Large Igneous Provinces from Gondwana and Adjacent Regions

Geological Society books refereeing procedures

This volume is published under an agreement between the Scientific Committee on Antarctic Research and the Geological Society of London and arises from the 12th International Symposium on Antarctic Earth Sciences (ISAES), Goa, July 2015.

GSL is the publisher of choice for books related to SCAR's geoscience activities, and SCAR receives a fee for all books published under this agreement.

Books published under this agreement are subject to the Society's standard rigorous proposal and manuscript review procedures.

It is recommended that reference to all or part of this book should be made in one of the following ways:

SENSARMA, S. & STOREY, B. C. (eds) 2018. *Large Igneous Provinces from Gondwana and Adjacent Regions*. Geological Society, London, Special Publications, **463**.

VIJAYA KUMAR, K., LAXMAN, M. B. & NAGARAJU, K. 2018. Mantle source heterogeneity in continental mafic Large Igneous Provinces: insights from the Panjal, Rajmahal and Deccan basalts, India. *In*: SENSARMA, S. & STOREY, B. C. (eds) *Large Igneous Provinces from Gondwana and Adjacent Regions*. Geological Society, London, Special Publications, **463**, 87–115. First published online July 11, 2017, https://doi.org/10.1144/SP463.5

GEOLOGICAL SOCIETY SPECIAL PUBLICATION NO. 463

Large Igneous Provinces from Gondwana and Adjacent Regions

EDITED BY

S. SENSARMA
University of Lucknow, India

and

B. C. STOREY
University of Canterbury, New Zealand

2018
Published by
The Geological Society
London

THE GEOLOGICAL SOCIETY

The Geological Society of London (GSL) was founded in 1807. It is the oldest national geological society in the world and the largest in Europe. It was incorporated under Royal Charter in 1825 and is Registered Charity 210161.

The Society is the UK national learned and professional society for geology with a worldwide Fellowship (FGS) of over 10 000. The Society has the power to confer Chartered status on suitably qualified Fellows, and about 2000 of the Fellowship carry the title (CGeol). Chartered Geologists may also obtain the equivalent European title, European Geologist (EurGeol). One fifth of the Society's fellowship resides outside the UK. To find out more about the Society, log on to www.geolsoc.org.uk.

The Geological Society Publishing House (Bath, UK) produces the Society's international journals and books, and acts as European distributor for selected publications of the American Association of Petroleum Geologists (AAPG), the Indonesian Petroleum Association (IPA), the Geological Society of America (GSA), the Society for Sedimentary Geology (SEPM) and the Geologists' Association (GA). Joint marketing agreements ensure that GSL Fellows may purchase these societies' publications at a discount. The Society's online bookshop (accessible from www.geolsoc. org. uk) offers secure book purchasing with your credit or debit card.

To find out about joining the Society and benefiting from substantial discounts on publications of GSL and other societies worldwide, consult www.geolsoc.org.uk, or contact the Fellowship Department at: The Geological Society, Burlington House, Piccadilly, London W1J 0BG: Tel. +44 (0)20 7434 9944; Fax +44 (0)20 7439 8975; E-mail: enquiries@geolsoc.org.uk.

For information about the Society's meetings, consult *Events* on www.geolsoc.org.uk. To find out more about the Society's Corporate Affiliates Scheme, write to enquiries@geolsoc.org.uk.

Published by The Geological Society from:
The Geological Society Publishing House, Unit 7, Brassmill Enterprise Centre, Brassmill Lane, Bath BA1 3JN, UK

The Lyell Collection: www.lyellcollection.org
Online bookshop: www.geolsoc.org.uk/bookshop
Orders: Tel. +44 (0)1225 445046, Fax +44 (0)1225 442836

British Library Cataloguing in Publication Data

A catalogue record for this book is available from the British Library.
ISBN 978-1-78620-325-0
ISSN 0305-8719

Distributors
For details of international agents and distributors see:
www.geolsoc.org.uk/agentsdistributors

Typeset by Nova Techset Private Limited, Bengaluru & Chennai, India
Printed and bound by CPI Group (UK) Ltd, Croydon CR0 4YY

Contents

Acknowledgements

Special thanks to the reviewers of the manuscripts submitted to this volume for their tireless, thorough and incisive comments that considerably helped to improve the manuscripts: Andrea Marzoli, Rajneesh Bhutani, N.V. Chalapathi Rao, Phil T. Leat, Phil Kyle, Teal Riley, Saibal Gupta, Peter C. Lightfoot, Talat Ahmad, M.L. Dora, Erfan Ali Mondal, Ramananda Chakrabarti, Dwijesh Ray, Hidehisa Mashima, Debajyoti Paul, Patricia Craig, Sujoy Kanti Ghosh, Keith Bell, Nitin R. Karmalkar, Gautam Sen, Saemi Halldorsson, R.W. Kent, Nilanjan Chatterjee, Raymond Duraiswami, Yong-Fei Zheng, Ashima Saikia and an anonymous reviewer. We also thank Angharad Hills, Tamzin Anderson, Sarah Gibbs and Rachael Kriefman at the Geological Society for thorough and professional editorial handling of the book.

Vivek P. Malviya (India) deserves a special mention for his outstanding and critical help on the Editorial desk as Editorial Assistant.

Gondwana Large Igneous Provinces (LIPs): distribution, diversity and significance

SARAJIT SENSARMA[1*], BRYAN C. STOREY[2] & VIVEK P. MALVIYA[3]

[1]*Centre of Advanced Study in Geology, University of Lucknow, Lucknow, Uttar Pradesh 226007, India*

[2]*Gateway Antarctica, University of Canterbury, Private Bag 4800, Christchurch 8140, New Zealand*

[3]*24E Mayur Residency Extension, Faridi Nagar, Lucknow, Uttar Pradesh 226016, India*

**Correspondence: sensarma2009@gmail.com*

Abstract: Gondwana, comprising >64% of the present-day continental mass, is home to 33% of Large Igneous Provinces (LIPs) and is key to unravelling the lithosphere–atmosphere system and related tectonics that mediated global climate shifts and sediment production conducive for life on Earth. Increased recognition of bimodal LIPs in Gondwana with significant, sometimes subequal, proportions of synchronous silicic volcanic rocks, mostly rhyolites to high silica rhyolites (±associated granitoids) to mafic volcanic rocks is a major frontier, not considered in mantle plume or plate process hypotheses. On a $\delta^{18}O$ v. initial $^{87}Sr/^{86}Sr$ plot for silicic rocks in Gondwana LIPs there is a remarkable spread between continental crust and mantle values, signifying variable contributions of crust and mantle in their origins. Caldera-forming silicic LIP events were as large as their mafic counterparts, and erupted for a longer duration (>20 myr). Several Gondwana LIPs erupted near the active continental margins, in addition to within-continents; rifting, however, continued even after LIP emplacements in several cases or was aborted and did not open into ocean by coeval compression. Gondwana LIPs had devastating consequences in global climate shifts and are major global sediment sources influencing upper continental crust compositions. In this Special Publication, papers cover diverse topics on magma emplacements, petrology and geochemistry, source characteristics, flood basalt–carbonatite linkage, tectonics, and the geochronology of LIPs now distributed in different Gondwana continents.

The continents that were stitched together in late Neoproterozoic–early Cambrian times formed the continent Gondwana, which combined with Laurentia in the Carboniferous to form Pangea and progressively fragmented in the Mesozoic. For much of that time, Gondwana was the largest continent on Earth, covering more than 100 million km^2 in area (Fig. 1). Gondwana constituted the present-day continents of South America, Africa, most of Australia and Antarctica, the Indian subcontinent and Madagascar, and parts of Arabia. A Zealandia continent, 94% of which is now submerged under ocean water, was also part of Gondwana (Mortimer *et al.* 2017*a*). All these remnants of Gondwana, including Zealandia, constitute more than 64% of the present-day continental mass (Torsvik & Cocks 2013). The supercontinent Gondwana survived for more than 300 myr until several parts of it separated and drifted away. It is conceivable that the records of major and significant geological events that are central to understanding the origin and evolution of continents through entire geological time are likely to be preserved in the present-day Gondwana continents. Major igneous events that took place over geological time in Gondwana are therefore intrinsic to the understanding of the Earth's lithosphere–atmosphere system that finally shaped Earth's surface processes conducive for life on this planet.

Our understanding of Gondwana evolution is much improved in the last three decades of research. We have realized that one of the important aspects, as evidenced from the reconstruction of continents with kinematic continuity using the GPlates software, could be that the bulk of Gondwana and its constituent cratons may have been relatively weakly affected by Phanerozoic tectonics (Torsvik & Cocks 2013). We have also understood that associated regional-scale intrusive (e.g. dyke swarms) rocks represent a powerful tool in reconstructing the original size and extent of Large Igneous Provinces (LIPs) in order to assess their enormity, and to better understand crust–mantle systems and related

From: Sensarma, S. & Storey, B. C. (eds) 2018. *Large Igneous Provinces from Gondwana and Adjacent Regions*. Geological Society, London, Special Publications, **463**, 1–16.
First published online November 27, 2017, https://doi.org/10.1144/SP463.11

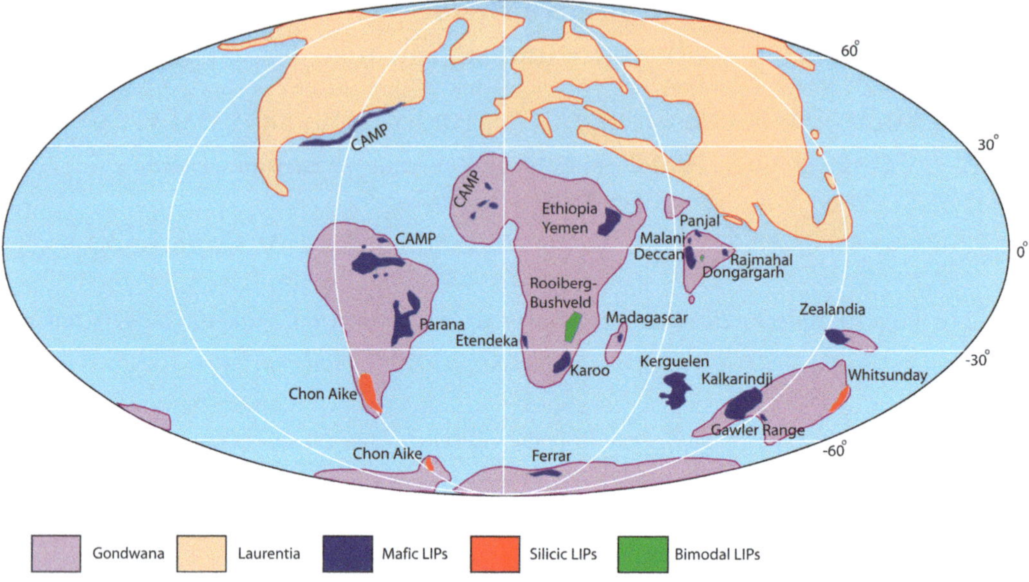

Fig. 1. Distribution of major Gondwana LIPs on a map after Gondwana break-up (*c.* 50.2 Ma). Already known bimodal LIPs are shown separately. The figure is modified after Coffin & Eldholm (1994), Bryan *et al.* (2002) and Paleomap Project, https://www.britannica.com/science (2001, C. R. Scotese, PALEOMAP Project).

tectonics. In many cases, remnants of single LIPs are now exposed in spatially separated continents. Despite our best efforts, however, understanding on their size, duration, frequency and link to continental break-up processes remains incomplete. The role of Gondwana LIPs in global climate and environmental shifts, and large-scale sediment production that influence even present-day upper continental crustal compositions are also of great interest.

This introductory paper to this Special Publication puts forward some perspectives as to how Gondwana large igneous events and related tectonics are important in understanding crust–mantle processes and their role in meditating global climate changes and sediment production over time. We also present an overview of the volume in the context of our understanding of Gondwana LIPs, highlighting the key points discussed in the included papers. For the benefit, in particular, of students and upcoming researchers, a note on possible issues concerning Gondwana LIP events that confront us at this time is briefly discussed at the end of this article.

Size and distribution of Gondwana LIPs

Large Igneous Provinces (LIPs) represent the largest volcanic events on Earth. The volcanic rocks in LIPs may cover huge areas of variable extent from as large as >100 000 km^2 (Bryan & Ernst 2007) to at least >50 000 km^2 (Sheth 2007). Gondwana is home to many of the largest LIPs globally. On the basis of a LIP inventory given in Ernst (2014), and different LIP webpages (largeigneousprovinces.org; mantleplume.org), the tentative size and distributions of the Gondwana LIPs are listed in Table 1. The compilation suggests that Gondwana LIPs cover nearly 58% of the surface area occupied by all LIPs on the present-day Earth (Gondwana and Laurentia LIPs put together). Amongst the LIPs, approximately 33% now occur in the Gondwana continents only.

Kalkarindji LIP (Australia), the oldest Phanerozoic LIP (*c.* 512–509 Ma), covers >2000 000 km^2 at present (Jourdan *et al.* 2014). The Paraná–Etendeka province (138–129 Ma), one of the largest LIPs ever erupted on Earth, covers presently at least 2 000 000 km^2 area (Ernst 2014), remnants of which are now exposed in Brazil, Paraguay and Africa across the Atlantic. The Afar province in Yemen–Ethiopia–Sudan–Egypt–Saudi Arabia at present covers an area as large as 2 000 000 km^2. The Panjal Traps (*c.* 290 Ma), presumably linked to the LIPs in the Himalayan magmatic province (HMP) emplaced during *c.* 290–270 Ma, is estimated to cover >2 000 000 km^2 in northern India–Pakistan–Tibet–Nepal. Another huge Gondwana LIP includes the Cretaceous (66–61 Ma) Deccan LIP (India) covering 600 000 km^2; on inclusion of part of the Madagascar province and the Seychelles, believed to be a detached remnant of the Deccan, the size of the

Deccan LIP may go up to >850 000 km^2. Other notable LIPs in Gondwana are the 188–178 Ma Karoo–Ferrar (nearly 150 000 km^2) in South Africa and Antarctica. With the Chon Aike province in Patagonia, recently recognized to be linked to the Karoo–Ferrar event (Pankhurst *et al.* 1998; Storey *et al.* 2013), the original size of the Karoo–Ferrar would obviously be larger than presently estimated. The 132–95 Ma Whitsunday province in eastern Australia covers nearly 200 000 km^2 (Bryan *et al.* 2000, 2012). Zealandia hosts volcanic rocks with an estimated volume of 4.9 million km^3 (Luyendyk 1995; Mortimer *et al.* 2017*a*). The total area covered by 116–95 Ma Rajmahal-Sylhet LIP (India) is *c.* 250 000 km^2 (Baksi 1995).

Precambrian LIPs, although still not adequately studied or known, are increasingly better identified and recognized in the Gondwana continents. The enormity of the Precambrian LIPs could be overwhelming as well (Ernst *et al.* 2013). The 2.19–2.10 Ga LIP event recognized in the West African Craton covers nearly 2000 000 km^2 at present, despite having been subjected to the prolonged action of secondary processes such as deformation, metamorphism, weathering and erosion. One of the major Precambrian LIPs is identified in the Yilgarn Craton in Australia. Despite prolonged weathering and erosion, several Precambrian LIPs located in Gondwana cover nearly 50 000–100 000 km^2: for example, the 1.9 Ga Great Dyke of the Zimbabwe Craton (Ernst 2014), the *c.* 2.5 Ga Dongargarh province (India: Sensarma 2007), the *c.* 2.06 Ma Rooiberg–Bushveld province (South Africa: Lenhardt & Eriksson 2012) and the *c.* 750 Ma Malani province (India: Sharma 2004) to name a few, implying that these provinces must have originally covered much larger areas. However, the effects of deformation and metamorphism, and other secondary processes make it difficult to identify and reconstruct matching details in the LIP remnants across continents. The lack of precise U–Pb ages in Precambrian LIPs further compounds the problem (Ernst *et al.* 2005, 2013).

The crust–mantle system of Gondwana LIPs and the tectonic context

The origin of LIPs is a first-order problem in Earth science. Neither the plume hypothesis nor plate tectonic processes adequately explain all observations in an integrated fashion for all LIPs. In some cases, none of the hypotheses is favoured. One of the emerging frontiers of LIP research today is recognition of the almost ubiquitous presence of significant amounts of silicic volcanic rocks, mostly rhyolites to high-silica rhyolites (and associated granitoids) in mafic LIPs. Indeed, in addition to mafic and silicic LIPs, bimodal LIPs do exist with substantial to subequal volumes of near-synchronous silicic and mafic rocks (Foulger 2007 and references therein). It is also known now that silicic LIP events could be as large as their mafic counterparts (Bryan & Ferrari 2013). Intrinsic to this compositional and petrological problem is the fact that the link between LIPs and continental break-up remains enigmatic. The break-up in many cases was aborted and did not culminate in the formation of an ocean because of coeval regional compression.

Increasing recognition of the substantial presence of silicic volcanic and associated plutonic rocks in continental mafic LIPs (e.g. Paraná–Etendeka (Harris & Milner 1997), Karoo–Ferrar–Chon Aike (Storey *et al.* 2013), Rajmahal (Ghose *et al.* 2016) and the Rooiberg–Bushveld province (Lenhardt & Eriksson 2012)) is a major advancement in contemporary LIP research. In the Karoo, the estimated volume of rhyolite is 35 000 km^3, which was emplaced after the main pulse (183–182 Ma) and is interstratified with basaltic lava (Cleverly *et al.* 1984). The Paraná–Etendeka province is dominantly mafic, but contains at least 20 000 km^3 of silicic volcanic rocks spread over 170 000 km^2 (Harris & Milner 1997). It is suggested that both mafic and silicic units in the Etendeka region may have originated from the same eruptive centres around the same time and are thus coeval (Milner *et al.* 1995). In the North Atlantic Igneous Province (NAIP) (62–54 Ma), a LIP not far from the Gondwana continent, substantial mafic–felsic rocks are present (Meade *et al.* 2014). In the Kalkarindji province, Jourdan *et al.* (2014) estimated the presence of a 15 000 km^{2}and 70 m-thick silicic volcanic breccia of explosive origin. New high-resolution U–Pb isotope dilution-thermal ionization mass spectrometry (ID-TIMS) geochronology on zircon and baddeleyite from both the *c.* 6000 km^3 layered silicic glassy hypabyssal intrusion, the Butcher Ridge Igneous Complex (BRIC), and from Ferrar mafic sills confirm that the BRIC magmatism occurred during the main phase of Ferrar LIP magmatism (184–182 Ma) (Nelson *et al.* 2015). It is also interesting to note that rhyolites to high-silica rhyolites in the LIPs, like Yemen–Ethiopian (500 km^3: Pankhurst *et al.* 2011), Dongargarh (8000 km^3: Sensarma *et al.* 2004; Sensarma 2007), the Sylhet Traps (Talukdar & Murthy 1971), Rooiberg–Bushveld (Lenhardt & Eriksson 2012) and Whitsunday (Bryan *et al.* 2000), are of high-temperature (800–1000°C) origin, compared to crust-derived normal magma temperatures (650–700°C).

Not only the compositional bimodality, but also the presence of substantial volumes of silicic volcanic rocks (plus silicic plutonic rocks), thus seem integral to several Gondwana mafic LIPs. The presence of a large volume of rheomorphic ignimbrites, ignimbrites, welded tuff, volcanic breecia/breccio-conglomerate and other primary pyroclastic deposits

Table 1. *Compilation of major Large Igneous Provinces (LIPs) in Gondwana with their respective area, age and duration*

SN	Name of LIP	Location	Covered area	Age (Ma)	Duration (Ma)	References
1	Afar Event	Yemen, Ethiopia, Djibouti, Saudi Arabia, Sudan, Egypt	2 000 000 km^2	31–29	3	Menzies *et al.* (1997); George *et al.* (1998); Hofmann *et al.* (1997)
2	NAVP (North Atlantic Volcanic Province) Event	UK and Greenland	1 300 000 km^2	62–58	4	Saunders *et al.* (1997)
3	Deccan Event	India and Seychelles	600 000 km^2, 1 800 000 km^2 (original area), 8 600 000 km^3	60.4–68	7.6	Eldholm & Coffin (2000)
4	Madagascar Event	Madagascar	260 000 km^2 (continental flood basalt portion only), 1 600 000 km^2, 4 400 000 km^3	90–84	6	Storey *et al.* (1997); Eldholm & Coffin (2000)
5	Kerguelen–Rajmahal Event	Indian Ocean, eastern India	6 000 000 km^3 (for South Kerguelen), 9 100 000 km^3 (for central Kerguelen–Brorken Ridge)	110–86	24	Mahoney *et al.* (1995); Kent *et al.* (1997); Eldholm & Coffin (2000); Frey *et al.* (2000)
6	Paraná–Etendeka Event	South America (Brazil, Paraguay), Africa (Namibia, Angola)	2 000 000 km^2, 1 800 000 km^2 (South America portion), 250 000 km^2 (African portion)	134–129, 138–135	9	Peate (1997)
7	Comei-Bunbury Event		*c.* 40 000 km^2	134–130	4	Zhu *et al.* (2009)
8	Karoo-Ferrar Event	Karoo (Africa), Ferrar (Antartica) provinces	Originally 5 000 000 km^3, originally (including underplating) 10 000 000 km^3, Karoo portion is 140 000 km^2, probably originally 1 000 000 km^2 and 2 500 000 km^3, Ferrar province is 500 000 km^3	183–179	16	Marsh *et al.* (1997); Storey & Kyle (1997); White (1997)
9	Chon Aike Event	Chon Aike silicic (South America) provinces	100 000 km^2	153–188	35	Pankhurst *et al.* (1998); Storey *et al.* (2013)
10	Whitsunday Event	Eastern Antartica	>500 000 km^2	132–95	37	Bryan *et al.* (2000, 2012)

11	CAMP (Central Atlantic Magmatic Province) Event	USA, South America, Africa	7 000 000 km^2, North America portion (400 000 km^2), South America portion (2 900 000 km^2), Africa portion (1 200 000 km^2)	204–191	13	May (1971); Courtney & White (1986); White & McKenzie (1989); Morgan *et al.* (1995); Burke (1996); Ernst & Buchan (1997); Marzoli *et al.* (1999)
12	Himalaya Neotethys Event (Panjal)	Northern India, Pakistan, Nepal, Tibet	*c.* 200 000 km^2	290–269	21	Garzanti *et al.* (1999)
13	Kalkarindji Flood Basalt	–	>2 000 000 km^2	508 ± 5–511.8 ± 1.9	4	Jourdan *et al.* (2014)
14	Malani Event	Aravali	*c.* 50 000 km^2	730–800 ± 50	20–80	Bhushan (2000); Sharma (2004)
15	Mundine Well Event	NW Australia (Pilbara, Yilgarn cratons)	180 000 km^2	758–752	6	http://www.largeigneousprovinces.org
16	Makuti Event	Zimbabwe, Zambia (Zambezi Belt)	–	880–740	6	http://www.largeigneousprovinces.org
17	Arabian–Nubian Shield Event	NE Africa and the Middle East (Arabian–Nubian Shield)	–	900–870	3	http://www.largeigneousprovinces.org
18	Lower Katangan Event(s)	Central Africa (Congo Craton)	140 000 km^2	1000–700	300	http://www.largeigneousprovinces.org
19	Mutare Event	Zimbabwe	20 000 km^2	1000–600	400	http://www.largeigneousprovinces.org
20	Gawler Range Event	Southern Australia (Gawler Craton)		1600–1500	100	http://www.largeigneousprovinces.org
21	Kolar Event	India	140 000 km^2	1800–1600	200	http://www.largeigneousprovinces.org
22	Bababudan Event	India (Western Dharwar Block)	30 000 km^2	3000–2500	500	http://www.largeigneousprovinces.org
23	Pietersburg–Giyani (O) Event	South Africa (north Kaapvaal Craton)	–	3500–3200	700	http://www.largeigneousprovinces.org
24	Rooiberg–Bushveld Event	Kaapvaal Craton	>200 000 km^2	2052 ± 48–2061 ± 2	37–55	Lenhardt & Eriksson (2012) and references therein
25	Dongargarh Event	Baster Craton		2506 + 4–2525 + 15; *c.* 2540–2467	*c.* 38–73	Sensarma (2007)

with/without rhyolitic lava flows in the silicic part of the LIP sequences (Chon–Aike–Karoo, Dongargarh, Whitsunday, Rajamahal, Etendeka and NAIP) and associated granitoids (crustal mush) may possibly suggest an association with large caldera structures. The largest silicic eruptions can sometimes be as large as flood basalt events as well (Bryan & Ferrari 2013).

The presence of large silicic rocks in so many Gondwana LIPs does not fit into the classic mantle plume model for large-volume mafic magma generation as the mantle plume model by itself does not produce significant volumes of silicic magma at any stage. Nor could the high-temperature (800–1000°C) large-volume silicic magmas be related to basaltic magma through fractionation. The presence of substantial silicic rocks within mafic LIPs is also not part of the models involving plate processes either. Obviously, the production of silicic rocks is linked to variable extents of crustal melting and interactions of crustal partial melts and mantle partial melts. A plausible scenario could be that the plume melting produces high-temperature magmas that impinge the crust to produce the crustal partial melts, and thereby promotes interactions of crustal and mantle partial melts. Of course, it is also possible that large volumes of mantle-derived primary basaltic magma are produced in an extensional tectonic setting independent of a mantle plume, which some workers ascribed to plume-induced rifting of the lithosphere (e.g. Samom *et al.* 2017).

Because continental crust is enriched in highly incompatible elements and thereby has a distinct radiogenic isotopic signature, incorporation of crustal components or interactions of crustal partial melts with the primary mantle-derived magma may lead to large-scale compositional heterogeneity in the final eruptive products that is now increasingly encountered in several Gondwana LIPs. For example, geochemical mass-balance calculations show that a basaltic komatiite melt may have dissolved about 20% silicic volcanics and erupted as a siliceous high-Mg basalt (SHMB) in the 2.5 Ga Dongargarh LIP (Sensarma *et al.* 2002). In the NAIP (62–54 Ma), widespread evidence of mixing and mingling of felsic (crustally derived) and mafic (mantle-derived) melts are manifested in textures, and in major, trace and high-resolution radiogenic (Sr-, Nd- and Pb-) isotope results (Meade *et al.* 2014), as also in the Panjal, to give rise to intermediate andesite rocks (Shellnutt 2017).

The low $\delta^{18}O$ value (<4–5‰) of rhyolites in some LIPs (e.g. Karoo: Miller & Harris 2007; and Dongargarh: Sensarma *et al.* 2004; Sensarma 2009) strongly argues for the involvement of hydrothermally altered crust in the generation of these rhyolites (Harris & Milner 1997; Colón *et al.* 2015). Nevertheless, the generation of LIP rhyolites call for various extents of crust–mantle interaction during emplacement of LIPs. The point is illustrated here by compiling and plotting $\delta^{18}O$ (whole rock/mineral separates as the case may be) v. initial $^{87}Sr/^{86}Sr$ of rhyolites from different Gondwana LIPs (Fig. 2). It is remarkable that rhyolites to high-silica rhyolites in Gondwana LIPs in practically all cases show a remarkable spread in-between continental crust ($\delta^{18}O$ *c.* 10–11‰) and mantle values (*c.* 5.5‰), or even 'mantle-like' values, and thus may plausibly have variable contributions of crust and mantle in their origin. Extremely high initial $^{87}Sr/^{86}Sr$ of rhyolites in some cases (e.g. Deccan) may be attributed to higher initial $^{87}Sr/^{86}Sr$ of the regional crust. Therefore, the composition (including isotopic) of the silicic component may adequately explain the trace element and isotopic variability in LIPs, and the presence of subducted crust in the plume source need not necessarily be invoked.

An interesting observation in many Gondwana mafic LIPs is the relative timing of the mafic and silicic phases; there is neither a consistent pattern to the eruption of silicic magmas, which may either precede (e.g. Ferrar, Panjal, Rajmahal, Rooiberg–Bushveld, Dongargarh), be contemporaneous with (e.g. Paraná–Etendeka) or post-date (Karoo–Lebombo) the peak of the basaltic flood volcanism.

Although the main phase of basalt erupted during a time period of *c.* 1–5 myr, occasionally extended to 1–10 myr in duration, there is significant departure too as longer time durations (30 myr for Karoo–Ferrar–Chon Aike, *c.* 24 myr for the Kerguelen–Rajmahal event, *c.* 70 myr for Dongargarh) is recorded in a couple of provinces (Fig. 3; also see Table 1). In fact, siliceous LIPs (e.g. 37 myr for Whitsunday, 20–80 myr for Malani) are known to have formed over a longer time period (40–60 myr: Bryan *et al.* 2000, 2013), However, more precise U–Pb single-zircon dating is required in several provinces to make the database robust.

The links between the emplacement of LIPs, rifting, and the process of the gradual break-up of continents and disintegration of Gondwana remain equally enigmatic (Storey 1995). There are several Mesozoic Gondwana LIPs (e.g. Deccan, Rajmahal, Karoo–Ferrar, Paraná and so on) that are plausibly linked to Gondwana rifting. However, in continental LIP provinces, whether the break-up was always concomitant to LIP emplacement and/or preceding it, or continued further in all cases even after LIP emplacement, is not adequately understood. This is important as Antarctica, even after it rifted from Gondwana, broke-up in the Mesozoic and moved to its present position in the South Pole, continued to rift further (Storey *et al.* 2013). Also, the rifting in many cases was aborted and did not finally break-up to open into an ocean because of coeval regional compression. Another important aspect is

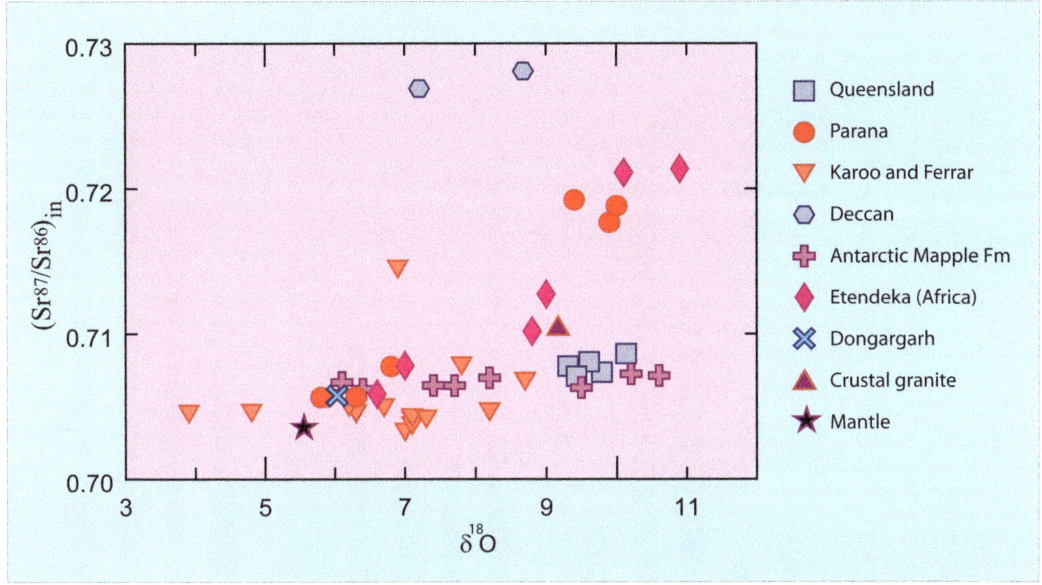

Fig. 2. $\delta^{18}O$ v. $(^{87}Sr/^{86}Sr)_{in}$ for silicic volcanic rocks (rhyolites, rhyodacites) occurring mostly in Gondwana mafic LIPs. The data spread between mantle and crustal values along a trend suggests contributions of both mantle and crust into the origin of silicic volcanic rocks in LIPs. Higher values in both oxygen- and Sr-isotopes in Paraná and Etendeka may suggest limited assimilation of supracrustals. $\delta^{18}O$ values are taken for mineral/whole-rock data as available in the literature. Data source: Queensland (Ewart 1982); Paraná (Harris *et al.* 2010); Karoo and Ferrar (Miller & Harris 2007); Deccan (Paul *et al.* 1977); Antarctic Mapple Formation (Riley *et al.* 2001); Etendeka (Harris *et al.* 2010); Dongargarh (Sensarma *et al.* 2004); crustal granite (Kempton & Harmon 1992); mantle (Rollinson 1993).

that many Gondwana LIPs (e.g. Chon Aike, Whitsunday) are close to once active plate boundaries and continental margins developed as part of plate processes. On the basis of lava geometry, the absence of domal uplift, a lack of large-volume eruption of high- to very-high-temperature (>1400°C) mantle partial melts (high Mg-rich basalts/picrites) and the enriched character of tholeiitic flood basalts, the Deccan Traps are also suggested to be intrinsically linked to plate processes by some researchers (e.g. Sheth 2005). The role of fluid in LIP genesis is also a matter of contemporary interest, particularly when nominally anhydrous minerals (e.g. olivine, pyroxene, plagioclase, garnet) could be a potential source of fluid during mantle melting even in within-plate settings. On the basis of the H_2O/Ce ratios of different mantle reservoirs, it is suggested that H_2O concentrations in plume sources may range from 300 to 1000 ppm (Hirschmann 2006 and references therein). It is argued that fluid released from mantle-plume induced subcontinent lithospheric metasomatism caused the latter (lithospheric mantle) to be enriched and fertile enough for the production of large volumes of flood basalts at shallower levels.

Effects of Gondwana LIPs on climate and sediment production

The present LIP inventory indicates that most of the LIP events occurred in Gondwana and adjacent regions. Consequently, a study of Gondwana LIP emplacement would be critical to understanding global climate changes and habitat. Many researchers suggest that LIP formation has severely influenced global climate, environmental changes and, perhaps, mass extinctions (Saunders 2005; Wignall 2001, 2005). LIP eruptions are temporally associated with climate/environmental changes that include rapid global warming and cooling, perturbations in p_{CO_2}, CH_4, SO_2 and halogens, sea-level changes, oceanic anoxia, calcification crises, mass extinctions, evolutionary radiations, and the release of gas hydrates (Storey *et al.* 2013 and references therein). The connection between the Karoo–Ferrar LIP eruptions and their effects on global climate, both on continents and in the marine realm, has been presented by Storey *et al.* (2013). Some workers consider that >11 km^3 of Deccan basalt eruptions in a short interval of 750 kyr were directly linked to an environmental shift by emitting poisonous

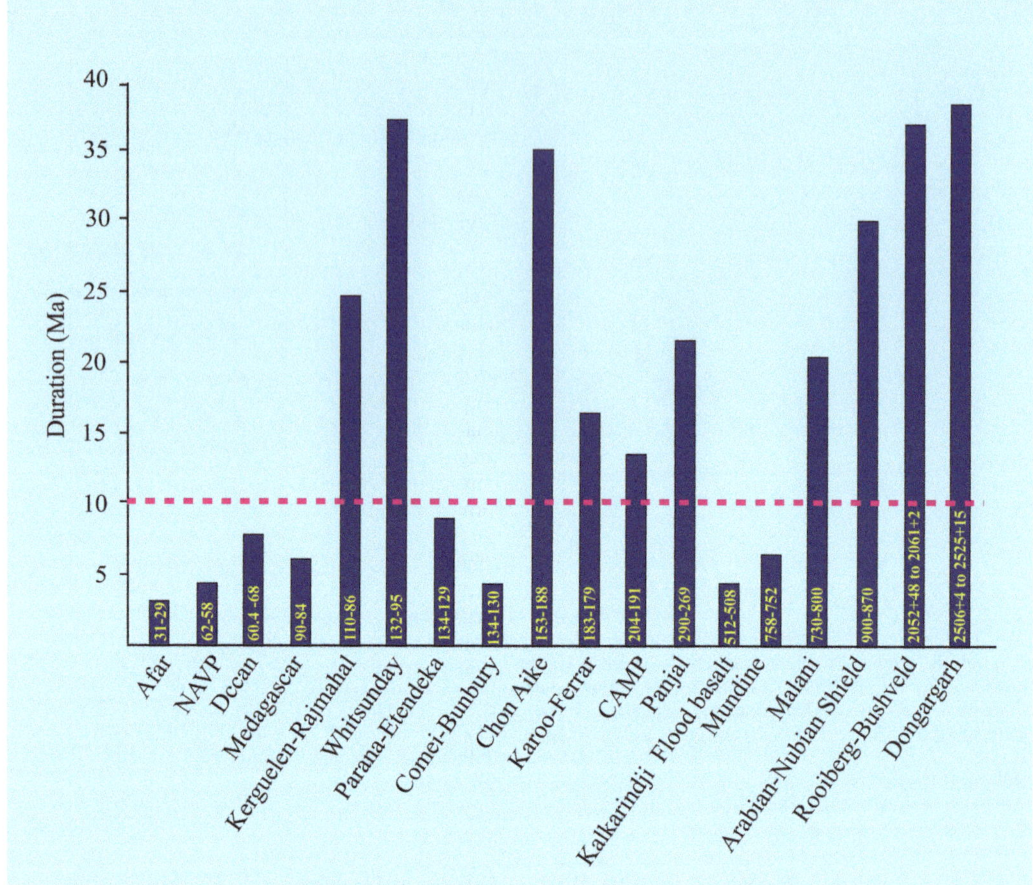

Fig. 3. The duration of LIP events in Gondwana and adjacent regions. Although the durations of several LIPs are within 1–10 myr, as expected in the mantle plume model, larger-duration (up to *c.* 40 myr) LIP events, including a few mafic LIPs, are there in Gondwana. For the data sources, see Table 1.

gases into the atmosphere with consequent adverse effects on the food chain in the very late Cretaceous, causing mass extinction of non-avial dinosaurs and ammonoids, and major biotic turnovers for invertebrates (corals, foraminiferas), reptiles and mammals (Schoene *et al.* 2015). Recent high-precision dating establishes the temporal synchronicity between the eruption of the Kalkarindji LIP (*c.* 512–509 Ma), a climatic shift and the early–middle Cambrian mass extinction in the beginning of the Phanerozoic Eon (Jourdan *et al.* 2014). On the other hand, some believe that LIP emplacement may have had a climatic impact, but did not necessarily directly cause mass extinction (Shellnutt *et al.* 2015). It may be equally possible that the biotic crisis may not be linked to the entire history of a LIP event. For example, the occurrence of repeated soft-sediment deformation structures (seismites), which are likely to be indicative of widespread seismicity, has been linked to end-Triassic mass extinction and the initial stages of the Central Atlantic Magmatic Province (CAMP) LIP emplacement (Lindström *et al.* 2015). Although the CAMP emplacement continued until the early Jurassic, there is no evidence of seismite up the section, while biotic recovery was on its way. This implies that the major igneous events on Gondwana are important links to understand the global climate and environmental changes, and associated biotic crisis. However, large caldera-forming events, as inferred from the large volume of high-temperature welded ignimbrites, welded tuff and other pyroclastic deposits within the LIP sequences, are known for their devastating consequences in global climate shifts.

The increased CO_2 emissions associated with Gondwana LIP emplacements must have caused enhanced CO_2 sequestering through increased silicate weathering and consequent sediment production.

In general, the upper continental crust (UCC) compositions closely approximate the intermediate granodioritic composition (SiO_2 55–66 wt%) believed to be forming typically at present-day convergent margins (Taylor & McLennan 1985). The weathering of rocks and erosion of sediments determine the mobility and distribution of chemical elements in different spheres of the Earth. The similarity in overall composition of global UCC and magmatic rocks of granodioritic compositions predominantly forming at the present-day convergent margins led to the suggestion that the subduction zone magmatic rocks are the principal or, perhaps, the sole global source for sediment production (see Taylor & McLennan 1985).

However, there are discrepancies to a factor of 2 or even ≥3 in the concentrations of certain trace elements such as Ni, Cr and Co in considering intermediate granodioritic rocks as the provenance (McLennan 2001). Continental flood basalt provinces (LIPs) are enriched in mafic/ultramafic igneous rocks and have a relatively high concentration of Cu, Co, Cr, etc. (Brügmann *et al.* 1987). In a recent study on the mineralogy and geochemistry of the river sediments in west-central India, Sharma *et al.* (2013) demonstrated that the discrepancy observed in the concentration of certain trace elements (e.g. Ni, Cr and Co) is better explained with the Deccan LIP as one of the sources (contributing more than 30%), in addition to the Archaean granitoids contributing the remaining (70%) to the bulk sediment compositions. Mafic LIPs, covering ≥50 000–100 000 km^2 in area, contain significant amounts of mafic rocks distributed across Gondwana, and therefore stand out as one of the major and potential global sediment sources. Also, it is suggested that at least 20% of Mg is lost from continents by chemical weathering (Mg being extremely mobile during weathering), which is one of the primary controls for finally leaving behind a Si-rich and Mg-poor continental mass, thereby decreasing crustal recycling by subduction (Lee *et al.* 2008), important for continental survival. So LIPs are important in better explaining the discrepancy in global UCC compositions and continent survival. Today *c.* 150 LIPs are known (Bryan & Ferrari 2013) and the inventory of LIPs is increasing. Hence, the identification and recognition of more LIPs of all ages in the geological record has wider implications.

Also, silicic LIPs constitute a significant part of the present-day UCC, and thus can act as a potential felsic end member, in addition to widespread granitoids, in calculating global UCC compositions. It is therefore pertinent to consider Gondwana LIPs, which cover a significant area on Earth's surface (see the earlier section on 'Size and distribution of Gondwana LIPs') as an important provenance in the models of UCC compositions and its chemical evolution. If the sediment inventory produced through Gondwana LIPs is better constrained in future, it is likely that the discrepancies observed in certain elemental abundances in UCC compositions, as mentioned above (McLennan 2001), could be better addressed and reconciled.

Overview of the volume

In the context of the ongoing studies on LIPs and their significance in better understanding Gondwana as a whole, the volume was conceived. Altogether 11 papers, including this one, have been included in the volume covering diverse topics on magma emplacements, petrology and geochemistry, source characteristics, flood basalt linkage to carbonatite magmatism, tectonics, and, to some extent, on geochronology of LIPs now distributed in different Gondwana continents.

Svensen *et al.* (2017) in an elegant and very timely review offer an overall perspective on the LIPs in Gondwana, including updated plate reconstructions using the GPlates and GMAP software from the key intervals of LIP magmatism in the history of Gondwana. This paper integrates information from several LIPs, such as the Kalkarindji, CAMP, Karoo–Ferrar and the Paraná–Etendeka, that are believed to have erupted within the time frame of the Gondwana supercontinent existence and recognizes the plume–crust interactions in the origin of LIPs. Importantly, they include LIP sills to estimate the volume of magmatism associated with the specific LIP events; thus, revising estimates of magma volume by including subvolcanic (dyke and sill) components, in addition to the lava flows, emplaced in adjacent sedimentary basins. The key role played by subvolcanic parts of LIPs in mediating climate and mass extinctions is also discussed in the paper.

The Ferrar LIP (Antarctica) has a linear outcrop pattern (length 3250 km) that is uncharacteristic for LIPs, although the extent that may be covered under thick ice cover is not clearly known. The Rajmahal province is also an north–south-trending elongate belt in eastern India (see Ghose *et al.* 2016). The mechanism of large volumes of magma emplacements in LIPs is of interest. The key lies in resolving the chronology of emplacements of different magmatic elements, such as lava and associated intrusive rocks, related to the same magmatic system. Subaerial lava flows can travel for more than 1000 km, depending on the eruption rate and topography. One of the suggested hypotheses in large magmatic systems is that crystal-free less-dense lava would be erupted earlier through a system of intrusions (called density-driven hypothesis). Obviously, eruption of the most evolved magma takes place first and subsequent magmas are increasingly richer in MgO content up the lava section. In the alternative 'cracked-lid' hypothesis, a sill-fed dyke network is overlain by lavas supplied by the randomly orientated underlying dyke field. **Elliot & Fleming (2017)** assess both the mechanism of magma emplacement in the Ferrar province by integrating the order of emplacement of the basaltic lava pile, and the intrusive (sill) rocks through field relationships and major element and selected trace element concentrations. They find that the geochemistry of lava and intrusive rocks is partially or marginally consistent, or sometimes drastically inconsistent, in different sections with the expectations of the 'density-driven' model. The cracked-lid sill-fed hypothesis implies that sills are intruded first and, progressively with time, magmas migrate upwards forming younger and stratigraphically higher sills, eventually breaking through to the surface as lava. Also, there should be some geochemical linkage between the lava flows and sills. The authors provide evidence and argue that none of the features of the magma emplacement hypothesis is favoured for the Ferrar LIP, and many factors such as density, lithostatic pressure, magma overpressure, composition and physical properties might have controlled the emplacement of magmas, and these depend on the timing and the place where the source was being tapped, whether as sills or lava flows.

Five major LIPs were emplaced at or near the Gondwana margin, including the Himalayan magmatic province (*c.* 290–270 Ma) which was emplaced along the Tethyan margin of Gondwana (Wang *et al.* 2014 and references therein), the Emishan LIP (China) and the Siberian Traps (*c.* 251 Ma). **Shellnutt (2017)**, in a succinct review of the Panjal Traps (*c.* 290 Ma), an important component of the Himalayan magmatic province, discusses the synchronous nature of basalts, that chemically range from continental tholeiite to ocean-floor basalt, and crustally derived silicic volcanic rocks (rhyolites and trachytes: $^{206}Pb/^{238}U$ zircon *in situ* age of 289 ± 3 Ma), that developed in a shallow lithospheric rift with significant mingling between crustal melts and mafic magmas. This implies that crust and mantle melting are linked to a common thermal perturbation, but not necessarily to a mantle plume. The silicic volcanic rocks are associated with silicic plutonic rocks, although the time bracket of the latter is not clearly known. One of the interesting aspects is that silicic eruption preceded the main basalt eruption in the Panjal. The Traps (India) have a thickness of <3000 m and cover an area of *c.* 100 000 km^2 that initially erupted within a continental rift setting and eventually transitioned into the opening of a nascent ocean basin that formed the Neotethys Ocean and the ribbon-like continent Cimmeria. The author also argues that the eruption of the Panjal Traps LIP during the Early Permian was not linked to the mid-Capitanian mass extinction or to post-Neotethys magmatism.

Phanerozoic Gondwana LIPs could be the windows to the mantle and its possible heterogeneity. **Vijaya Kumar *et al.* (2017)** provide a geochemical perspective using a global geochemical database taken from GEOROCK on the critical role of subcontinental lithospheric mantle with/without invoking a plume for the origin of three important LIPs: the Panjal Traps (*c.* 289 Ma), the Rajmahal (*c.* 117 Ma) and the Deccan Traps (*c.* 65 Ma). They show different degrees of melting; 5–20% of spinel peridotite for the Panjal, 10–20% for the Rajmahal and 1–20% for the Deccan Traps formed the primary basaltic to picritic basalts in the respective provinces. A spinel peridotite to a garnet-bearing source is suggested for the Panjal and Deccan basalts, respectively, with the Rajmahal pulse from some intermediate depth in-between these two. Their study emphasizes variable lithospheric mantle participation in the origin of these LIPs with no discernible input from normal mid-ocean ridge basalt (N-MORB)-type depleted mantle and continental crust. The estimated relative contributions of lithosphere and asthenosphere could be 67 and 33% for the Panjal, 52 and 48% in Rajmahal, and 28 and 72% for the Deccan, respectively. The apparent lack of alkaline rocks and carbonatite in the Panjal, in contrast to the Deccan and Rajmahal, basalts is ascribed to lithospheric thickness and fertility. Contrary to the assertion of **Shellnutt (2017)** that carbonatites occurring in the Koga and Jambil areas (NW Pakistan) nearly 300 km west of the Panjal Traps have a possible linkage to the Panjal basalts, this paper suggests that these carbonatites are of mantle-derived melts formed beneath a thick continental lithosphere far away from a rift, whereas the Panjal basalts are characteristically rift-related and formed underneath a thinned continental lithosphere; and, therefore, not linked. The authors suggested that the Indian subcontinental lithosphere fertility may have increased with decreasing age and is possibly related to Gondwana fragmentation. Alternatively, the subcontinental lithosphere mantle (SCLM) was variably enriched in different segments.

A snapshot of the sub-Deccan continental lithospheric mantle that participated in mantle melting during the Deccan Traps eruption is discussed by **Pandey *et al.* (2017)** from a mantle nodule, rarely fresh enough for study, caught up in *c.* 55 Ga lamprophre dyke within the Narmada Rift Zone in the Deccan province. The nodule (Ol + Opx + Grt), perhaps, bears the first conclusive evidence of modal mantle metasomatism in the Deccan, and has reacted with metasomatic fluid to give rise to an assemblage Ol + Cpx + Phol + Spl + Ap that equilibrated at a temperature of *c.* 1200°C and a pressure of *c.* 12 kb (*c.* 40 km depth). Olivine reported in the nodule is Fe-rich ($Fo_{85.34}$), more Fe-rich than off-craton (Fo_{88-92}) and on-craton (Fo_{91-94}) mantle xenoliths hitherto known, and may confirm it to be a xenolith of Fe-rich sub-Deccan lithospheric mantle from the garnet stability field. The Fo and Ni content is suggestive of a Fe-rich peridotite, rather than of a pyroxenitic source, and the presence of phlogopite and apatite provide compelling evidence of mantle metasomatism. This study clearly indicates a post-Deccan metasomatic layer that survived melting somewhat during the Deccan eruption because of the variable thickness of the underlying lithosphere; a situation encountered in several LIPs, including the Paraná–Etendeka.

Carbonatites are spatially associated with several LIPs, including those in Gondwana, but their temporal relationship is not well established, which may, indeed, provide insight into carbonatite genesis and its link with the LIPs. The possible genetic link of the *c.* 65 Ma main Deccan eruption event to the Ambar Dongar carbonatite ($^{40}Ar/^{39}Ar$ dating 65 ± 0.3 Ma: Ray & Pande 1999), the youngest Indian carbonatite occurrence, is thus of interest. One of the major difficulties in testing the LIP–carbonatite link is the nature of the mantle source for the primary carbonatite melt: that is, whether the parental carbonatite magma was generated from the SCLM or from deep mantle sources. **Chandra *et al.* (2017)** demonstrate through detailed mineralogical and geochemical compositions and geochemical modelling that the metasomatism of the SCLM below the Deccan was key to a possible petrogenetic link between the main Deccan eruption event and the Ambar Dongar carbonatite. A notable feature is that the Ambar Dongar carbonatite, both ferrocarbonatite and calciocarbonatite varieties, is extremely rich in REE (total REE 1025–31 117 ppm) with ferrocarbonatite richer in REE than calciocarbonatite. On the basis of trace element modelling, the authors proposed a model in which the contributions of CO_2-rich fluids and heat supplied by the Deccan plume metasomatized the SCLM (carbonated garnet lherzolite) at 100 km for the production of parental carbonated silicate magma that underwent liquid immiscibility at crustal depth for nephelinite and carbonatite production. However, the model could not explain the abundances of Rb, Ba, Th, U and the REE pattern adequately. Metasomatism of the SCLM below the Deccan Traps was possibly more common than envisaged.

Rifting and break-up of Gondwana led to the production of large-scale basalt magmatism, and the amalgamation of continents that eventually stitched together to form Gondwana is marked with a chain of granitoid plutons. Gondwana includes a >10 000 km-long chain of 530–500 Ma granitoid magmatism intruded during the assembly of continents along the northern margin (Veevers 2004). **Sadiq *et al.* (2017)** discussed the origin and complex hybridization process preserved in the 550 ± 15 Ma Nongpoh Granite in the Shillong Plateau in NE India that represents part of a large magmatic system where mafic injection during different stages of crystallization of granitoid magma and felsic–mafic magma interactions took place. Evidence of an earlier injection is manifested as mafic magmatic enclaves (MMEs) and syn-plutonic dykes representing the later phase. The 501 Ma chemical (U–Th–Pb) age of monazite in the MMEs represents similar tectonothermal events reported from other parts of India, SW Australia and Antarctica, and represents the global Neopalaeoproterozic–Cambrian Pan-African events during the amalgamation of Eastern Gondwana.

Whether rifting and break-up continued after the main break-up of Gondwana in the Mesozoic is of great interest. We have two papers in the volume addressing this question, although the datasets, methodologies adopted and line of treatment in these papers are different. **Mortimer *et al.* (2017*b*)** discuss new Ar/Ar and limited U–Pb ages and geochemistry (both trace element and Nd isotopic analyses) of widespread volcanism on the northern part of the Zealandia continent by using dredged samples obtained in different cruises and integrated with their earlier findings on the southern Zealandia in an attempt to give a more updated picture of magmatism across the whole Zealandia continent. This study showed that the lavas erupted both in northern and southern Zealandia are underlain by continental rather than oceanic crust, and record Late Cretaceous–Eocene intracontinental rifting both during and after continental break-up. Interestingly, Zealandia intraplate volcanic rocks share generally similar age ranges and composition with those of West Antarctica and Australia that were together before the break-up.

In the case of the Deccan Traps LIP, pre- and syn-magmatic rifting is widely reported. However, the question of whether rifting continued even after Deccan magmatism is not adequately studied or understood. The issue of possible post-magmatic rifting in the Deccan is dealt with by **Mitra *et al.* (2017)** by geochemical modelling of weathering of abundant tholeiitic basalt, as well as less common alkali basalt occurring in the far west of India (Kutch region) at elevated but variable atmospheric p_{CO_2} (p_{CO_2}: $10^{-2.5}$ and less), a condition that presumably prevailed in the end Cretaceous–early Paleocene time at about syn- to post-Deccan basalt emplacement. REACT, a module included in the software packages Geochemists' Workbench 10.0.4, is used to model thermodynamic equilibrium reaction paths for the weathering of basalt. The authors consider both open- and closed-system weathering in order to simulate basalt weathering with/without rainwater interactions characteristic of tropical climatic conditions that existed during or after Deccan eruption. More intense weathering of basalts to form kaolinite at the base of rift basins, and less weathering leading to formation of smectite in the rift flanks, is argued to have been controlled by rift-generated topography, and not mediated by variations in atmospheric p_{CO_2}. Post-magmatic tectonics, rather than climate, thereby controlled the weathering of basalts at the terminal to post-Deccan time.

A major challenge lies in identifying remnants of LIPs, particularly of Precambrian age, that occur now in spatial isolation in different continents or within the same landmass. In an attempt to reconstruct a Palaeoproterozoic LIP event of plume origin in the northern India Shield, **Samon *et al.* (2017)** discuss new geochemical and Sm–Nd isotopic constraints from a huge 2104 ± 23 Ma mafic sill that extended for more than 120 m in depth emplaced in an the Gwalior Basin in north central India. They integrate the results with other major synchronous mafic provinces in the northern Indian Shield that were part of the same LIP. The authors suggest that the record of intracratonic basin formation due to lithospheric stretching, plume-induced rift magmatism, large gabbro/doleritic sills and dykes emplacements between 2.5 and 2.1 Ga are part of a LIP event, and are associated with the break-up of the Kenorland supercontinent, the parts of which later amalgamated and became part of Gondwana.

The future

Based on the results of publications in this volume, we suggest that understanding the formation and age of the silicic plutonic rocks associated with silicic volcanic rocks in LIPs is a major frontier. The study of stratigraphic constraints supported by precise age dating with robust methods (e.g. U–Pb single-zircon data) and volume estimation of LIPs pose a great challenge in reconstructing the LIP events better. The definite linkage between LIP emplacement and biotic crisis need to be better established. More petrological constraints and the study of source characteristics are obviously required to reassess and refine the existing models of LIP genesis. One of the associated challenges before us is also to assess the possibility of a large eruption and associated risk. It is estimated that volcanism of the size of a LIP may happen once in 100 000 years. So there might be adequate signals and geological events, including earthquakes and other surficial processes, over a sufficient time period (years or decades) to provide a reasonable warning. Better resolution of seismic images might help in determining the amount of the eruptible magma and estimating the possibility of a large eruption. However, there are always large uncertainties in predicting such large eruptions. In order to lessen uncertainties in predicting LIP size eruptions, we need to know much more about the past LIPs, their ages and duration, what their true size was, and what types of deposits they produced. LIPs associated with continental margin settings in some cases pose a challenge to unravel whether LIP emplacement is always a within-plate phenomenon. Needless to say, the future study on post-magmatic rifting in LIP provinces and its implication for the underlying crust–mantle system in Gondwana will certainly add to our understanding of LIP tectonics.

Concluding remarks

Gondwana, comprising >64% of the present-day continental mass, is home to 33% of large igneous provinces (LIPs). The study of Gondwana LIPs is therefore critical to understanding the Earth's lithosphere–atmosphere system that finally shaped Earth's surface processes, thereby creating conditions conducive for life on this planet. Gondwana is home to many of the largest LIPs, each covering 200 000–2 000 000 km^2. The Precambrian LIPs in Gondwana are numerous and their size could also be overwhelming, although difficult to estimate because of the effects of deformation, metamorphism, and prolonged weathering and erosion. One of the major recent frontiers is the ubiquitous presence of significant to subequal amounts of synchronous silicic volcanic rocks in mafic LIPs, thereby suggesting a more common presence of bimodal LIPs than believed earlier. It is also now known that silicic LIP events could be as large as their mafic counterparts. Also, the origin of large silicic rocks in so many of the Gondwana LIPs by crust–mantle interactions is not part of the classic plume or plate tectonic model as it does not consider significant volumes of silicic magma production at any stage. Several Gondwana LIPs are truly linked to Gondwana rifting meditated directly or indirectly to mantle plume activity. However, whether the Gondwana break-up was syn- or pre-LIP emplacement, or continued even after LIP emplacement, is not adequately understood. Many LIPs, particularly silicic LIPs (SLIPs) and bimodal LIPs are also of longer duration (>20 myr), again not in tune with expectations of plume model (up to 10 myr). The few LIPs associated with a continental margin setting pose the question of whether LIP emplacement is strictly a within-plate phenomenon.

This would like to thank the organizers of the SCAR ISAES conference held in Goa in India in 2015 for facilitating the LIP session on which the papers within this volume are based. Detailed and critical reviews by Saibal Gupta and Phil Leat are gratefully acknowledged. S.S. thanks UGC (New Delhi) and University of Lucknow for support.

References

BAKSI, A.K. 1995. Petrogenesis and timing of volcanism in the Rajmahal flood basalt province, northeast India. *Chemical Geology*, **121**, 73–90.

BHUSHAN, S.K. 2000. MalaniRhyolies: a review. *Gondwana Research*, **3**, 65–77.

BRÜGMANN, G.E., ARNDT, N.T., HOFMANN, A.W. & TOBSCHALL, H.J. 1987. Noble metal abundances in komatiite suites from Alexo, Ontario and Gorgona Island, Colombia. *Geochimica et Cosmochimica Acta*, **51**, 2159–2169.

BRYAN, S.E. & ERNST, R. 2007. Revised definition of Large Igneous Provinces (LIPs). *Earth-Science Reviews*, **86**, 175–202.

BRYAN, S.E. & FERRARI, L. 2013. Large igneous provinces and silicic large igneous provinces: progress in our understanding over the last 25 years. *Geological Society of America Bulletin*, **125**, 1053–1078.

BRYAN, S.E., EWART, A., STEPHENS, C.J., PARIANOS, J. & DOWNES, P.J. 2000. The Whitsunday Volcanic Province, Central Queensland, Australia: lithological and stratigraphic investigations of a silicic-dominated large igneous province. *Journal of Volcanology and Geothermal Research*, **99**, 55–78.

BRYAN, S.E., RILEY, T.R., JERRAM, D.A., LEAT, P.T. & STEPHENS, C.J. 2002. Silicic volcanism: an under-valued component of large igneous provinces and volcanic rifted margins. *In*: MENZIES, M.A., KLEMPERER, S.L., EBINGER, C.J. & BAKER, J. (eds) *Magmatic Rifted Margins*. Geological Society of America, Special Papers, **362**, 99–120.

BRYAN, S.E., COOK, A.G., ALLEN, C.M., SIEGEL, C., PURDY, D.J., GREENTREE, J.S. & UYSAL, I.T. 2012. Early-mid Cretaceous tectonic evolution of eastern Gondwana: from silicic LIP magmatism to continental rupture. *Episodes*, **35**, 142–152.

BRYAN, S.E., OROZCO-ESQUIVEL, T., FERRARI, L. & LÓPEZ-MARTÍNEZ, M. 2013. Pulling apart the Mid to Late Cenozoic magmatic record of the Gulf of California: is there a Comondú Arc? *In*: GOMEZ-TUENA, A., STRAUB, S.M. & ZELLMER, G.F. (eds) *Orogenic Andesites and Crustal Growth*. Geological Society, London, Engineering Geology Special Publications, **385**, 389–407, https://doi.org/10.1144/SP385.8

BURKE, K. 1996. The African plate. *South African Journal of Geology*, **99**, 341–409.

CHANDRA, J., PAUL, D., VILADKAR, S.G. & SENSARMA, S. 2017. Origin of the Amba Dongar carbonatite complex, India and its possible linkage with the Deccan Large Igneous Province. *In*: SENSARMA, S. & STOREY, B.C. (eds) *Large Igneous Provinces from Gondwana and Adjacent Regions*. Geological Society, London, Special Publications, **463**. First published online July 10, 2017, https://doi.org/10.1144/SP463.3

CLEVERLY, R.W., BETTON, P.J., BRISTOW, J.W. & ERLANK, A.J. 1984. Geochemistry and petrogenesis of the Lebombo rhyolites. *In*: ERLANK, A.J. (ed.) *Petrogenesis of the Volcanic Rocks of the Karoo Province*. Geological Society of South Africa, Special Publications, **13**, 171–194.

COFFIN, M.F. & ELDHOLM, O. 1994. Large igneous provinces: crustal structure, dimensions, and external consequences. *Reviews of Geophysics*, **32**, 1–36.

COLÓN, D.P., BINDEMAN, I.N., ELLIS, B.S., SCHMITT, A.K. & FISHER, C.M. 2015. Hydrothermal alteration and melting of the crust during the Columbia River Basalt–Snake River Plain transition and the origin of low-$\delta^{18}O$ rhyolites of the central Snake River Plain. *Lithos*, **224–225**, 310–323.

COURTNEY, R.C. & WHITE, R.S. 1986. Anomalous heat flow and geoid across the Cape Verde Rise: evidence for dynamic support from a thermal plume in the mantle. *Geophysical Journal of the Royal Astronomical Society*, **87**, 815–867.

ELDHOLM, O. & COFFIN, M.F. 2000. Large igneous provinces and plate tectonics. *In*: RICHARDS, M., GORDON, R. & VAN DER HILST, R. (eds) *The History and Dynamics of Global Plate Motions*. American Geophysical Union, Geophysical Monographs, **121**, 309–326.

ELLIOT, D.H. & FLEMING, T.H. 2017. The Ferrar Large Igneous Province: field and geochemical constraints on supra-crustal (high-level) emplacement of the magmatic system. *In*: SENSARMA, S. & STOREY, B.C. (eds) *Large Igneous Provinces from Gondwana and Adjacent Regions*. Geological Society, London, Special Publications, **463**. First published online July 10, 2017, https://doi.org/10.1144/SP463.1

ERNST, R.E. 2014. *Large Igneous Provinces*. Cambridge University Press, Cambridge.

ERNST, R.E. & BUCHAN, K.L. 1997. Giant radiating dyke swarms: their use in identifying pre-Mesozoic large igneous provinces and mantle plumes. *In*: MAHONEY, J.J. & COFFIN, M.F. (eds) *Large Igneous Provinces: Continental, Oceanic, and Planetary Flood Volcanism*. American Geophysical Union, Geophysical Monographs, **100**, 297–333.

ERNST, R.E., BUCHAN, K.L. & CAMPBELL, I.H. 2005. Frontiers in large igneous province research. *Lithos*, **79**, 271–297.

ERNST, R.E., BLEEKER, W., SODERLUND, U. & KERR, A.C. 2013. Large Igneous Provinces and super continent: towards completing the plate tectonic revolution. *Lithos*, **174**, 1–14.

EWART, A. 1982. Petrogenesis of the tertiary anorogenic volcanic series of Southern Queensland, Australia, in the light of trace element geochemistry and O, Sr and Pb isotopes. *Journal of Petrology*, **23**, 344–382.

FOULGER, G.R. 2007. The 'plate' model for the genesis of melting anomalies. *In*: FOULGER, G.R. & JURDY, D.M. (eds) *Plates, Plumes, and Planetary Processes*. Geological Society of America, Special Papers, **430**, 1–28, https://doi.org/10.1130/2007.2430(01)

FREY, F.A., COFFIN, M.F. *ET AL.* 2000. Origin and evolution of a submarine large igneous province: the Kerguelen Plateau and Broken Ridge, southern Indian Ocean. *Earth and Planetary Science Letters*, **176**, 73–89.

GARZANTI, E., LE FORT, P. & SCIUNNACH, D. 1999. First report of Lower Permian basalts in South Tibet: tholeiitic magmatism during break-up and incipient opening of Neotethys. *Journal of Asian Earth Sciences*, **17**, 533–546.

GEORGE, R., ROGERS, N. & KELLEY, S. 1998. Earliest magmatism in Ethiopia: evidence for two mantle plumes in one flood basalt province. *Geology*, **26**, 923–926.

GHOSE, N.C., CHATTERJEE, N. & WINDLEY, B.F. 2016. Subaqueous early eruptive phase of the late Aptian Rajmahal volcanism, India: Evidence from volcaniclastic rocks, bentonite, black shales, and oolite. *Geoscience*

Frontiers, **8**, 809–822, https://doi.org/10.1016/j.gsf.2016.06.007

Harris, C. & Milner, S. 1997. Crustal origin for the Paraná rhyolites: discussion of 'description and petrogenesis of the Paraná rhyolites, southern Brazil' by Garland *et al.* (1995). *Journal of Petrology*, **38**, 299–302.

Harris, C., Whittingham, A.M., Milner, S.C. & Armstrong, R.A. 2010. Oxygen isotope geochemistry of the silicic volcanic rocks of the Etendeka-Paraná Province: source constraints. *Geology*, **18**, 1119–1121.

Hirschmann, M.M. 2006. Water, melting, and the deep Earth H_2O cycle. *Annual Review of Earth and Planetary Sciences*, **34**, 629–653.

Hofmann, C., Courtillot, V., Féraud, G., Rochette, P., Virgu, G., Ketefo, E. & Pik, R. 1997. Timing of the Ethiopian flood basalt event and implications for plume birth and global change. *Nature*, **389**, 838–841.

Jourdan, F., Hodges, K. *et al.* 2014. High-precision dating of the Kalkarindji large igneous province, Australia, and synchrony with the Early–Middle Cambrian (Stage 4–5) extinction. *Geology*, **42**, 543–546.

Kempton, P.D. & Harmon, R.S. 1992. Oxygen isotope evidence for large-scale hybridization of the lower crust during magmatic underplating. *Geochimica et Cosmochimica Acta*, **56**, 963–970.

Kent, W., Saunders, A.D., Kempton, P.D. & Ghose, N.C. 1997. Rajmahal basalts, eastern India: Mantle sources and melt distribution at a volcanic rifted margin. *In*: Mahoney, J.J. & Coffin, M.F. (eds) *Large Igneous Provinces: Continental, Oceanic, and Planetary Flood Volcanism*. American Geophysical Union, Geophysical Monographs, **100**, 145–182.

Lee, C.T.A., Morton, D.M., Little, M.G., Kistler, R., Horodyskyj, U.N., Leeman, W.P. & Agranier, A. 2008. Regulating continent growth and composition by chemical weathering. *Proceedings of the National Academy of Sciences of the United States of America*, **105**, 4981–4986.

Lenhardt, N. & Eriksson, P.G. 2012. Volcanism of the Palaeoproterozoic Bushveld Large Igneous Province: the Rooiberg Group, Kaapvaal Craton, South Africa. *Precambrian Research*, **214–215**, 82–94.

Lindström, S., Pedersen, G.K. *et al.* 2015. Intense and widespread seismicity during the end-Triassic mass extinction due to emplacement of a large igneous province. *Geology*, **43**, 387–390, https://doi.org/10.1130/G36444.1

Luyendyk, B.P. 1995. Hypothesis for Cretaceous rifting of east Gondwana caused by subducted slab capture. *Geology*, **23**, 373–376.

Mahoney, J.J., Jones, W.B., Frey, F.A., Salters, V.J.M., Pyle, D.G. & Davies, H.L. 1995. Geochemical characteristics of lavas from Broken Ridge, the Naturaliste Plateau and southernmost Kerguelen Plateau: Cretaceous plateau volcanism in the southeast Indian Ocean. *Chemical Geology*, **120**, 315–345.

Marsh, J.S., Hooper, P.R., Rehacek, J., Duncan, R.A. & Duncan, A.R. 1997. Stratigraphy and age of Karoo basalts of Lesotho and implications for correlations within the Karoo igneous province. *In*: Mahoney, J.J. & Coffin, M.F. (eds) *Large Igneous Provinces: Continental, Oceanic, and Planetary Flood Volcanism*. American Geophysical Union, Geophysical Monographs, **100**, 247–272.

Marzoli, A., Renne, P.R., Piccirillo, E.M., Ernesto, M., Bellieni, G. & De Min, A. 1999. Extensive 200-million-year-old continental flood basalts of the central Atlantic Magmatic Province. *Science*, **284**, 616–618.

May, P.R. 1971. Pattern of Triassic–Jurassic diabase dikes around the North Atlantic in the context of predrift position of the continents. *Geological Society of America Bulletin*, **82**, 1285–1292.

McLennan, S.M. 2001. Relationships between the trace element composition of sedimentary rocks and upper continental crust. *Geochemistry, Geophysics, Geosystems*, **2**, 1021, https://doi.org/10.1029/2000GC0 00109

Meade, F.C., Troll, V.R., Ellam, R.M., Freda, C., Font, L., Donaldson, C.H. & Klonowska, I. 2014. Bimodal magmatism produced by progressively inhibited crustal assimilation. *Nature Communications*, **5**, 4199, https://doi.org/10.1038/ncomms5199

Menzies, M., Baker, J., Chazot, G. & Al'kadasi, M. 1997. Evolution of the Red Sea volcanic margin, western Yemen. *In*: Mahoney, J.J. & Coffin, M.F. (eds) *Large Igneous Provinces: Continental, Oceanic, and Planetary Flood Volcanism*. American Geophysical Union, Geophysical Monographs, **100**, 29–43.

Miller Jodie, A. & Harris, C. 2007. Petrogenesis of the Swaziland and northern Natal rhyolites of the Lebombo rifted volcanic margin, South East Africa. *Journal of Petrology*, **48**, 185–218.

Milner, S.C., Duncan, A.R., Whittingham, A.M. & Ewart, A. 1995. Trans-Atlantic correlation of eruptive sequences and individual silicic volcanic units within the Paraná-Etendeka igneous province. *Journal of Volcanology and Geothermal Research*, **69**, 137–157.

Mitra, K., Mitra, S., Gupta, S., Bhattacharya, S., Chauhan, P. & Jain, N. 2017. Modelling basalt weathering at elevated CO_2 concentrations: implications for terminal to post-magmatic rifting in the Deccan Traps, Kachchh, India. *In*: Sensarma, S. & Storey, B.C. (eds) *Large Igneous Provinces from Gondwana and Adjacent Regions*. Geological Society, London, Special Publications, **463**. First published online July 17, 2017, https://doi.org/10.1144/SP463.8

Morgan, J.P., Morgan, W.J. & Price, E. 1995. Hotspot melting generates both hotspot volcanism and a hotspot swell? *Journal of Geophysical Research*, **100**, 8045–8062.

Mortimer, N., Campbell, H.J. *et al.* 2017*a*. Zealandia: Earth's hidden continent. *GSA Today*, **27**, 27–35, https://doi.org/10.1130/GSATG321A.1

Mortimer, N., Gans, P.B. *et al.* 2017*b*. Regional volcanism of northern Zealandia: post-Gondwana break-up magmatism on an extended, submerged continent. *In*: Sensarma, S. & Storey, B.C. (eds) *Large Igneous Provinces from Gondwana and Adjacent Regions*. Geological Society, London, Special Publications, **463**. First published online August 16, 2017, https://doi.org/10.1144/SP463.9

Nelson, D., Cottle, J., Barboni, M. & Schoene, B. 2015. Petrogenesis of the Butcher Ridge Igneous Complex, a unique layered glassy silicic intrusion within the Ferrar Large Igneous Province. Abstract 116 presented at the XII International Synposium on Antractic Earth Sciences, 13–17 July 2015, Goa, India.

Pandey, R., Rao, N.V.C., Pandit, D., Sahoo, S. & Dhote, P. 2017. Imprints of modal metasomatism in the post-Deccan subcontinental lithospheric mantle: petrological evidence from an ultramafic xenolith in an Eocene lamprophyre, NW India. *In*: Sensarma, S. & Storey, B.C. (eds) *Large Igneous Provinces from Gondwana and Adjacent Regions*. Geological Society, London, Special Publications, **463**. First published online July 5, 2017, https://doi.org/10.1144/SP463.6

Pankhurst, M.J., Schaefer, B.F. & Betts, P.G. 2011. Geodynamics of rapid voluminous felsic magmatism through time. *Lithos*, **123**, 92–101.

Pankhurst, R.J., Leat, P.T., Sruoga, P., Rapela, C.W., Márquez, M., Storey, B.C. & Riley, T.R. 1998. The Chon Aike province of Patagonia and related rocks in West Antarctica: a silicic large igneous province. *Journal of Volcanology and Geothermal Research*, **81**, 113–136.

Paul, D.K., Potts, P.J., Rex, D.C. & Beckinsale, R.D. 1977. Geochemical and petrogenetic study of the Girnar igneous complex, Deccan volcanic province, India. *Geological Society of America Bulletin*, **88**, 227–234.

Peate, D.W. 1997. The Paraná-Etendeka Province. *In*: Mahoney, J.J. & Coffin, M.F. (eds) *Large Igneous Provinces: Continental, Oceanic and Planetary Flood Volcanism*. American Geophysical Union, Geophysical Monographs, **100**, 217–245.

Ray, J.S. & Pande, K. 1999. Carbonatite alkaline magmatism associated with continental flood basalts at stratigraphic boundaries: cause of mass extinctions. *Geophysical Research Letters*, **26**, 1917–1920.

Riley, T.R., Leat, P.T., Pankhurst, R.J. & Harris, C. 2001. Origins of large volume rhyolitic volcanism in the Antarctic Peninsula and Patagonia by crustal melting. *Journal of Petrology*, **42**, 1043–1106.

Rollinson, H.R. 1993. *Using Geochemical Data: Evaluation, Presentation, Interpretation*. Longman, Harlow, UK.

Sadiq, M., Umrao, R.K., Sharma, B.B., Chakraborti, S., Bhattacharyya, S. & Kundu, A. 2017. Mineralogy, geochemistry and geochronology of mafic magmatic enclaves and their significance in evolution of Nongpoh granitoids, Meghalaya, NE India. *In*: Sensarma, S. & Storey, B.C. (eds) *Large Igneous Provinces from Gondwana and Adjacent Regions*. Geological Society, London, Special Publications, **463**. First published online July 6, 2017, https://doi.org/10.1144/SP463.2

Samom, J.D., Ahmad, T. & Choudhary, A.K. 2017. Geochemical and Sm–Nd isotopic constraints on the petrogenesis and tectonic setting of the Proterozoic mafic magmatism of the Gwalior Basin, central India: the influence of Large Igneous Provinces on Proterozoic crustal evolution. *In*: Sensarma, S. & Storey, B.C. (eds) *Large Igneous Provinces from Gondwana and Adjacent Regions*. Geological Society, London, Special Publications, **463**. First published online July 10, 2017, https://doi.org/10.1144/SP463.10

Saunders, A.D. 2005. Large Igneous Provinces: origin and environmental consequences. *Elements*, **1**, 259–263.

Saunders, A.D., Fitton, J.G., Kerr, A.C., Norry, M.J. & Kent, R.W. 1997. The North Atlantic Igneous Province. *In*: Mahoney, J.J. & Coffin, M.F. (eds) *Large Igneous Provinces: Continental, Oceanic, and Planetary Flood Volcanism*. American Geophysical Union, Geophysical Monographs, **100**, 45–93.

Schoene, B., Samperton, K.M., Eddy, M.P., Keller, G., Adatte, T., Bowring, S.A. & Gertsch, B. 2015. U–Pb geochronology of the Deccan Traps and relation to the end-Cretaceous mass extinction. *Science*, **347**, 182–184.

Sensarma, S. 2007. A bimodal large igneous province and the plume debate: the Paleoproterozoic Dongargarh Group, central India. *In*: Foulger, G.R. & Jurdy, D.M. (eds) *Plates, Plumes, and Planetary Processes*. Geological Society of America, Special Papers, **430**, **430**, 831–839.

Sensarma, S. 2009. Low-$\delta^{18}O$ rhyolites in the c. 2.5 Ga Dongargarh bimodal LIP, India: textural and chemical evidence for magma mixing and source characteristics. Paper presented at the Geological Society of America Penrose Conference on Low-δ18O Rhyolites and Crustal Melting: Growth and Redistribution of the Continental Crust, 9–13 September 2009, Twin Falls, Idaho, and Yellowstone National Park, Wyoming, USA.

Sensarma, S., Palme, H. & Mukhopadhyay, D. 2002. Crust–mantle interaction in the genesis of siliceous high magnesian basalts: evidence from the Early Proterozoic Dongargarh Supergroup, India. *Chemical Geology*, **187**, 21–37.

Sensarma, S., Hoernes, S. & Mukhopadhyay, D. 2004. Relative contributions of crust and mantle to the origin of the Bijli Rhyolite in a Palaeoproterozoic bimodal volcanic sequence (Dongargarh Group), central India. *Journal of Earth System Science*, **113**, 619–648.

Sharma, A., Sensarma, S., Kumar, K., Khanna, P.P. & Saini, N.K. 2013. Mineralogy and geochemistry of the Mahi River sediments in tectonically active western India: implications for Deccan large igneous province source, weathering and mobility of elements in a semi-arid climate. *Geochimica et Cosmochimica Acta*, **104**, 63–83.

Sharma, K.K. 2004. The Neoproterozoic Malani magmatism of the northwestern Indian shield: implications for crust-building processes. *Proceedings of the Indian Academy of Science (Earth Planetary Sciences)*, **113**, 795–807.

Shellnutt, J.G. 2017. The Panjal Traps. *In*: Sensarma, S. & Storey, B.C. (eds) *Large Igneous Provinces from Gondwana and Adjacent Regions*. Geological Society, London, Special Publications, **463**. First published online July 6, 2017, https://doi.org/10.1144/SP463.4

Shellnutt, J.G., Bhat, G.M., Wang, K.-L., Yeh, M.-W., Brookfield, M.E. & Jahn, B.-M. 2015. Multiple mantle sources of the early Permian Panjal Traps, Kashmir, India. *American Journal of Science*, **315**, 589–619.

Sheth, H.C. 2005. From Deccan to Réunion: no trace of a mantle plume. *In*: Foulger, G.R., Natland, J.H., Presnall, D.C. & Anderson, D.L. (eds) *Plates, Plumes, and Paradigms*. Geological Society of America, Special Papers, **388**, 477–501.

Sheth, H.C. 2007. Plume-related regional pre-volcanic uplift in the Deccan Traps: absence of evidence, evidence of absence: discussion. *In*: Foulger, G.R. & Jurdy, D.M. (eds) *Plates, Plumes, and Planetary Processes*. Geological Society of America, Special Papers, **430**, 803–813.

Storey, B.C. 1995. The role of mantle plumes in continental breakup: case histories from Gondwanaland. *Nature*, **377**, 301–308.

Storey, B.C. & Kyle, P.R. 1997. An active mantle mechanism for Gondwana breakup. *In*: Hatton, C.J. (ed.)

Special Issue on the Proceedings of the Plumes, Plates and Mineralisation '97 Symposium. South African Journal of Geology, **100**, 283–290.

Storey, B.C., Vaughan, A.P.M. & Riley, T.R. 2013. The link between large igneous provinces, continental break-up and environmental change: evidence reviewed from Antarctica. *Earth and Environmental Science Transactions of the Royal Society of Edinburgh*, **104**, 1–14.

Storey, M., Mahoney, J.J. & Saunders, A.D. 1997. Cretaceous basalts in Madagascar and the transition between plume and continental lithosphere mantle sources. *In*: Mahoney, J.J. & Coffin, M.F. (eds) *Large Igneous Provinces: Continental, Oceanic, and Planetary Flood Volcanism*. American Geophysical Union, Geophysical Monographs, **100**, 95–122.

Svensen, H.H., Torsvik, T.H. *et al.* 2017. Gondwana Large Igneous Provinces: plate reconstructions, volcanic basins and sill volumes. *In*: Sensarma, S. & Storey, B.C. (eds) *Large Igneous Provinces from Gondwana and Adjacent Regions*. Geological Society, London, Special Publications, **463**. First published online August 30, 2017, https://doi.org/10.1144/SP463.7

Talukdar, S.C. & Murthy, M.V.N. 1971. The Sylhet traps, their tectonic history, and their bearing on problems of Indian flood basalt provinces. *Bulletin Volcanologique*, **35**, 602–618.

Taylor, S.R. & McLennan, S.M. 1985. *The Continental Crust; Its Composition and Evolution. An Examination of the Geochemical Record Preserved in Sedimentary Rocks*. Blackwell, Oxford.

Torsvik, T.H. & Cocks, L.R.M. 2013. GR focus review: Gondwana from top to base in space and time. *Gondwana Research*, **24**, 999–1030.

Veevers, J.J. 2004. Gondwanaland from 650 to 500 Ma assembly through 320 Ma merger in Pangea to 185–100 Ma breakup: supercontinental tectonics via stratigraphy and radiometric dating. *Earth-Science Reviews*, **68**, 1–132.

Vijaya Kumar, K., Laxman, M.B. & Nagaraju, K. 2017. Mantle source heterogeneity in continental mafic Large Igneous Provinces: insights from the Panjal, Rajmahal and Deccan basalts, India. *In*: Sensarma, S. & Storey, B.C. (eds) *Large Igneous Provinces from Gondwana and Adjacent Regions*. Geological Society, London, Special Publications, **463**. First published online July 11, 2017, https://doi.org/10.1144/SP463.5

Wang, M., Li, C., Wu, Y.-W. & Xie, C.-M. 2014. Geochronology, geochemistry, Hf isotopic compositions and formation mechanism of radial mafic dykes in northern Tibet. *International Geology Review*, **56**, 187–205.

White, R.S. 1997. Mantle plume origin for the Karoo and Ventersdorp flood basalts, South Africa. *In*: Hatton, C.J. (ed.) *Special Issue on the Proceedings of the Plumes, Plates and Mineralisation '97 Symposium. South African Journal of Geology*, **100**, 271–282.

White, R.S. & McKenzie, D. 1989. Magmatism at rift zones: the generation of volcanic continental margins and flood basalts. *Journal of Geophysical Research*, **94**, 7685–7729.

Wignall, P.B. 2001. Large igneous provinces and mass extinctions. *Earth-Science Reviews*, **53**, 1–33.

Wignall, P.B. 2005. The link between large igneous province eruptions and mass extinctions. *In*: Saunders, A.D. (ed.) *Large Igneous Provinces: Origin and Environmental Consequences. Elements*, **1**, (5), 293–297.

Zhu, D.C., Chung, S.L., Mo, X.X., Zhao, Z.D., Niu, Y., Song, B. & Yang, Y.H. 2009. The 132 Ma Comei-Bunbury large igneous province: remnants identified in present-day southeastern Tibet and southwestern Australia. *Geology*, **37**, 583–586, https://doi.org/10.1130/G30001A.1

Gondwana Large Igneous Provinces: plate reconstructions, volcanic basins and sill volumes

H. H. SVENSEN[1]*, T. H. TORSVIK[1,2,3,4], S. CALLEGARO[1], L. AUGLAND[1], T. H. HEIMDAL[1], D. A. JERRAM[1,5], S. PLANKE[1,6] & E. PEREIRA[7]

[1]*Centre for Earth Evolution and Dynamics (CEED), University of Oslo, PO Box 1028, Blindern, 0316 Oslo, Norway*

[2]*Geodynamics, Geologiske Undersøkelse (NGU), Leiv Eirikssons Vei 39, N-7491 Trondheim, Norway*

[3]*School of Geosciences, University of Witwatersrand, Johannesburg, WITS 2050, South Africa*

[4]*GFZ German Research Centre for Geosciences, Telegrafenberg, 14473 Potsdam, Germany*

[5]*DougalEARTH Ltd, 31 Whitefields Crescent, Solihull B91 3NU, UK*

[6]*Volcanic Basin Petroleum Research (VBPR), Oslo Science Park, Gaustadalléen 21, N-0349 Oslo, Norway*

[7]*Department of Stratigraphy and Paleontology, Rio de Janeiro State University, Brazil*

**Correspondence: hensven@geo.uio.no*

Abstract: Gondwana was an enormous supertarrane. At its peak, it represented a landmass of about 100×10^6 km^2 in size, corresponding to approximately 64% of all land areas today. Gondwana assembled in the Middle Cambrian, merged with Laurussia to form Pangea in the Carboniferous, and finally disintegrated with the separation of East and West Gondwana at about 170 Ma, and the separation of Africa and South America around 130 Ma. Here we have updated plate reconstructions from Gondwana history, with a special emphasis on the interactions between the continental crust of Gondwana and the mantle plumes resulting in Large Igneous Provinces (LIPs) at its surface. Moreover, we present an overview of the subvolcanic parts of the Gondwana LIPs (Kalkarindji, Central Atlantic Magmatic Province, Karoo and the Paraná–Etendeka) aimed at summarizing our current understanding of timings, scale and impact of these provinces. The Central Atlantic Magmatic Province (CAMP) reveals a conservative volume estimate of 700 000 km^3 of subvolcanic intrusions, emplaced in the Brazilian sedimentary basins (58–66% of the total CAMP sill volume). The detailed evolution and melt-flux estimates for the CAMP and Gondwana-related LIPs are, however, poorly constrained, as they are not yet sufficiently explored with high-precision U–Pb geochronology.

The Gondwana supertarrane had a surface area of about 100×10^6 km^2, and was assembled in Late Neoproterozoic and Cambrian times. It later amalgamated with Laurussia to form Pangea in the Late Carboniferous. Gondwana (or Gondwanaland) was first named by Medlicott & Blanford (1879), and included most of South America, Africa, Madagascar, India, Arabia, East Antarctica and Western Australia (Torsvik & Cocks 2013). Gondwana was affected by several Large Igneous Provinces (LIPs) from the Cambrian to its final break-up in the Cretaceous. These LIPs were likely to have been sourced from mantle plumes, and had major impacts on deep and shallow crustal rheology and on the Earth's climate. In general, the structure of the LIPs is two-fold, with (1) surface lava flows and pyroclastic deposits, and (2) a network of subvolcanic sills and dykes. The latter were commonly emplaced in sedimentary basins, and thus led to a wide range of effects including metamorphism, devolatilization, porosity reduction, and phreatic and phreatomagmatic eruptions.

The aim of this paper is to present plate reconstructions of the evolution of Gondwana, and to assess the timing and consequences of LIP formation. This is of great interest since Gondwana continental LIPs are associated with both major mass extinctions and rapid climatic changes. We stress that our contribution is not meant as a comprehensive review but as an update on key periods of the

From: SENSARMA, S. & STOREY, B. C. (eds) 2018. *Large Igneous Provinces from Gondwana and Adjacent Regions*. Geological Society, London, Special Publications, **463**, 17–40.
First published online August 20, 2017, https://doi.org/10.1144/SP463.7

long Gondwana history (pertaining to the LIPs: Kalkarindji, Central Atlantic Magmatic Province, Karoo and the Paraná–Etendeka), and a useful summary of our current understanding. We put a special emphasis on the emplacement environments of the subvolcanic parts of the LIPs, in particular of the Central Atlantic Magmatic Province (CAMP) and the affected sedimentary basins, as this theme is poorly known outside the Brazilian geological community.

Methods

Plate reconstructions

We used GPlates (www.gplates.org; and GMAP (Torsvik & Smethurst 1999) for plate reconstructions and data analysis. Mesozoic reconstructions use a palaeomagnetic reference frame based on the zero Africa longitude method (Torsvik *et al.* 2012), whilst our Cambrian reconstruction is detailed in Torsvik *et al.* (2014). The lower mantle is dominated by two equatorial and antipodal regions of low seismic shear-wave velocities, referred to as the Large Low Shear-wave Velocity Provinces (LLSVPs), or using more friendly acronyms of Kevin Burke, TUZO (the LLSVP beneath Africa) and JASON (its Pacific counterpart). TUZO and JASON are prominent in all global shear-wave tomographic models, and most reconstructed Large Igneous Provinces (LIPs) and kimberlites over the past 300 myr – and perhaps much longer (Torsvik *et al.* 2014, 2016) – have erupted directly above their margins, termed the plume generation zones (PGZs: Burke *et al.* 2008). In order to relate deep-mantle processes to surface processes in a palaeomagnetic reference frame, we have counter-rotated the PGZs to account for true polar wander (TPW). Here we use the 0.9% slow contour in the tomographic shear-wave model (s10mean: Doubrovine *et al.* 2016) as the best approximation of the plume source region in the deep mantle. This is shown in Figures 1–3, where the location of the PGZ (thick red line) is corrected for TPW. In this way, we apply estimated TPW rotations (Torsvik *et al.* 2014) to the mantle to simultaneously visualize how the reconstructed continents and the underlying mantle structures would look like in the palaeogeographical reference frame (Torsvik & Cocks 2013)

Geochronology

Most of the LIPs described here (see Table 1) have been dated by high-precision zircon U–Pb chemical abrasion isotope dilution thermal ionization mass spectrometry (CA-ID-TIMS) geochronology, with the use of a common U–Pb tracer allowing direct interlaboratory data comparison (i.e. the EARTHTIME tracer: Condon *et al.* 2015). The CA-ID-TIMS data from the Paraná part of the Paraná–Etendeka LIP (Janasi *et al.* 2011; Florisbal *et al.* 2014) and the data from Svensen *et al.* (2012) from the Karoo used in-house U–Pb tracers, but the tracer used in the latter study has been calibrated to the EARTHTIME tracer (Svensen *et al.* 2015*a*; Corfu *et al.* 2016). There are no U–Pb data from the Etendeka part of the Paraná–Etendeka LIP, so only $^{40}Ar/^{39}Ar$ ages are reported (corrected to the latest standards). We have not included ion microprobe (SIMS/SHRIMP) and laser ablation inductively coupled plasma mass spectrometry (LA-ICP-MS) data as these methods do not yield ages with the required precision to evaluate potential correlations of short-lived geological events (e.g. LIP magmatism and mass extinction events). Furthermore, due to the relatively low level of precision of the data, potential Pb-loss is difficult to constrain, again leading to large uncertainties regarding the accuracy of ages. Similarly, baddeleyite ages are generally left out of this review – unless they are combined with zircon and, hence, can be evaluated relative to zircon data – as these are also prone to Pb loss if not abraded (for a discussion of potential problems with baddeleyite ages, see, e.g., Corfu *et al.* 2016). All the zircon U–Pb ID-TIMS ages considered are reported, including tracer calibration uncertainties and excluding decay constants uncertainties, and are shown in Figure 4. For comparison, $^{40}Ar/^{39}Ar$ age ranges (recalculated to conform to the Fish Canyon sanidine age of Kuiper *et al.* 2008) are also included in Figure 4, as these ages represent the majority of analyses for several of the LIPs and has, in some cases, been used to argue for longevity of magmatic activity (e.g. Jourdan *et al.* 2005).

Basin-scale sill volume estimates

Total sill volumes from the sedimentary basins are notoriously difficult to estimate. Our approach is similar to that of Svensen *et al.* (2015*b*) from the Karoo Basin, using the aerial extent of the relevant basins and borehole-based sill thicknesses. Data from the literature (chiefly, outcrop, borehole and seismic data) were then scouted and used to estimate the total cumulative thickness of sills in the basins. Thicknesses of the sills are likely to vary across the basins, commonly with maxima at the centres; thus, minimum, maximum and mean thicknesses were used to calculate the total volume. We argue that the mean thickness is the most reasonable value to be used for estimating sill volumes in sedimentary basins, but minimum and maximum volumes are also reported in order to show the total range of plausible values. We summarize the literature-based sill volume estimates in Table 2, and our new CAMP compilation in Table 3, followed by a separate CAMP section in the Results.

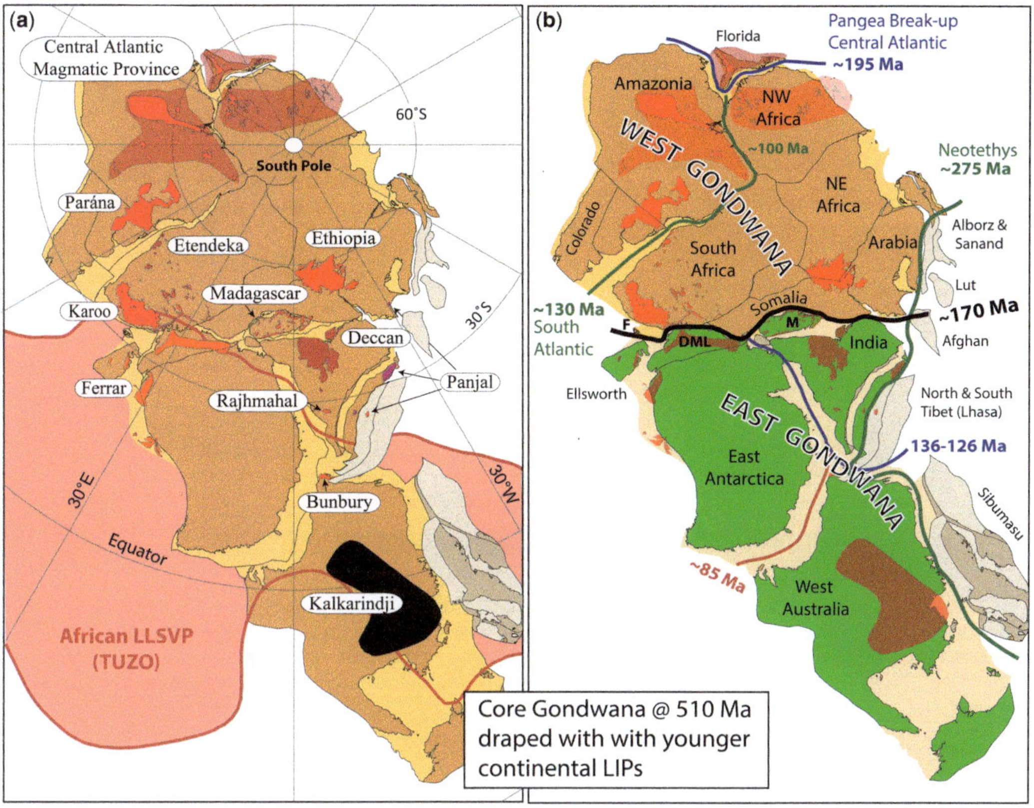

Fig. 1. (**a**) Reconstruction of core Gondwana at 510 Ma (Torsvik *et al.* 2014), around the time when the Kalkarindji LIP (black outline) erupted in Western Australia above the margin of TUZO (African Large Low Shear-wave Velocity Province; LLSVP). Also shown are all younger continental LIPs that affected Gondwana continental crust (see Table 1). (**b**) Similar to (a) but highlighting some important phases of Gondwana dispersal until the Early Cretaceous. The main break-up of Gondwana started at some time after 170 Ma when West and East Gondwana (green shading) separated. Peri-Gondwanan terranes that drifted off Gondwana (as part of Pangea) in the Early Permian (opening of Neotethys) are shaded in grey. F, Falkland; DML, Dronning Maud Land; M, Madagascar.

Results

The birth and demise of Gondwana

Unification of the many old cratons and terranes to form Gondwana began in the Late Neoproterozoic and continued into the Cambrian, but was largely over before the Middle Cambrian. The core of Gondwana included most of South America, Africa, Madagascar, India, Arabia, East Antarctica and Western Australia (Torsvik & Cocks 2013), and it reached a size of about 100×10^6 km^2 at its maximum (*c.* 64% of all land areas today). There were also many smaller units along its margins, such as Avalonia (including England), which had already drifted away during the Early Ordovician (with opening of the Rheic Ocean).

Gondwana merged with Laurussia (Laurentia, Baltica and Avalonia) in the Carboniferous at around 320 Ma to form Pangea. The disintegration of Gondwana had already started in the Early Ordovician but its demise after the Pangea assembly (Fig. 1) can be summarized as follows:

- Opening of the Neotethys from about 275 to 260 Ma, when a string of microcontinents and terranes such as Alborz and Sanand (Iran) through to Lut, Afghanistan, Tibet and Sibumasu moved away from the eastern rim of the Gondwanan sector of Pangea.
- Opening of the Central Atlantic at around 195 Ma. This led to the definite break between North and South Pangea, with the result that the core Gondwanan continent regained its independence for 20–30 myr. Florida was left behind with North America.
- Separation of East and West Gondwana at around 170 Ma; this is essentially the demise

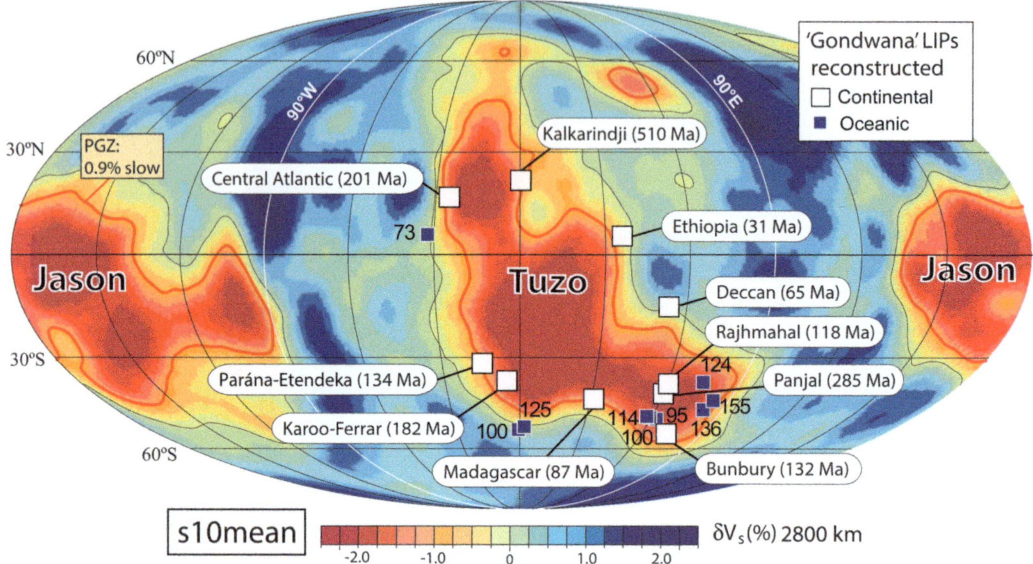

Fig. 2. Reconstruction of all 'Gondwana'-related Phanerozoic LIPs (31–510 Ma) using a hybrid reference frame (Torsvik *et al.* 2016) and draped on the s10mean tomographic model of Doubrovine *et al.* (2016). The plume generation zone (PGZ) in this model corresponds to the 0.9% slow contour. LIPs with white-squared symbols are continental LIPs, whilst those with blue-squared symbols are oceanic plateaus (LIP numbers are ages in Ma: see Table 1).

of Gondwana, but other important break-up phases that involved Gondwana continental crust included the East Gondwana break-up, starting with a separation between East Antarctica–Australia from India between 136 and 126 Ma, and the West Gondwana break-up, starting with opening of the South Atlantic at around 134 Ma.

Other younger but important post-Gondwana break-up phases include: (i) departure of India/Seychelles from Madagascar (opening of the Mascarene Basin) at around 84 Ma; (ii) East Antarctica and Australia separating at *c.* 85 Ma; (iii) India and Seychelles separating at 62–63 Ma; and (iv) departure of Arabia from Africa (opening of the Red Sea). This probably started with an early short-lived phase of seafloor spreading at around 26 Ma, and saw the onset of a second and still ongoing phase of seafloor spreading in the Pliocene, about 5 myr ago.

Constraining volumes: CAMP volcanism and sills in Brazil

Despite the growing amount of literature pertaining to the CAMP event, a thorough assessment of the preserved and original volumes of CAMP products is still lacking, with the only exception being a contribution by McHone (2003). In general, a total pre-erosional volume of 2.5×10^6 km^3 is addressed (e.g. Marzoli *et al.* 1999*b*). The total volume of magmatic products emplaced by the CAMP is within the same order of magnitude of those of other Large Igneous Provinces, but the general thickness (of outpoured material) is estimated as being smaller (Sebai *et al.* 1991; McHone 2003). Before we discuss the role of LIPs in the evolution and break-up of Gondwana, we would like to address the lack of constraint on the volumes associated with the CAMP event.

In general, intrusive and subvolcanic magmatism seems to constitute a large portion of the CAMP event. The only published estimates for volumes of CAMP products and relative degassing (McHone 2003) do not consider the existence of: (i) CAMP intrusions; (ii) Bolivian, Moroccan and Iberian CAMP occurrences (Knight *et al.* 2004; Bertrand *et al.* 2014; Callegaro *et al.* 2014); and (iii) an underestimate the extent of vast sills from Taoudenni–Hank–Reggane basins in Mali and Algeria (Verati *et al.* 2005; Charaf Chabou *et al.* 2010), from the Fouta Djalon Plateau (Deckart *et al.* 2005), and South America. Here we estimate the original volume of the CAMP starting from what is the present-day volume and surface of CAMP relicts. Particular emphasis is put on the sills and the intrusions, partly because the only available estimates to date are mainly focused on the lava piles, and partly because intrusive and subvolcanic bodies yield a high potential for degassing volatiles able to impact the global

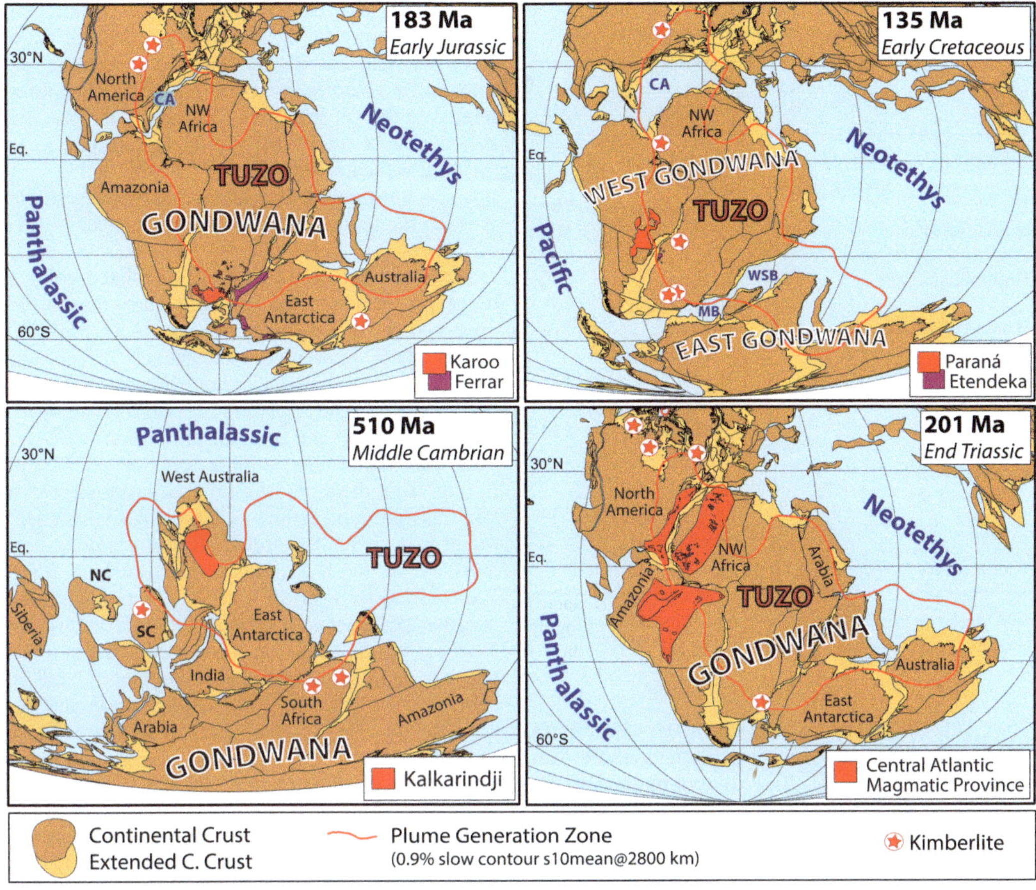

Fig. 3. Reconstructions at 510 Ma (Kalkarindji), 201 Ma (Central Atlantic Magmatic Province), 183 Ma (Karoo and Ferrar) and 135 Ma (134?: Paraná–Etendeka). These are palaeomagnetic reconstructions (Torsvik *et al.* 2012, 2014) but the plume generation zones has been counter-rotated to account for true polar wander. We have also reconstructed kimberlites within ±5 Ma for each reconstruction interval. CA, Central Atlantic; MB, Mozambique Basin; NC, North China; SC, South China; WSB, West Somali Basin.

environment (e.g. Svensen *et al.* 2004, 2009; Ganino & Arndt 2009; Aarnes *et al.* 2010), making them significant in the broader picture of understanding the relationship between CAMP and the end-Triassic mass extinction (Marzoli *et al.* 2004; Blackburn *et al.* 2013).

Products of extensive subvolcanic CAMP magmatism are present in northern Brazil, as thick doleritic sheets emplaced within the Palaeozoic sediments of the extensive Amazonas (*c.* 5×10^5 km^2), Solimões (*c.* $4 \times 10^5 - 6 \times 10^5$ km^2) and Acre basins (2.3×10^5 km^2: Milani & Zalán 1999; De Min *et al.* 2003) (Fig. 5). The stratigraphy of these deep (down to 5 km) intracontinental sedimentary basins includes three or four Palaeozoic supersequences of mainly siliciclastic rocks, along with thick (up to 1600 m: Milani & Zalán 1999) Carboniferous–Permian evaporitic and carbonate deposits. CAMP sills intruded Carboniferous–Permian evaporites and carbonates in the Solimões Basin, whereas, in the Amazonas Basin, sills are recorded both in Ordovician–Carboniferous siliciclastic sediments and in Carboniferous–Permian evaporites (Wanderley Filho *et al.* 2006). The maximum cumulative sill thickness (1038 m: Wanderley Filho *et al.* 2006) is reached in the Solimões Basin, and this decreases both eastwards towards the Amazonas Basin (between 100 and 809 m) (Fig. 5) and westwards towards the Acre Basin (Almeida 1986; De Min *et al.* 2003). The sill emplacement depth for the Brazilian basins typically varies between 1000 and 3500 m (Wanderley Filho *et al.* 2006) (Fig. 5). Affiliation to the CAMP of intrusive rocks found within the 600 000 km^2-wide Parnaíba sedimentary basin is questionable given the substantial spread in K–Ar ages shown by these rocks (Mizusaki *et al.*

Table 1. *Phanerozoic Large Igneous Provinces (LIPs) linked to 'Gondwana' and its dispersal history*

Continental	Age (Ma)	Oceanic plateaus	Age (Ma)
Ethiopia	31	Sierra Leone Rise	73
Deccan Traps	65	Broken Ridge	95
Madagascar	87	Central Kerguelen	100
Rajhmahal Traps	118	Agulhas Plateau	100
Bunbury Basalts–Cuvier–Gascoyne	132	Southern Kerguelen	114
Paraná–Etendeka	134	Wallaby Plateau	124
Argo Margin	155	Maud Rise	125
Karoo	182		
Central Atlantic Magmatic Province	201		
Panjal Traps	285		
Kalkarindji	510		

LIPs are separated into continental (see Fig. 1) and oceanic plateaus (Fig. 2).

2002). On the other hand, the presence of lava flows of clear CAMP age (Mosquito Formation: De Min *et al.* 2003; Merle *et al.* 2011) overlying the Palaeozoic sedimentary sequences in the Parnaíba Basin (the only known extrusive components of CAMP magmatism in Brazil: Marzoli *et al.* 1999*b*) may indicate that at least some of the dolerite sheets (Porto & Pereira 2014) recorded in the boreholes are CAMP related. The Parnaíba Basin does, however, host sills of undoubted Paraná–Etendeka affinity (Mizusaki *et al.* 2002; Porto & Pereira 2014) within its up to 3500 m-thick Palaeozoic sediments (siliciclastic and calcareous-evaporitic: Milani & Zalán 1999).

CAMP sills are often cropping out in restricted areas in sedimentary basins (outcrops cover *c.* 20% of the total), but borehole data from the basins generally intercept dolerites resting underneath most of the basin surface (cf., e.g., the Taoudenni Basin or the Brazilian basins: Milani & Zalán 1999; Verati *et al.* 2005). Boreholes and seismic data often reveal several sills stacked at different levels of the sedimentary pile (Fig. 5). We thus calculated the volume of CAMP sills by considering the extension of the basins that host them multiplied by the cumulative thickness of the doleritic sheets (Fig. 6; Table 3). Depending on the structure of the basin, the thickness of the sills can vary from the centre to the edges, tapering towards both ends of the basin (e.g. within the Amazon Basin: Fig. 5) or as a regional trend (e.g. thickness waning east and west of the Solimões Basin: De Min *et al.* 2003), resulting in high uncertainty for the thickness estimation. Therefore, we calculated the minimum, average and maximum volumes by considering the variability related to the sill thickness (Table 3). The total estimated volume of CAMP products approaches 1×10^6 km^3 if the average thickness of sills and in-basin flows is considered. If a less conservative estimate is used, the total subvolcanic volume exceeds 1.6×10^6 km^3.

It should be noted that these calculated volumes hinge on the extension of sedimentary basins and observation of the magmatic products preserved within them. Therefore, the better preservation of sills with respect to lava flows may lead to a disparity between intrusive and extrusive volumes that might not reflect the original proportion of CAMP products. While sills volume are well constrained through this method, lava-flow volumes should be taken as minimum estimates, since CAMP lava fields might have extended above all those regions presently cut by the dyke swarms (cf. McHone 2003). We decided to limit the calculations to basin areas because of the better-constrained information we have for these settings in the present: that is, being target to sedimentation, they are proven to be depocentres, capable of accommodating significant volumes of lava flows. Also, further potential CAMP occurrences are known to exist in remote areas of Africa and South America, but we here limit out assessment to the CAMP occurrences for which some (geochronological and/or geochemical) data have been published: that is, for which a CAMP affiliation was demonstrated. Therefore, we stress that these estimates are rather conservative and that future studies on the CAMP might result in larger volume estimates for this LIP.

Discussion

The major LIPs of Gondwana

Gondwana is not only unique for having been the largest unit of continental crust on Earth for more than 200 myr before its amalgamation with Laurussia in the Carboniferous to form Pangea, but also for hosting abundant LIPs. Around 30 Phanerozoic LIPs are commonly listed in global compilations, and 19 of these (11 continental and seven oceanic LIPs, Table 1) – 60% of all LIPs – affected Gondwanan continental lithosphere and its margins and oceanic lithosphere during Gondwana dispersal. Four case studies are discussed in this contribution, including the Cambrian Kalkarindji LIP in Western Australia, the end-Triassic CAMP that affected vast areas in North America, NW Africa and South America, the Early Jurassic Karoo–Ferrar LIP in South Africa and Antarctica (heralding the Jurassic

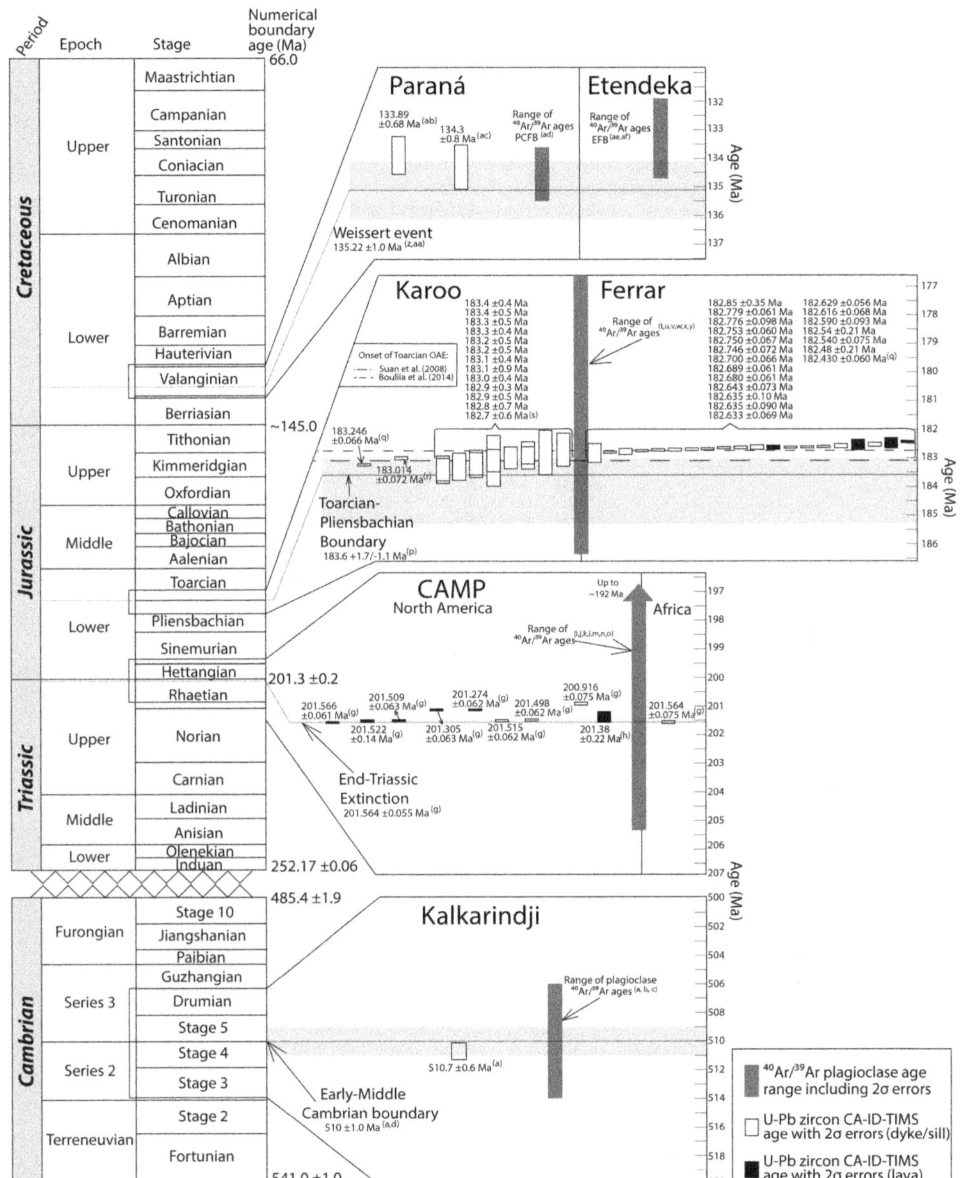

Fig. 4. Zircon U–Pb ID-TIMS and $^{40}Ar/^{39}Ar$ mineral data from the Central Atlantic Magmatic Province, the Karoo and Ferrar, and the Paraná–Etendeka Large Igneous Provinces. Uncertainties of all U–Pb ages reported include both the analytical uncertainty and the uncertainty derived from the tracer calibration, except for the data from Janasi *et al.* (2011) and Florisbal *et al.* (2014) where information on whether the tracer calibration uncertainties were included in the age calculations could not be obtained. The decay constant uncertainties are not included. Uncertainties on the reference events or boundaries in each panel are 2σ. The chronostratigraphic base chart is taken from Cohen *et al.* (2013; updated). The referred ages are taken from: (a) Jourdan *et al.* (2014); (b) Glass & Phillips (2006); (c) Evins *et al.* (2009); (d) Harvey *et al.* (2011); (e) Landing *et al.* (1998); (g) Blackburn *et al.* (2013); (h) Schoene *et al.* (2010); (i) Jourdan *et al.* (2009); (j) Nomade *et al.* (2007); (k) Hames *et al.* (2000); (l) Beutel *et al.* (2005); (m) Marzoli *et al.* (2011); (n) Knight *et al.* (2004); (o) Nomade *et al.* (2000); (q) Burgess *et al.* (2015); (r) Sell *et al.* (2014); (s) Svensen *et al.* (2012); (t) Jourdan *et al.* (2008); (u) Le Gall *et al.* (2002); (v) Jourdan *et al.* (2004); (w) Jourdan *et al.* (2005); (x) Jourdan *et al.* (2007*a*); (y) Jourdan *et al.* (2007b); (z) Martinez *et al.* (2015); (aa) Aguirre-Urreta *et al.* (2015); (ab) Florisbal *et al.* (2014); (ac) Janasi *et al.* (2011); (ad) Thiede & Vasconcelos (2010); (ae) Marzoli *et al.* (1999*a*); and (af) Renne *et al.* (1996). CAMP, Central Atlantic Magmatic Province; OAE, Oceanic Anoxic Event.

Table 2. *Major volcanic basins, Large Igneous Provinces (LIP), and environmental changes*

LIP	Comment	Correlated event	CIE*	Environmental changes	Biosphere changes	Sedimentary metamorphism	Sill age (Ma)	Sill volume (km^3)	References
Paraná–Etendeka	Paraná and Etendeka basins	Valanginian OAE	Positive	Cooling	limited	Sanstone/Shale	134		
	Paraná basin sills							112 000	4
	Etendeka sills and complexes							>10 000	This work
Karoo–Ferrar	Karoo Basin sills	Toarcian	Negative	Warming	Minor Extinction	Shale	182.6	370 000	1
	Antarctica sills					Sandstone + coal	182.6	170 000–230 000	2,3
	Tasmania sills							15 000	3
	All sills in the major basins							555 000–615 000	
CAMP	All affected basins	End-Triassic	Negative	Warming	Mass extinction	Evaporite + shale	201	700 000	This work
Siberian Traps	Tunguska Basin sills	End-Permian	Negative	Warming	Mass extinction	Evaporite + shale	252	780 000	5
Kalkarinji	Australia	Early–Middle Cambrian	Positive	OAE†	Mass extinction	Shale	510.7	?	

*CIE, Carbon Isotope Excursion.
†OAE, Oceanic Anoxic Event.
References: 1, Svensen *et al.* (2012); 2, Elliot & Fleming (2000); 3, Storey *et al.* (2013); 4, Frank *et al.* (2009); 5, Vasiliev *et al.* (2000).

Table 3. *Sedimentary basins with CAMP sills and their estimated volumes*

CAMP sills		Thickness (km)	Basin area (km²)	Average volume (km³)	Minimum volume (km³)	Maximum volume (km³)	References
North America	Newark Basin – Palisades (PA, NJ, NY)*	Up to 0.35	2400	480	240	840	1, 2, 3
	Culpeper Basin – Belmont, Rapidan sills (VA, MD)	0.37–0.6	2750	1380	550	1650	2, 4
	Gettysburg Basin York Haven sill (PA, MD)*	0.676	170	77	34	115	2, 4
	Deep River Basin – Sandford, Durham (NC, SC)	0.2	4000	800	200	1200	2, 3
	Dan River – Danville Basin (NC)	0.2	1500	300	75	450	2, 3
Africa	Anti Atlas (Morocco)	0.05–0.1	60 000	9000	6000	12 000	5
	Taoudenni (Mali)	0.2–0.4	100 000	30 000	20 000	40 000	5, 6
	Reggane, Tindouff, Hank (Mauritania, Algeria)	0.2–0.5	240 000	72 000	48 000	120 000	5
	Fouta Djalon (Guinea)	0.01–0.9	150 000	60 000	1500	135 000	7
South America	Acre (Brazil)	0.1–0.8	230 000	69 000	23 000	184 000	8
	Solimoes (Brazil)	Up to 1.038	400 000	200 000	40 000	400 000	8, 9
	Amazon (Brazil)	Up to 0.915	500 000	250 000	50 000	458 000	8, 9
	Tarabuco, Camiri (Bolivia)	0.03–0.14	30 000	2100	300	4200	10
	Devil's Island (French Guyana)	0.035	13	0	–	–	9
Europe	Pyrenees 'ophites' (Spain, France)	0.05–0.2	35 000	3500	1750	17 500	11, 12
			1 760 000	699 000	192 000	1 370 000	

*For the Gettysburg and Newark basins, the reported area is not that of the entire basins but only the area of extension of the sills.
Data from: 1, Puffer *et al.* (2009); 2, McHone (2003); 3, Luttrell (1989); 4, Woodruff *et al.* (1995); 5, Charaf Chabou *et al.* (2010); 6, Verati *et al.* (2005); 7, Deckart *et al.* (2005); 8, De Min *et al.* (2003); 9, Milani & Zalán (1999); 10, Bertrand *et al.* (2014); 11, Béziat *et al.* (1991); 12, Callegaro *et al.* (2014).

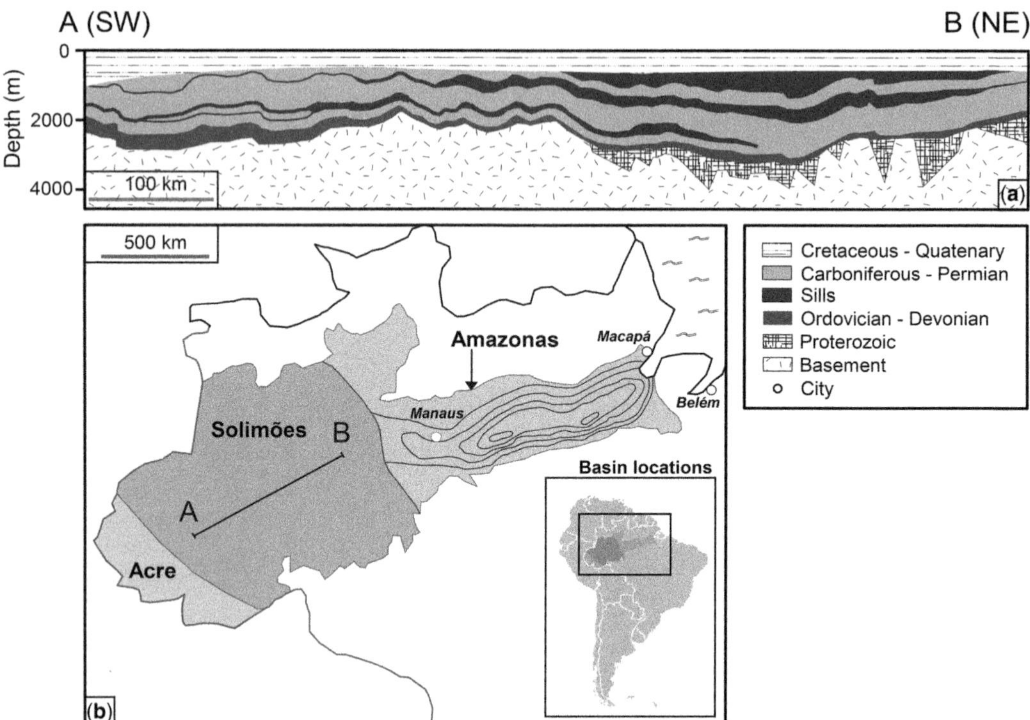

Fig. 5. (**a**) Qualitative SW–NE cross-section of the Solimões Basin (Brazil: A–B line marked in b highlights the presence of multiple thick sills encompassing the entire basin. (**b**) Outline of three Brazilian basins, Acre, Solimões and Amazon from west to east. Isopachs of cumulative sill thickness are marked for the Amazon Basin only (one line each 200 m in thickness, from 100 to 900 m). Adapted from Wanderley Filho *et al.* (2006).

break-up of Gondwana), and, finally, the Paraná–Etendeka LIP, which affected large areas in South America and SW Africa (e.g. Brazil, Namibia and Angola), and assisted the opening of the South Atlantic from around 134 Ma.

The Kalkarindji Large Igneous Province

In the early Palaeozoic, Gondwana stretched from the South Pole to the equator (West Australia) and the Kalkarindji continental flood basalt province erupted at equatorial latitudes (Figs 1 & 3). Kalkarindji is a relatively poorly known LIP located in the interior of Western Australia. The name of the province was originally proposed by Glass & Phillips (2006) to encompass the Early Cambrian basalts of the Antrim Plateau in northern Australia, and a number of smaller basalt and dolerite outcrops in the interior of the continent. Recent overviews of the province can be found in Ernst (2014) and www.largeigneousprovinces.org/LOM

The current extent of the Kalkarindji outcrops is about 55 000 km^2, with an estimated initial areal extent of 400 000 km^2 and a volume of 0.15×10^6 km^3 (Marshall *et al.* 2016 and references therein). The areal extent of the entire Kalkarindji province is estimated to be at least 2.1×10^6 km^2, but could be greater than 3×10^6 km^2 if igneous complexes in the Adelaide Fold Belt in South Australia are included as part of the province (Evins *et al.* 2009; Ernst 2014). The offshore extent of the province is currently unknown, but the Milliwindi dyke extends at least 70 km to the north of the Kimberly (Hanley & Wingate 2000). The volume of the erupted and associated shallow intrusive rocks is difficult to determine due to the scattered outcrops (see Fig. 7a), but is likely to be greater than 1.5×10^6 km^3 (Ernst 2014).

The igneous complexes of the Kalkarindji LIP consist dominantly of inflated, subaerial basalt flows and dolerite sheet intrusions. The basalts are dominantly low-Ti tholeiites and highly enriched in incompatible elements, suggesting continental contamination of the magmas (Glass & Phillips 2006; Ernst 2014). The most extensive and best-preserved basalt outcrops are located in the remote western part of the Antrim Plateau, reaching a thickness of at least 1.1 km. The lava flows are typically 20–60 m thick with massive interior and fractured or brecciated flow tops. Intraflow sedimentary units, including

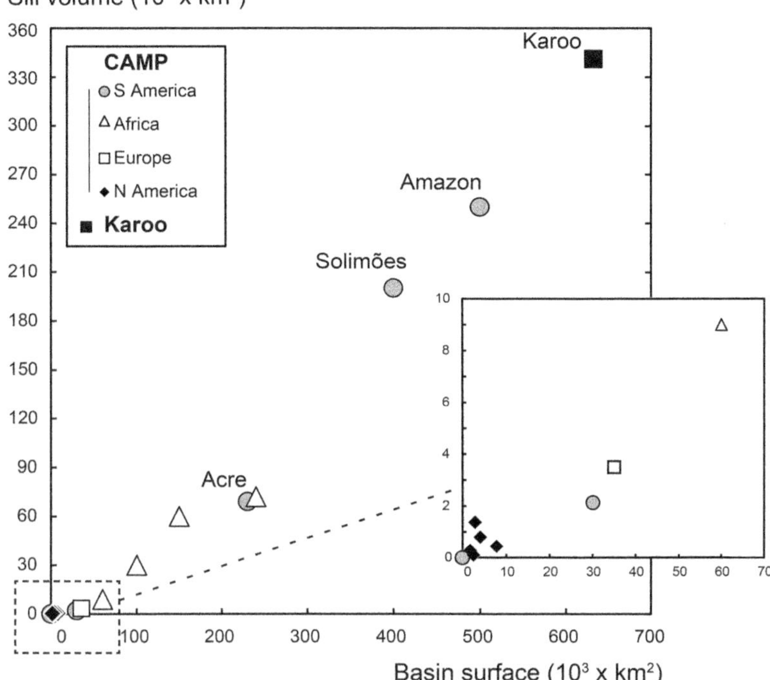

Fig. 6. The average volume of CAMP sills (in km^3) is plotted against the surface (in km^2) of the basins hosting them. Besides a clear positive correlation between the basin surface and total volume of the sills emplaced, the bigger volume of CAMP sills from the Brazilian basins compared to those from African, North American and European basins is visible. The inset is an enlarged detail of the smaller sills, and the volume of the Karoo sills is plotted for comparison.

aeolian sandstones, are locally present. A basaltic agglomerate, the Blackfell Rockhole Member, has been described as an explosive phreatomagmatic tephra deposit but was recently reinterpreted as 25–40 m-thick rubbly pahoehoe lava flow (Marshall *et al.* 2016). In the Antrim Plateau, the extrusive pile was emplaced above Proterozoic clastic sedimentary basins, and is overlain by stromatolitic limestones of Late Cambrian age.

Subvolcanic complexes and the igneous plumbing systems are poorly exposed. An eruptive centre in the central part of the Antrim Plateau has been proposed by Glass & Phillips (2006). The Milliwindi dolerite dyke, a NW-trending, more than 200-km-long intrusion in the NW Kimberly, has also been proposed as a potential feeder dyke for the extrusive basalts (Hanley & Wingate 2000).

An extensive Early Cambrian intrusive complex is present in the Officer Basin in central Australia (Jourdan *et al.* 2014). This Neoproterozoic intracratonic basin is more than 8 km deep with an areal extent of more than 300 000 km^2. It is filled with marine carbonates, clastic sediments and evaporites. The structuring of the basin is mainly due to halokinesis. Dolerite sills are exposed in the NW Akubra-Boondawari area, and are penetrated by several petroleum exploration boreholes (Grey *et al.* 2005). Exploratory fieldwork in the Antrim Plateau region to the north revealed no presence of intrusive sills or hydrothermal vent complexes in the Proterozoic basins stratigraphically below the basalts.

A U–Pb zircon age of 510.7 ± 0.6 Ma obtained from the coarse-grained Milliwindi dolerite dyke is considered a minimum age for the onset of magmatic activity in the Kalkarindji LIP (Jourdan *et al.* 2014). This age overlaps the Early–Middle Cambrian boundary (510.0 ± 1.0 Ma: Landing *et al.* 1998; Harvey *et al.* 2011) and the associated extinction event (Fig. 4). The four $^{40}Ar/^{39}Ar$ ages that exist from the Kalkarindji LIP come from three flows and a sill, and range from 509.0 ± 2.6 to 511.9 ± 1.9 Ma, consistent with the U–Pb age (Glass & Phillips 2006; Evins *et al.* 2009; Jourdan *et al.* 2014).

The Central Atlantic Magmatic Province

Pangea was straddling the equator in the Late Triassic (Fig. 3), and the Central Atlantic Magmatic

Fig. 7. (**a**) Two lava flows in a small quarry in Kalkarindji LIP, Australia. Photograph: S. Planke. (**b**) Planar sills in the Nico Malan Pass road section in the Karoo Basin, South Africa. Photograph: H. H. Svensen. (**c**) CAMP lava flows emplaced in a Mesozoic sedimentary basin crop out near Aouli village (close to Midelt) in the Moroccan Middle Atlas. Photograph: S. Callegaro. (**d**) Thick sill intrusion into aeolian sandstones, NW Namibia, Etendeka. Photograph: D.A. Jerram. (**e**) Panoramic view of a thick lava sequence south of the Huab River, NW Namibia. Section length is approximately 7 km, Etendeka. Photograph: D.A. Jerram. (**f**) Panorama of thick dolerite sills around 'Finger Mountain', Trans-Antartic Mountains, Antarctica. Outcrop length is approximately 6 km. Photograph: D.A. Jerram.

Province (CAMP) magmatic activity was located at equatorial–subtropical latitudes. As for many other LIPs, CAMP assisted a major plate tectonic reorganization that led to the opening of the central part of the Atlantic Ocean at around 195 Ma, and thus split the main area of the supercontinent into northern and southern Pangea, with the latter again often termed Gondwana. With a north–south extension exceeding 6000 km and laterally spread along more than 2500 km, CAMP is thus preserved (Marzoli *et al.*

1999*b*) on the eastern margin of North America (from Nova Scotia to Florida), western Europe (France and Iberian Peninsula), in West Africa (from Morocco to the Ivory Coast) and northern South America (French Guyana, Surinam, Brazil, and Bolivia). CAMP magmatism is expressed by lava flows from fissure eruptions fed by dykes and sills, with magma ponding in deep and shallow magma chambers, a few of which are preserved to date as layered mafic intrusions.

Most of the CAMP rocks are tholeiitic low-Ti ($TiO_2 < 2$ wt%: De Min *et al.* 2003) continental flood basalts or basaltic andesites, whereas CAMP high-Ti tholeiites are restricted to a narrow zone in NE South America (Suriname, French Guyana and northern Brazil: Deckart *et al.* 2005; Merle *et al.* 2011) and the southern margin of the West African craton (Liberia, Sierra Leone). A peculiar feature of this LIP compared to what is observed for others is the lack of alkaline and acidic magmatism (cf. e.g. Karoo or Deccan Traps; Marsh & Eales 1984; Parisio *et al.* 2016) and the very minor volume of high-Ti products (cf., e.g., magmatism of the Paraná–Etendeka LIP: Peate & Hawkesworth 1996). CAMP lava piles are mainly preserved within continental Mesozoic rift basins, often filled with red beds and evaporites. Lava-flow remnants crop out interdigitated among (or are buried between) fluvio-lacustrine sediments in basins of the Newark Supergroup in the eastern USA and Nova Scotia (Merle *et al.* 2014), of the High and Middle Atlas, Arganà and Meseta in Morocco (Knight *et al.* 2004), and the Algarve and Santiago do Caçem in Portugal (Callegaro *et al.* 2014). The volcanic piles may range in overall thicknesses between 50 and 600 m (e.g. Merle *et al.* 2014). Notably, other smaller basins spread all over the surface of the province do contain minor volumes of lava flows, such as in Bolivia (Bertrand *et al.* 2014) or in Algeria (Béchar: Charaf Chabou *et al.* 2010). Erosion wiped away a large portion of the outpoured tholeiites, but since geochemical observations show that CAMP lava piles were fed by the preserved coeval dykes (e.g. McHone 1996; Puffer *et al.* 2009; Merle *et al.* 2014), the original surface of CAMP lava fields may have corresponded to that presently enclosed by the dyke swarms (McHone 2003).

From the assessment presented here, we estimate that the total volume of CAMP products ranges within 1×10^6–1.6×10^6 km^3. It is clear from this analysis that the Brazilian basins host the majority of CAMP products, both in terms of sills (520 000 km^3, *c.* 70% of the total volume of CAMP sills) and of lava flows (64 000 km^3, *c.* 40% of the total volume estimated for basinal extrusive products). This assessment further highlights the fact that an important portion of the CAMP is constituted by intrusive and subvolcanic products, and that the majority of them are hosted by organic-rich sedimentary basins. Detailed studies of the magma–sediment interaction, coupled with high-precision U–Pb geochronology, will be the key for understanding the impact of CAMP on the global climate.

Only two high-precision zircon U–Pb CA-ID-TIMS studies have been published for the CAMP (Fig. 4) (Schoene *et al.* 2010; Blackburn *et al.* 2013). In addition, three air abrasion multigrain zircon and/or baddeleyite U–Pb ID-TIMS studies from two sills and one basalt in North America had previously been published (Dunning & Hodych 1990; Hodych & Dunning 1992; Schoene *et al.* 2006). Due to much larger uncertainties or unconstrained Pb-loss, these data are not included in Figure 4. The oldest U–Pb-dated CAMP basalt (the North Mountain Basalt; lowermost flow in the Fundy Basin) from North America has an age of 201.566 ± 0.061–201.52 ± 0.14 Ma (two different samples: Blackburn *et al.* 2013) or 201.38 ± 0.22 Ma (Schoene *et al.* 2010), and ages range up to 201.274 ± 0.062 for the youngest dated basalt (Blackburn *et al.* 2013). Three sills are also dated from the North American part of CAMP, giving ages between 201.515 ± 0.062 and 200.916 ± 0.075 Ma, the latter being the youngest CAMP sample dated (Blackburn *et al.* 2013). One sill from Morocco, representing the only high-precision U–Pb date from the African part of the CAMP, is dated at 201.564 ± 0.075 Ma (Blackburn *et al.* 2013). Blackburn *et al.* (2013) estimated the end-Triassic extinction to have occurred at 201.564 ± 0.055 Ma, demonstrating synchronicity between early CAMP magmatism and extinction. Available $^{40}Ar/^{39}Ar$ ages show a relatively large range from *c.* 204 to *c.* 192 Ma, but with a significant majority clustering tightly around 201–202 Ma, also indistinguishable from the mean age of all CAMP magmatism (Marzoli *et al.* 2011).

The Karoo–Ferrar Large Igneous Province and related basins

The bulk of Pangea was still rather intact by the Early Jurassic, with limited opening of the Central Atlantic. Karoo–Ferrar magmatic activity in southern Gondwana was centred on southerly latitudes (30–60°S: Fig. 3). The Karoo–Ferrar LIP covers a vast range of Gondwanaland areas. It is extensively preserved throughout southern Africa (e.g. South Africa, Namibia and Lesotho: du Toit 1920; Marsh *et al.* 1997; Neumann *et al.* 2011), in Tasmania and in Antarctica (e.g. Heimann *et al.* 1994; White *et al.* 2009). Invariably, the intrusive component of this LIP is emplaced within thick sedimentary sequences (e.g. the Karoo Supergroup in Africa and the Beacon Supergroup in Antarctica). In both the Antarctica sections and the Karoo sequences, the sill complexes can be mapped from the basement up through the main sedimentary succession (e.g. du Toit 1920; Marsh & Eales 1984; Galerne *et al.* 2008; Polteau *et al.* 2008; Jerram *et al.* 2010; Svensen *et al.* 2015*b*). By far, the most extensively studied parts of the Karoo–Ferrar LIP are in the Southern African portion and this provides us with an extensive geochronological framework expanded on below.

The Upper Carboniferous–Jurassic Karoo Supergroup in South Africa has a maximum cumulative thickness of 12 km and a preserved maximum thickness of 5.5 km (Tankard *et al.* 2009). The current area with Karoo sediments cropping out in South Africa is about 630 000 km^2 (Svensen *et al.* 2015*b*). The depositional environments range from marine to fluvial and, finally, aeolian (Catuneanu *et al.* 1998). The Karoo Basin is overlain by 1.3 km of volcanic rocks of the Drakensberg Group, consisting mainly of stacked basalt flows erupted into a continental and dry environment. The plumbing system of the flood basalts is a basin-scale sill complex consisting of sills and dykes (Marsh & Eales 1984; Chevallier & Woodford 1999). The thickest sill in the basin is about 220 m thick, most are in the 10–60 m range, and the sills were emplaced at about 182.6 Ma (Svensen *et al.* 2012, 2015*c*). The composition of the sills is mainly tholeitic, with a few more evolved intrusions (andesitic) (Marsh & Eales 1984; Neumann *et al.* 2011).

Hundreds of breccia pipes and hydrothermal vent complexes are rooted in the contact aureoles in the Karoo Basin (Svensen *et al.* 2006, 2007). Alexander du Toit suggested that these degassing pipes formed due to igneous gas release, but recent research has stressed the importance of sediment-derived volatiles in their generation (e.g. Jamtveit *et al.* 2004; Svensen *et al.* 2007; Aarnes *et al.* 2010, 2012). The devolatilization involved both organic and inorganic reactions, leading to the formation of high-temperature minerals and lowering of the organic carbon content in shales (Aarnes *et al.* 2012; Svensen *et al.* 2015*b*). The hydrothermal vent complexes commonly crop out in the uppermost 400–500 m of the basin, and are associated with sills in the subsurface (Svensen *et al.* 2006, 2015*b*). In the upper parts of the basin stratigraphy (e.g. in the Elliot and Clarens formations), magma–water interactions led to the formation of phreatomagmatic complexes. One of the best-studied complexes, the Sterkspruit Complex, represents a >45 km^2 explosion crater filled with a variety of lavas, tuffs, sediment breccias, hyaloclastites and various other pyroclastic rocks (McClintock *et al.* 2008).

In Antarctica, the volcanic and subvolcanic rocks are scattered across vast areas along the Transantarctic Mountains and Queen Maud Land (e.g. Elliot & Fleming 2000; Elliot & Hanson 2001; McClintock & White 2006; White *et al.* 2009; Muirhead *et al.* 2014). Along the Transantarctic Mountains, a large sedimentary basin and its basement are exposed (e.g. Barrett *et al.* 1986; Storey *et al.* 2013). The sedimentary rocks are classified as the Beacon Supergroup and include Devonian–Triassic clastic sediments, mainly sandstones, also including Permian coal seams. Early Jurassic sedimentary rocks include sandstones intermixed and interbedded with tuffs and peperites (e.g. McClintock & White 2006; Muirhead *et al.* 2014). One of the most prominent sills in the Ferrar is the Peneplain sill, estimated to cover 19 000 km^2 with a thickness of 250 m (4750 km^3) (Gunn & Warren 1962; White *et al.* 2009). Other thick sills include the Basement sill and the Asgard and Mount Fleming sills, all of which attain thickness in the hundreds of metres and have a large aerial extent (e.g. Marsh 2004; Jerram *et al.* 2010).

There are three high-precision U–Pb ID-TIMS papers published from Karoo (Svensen *et al.* 2012; Sell *et al.* 2014; Burgess *et al.* 2015) and one extensive study from Ferrar (including 19 samples from three different areas of Antarctica and one sample from Tasmania: Burgess *et al.* 2015) (Fig. 4). The ages from Karoo obtained by Svensen *et al.* (2012) were mainly obtained on air-abraded zircons, with only a few chemically abraded zircons, but the study includes an extensive dataset on 14 samples with concordant zircons showing a large degree of consistency and, thus, most probably representing the true ages of the dated sills within uncertainty. The zircon ages from Sell *et al.* (2014) and Burgess *et al.* (2015) were obtained by CA-ID-TIMS. Only sills have been dated from Karoo and the ages range from 183.4 ± 0.4 to 182.7 ± 0.6 Ma (Svensen *et al.* 2012; Corfu *et al.* 2016), with the two most precise ages being 183.246 ± 0.066 and 183.014 ± 0.075 Ma (Sell *et al.* 2014; Burgess *et al.* 2015). In Ferrar, 20 samples from 10 lavas, eight sills and two intrusions have been dated by CA-ID-TIMS (Burgess *et al.* 2015). The lava ages range from 182.779 ± 0.066 to 182.430 ± 0.066 Ma, and the intrusive rocks range in age from 182.85 ± 0.35 to 182.540 ± 0.075 Ma (Burgess *et al.* 2015). Different astrochronological models exist for the onset of the Toarcian Ocean Anoxic Event (TOAE), one model placing the onset of the TOAE at 183.1 Ma (Boulila *et al.* 2014) and another at 182.75 Ma (Suan *et al.* 2008; Burgess *et al.* 2015), both anchored at the radiometrically determined Pliensbachian–Toarcian boundary age of $183.6 + 1.7/-1.1$ (Pálfy *et al.* 2000; Sell *et al.* 2014). Whichever model is used, magmatism in Karoo overlaps the TOAE, while magmatism in Ferrar only overlaps with the TOAE using the model of Suan *et al.* (2008). Sills and lavas dated by $^{40}Ar/^{39}Ar$ show a relatively large spread (Fig. 4), but the majority of ages cluster between *c.* 182 and 186 Ma (Jourdan *et al.* 2008).

Paraná–Etendeka

By the Early Cretaceous, seafloor spreading in the Central Atlantic and the West Somali and Mozambique basins were well under way. The continents were spread from pole-to-pole and Paraná–Etendeka erupted at southerly subtropical latitudes (Fig. 3). The Cretaceous break-up of South America from Africa is a late-stage part of the Gondwana break-up and the Paraná–Etendeka LIP is the manifestation of this break-up; plume activity with volcanism started at 135–134 Ma, with large volumes of magma occurring in the first few million years up to, and including, the onset of break-up (e.g. Jerram & Widdowson 2005). This volcanism and the pre-volcanic stratigraphy help to correlate the Paraná (Brazil) with the Etendeka (Namibia). A large volume of the preserved volcanic rocks are found on the Paraná side.

The volcanic stratigraphy of the Paraná–Etendeka, as for many of the LIPs, has been mapped on a gross scale using chemostratigraphic relationships (e.g. Paraná: Peate 1997; Etendeka: Marsh *et al.* 2001). Increasingly detailed field-based volcanological correlations of the lava sequences have also identified disconformities within lava sequences of the same chemostratigraphic type, as well as detailed understanding of the volcanic evolution of the province (e.g. Jerram *et al.* 1999; Jerram & Stollhofen 2002; Waichel *et al.* 2008). Also, key correlations with the base basalt sequences and the immediate underlying and interbedded stratigraphy (Jerram & Widdowson 2005; Petry *et al.* 2007; Waichel *et al.* 2012) have been made. Correlations across the South Atlantic are possible using these stratigraphic and chemical relationships between key units. Volumes calculated with help from these correlations highlight large-volume individual events, with some of the largest individual silicic eruptions adding up to several thousands of km^3 (Bryan *et al.* 2010). Such events are manifested as both eruptions and as intrusions with big sill and volcanic centre complexes (Jerram & Bryan 2015).

The Paraná–Etendeka has also been linked with climatic changes as recorded by carbon isotope excursions, specifically the mid-Valanginian Weissert Event (e.g. Erba *et al.* 2004; Aguirre-Urreta *et al.* 2015; Martinez *et al.* 2015). However, the scale and impact of this event is somewhat small compared to events such as the end-Permian. The volcanic basin of the Paraná–Etendeka contains significant aeoilan deposits (Mountney *et al.* 1998; Jerram *et al.* 2000), and great interaction of the volcanics with these sandstones is observed (Jerram & Stollhofen 2002; Petry *et al.* 2007). As such, there may have been a dampening down of the effects of gasses given off during this event compared to other LIPs as the interaction can often result in a net sequestration of gasses, such as CO_2 in diagenetic cements (Jones *et al.* 2016).

The geochronological picture from the Paraná–Etendeka is incomplete. Some U–Pb ID-TIMS age data include a combined zircon (multigrain fraction; air abraded) and baddeleyite (multigrain) weighted mean age from a dacitic volcanic rock (Janasi *et al.* 2011), and a five-zircon (multigrain; CA) upper intercept age from a composite dyke (Florisbal *et al.* 2014). No tracer calibration uncertainties were reported for these data but, as the uncertainties on the calculated ages are rather large, this is not a critical factor for comparison with other ages. The interpreted ages for the dacite and the dyke, respectively, are 134.3 ± 0.8 and 133.9 ± 0.7 Ma (Janasi *et al.* 2011; Florisbal *et al.* 2014), but due to the number few analyses and the clear presence of Pb loss in the analysed zircons, the accuracy of the ages cannot be considered as highly robust. Thiede & Vasconcelos (2010) reviewed $^{40}Ar/^{39}Ar$ ages from the Paraná and compiled the most robust ages. They also re-analysed some of the samples that deviated significantly from the mean age, and showed that all robust ages cluster tightly and define an interval of magmatism of less than *c.* 1.2 Ma, with a mean age of 134.6 Ma. Recent palaeomagnetic data suggest a longer lived magmatic range of *c.* 4 Ma, within a similar time frame (Dodd *et al.* 2015).

On the Etendeka (Namibian) side, there are only $^{40}Ar/^{39}Ar$ ages reported (e.g. Renne *et al.* 1996; Jerram *et al.* 1999; Marzoli *et al.* 1999*a*). They are stated as slightly younger (133–132 Ma) but when recalculated using the latest techniques for Ar/Ar calibration (Kuiper *et al.* 2008; Renne *et al.* 2010, see also Dodd *et al.* 2015), these come in at *c.* 134 Ma, overlapping with ages from the Paraná Basin (Fig. 4). Magmatic activity in the Paraná–Etendeka Magmatic Province appears to be slightly younger – but overlapping within error – than the onset of the mid-Valanginian Weissert Event (carbon isotope shift), dated at 135.22 ± 1.0 Ma (e.g. Aguirre-Urreta *et al.* 2015; Martinez *et al.* 2015). The question of synchronicity and matching between the geochronological data available for the two events remains open until better age constraints are attained.

LIPs, plates and the big picture

The dawn of the Phanerozoic is exceptional in many ways: most continents were located in the southern hemisphere, the atmospheric CO_2 concentration was, perhaps, 5–10 times the current level, and a global sea-level rise with largely warm surface seawater temperatures appear to have characterized the Cambrian and much of the Ordovician (Fig. 8). Abrupt changes in the atmospheric concentration of greenhouse gases have occurred episodically through Earth's history, and many of these climatic and environmental perturbations show a causal relationship with LIP eruptions, which have been sourced by plumes from the margins of two main thermochemical provinces in the deep mantle, TUZO and JASON. The Gondwana LIPs were all sourced from TUZO (Figs 2 & 3). During the Palaeozoic, most of the continents moved northwards, Pangea formed in the Late Carboniferous, and by the Late Triassic Pangea was centred around the equator and overlying the TUZO LLSVP (Fig. 3). CAMP magmatic activity was located to equatorial–subtropical latitudes, and contemporaneous kimberlites are found in southern Africa and North America. CAMP and kimberlites were sourced by deep plumes from the western margin of TUZO. The CAMP marks the initial break-up of the Pangaea supercontinent, relicts of its extensive tholeiitic magmatism are presently preserved in four continents along both sides of the Atlantic Ocean (Marzoli *et al.* 1999*b*), and CAMP contributed to one of the 'big five' biotic extinction events in the Phanerozoic.

By the Early Cretaceous, seafloor spreading in the Central Atlantic and the West Somali and Mozambique basins were well underway. The continents were spread from pole-to-pole and Paraná–Etendeka erupted at southerly subtropical latitudes (Fig. 3) with contemporaneous kimberlite activity in South Africa, Namibia and NW Africa. As for the end-Triassic and Early Jurassic, this magmatic activity was sourced by deep plumes along the western margin of TUZO. This is also exemplified in Figure 9: in this schematic cartoon we envisage that long-lived subduction along the South American (Andean) margin was the triggering mechanism for a deep plume that eventually rose through the mantle and led to catastrophic upper mantle melting, forming the Paraná–Etendeka LIP.

LIPs may have caused or contributed to four of the 'big five' biotic extinction events in the Phanerozoic: the end-Devonian (Yakutsk in Siberia), the end-Permian (Siberian Traps), the end-Triassic (Central Atlantic Magmatic Province, CAMP) and the end-Cretaceous (K-Pg: Deccan Traps). In addition, the Middle Permian extinction (Capitanian) was causally linked to the Emeishan LIP (e.g. Bond *et al.* 2010; Jerram *et al.* 2016*b*), but formed outside Gondwana and is thus not included in this work. The K-Pg mass extinction event, however, is unique because it is coincident with the Chicxulub bolide impact. The oldest Phanerozoic extinction, and the third in importance, at the end of the Ordovician (Hirnantian) is not linked with any known LIP, even though widespread explosive volcanism is suggested to have played a role in the global cooling (Buggisch *et al.* 2010).

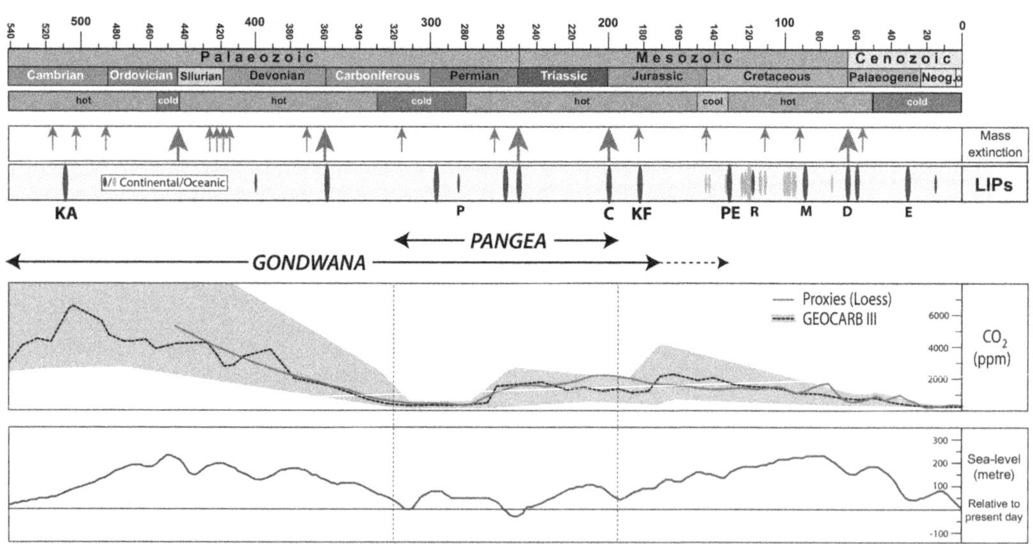

Fig. 8. Phanerozoic timescale, icehouse (cold) v. greenhouse (hot) conditions, extinction events (five major with larger arrows), global LIP events (continental and oceanic), atmospheric p_{CO_2} (Royer 2006) and global sea-level variations (Haq & Al-Qahtani 2005; Haq & Shutter 2008). Abbreviations for LIPs affecting Gondwana continental lithosphere are as follows: KA, Kalkarindji; P, Panjal; C, Central Atlantic Magmatic Province; KF, Karoo-Ferrar; PE, Parená-Etendeka; R, Rajhmahal; M, Madagascar; D, Deccan; E, Ethiopia. We have also indicated with arrows the life time of Gondwana and Pangea.

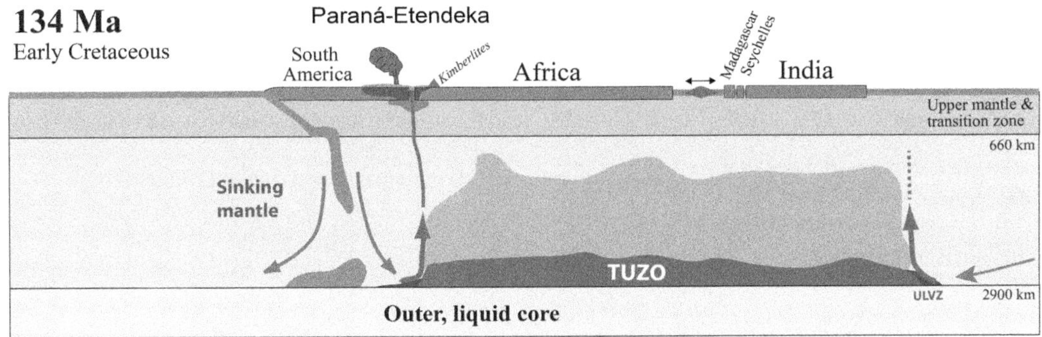

Fig. 9. Subducted lithospheric slabs may interact with the margins of TUZO and JASON, triggering the formation of plumes. The cartoon is a profile approximately through the reconstructed Paraná–Etendeka LIP, North Madagascar–Seychelles and India (cf. Fig. 3) at around 135 Ma. The triggering slabs for plumes sourcing the Paraná–Etendeka LIP and contemporaneous kimberlites in South Africa and Namibia must clearly have been linked to subduction along the western margin of South America. The South Atlantic opened shortly after the Paraná–Etendeka LIP.

The importance of subvolcanic intrusions and sill volumes

The timing of environmental changes and LIPs is summarized in Figure 8. Several processes are suggested as links between LIPs and environmental changes, including: (1) lava degassing of mantle volatiles, with or without a contribution from recycled continental crust; (2) degassing of volatiles derived from the sill-related contact aureoles of heated sedimentary rocks; and (3) degassing from sills and lavas contaminated by volatiles derived from the intruded sedimentary rocks or other crustal rocks (e.g. Svensen *et al.* 2004; Ganino & Arndt 2009; Sobolev *et al.* 2011). The actual extinction mechanisms are debated, and the suggestions include extreme temperatures, oceanic anoxia and pH reduction, sulphuric acid poisoning, and ozone layer destruction followed by extreme UV-B radiation (e.g. Visscher *et al.* 2004; Robock 2005; Bacon *et al.* 2013; Elliott-Kingston *et al.* 2014).

The importance of volcanic basins for the evolution of LIPs and the relationship to environmental changes has been stressed in recent years. During the formation of LIPs, magma is commonly emplaced as sills, dykes and igneous centres in the upper crust (Jerram & Bryan 2015). In cases where the volume of magma emplaced in sedimentary basins is high, these basins are commonly referred to as 'volcanic basins' (Fig. 10) (e.g. Jerram 2015; Abdelmalak *et al.* 2016; Jerram *et al.* 2016*a*). Volcanic basins are present along rifted continental margins and on lithospheric cratons (e.g. Coffin & Eldholm 1994), and represent vast basin areas that contain significant volumes of LIP-related igneous rocks and, in some cases, particularly where underplating has occurred and sill complexes exist, the intrusive component can be larger than the extrusive counterparts (basalt/lava) and pyroclastic deposits. Sill emplacement is characterized by the development of hydrothermal vent complexes that cut up through the basin and erupt at the surface through pipe structures (e.g. Fig. 10) (Svensen *et al.* 2006; Jerram *et al.* 2016*a*) that can be found both before and within the volcanic pile and which can indicate the relative timing of the sill emplacement (e.g. Angkasa *et al.* 2017).

Sill thicknesses can exceed 350 m, sandwiched between sedimentary sequences and associated contact metamorphic aureoles. Sills can also occur within the lava piles themselves (e.g. Hansen *et al.* 2011), but their identification is complicated within a similar host rock and also emphasizes a potential for the underestimate of the volume of the intrusive component. A selection of field photographs from several of the LIPs described in this work is shown in Figure 7. Contact metamorphism of the host sedimentary rocks cause devolatilization and the generation of H_2O, CO_2, CH_4 and SO_2 depending on the composition of the country rocks, including their content of organic carbon (e.g. Svensen *et al.* 2009). As a rule of thumb, the volume of heated sedimentary rocks is twice the sill volume, as suggested both by modelling studies and by mineral and maceral temperature proxies (e.g. Aarnes *et al.* 2011). These volatiles may reach the atmosphere via direct degassing in breccia pipes and hydrothermal vent complexes or by seepage through fractures and sediment permeability. In addition to the sediment degassing, basalt lava degassing releases a range of volatiles to the atmosphere, including CO_2 and SO_2 (e.g. Self *et al.* 2006).

A classic epitome of the LIP mass-extinction paradigm is the relationship between CAMP and end-Triassic mass extinction. Combined geochronological and stratigraphic data have evidenced that the

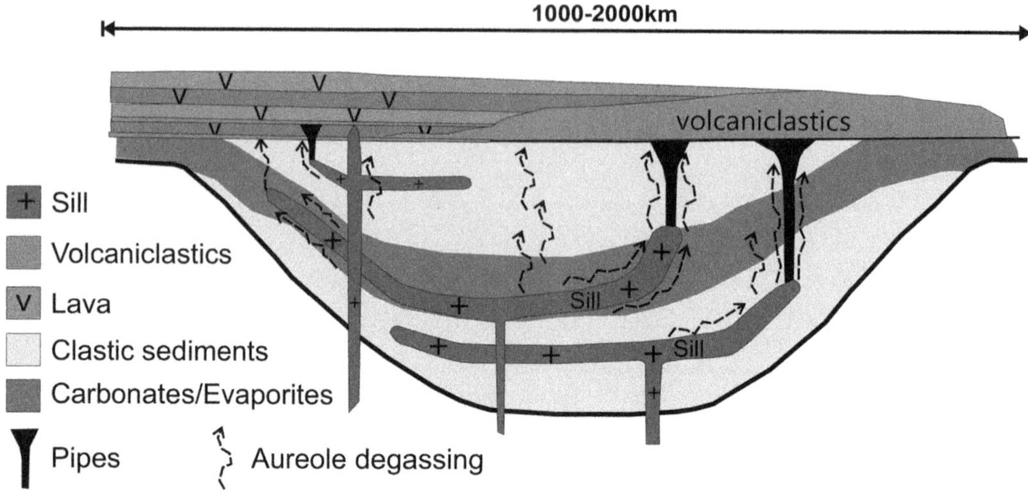

Fig. 10. Schematic cross-section of a volcanic basin, showing sills, dykes, pipes, and surface deposits of lava and pyroclastics. Modified from Jerram *et al.* (2016*a*).

two events coincide, but few constraints exist on how the CAMP played its forcing on the end-Triassic global environment. On the stratigraphic record, from worldwide localities, the end-Triassic biotic crisis is also associated to a negative carbon isotope excursion that reflects a disruption of the carbon cycle, for which the CAMP is considered the cause. Modelling shows that the sharp negative $\delta^{13}C$ excursion associated with the extinction can be obtained by rapid release of very isotopically light carbon (e.g. Bachan & Payne 2016). Since carbon released by flood basalt magmatism is isotopically too heavy to drive such negative excursions, the source of isotopically light carbon must be sought in the degassing from heated sediments. In this view, the study of subvolcanic intrusions and the estimation of sill volumes are fundamental because the degassing from organic-rich sediments in contact with such magmatic structures is, to the best of our knowledge, the most efficient way to transfer volatiles from the lithosphere to the Earth's atmosphere during LIP events. The volume estimation of CAMP magmas presented here further highlights that an important portion of the CAMP is constituted by intrusive and subvolcanic products, and that the majority of them are hosted by the organic-rich sedimentary basins in Brazil.

There are still unresolved aspects related to the age of sills and dykes, their volumes, and emplacement rate, even though recent work has made advances in several LIPs (e.g. Jourdan *et al.* 2005, 2014; Svensen *et al.* 2012; Blackburn *et al.* 2013; Burgess *et al.* 2015). These issues are challenging to resolve as sills and dykes are non-uniformly distributed in volcanic basins, and basin-scale 3D seismic data and boreholes are not always available. Still, we stress that current knowledge and data suggest that the subvolcanic intrusions in sedimentary basins hold a key to understand past environmental changes.

Summary

- In Palaeozoic and Early Mesozoic times, Gondwana LIPs and the majority of kimberlites were sourced by deep plumes from the margins of the African LLSVP (TUZO), one of two major stable thermochemical piles at the core–mantle boundary. Gondwana essentially existed from Cambrian to Jurassic times and in its lifetime three continental LIPs (Kalkarindji, CAMP and Karoo–Ferrar) directly affected Gondwanan continental crust at around 510, 201 and 183 Ma.
- Two of these LIPs (CAMP and Karoo–Ferrar) assisted the break-up of Gondwana and Pangea, as witnessed by the opening of the Central Atlantic Ocean (splitting Pangea) and the West Somali and Mozambique basins (separating West and East Gondwana) during the Jurassic.
- Soon after, the opening of the South Atlantic in the early Cretaceous from around 130 Ma took place only about 4–5 myr from the onset of the Paraná–Etendeka LIP, which affected vast areas in Brazil and parts of Namibia.
- The architecture of the major Gondwana LIPs consists of a subvolcanic and a volcanic part. Subvolcanic sills and dykes represent a significant part of the LIP igneous volume, commonly emplaced in sedimentary basins (i.e. volcanic

basins). Sills are commonly present throughout the basin stratigraphies and the sill volumes are thus scaled to the basin size.
- A new assessment of the CAMP LIP shows that the sills have a volume of about 700 000 km^3 (based on an average volume estimate – Table 3), more than half of it present in the Brazilian basins. This suggests that sill emplacement and the related contact metamorphism and devolatilization probably contributed to the climatic change and mass extinction at the end of the Triassic.

This work was supported by the Research Council of Norway through its Centres of Excellence funding scheme to the Centre of Earth Evolution and Dynamics (CEED), project number 223272. L. Parisio is thanked for fruitful insights and discussions over the extent of CAMP outcrops.

References

Aarnes, I., Svensen, H., Connolly, J.A.D. & Podladchikov, Y.Y. 2010. How contact metamorphism can trigger global climate changes: modeling gas generation around igneous sills in sedimentary basins. *Geochimica et Cosmochimica Acta*, **74**, 7179–7195.

Aarnes, I., Svensen, H., Polteau, S. & Planke, S. 2011. Contact metamorphic devolatilization of shales in the Karoo Basin, South Africa, and the effects of multiple sill intrusions. *Chemical Geology*, **281**, 181–194.

Aarnes, I., Podladchikov, Y. & Svensen, H. 2012. Devolatilization-induced pressure build-up: implications for reaction front movement and breccia pipe formation. *Geofluids*, **12**, 265–279, https://doi.org/10.1111/j.1468-8123.2012.00368.x

Abdelmalak, M.M., Planke, S., Faleide, J.I., Jerram, D. A., Zastrozhnov, D., Eide, S. & Myklebust, R. 2016. The development of volcanic sequences at rifted margins: new insights from the structure and morphology of the Vøring Escarpment, mid-Norwegian Margin. *Journal of Geophysical Research: Solid Earth*, **121**, 5212–5236, https://doi.org/10.1002/2015JB012788

Aguirre-Urreta, B., Lescano, M., Schmitz, M.D., Tunik, M., Concheyro, A., Rawson, P.F. & Ramos, V.A. 2015. Filling the gap: new precise Early Cretaceous radioisotopic ages from the Andes. *Geological Magazine*, **152**, 557–564.

Almeida, F.F.M. 1986. Distribuição regional e relações tectônicas do magmatismo pós-Palcozóico no Brasil. *Revista Brasileira de Geociências*., **16**, 325–349.

Angkasa, S.S., Jerram, D.A. *et al.* 2017. Mafic intrusions, hydrothermal venting, and the basalt-sediment transition: Linking onshore and offshore examples from the North Atlantic igneous Province. *Interpretation*, **5**, ISSN: 2324-8858, published 15 May 2017.

Bachan, A. & Payne, J.L. 2016. Modelling the impact of pulsed CAMP volcanism on p_{CO_2} and $\delta^{13}C$ across the Triassic–Jurassic transition. *Geological Magazine*, **153**, 252–270.

Bacon, K.L., Belcher, C.M., Haworth, M. & McElwain, J.C. 2013. Increased atmospheric SO_2 detected from changes in leaf physiognomy across the Triassic–Jurassic boundary interval of East Greenland. *PloS ONE*, **8**, e60614.

Barrett, P.J., Elliot, D.H. & Lindsay, J.F. 1986. The Beacon supergroup (Devonian–Triassic) and Ferrar Group (Jurassic) in the Beardmore glacier area, Antarctica. *Antarctic Research Series*, **36**, 339–428.

Bertrand, H., Fornari, M., Marzoli, A., García-Duarte, R. & Sempere, T. 2014. The Central Atlantic Magmatic Province extends into Bolivia. *Lithos*, **188**, 33–43.

Beutel, E.K., Nomade, S., Fronabarger, A.K. & Renne, P.R. 2005. Pangea's complex breakup: a new rapidly changing stress field model. *Earth and Planetary Science Letters*, **236**, 471, https://doi.org/10.1016/j.epsl.2005.03.021

Béziat, D., Joron, J.L., Monchoux, P., Treuil, M. & Walgenwitz, F. 1991. Geodynamic implications of geochemical data for the Pyrenean ophites (Spain–France). *Chemical Geology*, **89**, 243–262.

Blackburn, T.J., Olsen, P.E. *et al.* 2013. Zircon U–Pb geochronology links the end-Triassic extinction with the Central Atlantic Magmatic Province. *Science*, **340**, 941–945.

Bond, D.P.G., Hilton, J., Wignall, P.B., Ali, J.R., Stevens, L.G., Sun, Y.-D. & Lai, X.-L. 2010. The Middle Permian (Capitanian) mass extinction on land and in the oceans. *Earth-Science Reviews*, **102**, 100–116, https://doi.org/10.1016/j.earscirev.2010.07.004

Boulila, S., Galbrun, B., Huret, E., Hinnov, L., Rouget, I., Gardin, S. & Bartolini, A. 2014. Astronomical calibration of the Toarcian Stage: implications for sequence stratigraphy and duration of the early Toarcian OAE. *Earth and Planetary Science Letters*, **386**, 98–111.

Bryan, S.E., Ukstins Peate, I. *et al.* 2010. The largest volcanic eruptions on Earth. *Earth-Science Reviews*, **102**, 207–229.

Buggisch, W., Joachimski, M.M., Lehnert, O., Bergström, S.M., Repetski, J.E. & Webers, G.F. 2010. Did intense volcanism trigger the first late Ordovician icehouse? *Geology*, **38**, 327–330.

Burgess, S.D., Bowring, S.A., Fleming, T.H. & Elliot, D.H. 2015. High-precision geochronology links the Ferrar large igneous province with early-Jurassic ocean anoxia and biotic crisis. *Earth and Planetary Science Letters*, **415**, 90–99.

Burke, K., Steinberger, B., Torsvik, T.H. & Smethurst, M.A. 2008. Plume generation zones at the margins of large low shear velocity provinces on the core–mantle boundary. *Earth and Planetary Sciences*, **265**, 49–60.

Callegaro, S., Rapaille, C. *et al.* 2014. Enriched mantle source for the Central Atlantic magmatic province: new supporting evidence from southwestern Europe. *Lithos*, **188**, 15–32.

Catuneanu, O., Hancox, P.J. & Rubidge, B.S. 1998. Reciprocal flexural behaviour and contrasting stratigraphies: a new basin development model for the Karoo retroarc foreland system, South Africa. *Basin Research*, **10**, 417–439.

Charaf Chabou, M., Bertrand, H. & Sebaï, A. 2010. Journal of African Earth Sciences Geochemistry of the Central Atlantic Magmatic Province (CAMP) in south-western Algeria. *Journal of African Earth Sciences*, **58**, 211–219.

Chevallier, L. & Woodford, A. 1999. Morpho-tectonics and mechanism of emplacement of the dolerite rings and sills of the western Karoo, South Africa. *South African Journal of Geology*, **102**, 43–54.

COFFIN, M.F. & ELDHOLM, O. 1994. Large Ignoeus Provinces: crustal structure, dimensions and external consequences. *Reviews of Geophysics*, **32**, 1–36.

COHEN, K.M., FINNEY, S.C., GIBBARD, P.L. & FAN, J.-X. 2013. The ICS International Chronostratigraphic Chart. *Episodes*, **36**, 199–204.

CONDON, D.J., SCHOENE, B., MCLEAN, N.M., BOWRING, S.A. & PARRISH, R.R. 2015. Metrology and traceability of U–Pb isotope dilution geochronology (EARTHTIME Tracer Calibration Part I). *Geochimica et Cosmochimica Acta*, **164**, 464–480.

CORFU, F., SVENSEN, H. & MAZZINI, A. 2016. Comment to paper: evaluating the temporal link between the Karoo LIP and climatic–biologic events of the Toarcian Stage with high-precision U–Pb geochronology by Bryan Sell, Maria Ovtcharova, Jean Guex, Annachiara Bartolini, Fred Jourdan, Jorge E. Spangenberg, Jean-Claude Vicente, Urs Schaltegger in *Earth and Planetary Science Letters*, 408 (2014) 48–56. *Earth and Planetary Science Letters*, **434**, 349–352.

DECKART, K., BERTRAND, H. & LIÉGEOIS, J.-P. 2005. Geochemistry and Sr, Nd, Pb isotopic composition of the Central Atlantic Magmatic Province (CAMP) in Guyana and Guinea. *Lithos*, **82**, 289–314.

DE MIN, A., PICCIRILLO, E.M., MARZOLI, A., BELLIENI, G., RENNE, P.R., ERNESTO, M. & MARQUES, L.S. 2003. The Central Atlantic Magmatic Province (CAMP) in Brazil: petrology, geochemistry, ^{40}Ar/^{39}Ar ages, paleomagnetism and geodynamic implications. The central Atlantic magmatic province: insights from fragments of Pangea. *In*: HAMES, W., MCHONE, J.G., RENNE, P.C. & RUPPEL, C. (eds) *The Central Atlantic Magmatic Province: Insights from Fragments of Pangea*. American Geophysical Union, Geophysical Monograph Series, **136**, 91–128, https://doi.org/10.1029/136GM06

DODD, S.C., MAC NIOCAILL, C. & MUXWORTHYA, A.R. 2015. Long duration (>4 Ma) and steady-state volcanic activity in the early Cretaceous Paraná–Etendeka Large Igneous Province: new palaeomagnetic data from Namibia. *Earth and Planetary Science Letters*, **414**, 16–29.

DOUBROVINE, P.V., STEINBERGER, B. & TORSVIK, T.H. 2016. A failure to reject: testing the correlation between large igneous provinces and deep mantle structures with EDF statistics. *Geochemistry, Geophysics, Geosystems*, **17**, 1130–1163, https://doi.org/10.1002/2015GC006044

DUNNING, G.R. & HODYCH, J.P. 1990. U/Pb zircon and baddeleyite ages for the Palisades and Gettysburg sills of the northeastern United States: implications for the age of the Triassic/Jurassic boundary. *Geology*, **18**, 795, https://doi.org/10.1130/0091-7613(1990)0182.3.CO;2

DU TOIT, A.I. 1920. The Karoo dolerites of South Africa: a study in hypabyssal injection. *Transactions of the Geological Society of South Africa*, **23**, 1–42.

ELLIOT, D.H. & FLEMING, T.H. 2000. Weddell triple junction: the principal focus of Ferrar and Karoo magmatism during initial breakup of Gondwana. *Geology*, **28**, 539–542.

ELLIOT, D.H. & HANSON, R.E. 2001. Origin of widespread, exceptionally thick basaltic phreatomagmatic tuff breccia in the Middle Jurassic Prebble and Mawson Formations, Antarctica. *Journal of Volcanology and Geothermal Research*, **111**, 183–201.

ELLIOTT-KINGSTON, C., HAWORTH, M. & MCELWAIN, J.C. 2014. Damage structures in leaf epidermis and cuticle as an indicator of elevated atmospheric sulphur dioxide in early Mesozoic floras. *Review of Palaeobotany and Palynology*, **208**, 25–42.

ERBA, E., BARTOLINI, A. & LARSON, R.L. 2004. Valanginian Weissert oceanic anoxic event. *Geology*, **32**, 149–152.

ERNST, R.E. 2014. *Large Igneous Provinces*. Cambridge University Press, London.

EVINS, L., JOURDAN, F. & PHILIPS, D. 2009. The Cambrian Kalkarindji Large Igneous Province: extent and characteristics based on new ^{40}Ar/^{39}Ar and geochemical data. *Lithos*, **110**, 294–304, https://doi.org/10.1016/j.lithos.2009.01.014

FLORISBAL, L.M., HEAMAN, L.M., DE ASSIS JANASI, V.A. & DE FATIMA BITENCOURT, M. 2014. Tectonic significance of the Florianópolis dyke Swarm, Paraná–Etendeka Magmatic Province: a reappraisal based on precise U–Pb dating. *Journal of Volcanology and Geothermal Research*, **289**, 140–150.

FRANK, H.T., ELISA, M., GOMES, B., LUIZ, M. & FORMOSO, L. 2009. Review of the areal extent and the volume of the Serra Geral Formation, Paraná Basin, South America. *Pesquisas em Geosciences*, **36**, 49–57.

GALERNE, C.Y., NEUMANN, E.R. & PLANKE, S. 2008. Emplacement mechanisms of sill complexes: information from the geochemical architecture of the Golden Valley Sill Complex, South Africa. *Journal of Volcanology and Geothermal Research*, **177**, 425–440.

GANINO, C. & ARNDT, N.T. 2009. Climate changes caused by degassing of sediments during the emplacement of large igneous provinces. *Geology*, **37**, 323–326.

GLASS, L.M. & PHILLIPS, D. 2006. The Kalkarindji continental flood basalt province: a new Cambrian large igneous province in Australia with possible links to faunal extinctions. *Geology*, **34**, 461–464, https://doi.org/10.1130/G22122.1

GREY, K., HOCKING, R.M. ET AL. 2005. *Lithostratigraphic Nomenclature of the Officer Basin and Correlative Parts of the Paterson Orogen Western Australia*. Geological Survey of Western Australia, Report 93.

GUNN, B.M. & WARREN, G. 1962. Geology of Victoria Land between the Mawson and Murlock Glaciers, Antarctica. *New Zealand Geological Survey Bulletin*, **71**, 1–157.

HAMES, W.E., RENNE, P.R. & RUPPEL, C. 2000. New evidence for geologically instantaneous emplacement of earliest Jurassic Central Atlantic magmatic province basalts on the North 7 American margin. *Geology*, **28**, 859, https://doi.org/10.1130/0091-7613(2000)282.0.CO;2

HANLEY, L.M. & WINGATE, M.T.D. 2000. SHRIMP zircon age for an Early Cambrian dolerite dyke: an intrusive phase of the Antrim Plateau Volcanics of northern Australia. *Australian Journal of Earth Sciences*, **47**, 1029–1040, https://doi.org/10.1046/j.1440-0952.2000.00829.x

HANSEN, J., JERRAM, D.A., MCCAFFREY, K. & PASSEY, S.R. 2011. Early Cenozoic saucer-shaped sills of the Faroe Islands: an example of intrusive styles in basaltic lava piles. *Journal of the Geological Society, London*, **168**, 159–178, https://doi.org/10.1144/0016-7649 2010-012

HAQ, B.U. & AL-QAHTANI, A.M. 2005. Phanerozoic cycles of sea-level change on the Arabian Platform. *GeoArabia*, **10**, 127–160.

HAQ, B.U. & SHUTTER, S.R. 2008. A chronology of Paleozoic sea-level changes. *Science*, **322**, 64–68.

HARVEY, T.H.P., WILLIAMS, M. *ET AL.* 2011. A refined chronology for the Cambrian succession of southern Britain. *Journal of the Geological Society, London*, **168**, 705–716, https://doi.org/10.1144/0016-76492010-031

HEIMANN, A., FLEMING, T.H., ELLIOT, D.H. & FOLAND, K.A. 1994. A short interval of Jurassic continental flood basalt volcanism in Antarctica as demonstrated by $^{40}Ar/^{39}Ar$ geochronology. *Earth and Planetary Science Letters*, **121**, 19–41.

HODYCH, J.P. & DUNNING, G.R. 1992. Did the Manicouagan impact trigger end-of-Triassic mass extinction? *Geology*, **20**, 51–54, https://doi.org/10.1130/0091-7613(1992)0202.3.CO;2

JAMTVEIT, B., SVENSEN, H., PODLADCHIKOV, Y. & PLANKE, S. 2004. Hydrothermal vent complexes associated with sill intrusions in sedimentary basins. *In*: BREITKREUZ, C. & PETFORD, N. (eds) *Physical Geology of High-Level Magmatic Systems*. Geological Society, London, Special Publications, **234**, 233–241, https://doi.org/10.1144/GSL.SP.2004.234.01.15

JANASI, V.A., DE FREITAS, V.A. & HEAMAN, L.H. 2011. The onset of flood basalt volcanism, Northern Paraná Basin, Brazil: a precise U–Pb baddeleyite/zircon age for a Chapecótype dacite. *Earth and Planetary Science Letters*, **302**, 147–153.

JERRAM, D.A. 2015. *Hot Rocks and Oil: Are Volcanic Margins the New Frontier?* Geofacets. Elsevier, Amsterdam, https://www.elsevier.com/__data/assets/pdf_file/0008/84887/ELS_Geofacets-Volcanic-Article_Digital_r5.pdf

JERRAM, D.A. & BRYAN, S.E. 2015. Plumbing systems of shallow level intrusive complexes. *In*: BREITKREUZ, C. & ROCCHI, S. (eds) *Physical Geology of Shallow Magmatic Systems*. ISBN: 978-3-319-14083-4. Springer International Publishing.

JERRAM, D.A. & STOLLHOFEN, H. 2002. Lava/sediment interaction in desert settings; are all peperite-like textures the result of magma–water interaction? *Journal of Volcanology and Geothermal Research*, **114**, 231–249.

JERRAM, D.A. & WIDDOWSON, M. 2005. The anatomy of Continental Flood Basalt Provinces: geological constraints on the processes and products of flood volcanism. *Lithos*, **79**, 385–405.

JERRAM, D.A., MOUNTNEY, N., HOLZFÖRSTER, F. & STOLLHOFEN, H. 1999. Internal stratigraphic relationships in the Etendeka Group in the Huab Basin, NW Namibia: understanding the onset of flood volcanism. *Journal of Geodynamics*, **28**, 393–418.

JERRAM, D.A., MOUNTNEY, N., HOWELL, J., LONG, D. & STOLLHOFEN, H. 2000. Death of a Sand Sea: an active erg systematically buried by the Etendeka flood basalts of NW Namibia. *Journal of the Geological Society, London*, **157**, 513–516, https://doi.org/10.1144/jgs.157.3.513

JERRAM, D.A., DAVIS, G.R., MOCK, A., CHARRIER, A. & MARSH, B. 2010. Quantifying 3D crystal populations, packing and layering in shallow intrusions: a case study from the Basement Sill, Dry Valleys, Antarctica. *Geosphere*, **6**, 537–548, https://doi.org/10.1130/GES00538.1

JERRAM, D.A., SVENSEN, H.H., PLANKE, S., POLOZOV, A.G. & TORSVIK, T.H. 2016*a*. The onset of flood volcanism in the north-western part of the Siberian traps: explosive volcanism v. effusive lava flows. *Palaeogeography, Palaeoclimatology, Palaeoecology*, **441**, 38–50.

JERRAM, D.A., WIDDOWSON, M., WIGNALL, P.B., SUN, Y., LAI, X., BOND, D.P.G. & TORSVIK, T.H. 2016*b*. Submarine palaeoenvironments during Emeishan flood basalt volcanism, SW China: Implications for plume-lithosphere interaction during the Capitanian, Middle Permian ('end Guadalupian') extinction event. *Palaeogeography, Palaeoclimatology, Palaeoecology*. ISSN 0031-0182, **441**, 65–73. https://doi.org/10.1016/j.palaeo.2015.06.009

JONES, M.T., JERRAM, D.A., SVENSEN, H.H. & GROVE, C. 2016. The effects of large igneous provinces on the global carbon and sulphur cycles. *Palaeogeography, Palaeoclimatology, Palaeoecology*, **441**, 4–21.

JOURDAN, F., FÉRAUD, G. *ET AL.* 2004. The Karoo triple junction questioned: evidence from $^{40}Ar/^{39}Ar$ Jurassic and Proterozoic ages and geochemistry of the Okavango dike swarm (Botswana). *Earth and Planetary Science Letters*, **222**, 989–1006.

JOURDAN, F., FERAUD, G., BERTRAND, H., KAMPUNZU, A.B., TSHOSO, G., WATKEYS, M.K. & LE GALL, B. 2005. The Karoo large igneous province: brevity, origin, and relation with mass extinction questioned by new $^{40}Ar/^{39}Ar$ age data. *Geology*, **33**, 745–748.

JOURDAN, F., FÉRAUD, G., BERTRAND, H. & WATKEYS, M.K. 2007*a*. From flood basalts to the inception of oceanization: example from the $^{40}Ar/^{39}Ar$ high-resolution picture of the Karoo large igneous province. *Geochemistry, Geophysics, Geosystems*, **8**, 1–20, https://doi.org/10.1029/2006GC001392

JOURDAN, F., FÉRAUD, G., BERTRAND, H., WATKEYS, M.K. & RENNE, P.R. 2007*b*. Distinct brief major events in the Karoo large igneous province clarified by new $^{40}Ar/^{39}Ar$ ages on the Lesotho basalts. *Lithos*, **98**, 195–209.

JOURDAN, F., FÉRAUD, G., BERTRAND, H., WATKEYS, M.K. & RENNE, P.R. 2008. The $^{40}Ar/^{39}Ar$ ages of the sill complex of the Karoo large igneous province: implications for the Pliensbachian–Toarcian climate change. *Geochemistry, Geophysics, Geosystems*, **9**, 1–20, https://doi.org/10.1029/2008GC001994

JOURDAN, F., MARZOLI, A. *ET AL.* 2009. $^{40}Ar/^{39}Ar$ ages of CAMP in North America: implications for the Triassic–Jurassic boundary and the ^{40}K decay constant bias. *Lithos*, **110**, 167–180.

JOURDAN, F., HODGES, K. *ET AL.* 2014. High-precision dating of the Kalkarindji large igneous province, Australia, and synchrony with the Early–Middle Cambrian (Stage 4–5) extinction. *Geology*, **42**, 543–546.

KNIGHT, K.B., NOMADE, S., RENNE, P.R., MARZOLI, A., BERTRAND, H. & YOUBI, N. 2004. The Central Atlantic Magmatic Province at the Triassic–Jurassic boundary: paleomagnetic and $^{40}Ar/^{39}Ar$ evidence from Morocco for brief, episodic volcanism. *Earth and Planetary Science Letters*, **228**, 143–160.

Kuiper, K.F., Deino, A., Hilgen, F.J., Krijgsman, W., Renne, P.R. & Wijbrans, J.R. 2008. Synchronizing rock clocks of Earth history. *Science*, **320**, 500–504.

Landing, E., Bowring, S.A., Davidek, K.L., Westrop, S.R., Geyer, G. & Heldmaier, W. 1998. Duration of the Early Cambrian: U–Pb ages of volcanic ashes from Avalon and Gondwana. *Canadian Journal of Earth Sciences*, **35**, 329–338, https://doi.org/10.1139/e97-107

Le Gall, B., Tshoso, G. *et al.* 2002. $^{40}Ar/^{39}Ar$ geochronology and structural data from the giant Okavango and related mafic dike swarms, Karoo igneous province, Botswana. *Earth and Planetary Science Letters*, **202**, 595–606.

Luttrell, G.W. 1989. Stratigraphic nomenclature of the Newark supergroup of eastern North America. United States Geological Survey Bulletin 1572, 136 p.

Marsh, B.D. 2004. A magmatic mush column rosetta stone: the McMurdo Dry Valleys of Antarctica. *Eos, Transactions of the American Geophysical Union*, **85**, 497–508, https://doi.org/10.1029/2004EO470001

Marsh, J.S. & Eales, H.V. 1984. The chemistry and Petrogenesis of igneous rocks of the Karoo central area, southern Africa. *In*: Erlank, A.J. (ed.) *Petrogenesis of the Volcanic Rocks of the Karoo Province*. Geological Society of South Africa, Special Publications, **13**, 27–67.

Marsh, J.S., Hooper, P.R., Rehacek, J. & Duncan, R.A. 1997. Stratigraphy and age of the Karoo basalts of Lesotho and implications for correlations within the Karoo igneous province. *In*: Mahoney, J.J. & Coffin, M.F. (eds) *Large Igneous Provinces: Continental, Oceanic, and Planetary Flood Volcanism*. American Geophysical Union, Geophysical Monograph Series, **100**, 247–272.

Marsh, J.S., Ewart, A., Milner, S.C., Duncan, A.R. & Miller, R.McG. 2001. The Etendeka Igneous Province: magma types and their stratigraphic distribution with implications for the evolution of the Paraná–Etendeka flood basalt province. *Bulletin of Volcanology*, **62**, 464–486.

Marshall, P.E., Widdowson, M. & Murphy, D.T. 2016. The Giant Lavas of Kalkarindji: rubbly pāhoehoe lava in an ancient continental flood basalt province. *Palaeogeography, Palaeoclimatology, Palaeoecology*, **441**, 22–37.

Martinez, M., Deconinck, J.F., Pellenard, P., Riquier, L., Company, M., Reboulet, S. & Moiroud, M. 2015. Astrochronology of the Valanginian–Hauterivian stages (Early Cretaceous): chronological relationships between the Paraná–Etendeka large igneous province and the Weissert and the Faraoni events. *Global and Planetary Change*, **131**, 158–173.

Marzoli, A., Melluso, L. *et al.* 1999*a*. Geochronology and petrology of Cretaceous basaltic magmatism in the Kwanza basin (western Angola), and relationships with the Parana–Etendeka continental flood basalt province. *Journal of Geodynamics*, **28**, 341–356.

Marzoli, A., Renne, P.R., Piccirillo, E.M., Ernesto, M., Bellieni, G. & De Min, A. 1999*b*. Extensive 200-million-year-old continental flood basalts of the Central Atlantic Magmatic Province. *Science*, **284**, 616–618.

Marzoli, A., Bertrand, H. *et al.* 2004. Synchrony of the Central Atlantic magmatic province and the Triassic–Jurassic boundary climatic and biotic crisis. *Geology*, **32**, 973–976.

Marzoli, A., Jourdan, F. *et al.* 2011. Timing and duration of the Central Atlantic magmatic province in the Newark and Culpeper basins, eastern USA. *Lithos*, **122**, 175–188.

Marshall, P.E., Widdowson, M. & Murphy, D.T. 2016. The Giant Lavas of Kalkarindji: rubbly pāhoehoe lava in an ancient continental flood basalt province. *Palaeogeography, Palaeoclimatology, Palaeoecology*, **441**, 22–37, https://doi.org/10.1016/j.palaeo.2015.05.006

McClintock, M. & White, J.D.L. 2006. Large phreatomagmatic vent complex at Coombs Hills, Antarctica: wet. explosive initiation of flood basalt volcanism in the Ferrar–Karoo LIP. *Bulletin of Volcanology*, **68**, 215–239.

McClintock, M., White, J.D.L., Houghton, B.F. & Skilling, I.P. 2008. Physical volcanology of a large crater-complex formed during the initial stages of Karoo flood basalt volcanism, Sterkspruit, Eastern Cape, South Africa. *Journal of Volcanology and Geothermal Research*, **172**, 93–111.

McHone, G. 1996. Broad terrane Jurassic flood basalts across northeastern North America. *Geology*, **24**, 319–322.

McHone, J.G. 2003. Volatile emissions from Central Atlantic Magmatic Province Basalts: mass assumptions and environmental consequences. *In*: Hames, W., McHone, J.G., Renne, P.C. & Ruppel, C. (eds) *The Central Atlantic Magmatic Province: Insights from Fragments of Pangea*. American Geophysical Union, Geophysical Monograph Series, **136**, 241–254.

Medlicott, H.B. & Blanford, W.T. 1879. *Manual of the Geology of India Pt. 1*. Geological Survey Office, Calcutta.

Merle, R., Marzoli, A. *et al.* 2011. $^{40}Ar/^{39}Ar$ ages and Sr–Nd–Pb–Os geochemistry of CAMP tholeiites from Western Maranhão basin (NE Brazil). *Lithos*, **122**, 137–151.

Merle, R., Marzoli, A. *et al.* 2014. Sr, Nd, Pb and Os isotope systematics of CAMP tholeiites from Eastern North America (ENA): evidence of a subduction-enriched mantle source. *Journal of Petrology*, **55**, 133–180.

Milani, E.J. & Zalán, P.V. 1999. An outline of the geology and petroleum systems of the Paleozoic interior basins of South America. *Episodes*, **22**, 199–205.

Mizusaki, A.M.P., Thomaz-Filho, A., Milani, E.J. & De Césero, P. 2002. Mesozoic and Cenozoic igneous activity and its tectonic control in northeastern Brazil. *Journal of South American Earth Sciences*, **15**, 183–198.

Mountney, N., Howell, J., Flint, S. & Jerram, D. 1998. Aeolian and alluvial deposition within the Mesozoic Etjo Sandstone Formation, northwest Namibia. *Journal of African Earth Sciences*, **27**, 175–192.

Muirhead, J.D., Airoldi, G., White, J.D.L. & Rowland, J.V. 2014. Cracking the lid: sill-fed dikes are the likely feeders of flood basalt eruptions. *Earth and Planetary Science Letters*, **406**, 187–197.

Neumann, E.-R., Svensen, H., Galerne, C.Y. & Planke, S. 2011. Multistage evolution of dolerites in the Karoo Large Igneous Province, Central South Africa. *Journal of Petrology*, **52**, 959–984.

NOMADE, S., THÉVENIAUT, H., CHEN, Y., POUCLET, A. & RIGOLLET, C. 2000. Paleomagnetic study of French Guyana Early Jurassic dolerites: hypothesis of a multistage magmatic event. *Earth and Planetary Science Letters*, **184**, 155–168.

NOMADE, S., KNIGHT, K.B. *ET AL.* 2007. Chronology of the Central Atlantic Magmatic Province: implications for the Central Atlantic rifting processes and the Triassic–Jurassic biotic crisis. *Palaeogeography, Palaeoclimatology, Palaeoecology*, **244**, 326–344.

PÁLFY, J., SMITH, P.L. & MORTENSEN, J.K. 2000. A U–Pb and $^{40}Ar/^{39}Ar$ time scale for the Jurassic. *Canadian Journal of Earth Sciences*, **37**, 923–944.

PARISIO, L., JOURDAN, F. *ET AL.* 2016. $^{40}Ar/^{39}Ar$ ages of alkaline and tholeiitic rocks from the northern Deccan Traps: implications for magmatic processes and the K–Pg boundary. *Journal of the Geological Society, London*, **173**, 679–688, https://doi.org/10.1144/jgs2015-133

PEATE, D.W. 1997. The Parana-Etendeka Province. *In*: MAHONEY, J.J. & COFFIN, M.F. (eds) *Large Igneous Provinces*. American Geophysical Union, Washington, DC, 217–245.

PEATE, D.W. & HAWKESWORTH, C.J. 1996. Lithospheric to asthenospheric transition in Low-Ti flood basalts from southern Paranà & Brazil. *Chemical Geology*, **127**, 1–24.

PETRY, K., JERRAM, D.A., DEL PILAR, M.D. & ZERFASS, H. 2007. Volcanic–sedimentary features in the Serra Geral Fm., Paraná Basin, southern Brazil: examples of dynamic lava–sediment interactions in an arid setting. *Journal of Volcanology and Geothermal Research*, **159**, 313–325.

POLTEAU, S., FERRÉ, E., PLANKE, S., NEUMANN, E.-R. & CHEVALLIER, L. 2008. How are saucer-shaped sills emplaced? Constraints from the Golden Valley Sill, South Africa. *Journal of Geophysical Research*, **113**, B12104, https://doi.org/10.1029/2008JB005620

PORTO, A. & PEREIRA, E. 2014. Seismic interpretation of igneous intrusions and their implications for an unconventional petroleum system in southeastern Parnaiba Basin, northeastern Brazil. Abstract V51B presented at *American Geophysical Union Fall Meeting 2014*, 15–19 December 2014, San Francisco, CA, USA.

PUFFER, J.H., BLOCK, K.A. & STEINER, J.C. 2009. Transmission of flood basalts through a shallow crustal sill and the correlation of sill layers with extrusive flows: the Palisades intrusive system and the basalts of the Newark Basin, New Jersey, U. S. A. *Journal of Geology*, **117**, 139–155.

RENNE, P.R., GLEN, J.M., MILNER, S.C. & DUNCAN, A.R. 1996. Age of Etendeka flood volcanism and associated intrusions in southwestern Africa. *Geology*, **24**, 659–662.

RENNE, P.R., MUNDIL, R., BALCO, G., MIN, K. & LUDWIG, K. R. 2010. Joint determination of ^{40}K decay constants and $^{40}Ar^{*}/^{40}K$ for the Fish Canyon sanidine standard, and improved accuracy for $^{40}Ar/^{39}Ar$ geochronology. *Geochimica et Cosmochimica Acta*, **74**, 5349–5367.

ROBOCK, A. 2005. Cooling following large volcanic eruptions corrected for the effect of diffuse radiation on tree rings. *Geophysical Research Letters*, **32**, L06702.

ROYER, D.L. 2006. CO_2-forced climate thresholds during the Phanerozoic. *Geochimica et Cosmochimica Acta*, **70**, 5665–5675.

SCHOENE, B., CROWLEY, J.L., CONDON, D.J., SCHMITZ, M.D. & BOWRING, S.A. 2006. Reassessing the uranium decay constants for geochronology using ID-TIMS U–Pb data. *Geochimica et Cosmochimica Acta*, **70**, 426–445.

SCHOENE, B., GUEX, J., BARTOLINI, A., SCHALTEGGER, U. & BLACKBURN, T.J. 2010. Correlating the end- Triassic mass extinction and flood basalt volcanism at the 100 ka level. *Geology*, **38**, 387, https://doi.org/10.1130/G30683.1

SEBAI, A., FERAUD, G., BERTRAND, H. & HANES, J. 1991. ^{40}Ar–^{39}Ar dating and geochemistry of tholeiitic magamatism related to the early opening of the Central Atlantic rift. *Earth and Planetary Science Letters*, **194**, 455–472.

SELF, S., WIDDOWSON, M., THORDARSON, T. & JAY, A.E. 2006. Volatile fluxes during flood basalt eruptions and potential effects on the global environment: a Deccan perspective. *Earth and Planetary Science Letters*, **248**, 518–532.

SELL, B., OVTCHAROVA, M. *ET AL.* 2014. Evaluating the temporal link between the Karoo LIP and climatic–biologic events of the Toarcian Stage with high-precision U–Pb geochronology. *Earth and Planetary Science Letters*, **408**, 48–56.

SOBOLEV, S.V., SOBOLEV, A.V. *ET AL.* 2011. Linking mantle plumes, large igneous provinces and environmental catastrophes. *Nature*, **477**, 312–316.

STOREY, B.C., VAUGHAN, A.P.M. & RILEY, T.R. 2013. The links between large igneous provinces, continental break-up and environmental change: evidence reviewed from Antarctica. *Earth and Environmental Science, Transactions of the Royal Society of Edinburgh*, **104**, 17–30.

SUAN, G., PITTET, B., BOUR, I., MATTOILI, E., DUARTE, L.V. & MAILLIOT, S. 2008. Duration of the Early Toarcian carbon isotope excursion deduced from spectral analysis: consequence for its possible causes. *Earth and Planetary Science Letters*, **267**, 666–679, https://doi.org/10.1016/j.epsl.2007.12.017

SVENSEN, H., PLANKE, S. & MALTHE-SØRENSSEN, A. 2004. Release of methane from a volcanic basin as a mechanism for initial Eocene global warming. *Nature*, **429**, 3–6.

SVENSEN, H., JAMTVEIT, B., PLANKE, S. & CHEVALLIER, L. 2006. Structure and evolution of hydrothermal vent complexes in the Karoo Basin, South Africa. *Journal of the Geological Society, London*, **163**, 671–682, https://doi.org/10.1144/1144-764905-037

SVENSEN, H., PLANKE, S., CHEVALLIER, L., MALTHE-SØRENSSEN, A., CORFU, B. & JAMTVEIT, B. 2007. Hydrothermal venting of greenhouse gases triggering Early Jurassic global warming. *Earth and Planetary Science Letters*, **256**, 554–566.

SVENSEN, H., PLANKE, S., POLOZOV, A.G., SCHMIDBAUER, N., CORFU, F., PODLADCHIKOV, Y.Y. & JAMTVEIT, B. 2009. Siberian gas venting and the end-Permian environmental crisis. *Earth and Planetary Science Letters*, **277**, 490–500.

SVENSEN, H., CORFU, F., POLTEAU, S., HAMMER, Ø. & PLANKE, S. 2012. Rapid magma emplacement in the Karoo Large Igneous Province. *Earth and Planetary Science Letters*, **325–326**, 1–9. https://doi.org/10.1016/j.epsl.2012.01.015

SVENSEN, H.H., HAMMER, Ø. & CORFU, F. 2015*a*. Astronomically forced cyclicity in the Upper Ordovician and U–Pb ages of interlayered tephra, Oslo Region, Norway. *Palaeogeography, Palaeoclimatology, Palaeoecology*, **418**, 150–159.

SVENSEN, H.H., PLANKE, S. *ET AL*. 2015*b*. Sub-volcanic intrusions and the link to global climatic and environmental changes. *In*: BREITKREUZ, C. & ROCCHI, S. (eds) *Physical Geology of Shallow Magmatic Systems*. ISBN: 978-3-319-14083-4. Springer International Publishing.

SVENSEN, H., POLTEAU, S., CAWTHORN, G. & PLANKE, S. 2015*c*. Sub-volcanic intrusions in the Karoo Basin, South Africa. *In*: BREITKREUZ, C. & ROCCHI, S. (eds) *Physical Geology of Shallow Magmatic Systems*. ISBN: 978-3-319-14083-4. Springer International Publishing.

TANKARD, A., WELSINK, H., AUKES, P., NEWTON, R. & STETTLER, E. 2009. Tectonic evolution of the Cape and Karoo basins of South Africa. *Marine and Petroleum Geology*, **26**, 1379–1412.

THIEDE, D.S. & VASCONCELOS, P.M. 2010. Parana flood basalts: rapid extrusion hypothesis confirmed by new $^{40}Ar/^{39}Ar$ results. *Geology*, **38**, 747–750.

TORSVIK, T.H. & COCKS, L.R.M. 2013. GR focus review: Gondwana from top to base in space and time. *Gondwana Research*, **24**, 999–1030.

TORSVIK, T.H. & SMETHURST, M.A. 1999. Plate tectonic modeling: virtual reality with GMAP. *Computer & Geosciences*, **25**, 395–402.

TORSVIK, T.H., VAN DER VOO, R. *ET AL*. 2012. Phanerozoic polar wander, paleogeography and dynamics. *Earth-Science Reviews*, **114**, 325–368.

TORSVIK, T.H., VAN DER VOO, R. *ET AL*. 2014. Deep mantle structure as a reference frame for movements in and on the Earth. *Proceedings of the National Academy of Sciences of the United States of America*, **111**, 8735–8740.

TORSVIK, T.H., STEINBERGER, B., ASHWAL, L.D., DOUBROVINE, P.V. & TRØNNES, R.C. 2016. Earth evolution and dynamics – a tribute to Kevin Burke. *Canadian Journal of Earth Sciences Special Issue*, **53**, 1073–1087, https://doi.org/10.1139/cjes-2015-0228

VASILIEV, Y.R., ZOLOTUKHIN, V.V., FEOKTISTOV, G.D. & PRUSSKAYA, S.N. 2000. Evaluation of the volume and genesis of Permo-Triassic Trap magmatism on the Siberian Platform. *Russian Geology and Geophysics*, **41**, 1696–1705.

VERATI, C., BERTRAND, H. & FÉRAUD, G. 2005. The farthest record of the Central Atlantic Magmatic Province into West Africa craton: precise $^{40}Ar/^{39}Ar$ dating and geochemistry of Taoudenni basin intrusives (northern Mali). *Earth and Planetary Science Letters*, **235**, 391–407.

VISSCHER, H., LOOY, C.V., COLLINSON, M.E., BRINKHUIS, H., CITTERT, J.H.A.V.K.V., KURSCHNER, W.M. & SEPHTON, M.A. 2004. Environmental mutagenesis during the end-Permian ecological crisis. *Proceedings of the National Academy of Sciences of the United States of America*, **101**, 12 952–12 956.

WAICHEL, B.L., SCHERER, C.M.S. & FRANK, H.T. 2008. Basaltic lavas covering active Aeolian dunes in the Paraná Basin in Southern Brazil: features and emplacement aspects. *Journal of Volcanology and Geothermal Research*, **169**, 59–72.

WAICHEL, B.L., DE LIMA, E.F., VIANA, A.R., SCHERER, C.M., BUENO, G.V. & DUTRA, G. 2012. Stratigraphy and volcanic facies architecture of the Torres Syncline, Southern Brazil, and its role in understanding the Paraná–Etendeka Continental Flood Basalt Province. *Journal of Volcanology and Geothermal Research*, **215–216**, 74–82.

WANDERLEY FILHO, J.R., TRAVASSOS, W.A.S. & ALVES, D.B. 2006. O diabásio nas bacias paleozóicas amazônicas-herói ou vilão. [The diabase in the Amazonian Paleozoic basins: hero or villain?]. *Boletim de Geociências da Petrobrás*, **14**, 177–184.

WHITE, J.D.L., BRYAN, S.E., ROSS, P.-S., SELF, S. & THORDARSON, T. 2009. Physical volcanology of continental large igneous provinces: update and review. *In*: THORDARSON, T., SELF, S., LARSEN, G., ROWLAND, S.K. & HÖSKULDSSON, A. (eds) *Studies in Volcanology: The Legacy of George Walker*. IAVCEI, Special Publications, **2**, 291–321.

WOODRUFF, L.G., FROELICH, A.J., BELKIN, H.E. & GOTTFRIED, D. 1995. Evolution of tholeiitic diabase sheet systems in the eastern United States: examples from the Culpeper Basin, Virginia–Maryland, and the Gettysburg Basin, Pennsylvania. *Journal of Volcanology and Geothermal Research*, **64**, 143–169.

The Ferrar Large Igneous Province: field and geochemical constraints on supra-crustal (high-level) emplacement of the magmatic system

DAVID H. ELLIOT[1]* & THOMAS H. FLEMING[2]

[1]*School of Earth Sciences and Byrd Polar and Climate Research Center, Ohio State University, Columbus, OH 43210, USA*

[2]*Department of Earth Sciences, Southern Connecticut State University, New Haven, CT 06515, USA*

**Correspondence: elliot.1@osu.edu*

Abstract: The Ferrar Large Igneous Province forms a linear outcrop belt for 3250 km across Antarctica, which then diverges into SE Australia and New Zealand. The province comprises numerous sills, a layered mafic intrusion, remnants of extensive lava fields and minor pyroclastic deposits. High-precision zircon geochronology demonstrates a restricted emplacement duration (<0.4 myr) at *c.* 182.7 Ma, and geochemistry demonstrates marked coherence for most of the Ferrar province. Dyke swarms forming magma feeders have not been recognized, but locally have been inferred geophysically. The emplacement order of the various components of the magmatic system at supra-crustal levels has been inferred to be from the top-down lavas first, followed by progressively deeper emplacement of sills. This order was primarily controlled by magma density, and the emptying of large differentiated magma bodies from depth. An alternative proposal is that the magma transport paths were through sills, with magmas moving upwards to eventually reach the surface to be erupted as extrusive rocks. These two hypotheses are evaluated in terms of field relationships and geochemistry in the five regional areas where both lavas and sills crop out. Either scenario is possible in one or more instances, but neither hypothesis applies on a province-wide basis.

Supplementary material: The locations of samples, and trace element data and major element analyses of samples are available at: https://doi.org/10.6084/m9.figshare.c.3819454

Large Igneous Provinces (LIPs) are characterized by the eruption of large volumes of magma in a relatively short span of time (Coffin & Eldholm 1994; Eldholm & Coffin 2000). Some old provinces are represented mainly by intrusive rocks (e.g. Mackenzie Dyke Swarm and the Muskox Intrusion: LeCheminant & Heaman 1989; Baragar *et al.* 1996), whereas others, young in age, are primarily known only from extrusive lava flows (e.g. Columbia River Basalts: Reidel *et al.* 2013). Provinces associated with the break-up of Gondwana in Mesozoic time (Paraná: Piccirillo *et al.* 1990; Peate 1997; Karoo: Marsh *et al.* 1997; Neumann *et al.* 2011) provide a more complete cross-section of the supra-crustal magmatic architecture, which is represented by a mix of intrusive and extrusive components. These Mesozoic provinces offer the possibility of addressing the mechanisms of emplacement by examination of the various elements of the magmatic system in the context of the spatial, temporal and geochemical relationships. The order of emplacement of the various elements of the system can potentially be assessed by field and geochronological data. Whereas lavas have clear stratigraphic chronology, field relationships seldom provide compelling evidence for the order of sill emplacement, nor for the order between intrusive and extrusive magmatic phases. Geochronological data in complex provinces such as the Karoo (Jourdan *et al.* 2005, 2007, 2008) have yielded a wide range of $^{40}Ar/^{39}Ar$ ages (determined on plagioclase), with uncertainties that make interpretation of the order of emplacement difficult to assess. These uncertainties remain, even though U–Pb age determinations have been conducted on zircons from Karoo dolerites and Lebombo rhyolites (Riley *et al.* 2004; Svensen *et al.* 2012; Sell *et al.* 2014).

The Ferrar LIP, which crops out principally in Antarctica but extends to SE Australia and New Zealand (Fig. 1), is a relatively simple province, with a limited range of geochemical types. The purpose of this paper is to evaluate, through field and geochemical data, two alternative hypotheses for the mechanisms of emplacement of sills and lavas.

The Ferrar Large Igneous Province

The Ferrar Large Igneous Province (FLIP), in outcrop, has a narrow linear distribution for 3250 km

From: SENSARMA, S. & STOREY, B. C. (eds) 2018. *Large Igneous Provinces from Gondwana and Adjacent Regions*. Geological Society, London, Special Publications, **463**, 41–58.
First published online July 10, 2017, https://doi.org/10.1144/SP463.1

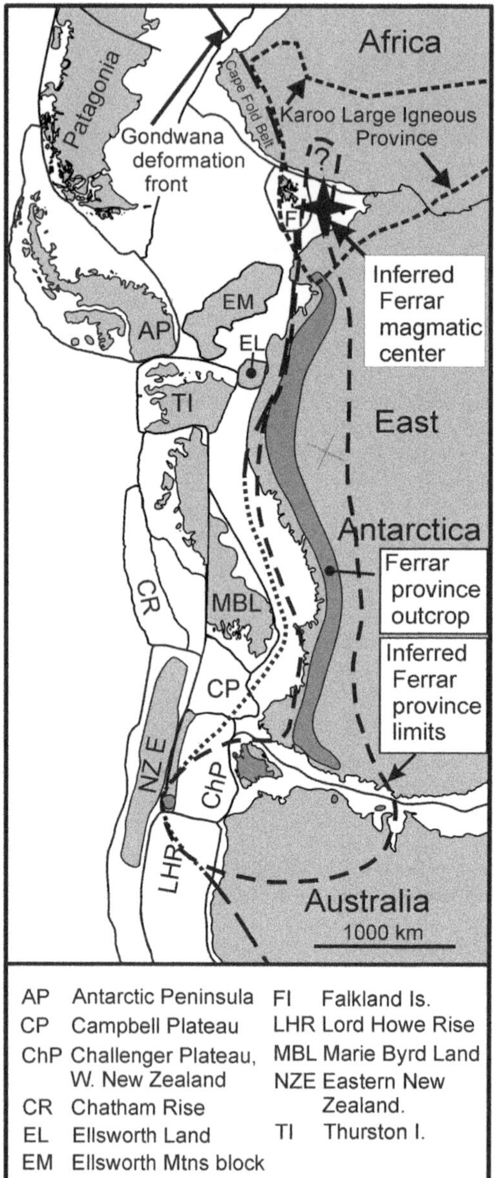

Fig. 1. Distribution of the Ferrar LIP in Gondwana (Gondwana modified from a reconstruction provided by the PLATES Project at the Institute of Geophysics at the University of Texas at Austin). The original extent of the Ferrar province beneath the East Antarctic Ice Sheet is speculative. The present-day South Pole is marked by a cross in East Antarctica.

across Antarctica from the Theron Mountains to north Victoria Land (Fig. 2), but then extends for another 500 km, and diverges to a width of about 1500 km to encompass New Zealand, Tasmania and SE Australia. The outcrop pattern in Antarctica is controlled mainly by Cenozoic uplift of the Transantarctic Mountains (see Elliot 2013). The distribution at the time of emplacement was undoubtedly much greater. Intrusive rocks comprise the Ferrar Dolerite sills and minor dykes (Elliot & Fleming 2004), and the Dufek layered mafic intrusion (Ford & Himmelberg 1991). The sills, with individual thicknesses generally between 100 and 200 m and cumulative thickness of 1500 m or more, occur predominantly in the 2.0–2.5 km-thick Devonian–Triassic Beacon Supergroup (Fig. 3) (Barrett 1991; Collinson *et al.* 1994; Bradshaw 2013). The sills, excluding outliers at and near Horn Bluff (longitude *c.* 150°E, about 700 km NW of the Mesa Range in Fig. 2), occur in a nearly continuous outcrop belt extending for about 2000 km from north Victoria Land to the Ohio Range, and intermittently thereafter for a further 1300 km to the Theron Mountains. Massive dolerite sills are particularly abundant south of the Mackay Glacier in south Victoria Land. Dykes are not common, but intrude both basement rocks and the Beacon sequence. In rare instances, dykes are observed cutting across Ferrar sills. The Dufek intrusion, as a whole, was first estimated to have a volume of approximately 50 000 km^3 and a thickness of 8–9 km (Behrendt *et al.* 1981; Ford & Himmelberg 1991), although later investigations led to the suggestion that it is two separate bodies with lesser thicknesses and a much smaller cumulative volume of approximately 6600 km^3 (Ferris *et al.* 1998). Extrusive rocks in the Ferrar province consist of basaltic pyroclastic rocks overlain by flood lavas of the Kirkpatrick Basalt, which attain a maximum extant thickness of approximately 900 m (Elliot & Fleming 2008). Intrusive rocks form 99% of the Ferrar outcrops and have an estimated volume of approximately 1.7×10^5 km^3, assuming a 150 km-wide original outcrop belt, whereas extrusive rocks, with an estimated extant volume of approximately 7000 km^3, are confined to widely spaced remnants of formerly more extensive lava fields (Fleming *et al.* 1995).

No strata younger than the Kirkpatrick Basalt lavas are present in the Transantarctic Mountains, except for Upper Cenozoic glacial deposits and alkaline volcanic rocks. A significant amount of stratigraphic section (Ferrar lavas and post-Ferrar sedimentary rocks) may be missing as a result of post-Early Jurassic erosion resulting mainly from uplift of the Transantarctic Mountains.

Geochemistry

The unifying characteristic of the FLIP is the distinctive geochemistry. Ferrar rocks are characterized by enriched Sr and Nd initial isotope ratios (Fig. 4), and crust-like trace element patterns (Fig. 5). The FLIP is

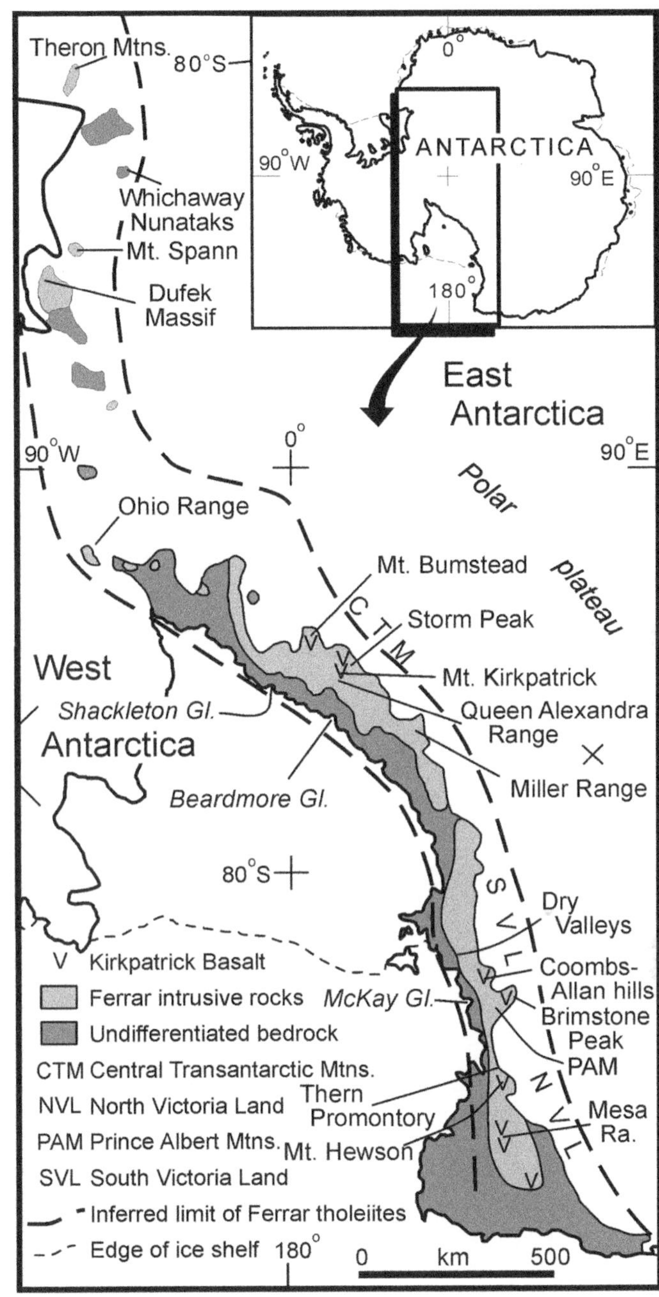

Fig. 2. Location map for the Ferrar LIP in Antarctica.

separated into the Mount Fazio Chemical Type (MFCT) and the Scarab Peak Chemical Type (SPCT) (Fleming *et al.* 1992) (Table 1). The MFCT comprises 99% of the province and includes the majority of all analysed rocks. The SPCT occurs as the stratigraphically youngest lava flow(s) in the central Transantarctic Mountains, and south and north Victoria Land, and also as sills in the Theron Mountains and the Whichaway Nunataks (Fig. 2) (Brewer *et al.* 1992; Leat *et al.* 2006). One sill from north Victoria Land has been reported to have SPCT composition (Hanemann & Viereck-Götte 2004), but

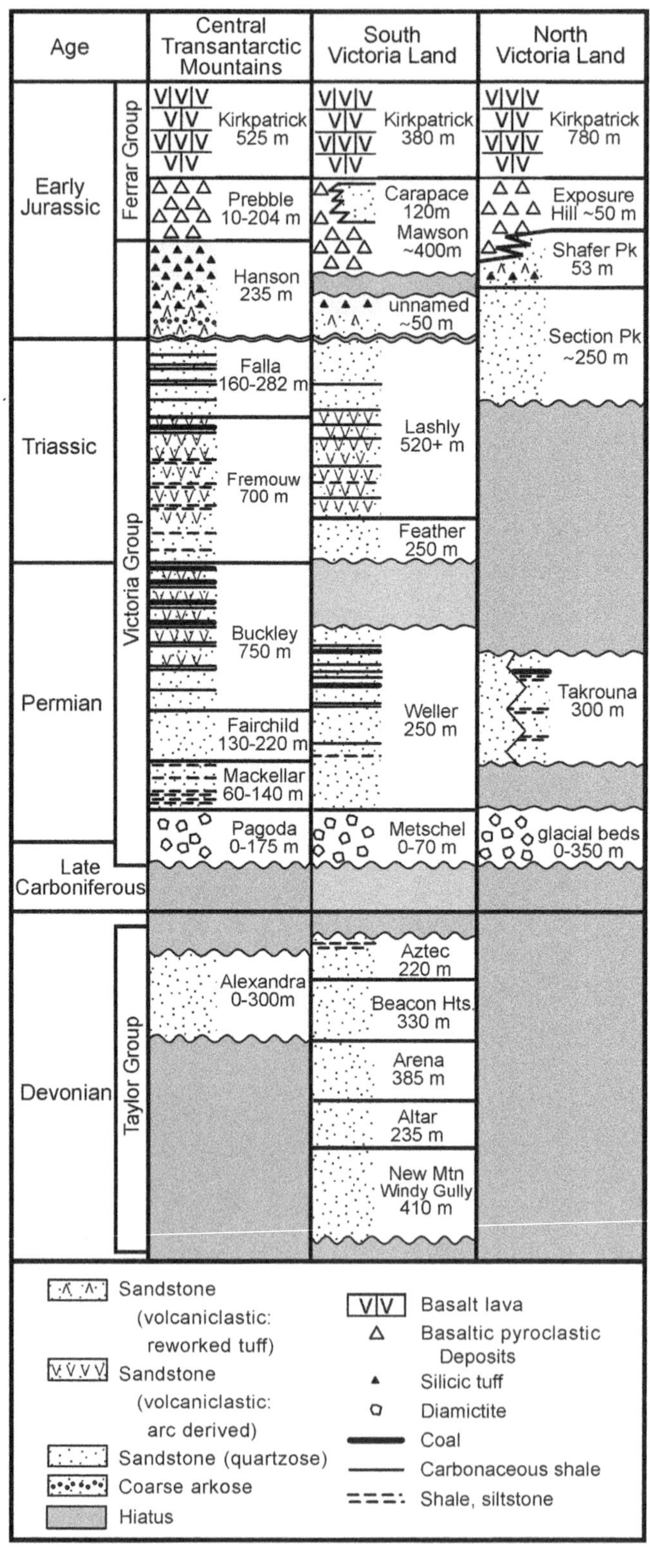
Age
Central Transantarctic Mountains
South Victoria Land
North Victoria Land
Early Jurassic
Triassic
Permian
Late Carboniferous
Devonian
Ferrar Group
Victoria Group
Taylor Group
Kirkpatrick 525 m
Prebble 10-204 m
Hanson 235 m
Falla 160-282 m
Fremouw 700 m
Buckley 750 m
Fairchild 130-220 m
Mackellar 60-140 m
Pagoda 0-175 m
Alexandra 0-300m
Kirkpatrick 380 m
Carapace 120m
Mawson ~400m
unnamed ~50 m
Lashly 520+ m
Feather 250 m
Weller 250 m
Metschel 0-70 m
Aztec 220 m
Beacon Hts. 330 m
Arena 385 m
Altar 235 m
New Mtn Windy Gully 410 m
Kirkpatrick 780 m
Exposure Hill ~50 m
Shafer Pk 53 m
Section Pk ~250 m
Takrouna 300 m
glacial beds 0-350 m
Sandstone (volcaniclastic: reworked tuff)
Sandstone (volcaniclastic: arc derived)
Sandstone (quartzose)
Coarse arkose
Hiatus
Basalt lava
Basaltic pyroclastic Deposits
Silicic tuff
Diamictite
Coal
Carbonaceous shale
Shale, siltstone

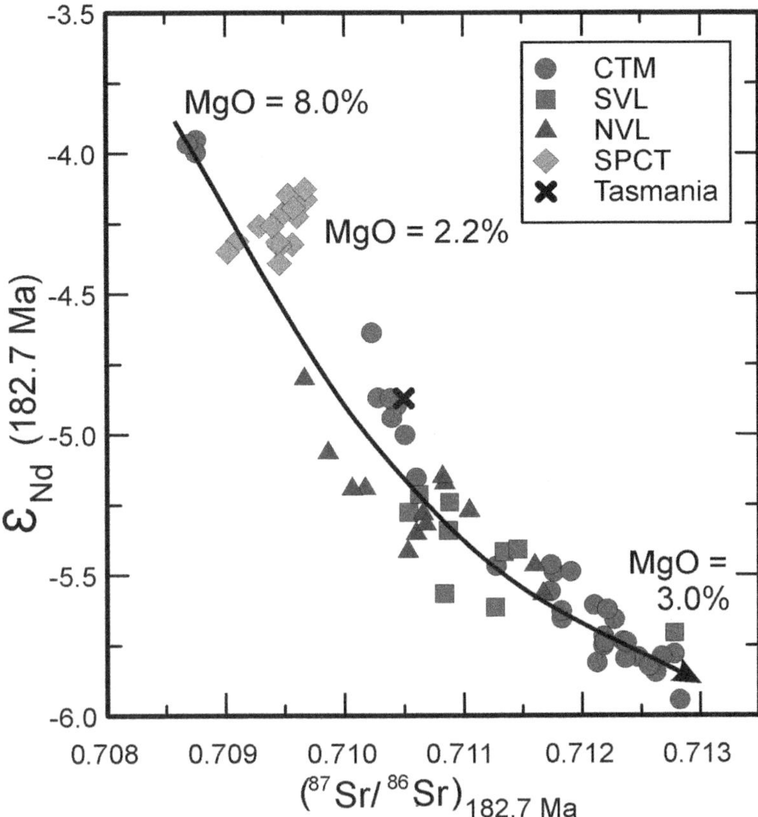

Fig. 4. ε_{Nd} v. $^{87}Sr/^{86}Sr$ at 183 Ma for Ferrar Large Igneous Province sills and lavas. Data sources: Tasmania – Hergt *et al.* (1989*b*); Transantarctic Mountains – Fleming *et al.* (1995); Elliot *et al.* (1999); Elliot & Fleming (2004).

here is regarded as a more highly evolved MFCT composition because it falls outside the restricted SPCT compositional range (having lower TiO_2, Y and Zr abundances, and higher Mg number).

The MFCT exhibits a range of geochemical compositions ($^{87}Sr/^{86}Sr_i$, *c.* 0.709–0.712; MgO, *c.* 9.3–2.6%; Zr, *c.* 60–175 ppm), which can be related by fractional crystallization accompanied by approximately 5% crustal assimilation (Figs 4–6) (Fleming *et al.* 1995). Although only data collected by the authors are illustrated here, analyses of Ferrar Dolerite sills in south Victoria Land (e.g. Hamilton 1965; Gunn 1966; Morrison & Reay 1995; Wilhelm & Wörner 1996; Antonini *et al.* 1999; Demarchi *et al.* 2001; Zieg & Marsh 2012) and north Victoria Land (e.g. Brotzu *et al.* 1988, 1992; Hornig 1993; Antonini *et al.* 1997; Hanemann & Viereck-Götte 2004) show that all belong to the MFCT, and they do not change the range of compositions except for samples from the Thern Promontory (Brotzu *et al.* 1988; Antonini *et al.* 1997), which include some compositions that are slightly more evolved. The SPCT has a distinct, evolved and restricted composition (Sr_i, *c.* 0.7095; MgO, *c.* 2.3%; Zr, *c.* 230 ppm), which lies off the chemical trends of the MFCT (Elliot *et al.* 1999).

Fig. 3. Simplified stratigraphic columns for the Taylor and Victoria groups of the Beacon Supergroup. Sources – Taylor and Victoria Groups in the central Transantarctic Mountains and south Victoria Land: Barrett (1991); Falla Formation: Elliot (1996); Hanson Formation: Elliot *et al.* (2017); Prebble Formation: Hanson & Elliot (1996); unnamed strata, south Victoria Land: Elliot & Grimes (2011); Mawson Formation: Ross *et al.* (2008); Prince Albert Mountains, south Victoria Land: Skinner & Ricker (1968); north Victoria Land glacial beds and the Takrouna Formation: Collinson *et al.* (1986); Section Peak and Shafer Peak formations: Schöner *et al.* (2011, 2007, respectively); Exposure Hill beds: Viereck-Götte *et al.* (2007); Kirkpatrick Basalt: Elliot & Fleming (2008). Note that the Victoria Group in south Victoria Land is half the thickness of that in the central Transantarctic Mountains.

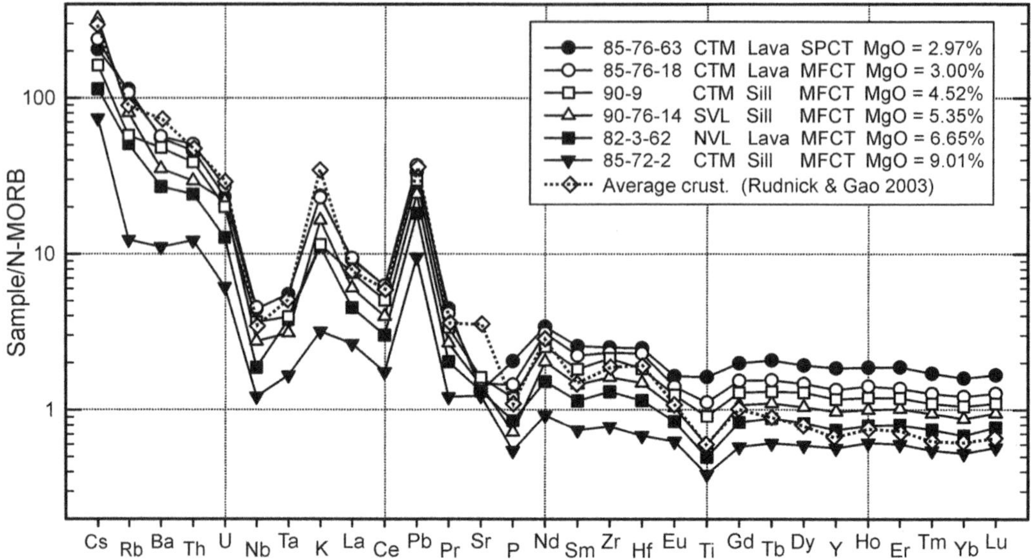

Fig. 5. MORB-normalized trace element diagram for selected samples of the Ferrar Group. The samples cover the entire range of MgO concentrations observed. Normalization factors from Sun & McDonough (1989). Data source: see the Supplementary material. Sample locations: 85-72-2 – Dawson Peak (Fig. 8); 82-3-62 – Haban Spur (Fig. 12); 90-76-14 – Pearse Valley (Fig. 10); 90-9 – Mt Black (Fig. 8); 85-76-18 & 63 – Storm Peak (Fig. 8).

Geochronology

Age determinations on the Ferrar LIP yielded a narrow range of dates by $^{40}Ar/^{39}Ar$ analysis of plagioclase crystals (Heimann *et al.* 1994; Fleming *et al.* 1997; Elliot *et al.* 1999), which established the duration of magmatic activity as relatively short (<2 myr) and close to the precision of the techniques. The accuracy of these dates, however, was compromised by issues related to uncertainty in the absolute age of international Ar reference materials (Renne *et al.* 1998). Subsequent multi-grain analyses of zircons, by the U–Pb thermal ionization mass spectrometry (TIMS) technique, provided a more accurate framework for the age of the Ferrar LIP (Encarnación *et al.* 1996; Minor & Mukasa 1997). More recently, zircons extracted from Ferrar samples have been analysed by the much more precise single-crystal chemical-abrasion isotope-dilution TIMS (CA ID-TIMS) method and have documented an even more restricted duration of emplacement (<0.4 myr), ranging from 182.779 ± 0.033 to 182.59 ± 0.079 Ma (2σ uncertainties) for the MFCT (Burgess *et al.* 2015). There is no clear temporal distinction between sills and lavas, although a limited number of analyses of SPCT rocks suggest that it is, permissibly, slightly younger (as young as 182.43 ± 0.036 Ma).

Source and mantle–lower crustal dispersal paths

Long-distance transport of Ferrar MFCT magmas from magmatic centres in the proto-Weddell Sea region was first proposed by Fleming *et al.* (1997) and Storey & Kyle (1997). Fleming *et al.* (1997) proposed that dykes, emanating from the magmatic centre, formed the feeders for the FLIP. Subsequently, the coherence of the SPCT geochemistry was interpreted to provide compelling support for the long-distance dyke transport proposal (Elliot *et al.* 1999; Elliot & Fleming 2000). Storey & Kyle (1997) suggested that the magmas originated at a mantle plume, migrated into large magma chambers with associated thermal bulges or topographical highs, and then flowed as sills and lavas along the length of the Transantarctic Mountains. Later, Ferris *et al.* (2003) advocated that Ferrar magmas entered the supra-crustal sedimentary sequences at the Dufek intrusion, which then formed the point source for long-distance magma transport through sills. Leat (2008), following Storey & Kyle (1997), also advocated long-distance transport through sills; however, long-distance sill transport is regarded as improbable because it requires magmas to cross a palaeotopographical high (the Ross High of Collinson *et al.* 1994) and then burrow down through the Taylor Group and penetrate the

Table 1. *Major element analyses of representative Ferrar sills and lavas*

Sample	85-72-2	11-5-1	96-73-16	85-51-20	85-51-41	04-01-01	02-09-58	02-11-64	90-75-5
Type	MFCT Sill	MFCT Sill	MFCT Sill	MFCT Lava	SPCT Lava	MFCT Sill	MFCT Sill	MFCT Sill	MFCT Lava
Region	CTM	CTM	CTM	CTM	CTM	SVL	SVL	SVL	SVL
SiO_2	52.05	54.60	56.90	58.33	56.85	55.12	57.32	58.41	57.56
TiO_2	0.49	0.63	1.13	1.59	1.97	0.63	0.94	1.09	0.99
Al_2O_3	16.37	15.06	13.84	12.81	12.07	14.77	14.44	14.51	13.70
Fe_2O_3	1.12	1.21	1.44	1.68	1.98	1.20	1.37	1.46	1.48
FeO	7.48	8.09	9.63	11.18	13.17	7.98	9.15	9.77	9.90
MnO	0.17	0.18	0.18	0.19	0.21	0.17	0.17	0.17	0.19
MgO	9.04	6.90	4.41	2.78	2.20	6.81	4.38	3.97	4.02
CaO	11.53	10.82	9.11	7.42	7.05	11.00	9.03	7.74	8.63
Na_2O	1.50	1.63	2.19	2.34	2.38	1.42	1.93	2.55	2.04
K_2O	0.20	0.78	1.03	1.47	1.83	0.84	1.15	0.17	1.33
P_2O_5	0.06	0.10	0.14	0.21	0.30	0.06	0.12	0.16	0.15
Total	100.00	100.00	100.00	100.00	100.00	100.00	100.00	100.00	100.00
LOI	1.93	0.74	0.76	0.59	0.33	-0.09	1.08	1.98	0.95
Mg#	68.3	60.3	44.9	30.7	22.9	60.3	46.1	42.0	42.0
Sample	97-62-12	97-68-3	97-57-2	97-51-6	97-51-52	82-9-2	82-10-2	81-2-30	82-2-5
Type	MFCT Sill	MFCT Sill	MFCT Sill	MFCT Lava	SPCT Lava	MFCT Sill	MFCT Sill	MFCT Lava	SPCT Lava
Region	SVL	SVL	SVL	SVL	SVL	NVL	NVL	NVL	NVL
SiO_2	53.34	54.97	56.41	56.07	57.02	56.37	58.25	55.34	57.08
TiO_2	0.51	0.63	0.73	0.82	1.96	0.73	1.04	0.65	1.99
Al_2O_3	14.70	14.59	14.59	14.03	11.91	14.40	13.84	14.71	12.00
Fe_2O_3	1.18	1.20	1.25	1.35	1.98	1.29	1.38	1.22	1.98
FeO	7.89	8.03	8.32	8.98	13.18	8.59	9.19	8.13	13.22
MnO	0.17	0.17	0.18	0.18	0.19	0.16	0.19	0.18	0.22
MgO	9.84	6.88	5.75	5.54	2.26	5.60	3.75	6.49	2.21
CaO	10.36	10.63	9.77	10.04	7.02	9.57	8.47	10.98	7.04
Na_2O	1.46	2.07	2.04	1.89	2.44	2.01	2.13	1.79	2.15
K_2O	0.49	0.74	0.85	0.98	1.78	1.16	1.57	0.40	1.82
P_2O_5	0.07	0.09	0.12	0.12	0.26	0.12	0.18	0.09	0.28
Total	100.00	100.00	100.00	100.00	100.00	100.00	100.00	100.00	100.00
LOI	0.63	0.34	0.03	0.17	0.16	1.54	1.35	1.42	0.38
Mg#	69.0	60.4	55.2	52.4	23.4	53.73	42.1	58.7	23.0

Analyses were performed by EDXRF at the Southern Connecticut State University and WDXRF at New Mexico Tech using Li-metaborate fusion methodologies modified after Norrish & Hutton (1969). Major element analyses given in wt% and normalized to 100% loss-free with $Fe_2O_3/FeO = 0.15$. CTM: central Transantarctic Mountains; NVL: north Victoria Land; SVL: south Victoria Land; LOI, loss on ignition; Mg#, Mg number. Sample locations are given in the Supplementary material.

basement rocks in the Dry Valleys to form the thick Basement Sill (Marsh 2004). That sill has been studied exhaustively in Wright Valley and is part of a more complex intrusion system, which includes a thick body with well-defined mineral layering at The Dais in Wright Valley (Marsh 2004; Marsh *et al.* 2005; Bédard *et al.* 2007). A thick tongue of orthopyroxene crystals, formed by flow differentiation and entrained in the Basement Sill during its emplacement, thins away from the region of Bull Pass; the feeder itself, an approximately 10 km-diameter funnel, is centred in the mountains immediately to the east of Bull Pass (Marsh *et al.* 2005) and implies vertical transport from reservoirs at depth. Although dykes are very sparse and widely distributed, several transect basement rocks, indicating the transport of magmas from depth at those sites, and not by transport through supra-crustal sills. Nevertheless, it is clear that magmas were dispersed laterally in sills for at least tens of kilometres, epitomized by the Basement Sill which has been identified over an area of about 10 000 km^2 (Marsh 2007).

The Muskox Intrusion, Coppermine River lavas and the Mackenzie Dyke Swarm, the latter known to extend over 2500 km (Baragar *et al.* 1996), are considered to be the best analogues for the FLIP but represent deeper levels of erosion of a possibly similar magmatic system. The Mackenzie Dyke Swarm, as observed today, occurs in rocks that were at mid-crustal depths.

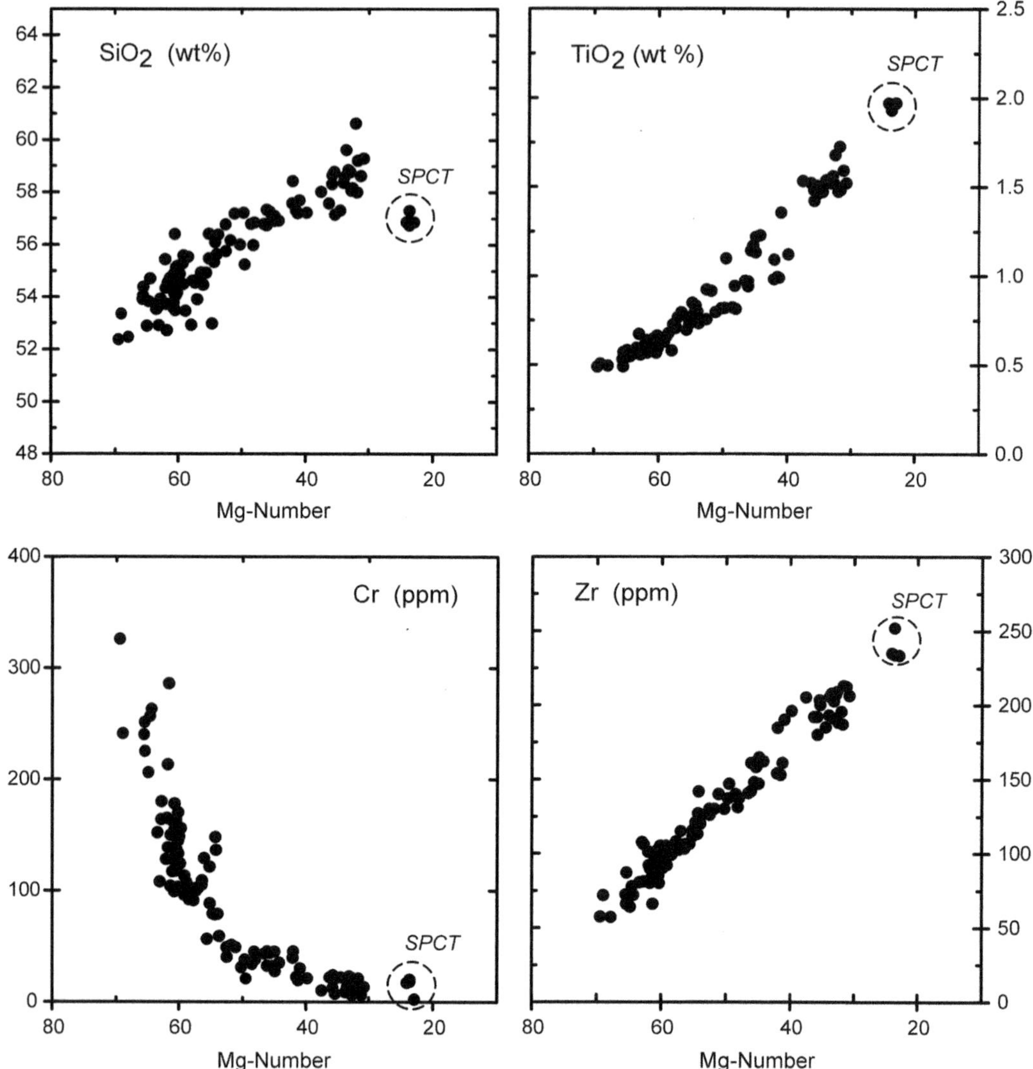

Fig. 6. Variation diagrams for selected major and trace elements for Ferrar Group lavas and sills to illustrate the geochemical coherence of the MFCT and the restricted and different composition of the SPCT. Data from Elliot *et al.* (1995, 1999) and Fleming *et al.* (1995, 1997).

With respect to the possibility of long-distance surface dispersal of lava flows, the Pomona flow of the Columbia River Basalt LIP has been traced for 600 km (Hooper 1997). Further, Self *et al.* (2008) correlated the Rajahmundry Trap lavas on the east coast of the Indian Peninsula with specific Deccan lava formations, and proposed that subaerial lava flow of more than 1000 km is probable given the right combination of magma eruption rates and topography. The SPCT composition occurs as a single lava flow at all outcrops (Elliot *et al.* 1999), except in north Victoria Land where multiple flows, or possibly flow lobes from a single eruptive event, are present (Mensing *et al.* 1991). These lava-flow outcrops, spread out over about 1500 km, are unlikely to represent surface transport from a single source (such as the Dufek intrusion, a further 1100 km distant) given that the flow path would have been along a rift valley system with inferred topographical lows (Elliot & Fleming 2004, 2008). Nevertheless, a regional gradient from a topographical high over a possible plume head in the proto-Weddell Sea region across Antarctica cannot be totally discounted.

Controls on shallow-level (supra-crustal) emplacement

Marsh (2004) examined the sills in the Dry Valleys region of south Victoria Land and proposed that density provided the fundamental control on the emplacement level of Ferrar magmas. The primary control on density was interpreted to be principally the proportion of phenocrysts (mainly orthopyroxene) carried by the magmas. In this hypothesis, the least dense, crystal-free magmas (proxied by MgO content) were initially emplaced as lavas through a system of regional dykes. With time, the magmatic system delivered increasingly dense magmas from depth. A point was reached at which the magmas became too dense for eruption and they were then emplaced as sills; with progressively increasing density (MgO content), the magmas were intruded at progressively greater depths in the Beacon sedimentary sequence. Marsh (2004) cited the lavas at Coombs Hills and Allan Hills, and the sills in the Dry Valleys, as exemplifying this process. Thus, in an ideal situation, the lavas should exhibit increasing MgO upsection and the sills increasing MgO down through the sill complex in the Beacon strata and basement rocks.

Airoldi *et al.* (2012) and Muirhead *et al.* (2012, 2014) examined dykes, inclined sheets and sills in the Coombs Hills, Allan Hills and the Dry Valleys region, and proposed an entirely different emplacement mechanism. They argued, on the basis of structural and magnetic studies, that the Ferrar dykes and inclined sheets were intruded into a neutral stress field, not an extensional stress field, which has long been accepted for dyke emplacement (Anderson 1951). In this hypothesis, the random orientation of dykes observed at Allan Hills reflects the neutral stress field, and dykes together with inclined sheets form the magma pathways from one sill to an overlying sill and eventually to the surface for eruption as lavas. Thus, the neutral stress field and sill inflation play a critical role in magma ascent, leading to a sill-fed dyke network overlain by lavas supplied by the underlying dyke field. This scenario has been referred to as the 'cracked-lid' hypothesis for flood basalt eruption (Muirhead *et al.* 2014). Nevertheless, this scenario invokes magma transport in dykes through crustal rocks, prior to invasion of supra-crustal rocks; this necessarily implies a regional stress field affecting the igneous and metamorphic crustal rocks. This scenario also suggests that there should be a link between lava compositions and those of underlying sills.

Stratigraphic and geochemical relationships

To assess the possible relationships between sill and lava compositions, the five areas with extant extrusive and intrusive phases, are examined (Figs 7–12). The stratigraphic distribution of the illustrated sills is based on their occurrence through the Beacon Supergroup and do not represent a simple vertical stack of sills at any one locality. Progressively younger strata of the Beacon Supergroup, which in the Shackleton Glacier region, the Dry Valleys of south Victoria Land and the Prince Albert Mountains (Fig. 2) have a low dip to the west (toward the ice-covered polar plateau), crop out at increasing elevations back from the mountain front. In the Queen Alexandra Range, the Beacon Supergroup describes a broad, shallow syncline, whereas, in north Victoria Land, the Permian strata fill a possible trough or rift with the Triassic Section Peak Formation cropping out on the inboard shoulder (polar plateau flank). Thus, the stratigraphic succession of sills may be spread out geographically over a lateral distance of 100 km or more, as is the case for the Shackleton Glacier region, the Dry Valleys of south Victoria Land and north Victoria Land, whereas, in the Queen Alexandra Range and the Prince Albert Mountains, the lavas and sills are in somewhat closer association. Further, sills may terminate and connect via thin dykes to other sills at higher stratigraphic levels, exchange stratigraphic positions and form climbing sheets connecting to other sills (Hamilton 1965; Elliot & Fleming 2004; Muirhead *et al.* 2012).

It should be noted that the thickest Devonian Taylor Group succession crops out in south Victoria Land, whereas the thickest Permo-Triassic Victoria Group succession is displayed in the central Transantarctic Mountains. The Beacon Supergroup in the Prince Albert Mountains is not well studied and its age is uncertain (Skinner & Ricker 1968); however, the occurrence of carbonaceous debris in the sequence suggests that it is part of the Victoria Group. In north Victoria Land, the Beacon strata are geographically scattered, and the composite stratigraphic column illustrated is less definitive than the columns for south Victoria Land and the central Transantarctic Mountains.

The MgO concentrations of chilled margins of sills are plotted adjacent to the stratigraphic columns. Those of fine-grained samples of lava flows in the same regions are plotted in the same figures but at a different scale (note that the SPCT magma type is indicated where present). In the Shackleton Glacier region (Figs 7a & 8), the MgO concentrations of the sills have a range of MgO = 4.3–7.3%, without any stratigraphic trend. The lavas at Mt Bumstead, above the lowest exposed flow, which

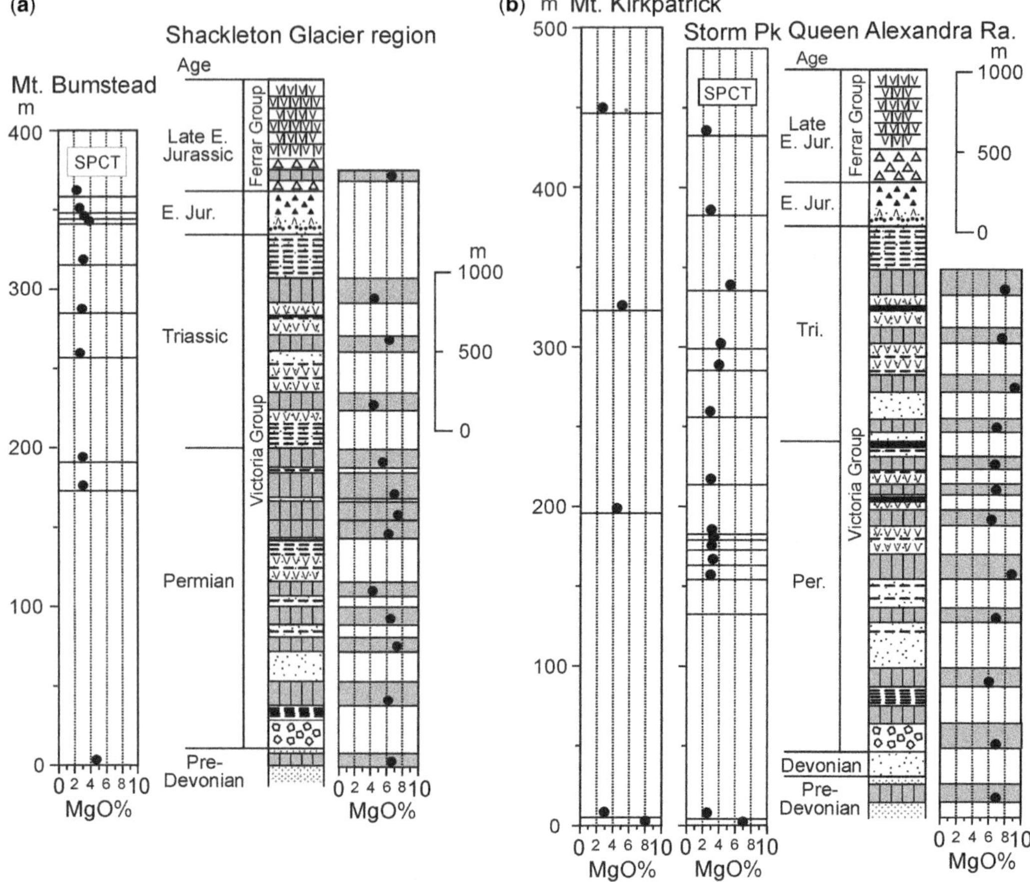

Fig. 7. (**a**) Schematic column to show the distribution of Ferrar Dolerite sills in basement rocks and the Victoria Group of the Beacon Supergroup in the Shackleton Glacier region, and a stratigraphic column for Kirkpatrick Basalt lavas at Mt Bumstead (see Fig. 8 for the locations). (**b**) Schematic column to show the distribution of Ferrar Dolerite sills in the Victoria Group of the Beacon Supergroup in the Queen Alexandra Range, and stratigraphic columns for Kirkpatrick Basalt lavas at Mt Kirkpatrick and Storm Peak (see Fig. 8 for the locations). The plotted points are the MgO concentrations of chilled margins of sills and of fine-grained lavas. In this and other figures, sills are projected onto the stratigraphic column and do not represent a single vertical sequence. Data sources: Hergt *et al.* (1989*a*: Portal Peak sill, third from the base in Fig. 7b); see the Supplementary material for other sills and lavas.

has MgO = 4.5%, are uniformly highly evolved (MgO = 2.7–3.4%). In the Queen Alexandra Range (Figs 7b & 8), the sills are relatively higher in MgO (6.0–9.4%), but again with no trend, and with the most MgO-rich olivine dolerite sills occurring in the middle and upper part of the Buckley Formation. Two lava columns are illustrated, the localities being 38 km apart. Both begin with a very thin flow with MgO > 7.0%. At Storm Peak overlying lavas are much less evolved (MgO = 2.6–5.7%) and without any clear trend. At Mt Kirkpatrick, there are only four very thick flows above the basal lava and they show an initial trend to less MgO-poor compositions. The lava sequence at Mt Falla, only 15 km distant from Mt Kirkpatrick, is similar to Storm Peak, with at least two correlative thick flows (second flow and capping flow: Barrett *et al.* 1986).

In the Dry Valleys of south Victoria Land (Figs 9a & 10), the sills are progressively higher in MgO downsection. The nearest lava flows occur about 100 km to the NNW at Carapace Nunatak (Fig. 10), and those flows, with the exception of the summit flow, show increasing MgO upsection (note that a flow with the SPCT composition is absent, and the sequence is much thinner than at all the major lava outcrop regions). Dolerite plugs and a thin sill invade the Mawson Formation at Coombs Hills and Allan Hills (Ross *et al.* 2008);

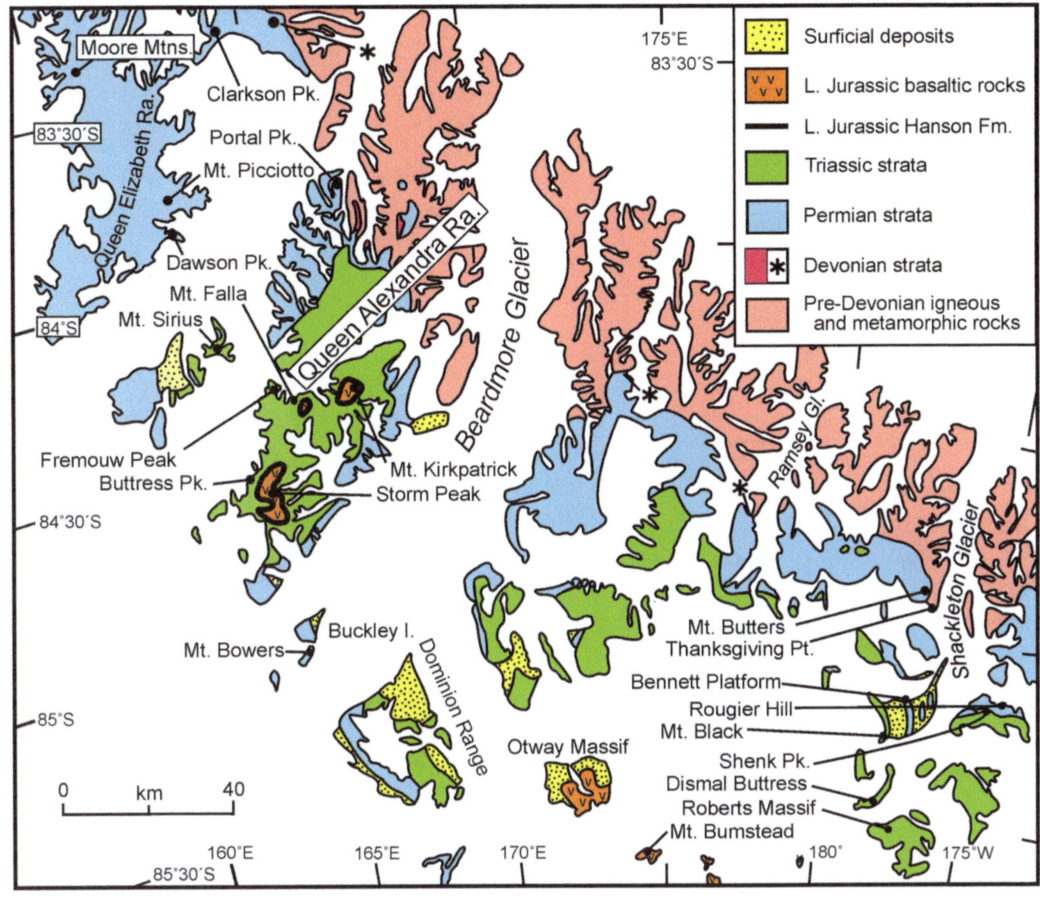

Fig. 8. Simplified geological map of the central Transantarctic Mountains to show the geographical distribution of Kirkpatrick Basalt lava flows, and Beacon Supergroup strata and associated Ferrar Dolerite sills.

all have MgO > 6.3%. In the Prince Albert Mountains (Figs 9b & 10), the sill and lava compositions overlap (3.9–7.3% MgO) and no clear trends are evident.

In north Victoria Land (Figs 11 & 12), the Victoria Group crops out over a wide area (250 km linear distribution) with only Triassic–Jurassic beds in close proximity to the lavas. To the south and west of the Mesa Range, sills are confined to Triassic–Jurassic quartzose strata; to the north, however, sills have been mapped in the Permian Takrouna Formation. North of the Mesa Range, chilled margin compositions (MgO = 7.1–7.2%) are available for only three sills (Hornig 1993; Hanemann & Viereck-Götte 2004: note that the Hanemann & Viereck-Götte 2004 sample from Mt Apolotok is reported to be a sill but is listed in their table as a dyke margin); one sill is documented to be 300 m thick (Hornig 1993) and is possibly at or near the top of the Permian section, and is illustrated as such in Figure 11; the sill at Mt Bower must be close to or at the base of the Section Peak Formation (Schöner *et al.* 2011). Several sills in this quartzose Triassic–Jurassic sequence are exposed in a nunatak (Exposure Hill) adjacent to the lavas in the southern Mesa Range (Elliot *et al.* 1986) and these have MgO = 3.7–5.6%. These sills are probably high in the Section Peak Formation, but a fault must separate Exposure Hill from the adjacent Mesa Range lavas, and neither the top nor the base of the formation is exposed there. The analysed sill samples to the south of the Mesa Range all have MgO < 4.5% (Brotzu *et al.* 1988). The Mesa Range Kirkpatrick lavas are predominantly more MgO-rich (mainly 6–8% MgO, with only two as low as 5% MgO) than sills south of the Mesa Range and at Exposure Hill; the lava sequence does not exhibit any trend upsection.

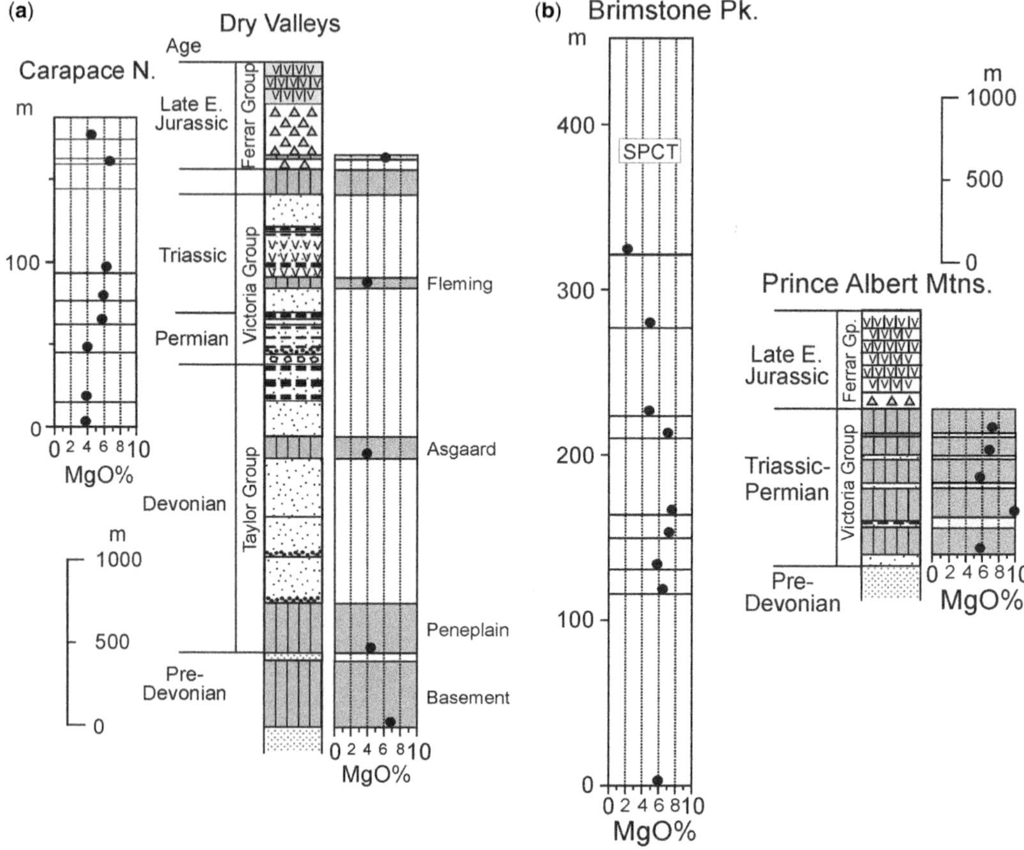

Fig. 9. (**a**) Schematic column to show the distribution of Ferrar Dolerite sills in the basement rocks and the Beacon Supergroup (Taylor and Victoria groups) in the Dry Valleys of south Victoria Land and a stratigraphic column for Kirkpatrick Basalt lavas at Carapace Nunatak (see Fig. 10 for the locations). (**b**) Schematic column to show the distribution of Ferrar Dolerite sills in the Beacon Supergroup (assumed to be the Victoria Group because of the presence of carbonaceous material) in the Prince Albert Mountains and a stratigraphic column for Kirkpatrick Basalt lavas at Brimstone Peak (see Fig. 10 for the locations). The plotted points are the MgO concentrations of chilled margins of sills and of fine-grained lavas. Data sources: Marsh (2007, chilled margin of the Fleming sill); Ross *et al.* (2008, sample 73160, stratigraphically highest sill); see also the Supplementary material.

The Mesa Range lavas overlie Lower Jurassic, silicic pyroclastic rocks, which have a regional thickness of about 50 m (Schöner *et al.* 2007, 2011). These overlie the 250 m-thick quartzose Section Peak Formation, which is intruded by two thick sills at Section Peak and along the escarpment that extends south to the Deep Freeze Range region (Fig. 12) where another section of Kirkpatrick Basalt lavas crops out at and near Mt Hewson (Brotzu *et al.* 1988). The sills at Section Peak are thick and uniform, and, together with the sills at Exposure Hill, show no sign of shallow intrusion such as vesicularity. This implies that the lavas, which generally have higher MgO than the sills, were erupted first in order for sufficient overburden to be present for sills to lack signs of degassing.

Discussion

In the density-driven vertical-emplacement hypothesis (Marsh 2004), eruption of the most evolved magma occurs first and subsequent magmas are increasingly MgO-rich up lava section; a point is reached at which magmas are no longer erupted at the surface, but are intruded sequentially, and with increasing MgO, downsection through the Beacon Supergroup. This relationship is not observed in north Victoria Land. There is no clear pattern in the Prince Albert Mountains that would support the hypothesis, whereas in the Dry Valleys region, the geochemistry, in general, is not inconsistent with it. In the Queen Alexandra Range, apart from the high-MgO basal flow, the lavas at Mt Kirkpatrick partially support the hypothesis,

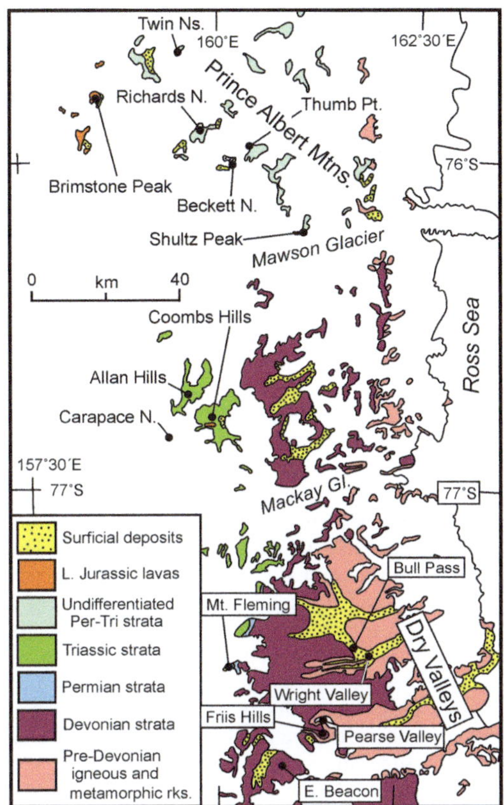

Fig. 10. Simplified geological map of south Victoria Land to show the geographical distribution of Kirkpatrick Basalt lava flows, and Beacon Supergroup strata and associated Ferrar Dolerite sills. Note that basaltic pyroclastic rocks, erupted prior to flood basalt effusion, occur in the western Coombs Hills and southern Allan Hills, and form small outcrops elsewhere but are here included with Triassic beds.

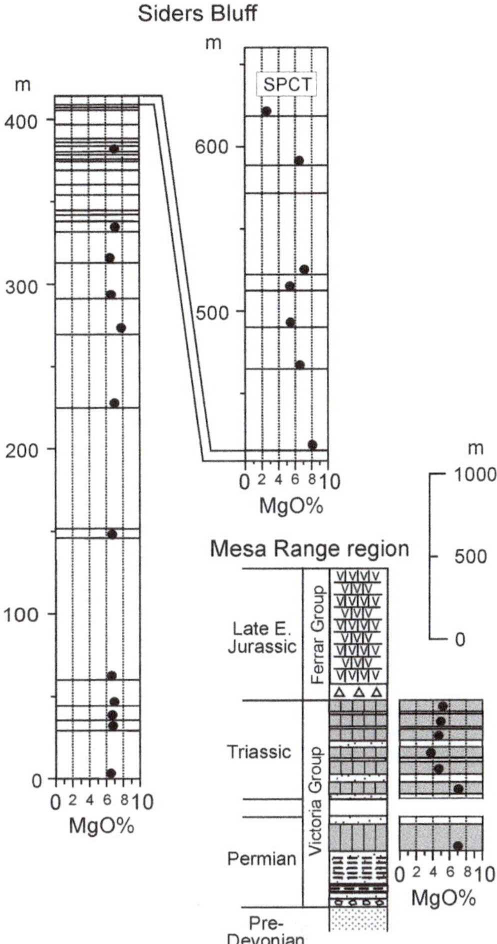

Fig. 11. Schematic column to show the distribution of Ferrar Dolerite sills in the Beacon Supergroup in the Mesa Range region of north Victoria Land and a stratigraphic column for Kirkpatrick Basalt lavas at Siders Bluff, Mesa Range (see Fig. 12 for the locations). The plotted points are the MgO concentrations of chilled margins of sills and of fine-grained lavas. The stratigraphic column is broken because there is no continuity between the Permian and Triassic strata. Data sources: Hornig (1993, sample FD004, Boggs Valley); Hanemann & Viereck-Götte (2004, sample RH54a, Mt Bower); see also the Supplementary material.

whereas the Storm Peak flows marginally support it. The sills, however, show no pattern of increasing MgO with depth. Similarly in the Shackleton Glacier region, apart from the basal very thick flow at Mt Bumstead, the lavas are uniformly strongly evolved. In contrast, the sills are higher in MgO except for three that are relatively evolved (*c.* 4.5%). In summary, the geochemistry is not inconsistent with the density-driven hypothesis in the Dry Valleys, but appears to negate the hypothesis in north Victoria Land and shows inconsistent patterns in the three other regions. It is clear that factors other than density play a significant role in the emplacement of magmas either at depth or at the surface.

In the cracked-lid sill-fed hypothesis (Muirhead *et al.* 2014), in which lavas are fed directly from underlying intrusions, there should be some geochemical correlation between the lava flows and the sills. In addition, this hypothesis implies that sills are intruded first and, with time, magmas migrate upwards forming younger and stratigraphically higher sills, eventually breaking through to the surface; lava flows cannot be erupted without corresponding sill emplacement at depth. Further, this hypothesis suggests dykes somewhere should be observed cutting strata immediately underlying the lavas and penetrating the lava succession, but no feeder dykes for

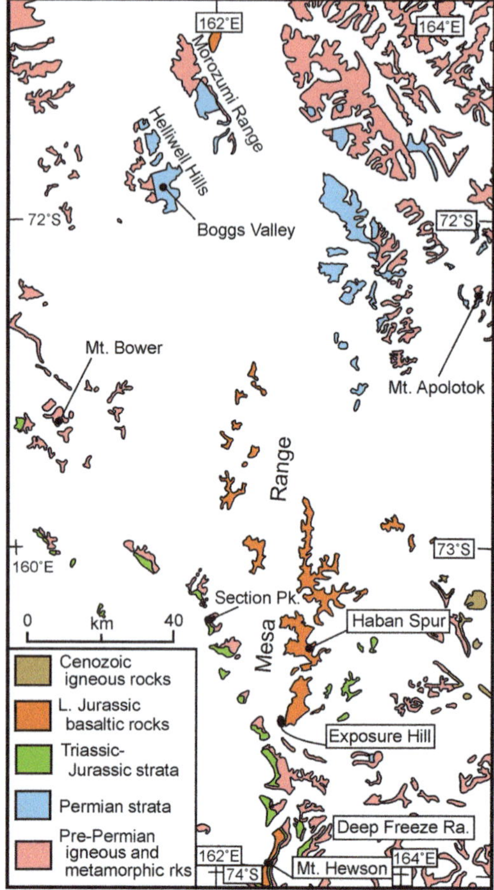

Fig. 12. Simplified geological map of north Victoria Land to show the geographical distribution of Kirkpatrick Basalt lava flows, and Beacon Supergroup strata and associated Ferrar Dolerite sills.

the lavas have been positively identified so far. There are instances of thin dykes connecting sills in the Shackleton Glacier region (Elliot & Fleming 2004), and examples of thicker inclined sheets and massive dolerite plugs with connecting dykes and sills in south Victoria Land (Hamilton 1965; Morrison & Reay 1995; Elliot & Fleming 2004; Muirhead *et al.* 2012). If the locations of lavas reflect sites of eruption associated with palaeotopographical lows in a Jurassic rift system (Elliot & Fleming 2008), there should be evidence for feeders. In the Coombs Hills–Allan Hills region (Ross *et al.* 2008; Muirhead *et al.* 2012) and at the Otway Massif (Elliot & Hanson 2001), massive dolerite bodies are present in association with, and intruded into, cauldron-like phreatomagmatic deposits (White & McClintock 2001) measuring as much as 350 m in thickness, and all overlain by Kirkpatrick Basalt lavas. These were sites of major extrusive activity; yet, the absence of reported vesicularity in the sills implies significant overburden at the time of intrusion (i.e. the lava column already existed at the time of dolerite emplacement).

In evaluating these hypotheses, uncertainties arise because sills and lavas in any one area assessed in this study crop out as much as 100 km apart. In the Shackleton Glacier region, given that some of the sills have higher MgO than the lava flows, it is plausible that the evolved lavas are the products of magma evolution in sill bodies. However, most sills are too thin to show significant *in situ* differentiation and no sill in this region shows any evidence (layering and accumulation of plagioclase and mafic minerals) for the extended fractional crystallization required to produce the evolved lavas, nor are there any sills with the evolved character of the lavas that could be considered constituent parts of the conduit systems. Further, an orthopyroxene mush comparable to that of the Basement Sill in south Victoria Land, a critical factor in the density-driven hypothesis, is not known to be present in any sill elsewhere. In the Queen Alexandra Range, the situation is similar. The very thick (up to nearly 200 m) evolved flows at Mt Kirkpatrick would require large reservoirs of cumulates, which have not been observed in any sill, and no sill has a composition suggesting that it might be part of the conduit system. In south Victoria Land, the range of lava and sill compositions is compatible with direct connections between them, and the same is the case in the Prince Albert Mountains. In north Victoria Land, the geochemistry is marginally consistent with a connection between the lava flows and sills north of the Mesa Range,

Many factors affect the emplacement of magmas, whether as sills or lava flows. These include density, lithostatic pressures, magma overpressures required for lateral emplacement, wall-rock composition and physical properties, and when and where the evolving source was tapped. It is interesting to note that thick (100+ m) flows are present at Mt Kirkpatrick where there are three such flows, one at Storm Peak that is highly evolved and quenched, and one at Mt Bumstead (and others at the adjacent Otway Massif). The margin of the thick flows at Mt Kirkpatrick have MgO = 2.5–4.5%, and at Mt Bumstead MgO = 4.5%; the lowest two flows at Mt Kirkpatrick have particularly evolved compositions. There are no suitable counterparts, such as thick cumulate zones indicating *in situ* fractional crystallization and evolution, in the underlying sills for the evolved flows, if the flows were to have originated via migration of magma into, and evolution within, the sills.

The SPCT magma composition provides an important constraint on the relationship between the intrusive and extrusive elements of the Ferrar magmatic system. Among the lavas, it is represented only by the uppermost flow(s) in the lava sequences. It is not represented by any analysed sill in the Transantarctic Mountains between Mt Spann (Fig. 2) and north Victoria Land (nor in SE Australasia), although it is represented in the Whichaway Nunataks and Theron Mountains (Fig. 2), which are 1250 km from the nearest extant lava in the Shackleton Glacier region. The SPCT lavas share chemical characteristics with the MFCT rocks, such as enriched initial isotope ratios of Sr and Nd, and crust-like trace element patterns. However, they clearly must have had a different evolutionary path because their evolved major and trace element geochemistry (Figs 5 & 6) is accompanied by Sr and Nd initial isotope ratios that lie at the least evolved end of the Ferrar province range (Fig. 4). The SPCT rocks could be compatible with the density-driven hypothesis, but as an entirely separate event. However, they are incompatible with the sill-fed cracked-lid hypothesis (Muirhead *et al.* 2014), and clearly demand a different emplacement path. SPCT lateral transport may have been predominantly at lower-crustal depths with ascent into supra-crustal rocks and to the surface outside the existing Ferrar outcrop belt. The SPCT plumbing system was not rooted in the subjacent extant supra-crustal Beacon sedimentary succession. This also opens the possibility that emplacement of at least some of the MFCT rocks may have included such transport from outside the area of existing outcrop.

Conclusions

The geochemical coherence of the principal geochemical type recognized in the Ferrar Large Igneous Province (99% of the province) suggests that it is a single episode of magmatism, which occurred over a limited time span. The mode and relative timing of emplacement of various elements of the magmatic architecture, however, are not entirely clear. No large dyke swarms or feeders have been recognized, leading to the suggestion, based on geochemical coherence, of a single source in the proto-Weddell Sea region of Gondwana, followed by lower-crustal transport and migration to the surface at a limited number of sites. Two scenarios have been proposed to explain the shallow emplacement history: one driven by magma physical properties (density) (Marsh 2004), and the other based on structural and field evidence (Muirhead *et al.* 2012, 2014). The contrasting hypotheses have been evaluated on field and geochemical data. Evidence from the Dry Valleys and Carapace Nunatak, south Victoria Land, is not inconsistent with the density-driven emplacement mechanism (Marsh 2004). At the remaining localities, field and geochemical data are inconsistent with the hypothesis. Field and geochemical data do not support the sill-fed hypothesis (Muirhead *et al.* 2014) as a geographically widespread scenario. Each process may be of local importance but neither hypothesis, based on field and geochemical data, applies province-wide.

The authors wish to acknowledge support over many years from the Office of Polar Programs, National Science Foundation. Two anonymous reviews are much appreciated and have significantly improved the manuscript. This is Byrd Polar and Climate Research Center contribution No. 1551.

References

Airoldi, G., Muirhead, J.D., Zanella, E. & White, J.D.L. 2012. Emplacement process of Ferrar Dolerite sheets at Allan Hills (South Victoria Land, Antarctica) inferred from magnetic fabric. *Geophysical Journal International*, **188**, 1046–1060, https://doi.org/10.1111/j.1365-246X.2011.05334.x

Anderson, E.M. 1951. *The Dynamics of Faulting and Dyke Formation with Application to Britain*. Oliver & Boyd, Edinburgh.

Antonini, P., Demarchi, G., Piccirillo, E.M. & Orsi, G. 1997. Distinct magma pulses in the Ferrar tholeiites of Thern Promontory (Victoria Land, Antarctica). *Terra Antarctica*, **4**, 33–39.

Antonini, P., Piccirillo, E.M., Petrini, R., Civetta, L., D'Antonio, M. & Orsi, G. 1999. Enriched mantle – Dupal signature in the genesis of the Jurassic Ferrar tholeiites from Prince Albert Mountains (Victoria Land, Antarctica). *Contributions to Mineralogy and Petrology*, **136**, 1–19.

Baragar, W.R.A., Ernst, R.E., Hulbert, L. & Peterson, T. 1996. Longitudinal petrochemical variation in the Mackenzie dyke swarm, northwestern Canadian Shield. *Journal of Petrology*, **37**, 317–359.

Barrett, P.J. 1991. The Devonian to Triassic Beacon Supergroup of the Transantarctic Mountains and correlatives in other parts of Antarctica. *In*: Tingey, R.J. (ed.) *The Geology of Antarctica*. Oxford Monographs on Geology and Geophysics, **17**. Oxford University Press, Oxford, 120–152.

Barrett, P.J., Elliot, D.H. & Lindsay, J.F. 1986. The Beacon Supergroup (Devonian–Triassic) and Ferrar Group (Jurassic) in the Beardmore Glacier area, Antarctica. *In*: Turner, M.D. & Splettstoesser, J.F. (eds) *Geology of the Central Transantarctic Mountains*. American Geophysical Union, Antarctic Research Series, **36**, 339–428.

Bédard, J.J., Marsh, B.D., Hersum, T.G., Naslund, H.R. & Mukasa, S.B. 2007. Large-scale mechanical redistribution of orthopyroxene and plagioclase in the basement sill, Ferrar Dolerites, McMurdo Dry Valleys, Antarctica: Petrological, mineral-chemical and field evidence for channelized movement of crystals and melt. *Journal of Petrology*, **48**, 2289–2326.

Behrendt, J.C., Drewry, D.J., Jankowski, E. & Grim, M.S. 1981. Aeromagnetic and radio echo ice-sounding measurements over the Dufek intrusion, Antarctica. *Journal of Geophysical Research*, **86**, (B4), 3014–3020.

Bradshaw, M.A. 2013. The Taylor Group (Beacon Supergroup): the Devonian sediments of Antarctica. *In*: Hambrey, M.J., Barker, P.F., Barrett, P.J., Bowman, V., Davies, B., Smellie, J.L. & Tranter, M. (eds) *Antarctic Palaeoenvironments and Earth-Surface Processes*. Geological Society, London, Special Publications, **381**, 67–97, https://doi.org/10.1144/SP381.23

Brewer, T.S., Hergt, J.M., Hawkesworth, C.J., Rex, D. & Storey, B.C. 1992. Coats Land dolerites and the generation of Antarctic continental flood basalts. *In*: Storey, B.C., King, E.C. & Livermore, R.A. (eds) *Magmatism and the Causes of Continental Break-Up*. Geological Society, London, Special Publications, **68**, 185–208, https://doi.org/10.1144/GSL.SP.1992.068.01.12

Brotzu, P., Capaldi, G., Civetta, L., Melluso, L. & Orsi, G. 1988. Jurassic Ferrar dolerites and Kirkpatrick basalts in northern Victoria Land (Antarctica): stratigraphy, geochronology and petrology. *Memorie della Societa Geologica Italiana*, **43**, 97–116.

Brotzu, P., Capaldi, G., Civetta, L., Orsi, G., Gallo, G. & Melluso, L. 1992. Geochronology and geochemistry of Ferrar rocks from north Victoria Land, Antarctica. *European Journal of Mineralogy*, **4**, 605–617.

Burgess, S.D., Bowring, S.A., Fleming, T.H. & Elliot, D. H. 2015. High precision geochronology links the Ferrar Large Igneous Province with early Jurassic ocean anoxia and biotic crisis. *Earth and Planetary Science Letters*, **415**, 90–99, https://doi.org/10.1016/j.epsl.2015.01.037

Coffin, M.F. & Eldholm, O. 1994. Large igneous provinces: crustal structure, dimensions, and external consequences. *Reviews of Geophysics*, **32**, 1–36.

Collinson, J.W., Pennington, D.C. & Kemp, N.R. 1986. Stratigraphy and petrology of Permian and Triassic fluvial deposits in northern Victoria Land, Antarctica. *In*: Stump, E. (ed.) *Geological Investigations in Northern Victoria Land*. American Geophysical Union, Antarctic Research Series, **46**, 211–242.

Collinson, J.W., Elliot, D.H., Isbell, J.L. & Miller, J.M. G. 1994. Permian–Triassic Transantarctic Basin. *In*: Veevers, J.J. & Powell, C.McA. (eds) *Permian–Triassic Pangaean Basins and Foldbelts Along the Panthalassan Margin of Gondwanaland*. Geological Society of America, Memoirs, **184**, 173–222.

Demarchi, G., Antonini, P., Piccirillo, E.M., Orsi, G., Civetta, L. & D'Antonio, M. 2001. Significance of orthopyroxene and major element constraints on the petrogenesis of Ferrar tholeiites from southern Prince Albert Mountains, Victoria land, Antarctica. *Contributions to Mineralogy and Petrology*, **142**, 127–146.

Eldholm, O. & Coffin, M.F. 2000. Large igneous provinces and plate tectonics. *In*: Richards, M., Gordon, R.G. & Van Der Hilst, R.D. (eds) *The History and Dynamics of Global Plate Motions*. American Geophysical Union, Geophysical Monographs, **121**, 309–326.

Elliot, D.H. 1996. The Hanson Formation: a new stratigraphical unit in Transantarctic Mountains, Antarctica. *Antarctic Science*, **8**, 389–394.

Elliot, D.H. 2013. The geological and tectonic evolution of the Transantarctic Mountains: a review. *In*: Hambrey, M.J., Barker, P.F., Barrett, P.J., Bowman, V., Davies, B., Smellie, J.L. & Tranter, M. (eds) *Antarctic Palaeoenvironments and Earth-Surface Processes*. Geological Society, London, Special Publications, **381**, 7–35, https://doi.org/10.1144/SP381.14

Elliot, D.H. & Fleming, T.H. 2000. Weddell triple junction: the principal focus of Ferrar and Karoo magmatism during initial breakup of Gondwana. *Geology*, **28**, 539–542.

Elliot, D.H. & Fleming, T.H. 2004. Occurrence and dispersal of magmas in the Jurassic Ferrar Large Igneous Province, Antarctica. *Gondwana Research*, **7**, 223– 237.

Elliot, D.H. & Fleming, T.H. 2008. Physical volcanology and geological relationships of the Ferrar Large Igneous Province, Antarctica. *Journal of Volcanology and Geothermal Research*, **172**, 20–37.

Elliot, D.H. & Grimes, C.G. 2011. Triassic and Jurassic strata at Coombs Hills, south Victoria Land: stratigraphy, petrology and cross-cutting breccia pipes. *Antarctic Science*, **23**, 268–280, https://doi.org/10.1017/S0954102010000994

Elliot, D.H. & Hanson, R.E. 2001. Origin of widespread, exceptionally thick basaltic phreatomagmatic tuff breccia in the Middle Jurassic Prebble and Mawson formations, Antarctica. *Journal of Volcanology and Geothermal Research*, **111**, 183–201.

Elliot, D.H., Haban, M.A. & Siders, M.A. 1986. The exposure hill formation, mesa range. *In*: Stump, E. (ed.) *Geological Investigations in Northern Victoria Land*. American Geophysical Union, Washington, DC, Antarctic Research Series, **46**, 267–278.

Elliot, D.H., Fleming, T.H., Haban, M.A. & Siders, M.A. 1995. Petrology and Mineralogy of the Kirkpatrick Basalt and Ferrar Dolerite, Mesa Range region, north Victoria Land, Antarctica. *In*: Elliot, D.H. & Blaisdell, G.L. (eds) *Contributions to Antarctic Research IV*. American Geophysical Union, Antarctic Research Series, **67**, 103–141.

Elliot, D.H., Fleming, T.H., Kyle, P.R. & Foland, K.A. 1999. Long Distance Transport of Magmas in the Jurassic Ferrar Large Igneous Province, Antarctica. *Earth and Planetary Science Letters*, **167**, 87–104.

Elliot, D.H., Larsen, D., Fanning, C.M., Fleming, T.H. & Vervoort, J.D. 2017. The Lower Jurassic Hanson Formation of the Transantarctic Mountains: implications for the Antarctic sector of the Gondwana plate margin. *Geological Magazine*, **154**, 777–803, https://doi.org/10.1017/S0016756816000388

Encarnación, J., Fleming, T.H., Elliot, D.H. & Eales, J.V. 1996. Synchronous emplacement of Ferrar and Karoo dolerites and the early breakup of Gondwana. *Geology*, **24**, 535–538.

Ferris, J.K., Johnson, A. & Storey, B.C. 1998. Form and extent of the Dufek intrusion, Antarctica, from newly compiled aeromagnetic data. *Earth and Planetary Science Letters*, **154**, 185–202.

Ferris, J.K., Storey, B.C., Vaughan, A.P.M., Kyle, P.R. & Jones, P.C. 2003. The Dufek and Forrestal intrusions, Antarctica: a centre for Ferrar Large Igneous Province dike emplacement. *Geophysical Research Letters*, **30**, 1348, https://doi.org/10.1029/2002GL016719

Fleming, T.H., Elliot, D.H., Jones, L.M., Bowman, J.R. & Siders, M.A. 1992. Chemical and isotopic variations in an iron-rich lava flow from North Victoria Land,

Antarctica: implications for low-temperature alteration and the petrogenesis of Ferrar magmas. *Contributions to Mineralogy and Petrology*, **111**, 440–457.

FLEMING, T.H., FOLAND, K.A. & ELLIOT, D.H. 1995. Isotopic and chemical constraints on the crustal evolution and source signature of Ferrar magmas, North Victoria Land, Antarctica. *Contributions to Mineralogy and Petrology*, **121**, 217–236.

FLEMING, T.H., HEIMANN, A., FOLAND, K.A. & ELLIOT, D.H. 1997. $^{40}Ar/^{39}Ar$ geochronology of Ferrar Dolerite sills from the Transantarctic Mountains, Antarctica: implications for the age and origin of the Ferrar magmatic province. *Geological Society of America Bulletin*, **109**, 533–546.

FORD, A.B. & HIMMELBERG, G.R. 1991. Geology and crystallization of the Dufek intrusion. *In*: TINGEY, R.J. (ed.) *The Geology of Antarctica*. Oxford Monographs on Geology and Geophysics, **17**. Oxford University Press, Oxford, 175–214.

GUNN, B.M. 1966. Modal and element variation in Antarctic tholeiites. *Geochimica et Cosmochimica Acta*, **30**, 881–920.

HAMILTON, W.B. 1965. *Diabase Sheets of the Taylor Glacier Region, Victoria Land, Antarctica*. United States Geological Survey, Professional Papers, **456-B**.

HANEMANN, R. & VIERECK-GÖTTE, L. 2004. Geochemistry of Jurassic Ferrar lava flows, sills and dikes sampled during the joint German–Italian Antarctic Expedition 1999–2000. *Terra Antarctica*, **11**, 39–54.

HANSON, R.E. & ELLIOT, D.H. 1996. Rift-related Jurassic basaltic phreatomagmatic volcanism in the central Transantarctic Mountains: precursory stage to flood-basalt effusion. *Bulletin of Volcanology*, **58**, 327–347.

HEIMANN, A., FLEMING, T.H., ELLIOT, D.H. & FOLAND, K.A. 1994. A short interval of Jurassic continental flood basalt volcanism in Antarctica as demonstrated by $^{40}Ar/^{39}Ar$ geochronology. *Earth and Planetary Science Letters*, **121**, 19–41.

HERGT, J.M., CHAPPELL, B.W., FAURE, G. & MENSING, T.M. 1989*a*. The geochemistry of Jurassic dolerites from Portal Peak, Antarctica. *Contributions to Mineralogy and Petrology*, **102**, 298–305.

HERGT, J.M., CHAPPELL, B.W., MCCULLOCH, T.M., MCDOUGALL, I. & CHIVAS, A.R. 1989*b*. Geochemical and isotopic constraints on the origin of the Jurassic dolerites of Tasmania. *Journal of Petrology*, **30**, 841–883.

HOOPER, P.R. 1997. The Columbia river flood Basalt Province: current status. *In*: MAHONEY, J.J. & COFFIN, M.F. (eds) *Large Igneous Provinces. Continental, Oceanic, and Planetary Volcanism*. American Geophysical Union, Geophysical Monographs, **100**, 1–27.

HORNIG, I. 1993. High-Ti and low-Ti tholeiites in the Jurassic Ferrar Group, Antarctica. *Geologisches Jahrbuch*, **E-47**, 335–369.

JOURDAN, F., FÉRAUD, G., BERTRAND, H., KAMPUNZU, A.B., TSHOSO, G., WATKEYS, M.K. & LE GALL, B. 2005. Karoo large igneous province: brevity, origin, and relation to mass extinction questioned by new $^{40}Ar/^{39}Ar$ data. *Geology*, **33**, 745–748, https://doi.org/10.1130/G21632.1

JOURDAN, F., FÉRAUD, G., BERTRAND, H. & WATKEYS, M.K. 2007. From flood basalts to the inception of oceanization: example from the $^{40}Ar/^{39}Ar$ high-resolution picture of the Karoo large igneous province. *Geochemistry, Geophysics, Geosystems*, **8**, Q02002, https://doi.org/10.1029/2006GC001392

JOURDAN, F., FÉRAUD, G., BERTRAND, H., WATKEYS, M.K. & RENNE, P.R. 2008. The $^{40}Ar/^{39}Ar$ ages of the sill complex of the Karoo large igneous province: implications for the Pliensbachian–Toarcian climate change. *Geochemistry, Geophysics, Geosystems*, **9**, Q06009, https://doi.org/10.1029/2008GC001994

LEAT, P.T. 2008. On the long-distance transport of Ferrar magmas. *In*: THOMSON, K. & PETFORD, N. (eds) *Structure and Emplacement of High-Level Magmatic Systems*. Geological Society, London, Special Publications, **302**, 45–61, https://doi.org/10.1144/SP302.4

LEAT, P.T., LUTTINEN, A.V., STOREY, B.C. & MILLAR, I.L. 2006. Sills of the Theron Mountains, Antarctica: evidence for long distance transport of mafic magmas during Gondwana break-up. *In*: HANSKI, E., MERTANEN, S., RÄMÖ, T. & VUOLLO, J. (eds) *Dyke Swarms: Markers of Crustal Evolution*. Taylor & Francis, Abingdon, UK, 183–199.

LECHEMINANT, A.N. & HEAMAN, L.M. 1989. Mackenzie igneous events, Canada: middle Proterozoic hotspot magmatism associated with ocean opening. *Earth and Planetary Science Letters*, **96**, 38–48.

MARSH, D.B. 2004. A magmatic mush column Rosetta Stone: the McMurdo Dry Valleys of Antarctica. *Eos*, **86**, 497, 502.

MARSH, B.D. 2007. Magmatism, magma, and magma chambers. *In*: SCHUBERT, G. (ed.) *Crustal and Lithosphere Dynamics*. Treatise on Geophysics, **6**. Elsevier, Amsterdam, 276–333, https://doi.org/10.1016/B978-0-444-53802-4.00116-0

MARSH, B.D., HERSUM, T.G., SIMON, A.C., CHARRIER, A.D. & SOUTER, B.J. 2005. Discovery of a funnel-like deep feeder zone for the Ferrar Dolerites, McMurdo Dry Valleys, Antarctica. *Abstract V14C-03 presented at the American Geophysical Union Meeting*, 5–9 December 2005, San Francisco, CA, USA.

MARSH, J.S., HOOPER, P.R., REHACEK, J., DUNCAN, R.A. & DUNCAN, A.R. 1997. Stratigraphy and age of Karoo basalts of Lesotho and implications for correlations within the Karoo Igneous Province. *In*: MAHONEY, J.J. & COFFIN, M.F. (eds) *Large Igneous Provinces: Continental, Oceanic, and Planetary Flood Volcanism*. American Geophysical Union, Geophysical Monographs, **100**, 247–272.

MENSING, T.M. & FAURE, G., JONES, L.M. & HOEFS, J. 1991. Stratigraphic correlation and magma evolution of the Kirkpatrick Basalt in the Mesa Range, northern Victoria Land, Antarctica. *In*: ULBRICH, H. & ROCHA CAMPOS, A. C. (eds) *Gondwana Seven Proceedings*. Instituto de Geosciências, Universidade de São Paulo, São Paulo, Brazil, 653–667.

MINOR, D. & MUKASA, S. 1997. Zircon U–Pb and hornblende $^{40}Ar–^{39}Ar$ ages for the Dufek layered mafic intrusion, Antarctica: implications for the age of the Ferrar large igneous province. *Geochimica et Cosmochimica Acta*, **61**, 2497–2504.

MORRISON, A.D. & REAY, A. 1995. Geochemistry of Ferrar Dolerite sills and dykes at Terra Cotta Mountains, south Victoria Land, Antarctica. *Antarctic Science*, **7**, 73–85, https://doi.org/10.1017/S0954102095000113

MUIRHEAD, J.D., AIROLDI, J., ROWLAND, J.V. & WHITE, J.D. L. 2012. Interconnected sills and inclined sheet intrusions control shallow magma transport in the Ferrar large igneous province, Antarctica. *Bulletin of the Geological Society of America*, **124**, 162–180, https://doi.org/10.1130/B30455.1

MUIRHEAD, J.D., AIROLDI, G., WHITE, J.D.L. & ROWLAND, J. V. 2014. Cracking the lid: sill-fed dikes are the likely feeders of flood basalt eruptions. *Earth and Planetary Science Letters*, **406**, 187–197, https://doi.org/10.1016/j.epsl.2014.08.036

NEUMANN, E.-R., SVENSEN, H., GALERNE, C.Y. & PLANKE, S. 2011. Multistage evolution of dolerites in the Karoo Large Igneous Province, central South Africa. *Journal of Petrology*, **52**, 959–984.

NORRISH, K. & HUTTON, J.T. 1969. An accurate X-ray spectrographic method for the analysis of a wide range of geological samples. *Geochimica et Cosmochimica Acta*, **33**, 431–453.

PEATE, D.W. 1997. The Paraná-Etendeka Province. *In*: MAHONEY, J.J. & COFFIN, M.F. (eds) *Large Igneous Provinces: Continental, Oceanic, and Planetary Flood Volcanism*. American Geophysical Union, Geophysical Monographs, **100**, 217–245.

PICCIRILLO, E.M., BELLIENI, G. ET AL. 1990. Lower Cretaceous tholeiitic dyke swarms from the Ponta Grossa arch (southeast Brazil): petrology, Sr–Nd isotopes and genetic relationships with the Parana flood volcanics. *Chemical Geology*, **89**, 19–48.

REIDEL, S.P., CAMP, V.E., ROSS, M.E., WOLFF, J.A., MARTIN, B.S., TOLAN, T.L. & WELLS, R.E. 2013. The Columbia River flood basalt province; stratigraphy, areal extent, volume, and physical volcanology. *In*: REIDEL, S.P., CAMP, V.E., ROSS, M.E., WOLFF, J.A., MARTIN, B.S., TOLAN, T.L. & WELLS, R.E. (eds) *The Columbia River Flood Basalt Province*. Geological Society of America, Special Papers, **497**, 1–43, https://doi.org/10.1130/2013.2497(01)

RENNE, P.R., SWISHER, C.C., DEINO, A.L., KARNER, D.B., OWENS, T.L. & DEPAOLO, D.J. 1998. Intercalibration of standards, absolute ages and uncertainties in Ar-Ar dating. *Chemical Geology*, **145**, 117–152.

RILEY, T.R., MILLAR, I.L., WATKEYS, M.K., CURTIS, M.L., LEAT, P.T., KLAUSEN, B. & FANNING, C.M. 2004. U–Pb zircon (SHRIMP) ages for Lebombo rhyolites, South Africa: refining the duration of Karoo volcanism. *Journal of the Geological Society, London*, **161**, 547–550, https://doi.org/10.1144/0016-764903-181

ROSS, P.-S., WHITE, J.D.L. & MCCLINTOCK, M.K. 2008. Physical volcanology of mafic volcaniclastic deposits and lavas in the Coombs–Allan Hills area, Ferrar large igneous province, Antarctica. *Journal of Volcanology and Geothermal Research*, **172**, 38–60.

RUDNICK, R.L. & GAO, S. 2003. Composition of the continental crust. *In*: HOLLAND, H.D. & TUREKIAN, K.K. (eds) *The Crust*. Treatise on Geochemistry, **3**. Elsevier, Amsterdam, 1–64.

SCHÖNER, R., VIERECK-GÖTTE, L., SCHEIDNER, J. & BOMFLEUR, B. 2007. Triassic–Jurassic sediments and multiple volcanic events in north Victoria Land, Antarctica: A revised stratigraphic model. *In*: COOPER, A.K., RAYMOND, C.R. ET AL. (eds) *Antarctica: A Keystone in a Changing World. On Line Proceedings of the 10th ISAES*. United States Geological Survey, Open-File Report, **2007-1047**, Short Research Paper 102, https://doi.org/10.3133/of2007-1047.srp102

SCHÖNER, R., BOMFLEUR, B., SCHEIDNER, J. & VIERECK-GÖTTE, L. 2011. A systematic description of the Section Peak Formation in north Victoria Land. *Polarforschung*, **80**, 71–87.

SELF, S., JAY, A.E., WIDDOWSON, M. & KESZTHELYI, L.O. 2008. Correlation of the Deccan and Rajahmundry Trap lavas; are these the longest and largest lava flows on Earth? *Journal of Volcanology and Geothermal Research*, **172**, 3–19.

SELL, B., OVTCHAROVA, M. ET AL. 2014. Evaluating the temporal link between the Karoo LIP and climatic–biologic events of the Toarcian Stage with high-precision U–Pb geochronology. *Earth and Planetary Science Letters*, **408**, 48–56, https://doi.org/10.1016/j.epsl.2014.10.008

SKINNER, D.N.B. & RICKER, J. 1968. The geology of the region between the Mawson and Priestley Glaciers, north Victoria Land. *New Zealand Journal of Geology and Geophysics*, **11**, 1041–1075.

STOREY, B.C. & KYLE, P.R. 1997. An active mantle mechanism for Gondwana break-up. *Journal of African Earth Sciences*, **100**, 283–290.

SUN, S.-S. & MCDONOUGH, W.F. 1989. Chemical and isotopic systematics of oceanic basalts: implications for mantle composition and processes. *In*: SAUNDERS, A. D. & NORRY, M.J. (eds) *Magmatism in the Ocean Basins*. Geological Society, London, Special Publications, **42**, 313–345, https://doi.org/10.1144/GSL.SP.1989.042.01.19

SVENSEN, H., CORFU, F., POLTEAU, S., HAMMER, Ø. & PLANKE, S. 2012. Rapid magma emplacement in the Karoo Large Igneous Province. *Earth and Planetary Science Letters*, **325–326**, 1–9, https://doi.org/10.1016/j.epsl.2012.01.015

VIERECK-GÖTTE, L., SCHÖNER, R., BOMFLEUR, B. & SCHNEIDER, J. 2007. Multiple shallow level sill intrusions coupled with hydromagmatic explosive eruptions marked the initial phase of Ferrar large igneous province magmatism in northern Victoria Land. *In*: COOPER, A.K., RAYMOND, C.R. ET AL. (eds) *Antarctica: A Keystone in a Changing World. On Line Proceedings of the 10th ISAES*. United States Geological Survey, Open-File Report, **2007-1047**, Short Research Paper 104, https://doi.org/10.3133/of2007-1047.srp104

WHITE, J.D.L. & MCCLINTOCK, M.K. 2001. Immense vent complex marks flood-basalt eruption in a wet, failed rift: Coombs Hills, Antarctica. *Geology*, **29**, 935–938.

WILHELM, S. & WÖRNER, G. 1996. Crystal size distribution in Jurassic Ferrar flows and sills (Victoria Land, Antarctica): evidence for processes of cooling, nucleation and crystallization. *Contributions to Mineralogy and Petrology*, **125**, 1–15.

ZIEG, M.J. & MARSH, D.B. 2012. Multiple reinjections and crystal-mush compaction in the Beacon sill, McMurdo Dry Valleys, Antarctica. *Journal of Petrology*, **53**, 2567–2591.

The Panjal Traps

J. GREGORY SHELLNUTT

Department of Earth Sciences, National Taiwan Normal University, 88 Tingzhou Road Section 4, Taipei 11677, Taiwan

jgshelln@ntnu.edu.tw

Abstract: The Early Permian (290 Ma) Panjal Traps are the largest contiguous outcropping of volcanic (basaltic, andesitic and silicic) rocks within the Himalaya that are associated with the Late Palaeozoic break-up of Gondwana. The basaltic Panjal Traps have compositional characteristics that range from continental tholeiite to ocean-floor basalt but it is clear that crustal contamination has played a role in their genesis. The basalts that show limited evidence for contamination have Sr–Nd isotopes ($^{87}Sr/^{86}Sr_i$ = 0.7043–0.7073; $\varepsilon_{Nd}(t) = 0 \pm 1$) similar to a chondritic (subcontinental lithospheric mantle) source, whereas the remaining basaltic rocks have a wide range of Nd ($\varepsilon_{Nd}(t)$ = −6.1 to +4.3) and Sr ($^{87}Sr/^{86}Sr_i$ = 0.7051–0.7185) isotopic values. The primary melt composition of the low-Ti Panjal Traps is picritic with mantle potential temperatures (T_P = 1400°C to 1450°C) similar to ambient mantle. The silicic volcanic rocks were derived by partial melting of the crust, whereas the andesitic rocks were derived by mingling between crustal and mantle melts. The Panjal Traps initially erupted within a continental rift setting. The rift eventually transitioned into a nascent ocean basin that led to seafloor spreading and the formation of the Neotethys Ocean and the ribbon-like continent Cimmeria.

The growth and development of mafic continental Large Igneous Provinces (MLIPs) is a first-order problem in the Earth Sciences, and requires a multidisciplinary approach to fully understand and characterize the thermal, biological and geological consequences of their formation (Wignall 2001; Jerram & Widdowson 2005; Campbell 2007; Bryan & Ernst 2008; Bryan & Ferrari 2013; Bond & Wignall 2014). MLIPs are voluminous, spatially contiguous regions of the continental crust where mantle-derived volcanic rocks and their intrusive equivalents were emplaced over a short period (*c.* 10 myr) of time (Campbell 2007; Bryan & Ernst 2008; Bryan & Ferrari 2013). The formation of LIPs is debated as many are considered to be derived by a diapiric upwelling of mantle material (mantle plume) possibly originating from the core–mantle boundary, whereas others are considered to be related to decompressional mantle melting associated with tensional plate stress – also known as the 'plate' model (Richards *et al.* 1989; Campbell & Griffiths 1990; Griffiths & Campbell 1990; Ernst & Buchan 2003; Anderson & Natland 2005; Ernst *et al.* 2005; Bryan & Ernst 2008; Foulger 2010; Ernst 2014). The rapid emplacement of mostly mafic magmas of LIPs is often, but not always, associated with continental break-up and mass extinctions. Therefore, the study of Phanerozoic LIPs has tremendous significance in understanding not only the secular thermal conditions of the lithosphere and plate tectonics but also how ecosystems respond to the effects of elevated rates of magmatism (Rampino & Stothers 1988; Coffin & Eldholm 1994; Courtillot *et al.* 1999; Wignall 2001; Jerram & Widdowson 2005; Saunders *et al.* 2007; Foulger 2010; van de Schootbrugge & Wignall 2016).

The Late Palaeozoic (*c.* 300–251 Ma) was an age of large polar glaciations, the zenith of Pangaea and two mass extinctions (Wignall 2001). Moreover, at least five major MLIPs were emplaced during this period. The Skagerrak-Centered LIP (*c.* 300 Ma) in central Laurasia (Torsvik *et al.* 2008; Timmerman *et al.* 2009), the Himalayan magmatic province (*c.* 290–270 Ma) along the Tethyan margin of Gondwana (Ernst & Buchan 2001; Zhu *et al.* 2010; Shellnutt *et al.* 2014; Wang *et al.* 2014), the Tarim LIP (*c.* 290–270 Ma) near the Tethyan margin of Laurasia (Yang *et al.* 2013; Xu *et al.* 2014), the Emeishan LIP (*c.* 260 Ma) of the South China Block (Shellnutt 2014) and the Siberian Traps (*c.* 251 Ma) of NE Laurasia (Saunders & Reichow 2009) collectively cover an area of >7× 10^6 km^2, and consist primarily of mafic volcanic and intrusive rocks. Unlike their Mesozoic and Cenozoic counterparts, the Late Palaeozoic MLIPs are almost all unrelated to plate separation (Coffin & Eldholm 1992; Torsvik *et al.* 2008; Shellnutt 2014; Xu *et al.* 2014). The only Late Palaeozoic LIP temporally associated with plate separation is the poorly constrained, ill-defined Himalayan magmatic province (HMP).

The HMP is an assortment of volcanic and plutonic sequences and mafic dykes that are broadly contemporaneous with the rifting of microcontinental terranes from the Tethyan margin of Gondwana

From: Sensarma, S. & Storey, B. C. (eds) 2018. *Large Igneous Provinces from Gondwana and Adjacent Regions*. Geological Society, London, Special Publications, **463**, 59–86.
First published online July 6, 2017, https://doi.org/10.1144/SP463.4

(Bhat *et al.* 1981; Honegger *et al.* 1982; Bhat 1984; Papritz & Rey 1989; Garzanti *et al.* 1999; Ernst & Buchan 2001; Yan *et al.* 2005; Zhu *et al.* 2010; Zhai *et al.* 2013; Shellnutt *et al.* 2014; Wang *et al.* 2014; Xu *et al.* 2016). The precise petrogenetic relationship between the various volcanic and plutonic rocks of the HMP remains uncertain. It is entirely possible that the HMP simply represents a collection of mutually exclusive disjointed units that were emplaced within the same regional-scale tensional regime along the Tethyan margin of Gondwana during the first 30 myr of the Permian (Fig. 1). The Permian volcanic sequences of the central and eastern Himalaya, such as the Abor, Nar-Tsum, Bhote Kosi, Selong and Mojiang volcanic groups, and the Qiangtang mafic dykes, are amongst the many occurrences of basaltic rocks that are considered to be related to the formation of the Neotethys Ocean and the ribbon-like continent Cimmeria (Bhat 1984; Sengör 1987; Yan *et al.* 2005; Zhu *et al.* 2010; Ali *et al.* 2012; Zhai *et al.* 2013; Wang *et al.* 2014). The Panjal Traps located in the western Himalaya of India and Pakistan form the single largest outcrop of the HMP but its origin and formation are relatively unconstrained (Bhat *et al.* 1981; Chauvet *et al.* 2008; Shellnutt *et al.* 2011*a*, 2014, 2015).

The Panjal Traps were first documented in the nineteenth century within the Pir Panjal and Zanskar mountain ranges of Kashmir with subsequent studies identifying their correlative units within other regions of the western Himalaya (Lydekker 1883; Middlemiss 1910; Wadia 1934, 1961). The Traps were initially considered to be Upper Palaeozoic in age by Middlemiss (1910), which was subsequently confirmed by *in situ* zircon U–Pb radioisotopic dating (Shellnutt *et al.* 2011*a*). Geochemical studies of the Traps have shown that they are generally similar to within-plate continental tholeiites, and can be correlated with the opening of the Neotethys Ocean and the formation of Cimmeria (Nakazawa & Kapoor 1973; Pareek 1976; Bhat & Zainuddin 1978, 1979; Bhat *et al.* 1981; Chauvet *et al.* 2008; Shellnutt *et al.* 2012, 2014, 2015).

Investigating the Panjal Traps has significant importance for unravelling pre-India–Eurasia collision tectonics of Gondwana and the formation of one of the few Late Palaeozoic ‘plate-separation’ mafic continental flood basalt provinces. Consequently, understanding the Panjal Traps can help to constrain issues such as the geodynamic and tectonomagmatic evolution of the Late Permian Tethyan margin of Gondwana, as well as offer an insight into the formation of LIPs with specific relevance to the ‘plate hypothesis’ (passive extension) v. ‘mantle plume theory’ (active extension) of LIP formation. The purpose of this paper is to provide a review on the current state of knowledge of the Panjal Traps and its petrogenetic relationship with the other Permian volcanic sequences of the Himalaya. The paper is divided into the following sections: (1) geological background; (2) age and duration of volcanism; (3) geochemical characteristics of the volcanic and

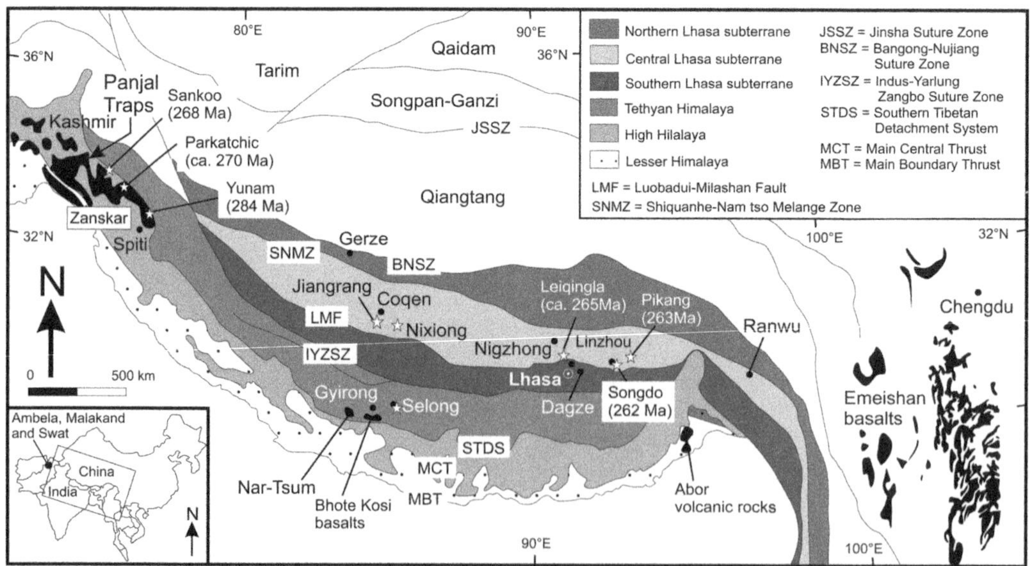

Fig. 1. Simplified geological map showing the distribution of Permian volcanic rocks within the Himalaya (based on Zhu *et al.* 2010). The inset map shows the locations of the Late Palaeozoic Ambela, Malakand and Swat granitic rocks.

plutonic rocks; (4) correlation with ultramafic volcanic rocks; (5) tectonomagmatic evolution; (6) mineral deposit potential; and (7) a regional examination of HMP rocks. The final section offers a synthesis of the Panjal Traps and suggestions for future research initiatives.

Geological background

The Tethyan domain of the western Indian Himalaya is composed of Precambrian–Late Palaeozoic rocks that form part of the Higher Himalaya. The Precambrian central crystalline complex consists of augen granite gneisses, nebulitic migmatites, and grey and dark paragneisses, and is the basement to the Tethyan passive margin sedimentary sequences. Overlying the basement is series of Cambrian–Lower Carboniferous sedimentary formations (Phe, Karsha, Kurgiakh, Thaple, Muth, Lipak and Po) that consist mainly of sandstone, shale, siltstone, arkose, carbonate and evaporite rocks (Gaetani *et al.* 1986; Garzanti *et al.* 1992, 1994, 1996*a*, *b*; Fuchs 1987; Myrow *et al.* 2006; Brookfield *et al.* 2013). Above the Lower Carboniferous units is the Middle–Upper Carboniferous fossiliferous (*Spirifer varuna* and *Camarophoria*-bearing) Fenestella shale, which is followed by the Agglomeratic slate consisting of siliciclastic and volcaniclastic material that may represent the first eruptive unit of the Panjal Traps.

The main volcanic sequence of the Panjal Traps erupted on the Upper Permian siliciclastic Nishatbagh beds, which were deposited on the Agglomeratic slate (Fig. 2) (Nakazawa *et al.* 1975; Garzanti *et al.* 1998; Wopfner & Jin 2009). The reported total thickness of the volcanic rocks in the Pir Panjal Range (western Kashmir) is approximately 3000 m, whereas in the Zanskar Range (eastern Kashmir) it is ≤300 m (Wadia 1934, 1961; Nakazawa & Kapoor 1973; Nakazawa *et al.* 1975; Chauvet *et al.* 2008; Wopfner & Jin 2009). Typically, individual flow units within the volcanic pile are 30 m thick. Deposited on top of the Panjal Traps are the Gangamopteris beds. The presence of *Gangamopteris kasmirensis* within sedimentary rocks on top of the volcanic rocks constrains the eruption age of the Panjal Traps to the Upper Permian (Fig. 2). The Gangamopteris beds are followed by the Zewan Formation (i.e. sandstones and carbonates) and the uppermost Permian–Lower Triassic Khunamuh Formation (shales) (Nakazawa *et al.* 1975; Wopfner & Jin 2009; Brookfield *et al.* 2013).

The Panjal Traps cover an area of approximately 10 000 km^2, exposed primarily around the Kashmir Valley along the Pir Panjal and Zanskar ranges (Fig. 3). The Traps are mostly basalt but there are volumetrically minor flows of basaltic andesite, andesite, rhyolite and dacite (Ganju 1944; Nakazawa & Kapoor 1973; Pareek 1976; Bhat *et al.* 1981; Shellnutt *et al.* 2012, 2014). The Traps show evidence of subaerial (columnar joints) and subaqueous (pillow structures) eruptive environments that Nakazawa & Kapoor (1973) interpreted as indicative of a near-shore, transgressive shallow-marine environment (Fig. 4). Intertrappean limestones, shales and slates are reported near Gulmarg and Bren (Pir Panjal Range), suggesting there were intermittent pauses during volcanism.

Chemostratigraphic variations of the basalts are identified within some profiles as TiO_2 (wt%) shows distinct changes from top to bottom of some volcanic sequences (Shellnutt *et al.* 2014). The volcanic centre of the Panjal Traps is inferred to be located in western Kashmir (Kashmir Valley)

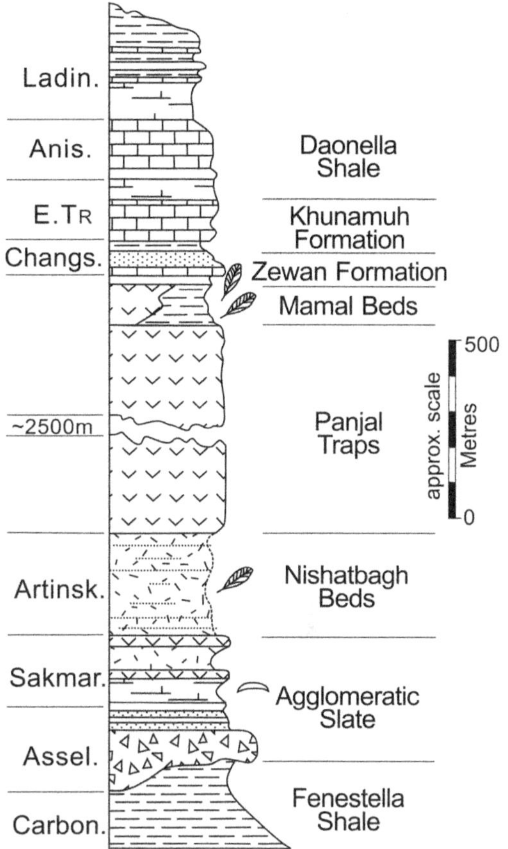

Fig. 2. Generalized Late Palaeozoic stratigraphy of the Pangaea Megasequence of Kashmir from Wopfner & Jin (2009). Carbon., Carboniferous; Assel., Asselian; Sakmar., Sakmarian; Artinsk., Artinskian. Changs., Changhsingian; ETR., Early Triassic; Anis., Anisian; Ladin., Landinian.

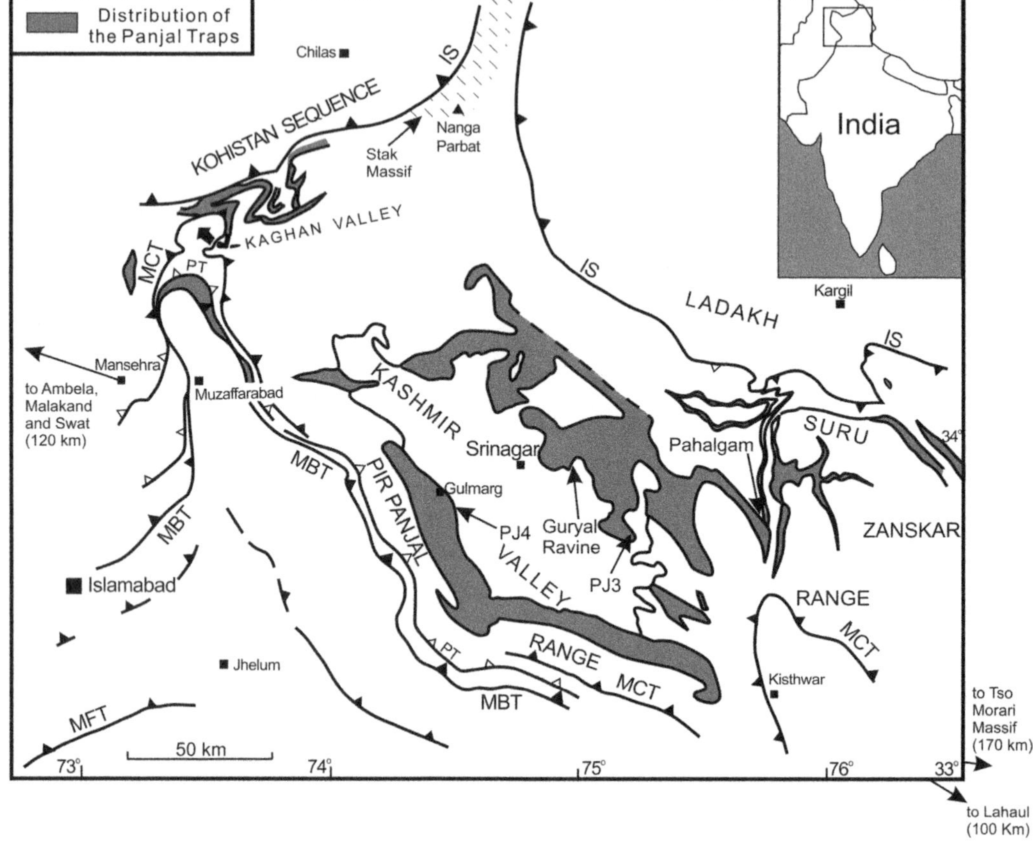

Fig. 3. Distribution of the Panjal Traps and major faults of the western Himalaya (based on Papritz & Rey 1989). IS, Indus Suture; MBT, Main Boundary Thrust; MCT, Main Central Thrust; MFT, Salt Range Main Frontal Thrust; PT, Panjal Thrust.

because the total flow thickness (*c.* 3 km) is greatest in that region (Nakazawa & Kapoor 1973). Andesitic rocks are located at specific eruptive horizons within the main basaltic sequences, whereas the silicic volcanic rocks appear to be below the initial basalt flows of the Kashmir Valley (Ganju 1944; Pareek 1976; Shellnutt *et al.* 2012). Permian komatiitic and basaltic rocks from Chhongtash in northern Ladakh are tentatively correlated with the Panjal Traps (Rao & Rai 2007).

Age and duration of volcanism

The precise timing and duration of the Panjal Traps was, until recently, only inferred based on the geological record. Initially, the Traps were considered to be Lower Palaeozoic in age. However, Middlemiss (1910), based on the presence of ammonites from the genus *Ophiceras* in overlying sediments, interpreted their age as Upper Palaeozoic. Subsequent interpretations suggested that the Traps erupted during the Late Palaeozoic–Early Mesozoic (Late Carboniferous–Early Triassic) but detailed structural and sedimentological studies by Nakazawa *et al.* (1975) and Kapoor (1977) showed that the lava erupted after the deposition of the Late–Middle Carboniferous Fenestella shale and before the deposition of the Gangamopteris beds. The fact that the Panjal Traps erupted before the deposition of the Gangamopteris beds constrains their emplacement age to no older than the Late Carboniferous but there are reports that indicated the Traps were underlain by the Gangamopteris beds (Nakazawa & Kapoor 1973; Nakazawa *et al.* 1975; Pareek 1976; Tripathi & Singh 1985, 1987; Wopfner & Jin 2009). Nakazawa *et al.* (1975) suggested that the Panjal Traps are definitively constrained to the Early–Middle Permian (Sakmarian–Artinskian) and that reports of an Early Triassic age are incorrect.

More recently, there were suggestions that the Panjal Traps erupted during the Middle–Late

Fig. 4. Field photographs of the Panjal Traps showing: (**a**) the contact between novaculite and the uppermost basalts at Guryal Ravine; (**b**) pillow basalt (near Awantipora); (**c**) the tectonic contact with Triassic carbonates (near Chandanwadi); and (**d**) relict columnar joint structures near Chandanwadi.

Permian. White (2002) and White & Saunders (2005) indicated that the Panjal Traps could be *c.* 260 Ma and possibly linked to the mid-Capitanian mass extinction, whereas Veevers & Tewari (1995) suggested that the Traps are correlative with Late Permian (*c.* 250 Ma) magmatic rocks found within the Gondwana margin after the opening of the Neotethys Ocean. Furthermore, Vannay & Spring (1993) and Noble *et al.* (2001) suggested that the Traps are likely to have erupted during the Kungutian–Artinskian (275–258 Ma).

The first radioisotopic age of the Panjal Traps was analysed from zircons collected from the silicic Traps near Srinagar (Shellnutt *et al.* 2011*a*). The *in situ* zircon laser ablation–inductively coupled plasma–mass spectrometry (LA-ICP-MS) U–Pb dating from a rhyolite near Srinagar yielded a weighted mean $^{206}Pb/^{238}U$ age of 289 ± 3 Ma and is consistent with the geological interpretations of Nakazawa *et al.* (1975). The age definitively demonstrates that the Traps erupted during the Early Permian and were not associated with the mid-Capitanian mass extinction or post-Neotethys magmatism.

Prior to the study of Shellnutt *et al.* (2011*a*), the Yunam A-type granite located within the Zanskar Valley was dated using the zircon U–Pb thermal ionization mass spectrometry (TIMS) technique and yielded an age of 284 ± 1 Ma (Spring *et al.* 1993). However, Spring *et al.* (1993) suggested there was no petrogenetic association between the Panjal Traps and the Yunam granite. In light of the new age data from the silicic volcanic rocks, the two units are probably related to the same tectonothermal system. Additionally, an Early Permian (284 ± 4.4 Ma) *in situ* zircon LA-ICP-MS U–Pb age of a granite from the Zanskar Valley is reported by Horton (2011) suggesting there was widespread plutonic magmatism that accompanied the Panjal volcanism. There are other Carboniferous–Permian granitic rocks reported in the region, such as Malakand, Ambela, Tarbela, Swat, Parkatchic and Sankoo, some of which are considered to be crustal melts associated with the injection of the mafic Panjal magma (Table 1) (Le Bas *et al.* 1987; Zeitler 1988; Spring *et al.* 1993; Smith *et al.* 1994; Noble *et al.* 2001; Ahmad *et al.* 2013).

Table 1. *Summary of Late Carboniferous–Middle Permian age dates from the western Himalaya*

Locality	Rock type	Method	Material	Age (Ma)	References
Srinagar	Rhyolite	$^{206}Pb/^{238}U$	Zircon	289 ± 3	Shellnutt *et al.* (2011*a*)
Yunam	Granite	$^{206}Pb/^{238}U$	Zircon	284 ± 1	Spring *et al.* (1993)
Zanskar	Granite	$^{206}Pb/^{238}U$	Zircon	284 ± 4.4	Horton (2011)
Zanskar	Leucogranite	$^{206}Pb/^{238}U$	Zircon	305 ± 11	Horton (2011)
Ambela (Koga)	Syenite	$^{87}Rb/^{86}Sr$	Rock	315 ± 15	Le Bas *et al.* (1987)
Ambela (Koga)	Ijolite	$^{87}Rb/^{86}Sr$	Rock	297 ± 4	Le Bas *et al.* (1987)
Ambela (Babaji)	Syenite	$^{206}Pb/^{238}U$	Zircon	280 ± 15	Smith *et al.* (1994)
Malakand	Granite	$^{206}Pb/^{238}U$	Zircon	279 ± 11	Smith *et al.* (1994)
Malakand	Granite	$^{206}Pb/^{238}U$	Zircon	271 ± 11	Zeitler (1988)
Peshawar	Microporphyry	$^{206}Pb/^{238}U$	Zircon	282 ± 5	Ahmad *et al.* (2013)
Peshawar	Granite	$^{206}Pb/^{238}U$	Zircon	278 ± 10	Ahmad *et al.* (2013)
Zanskar	Granite	$^{206}Pb/^{238}U$	Zircon	268 ± 5	Noble *et al.* (2001)
Zanskar	Granite	$^{206}Pb/^{238}U$	Zircon	270	Noble *et al.* (2001)
Tarbela	Syenite	$^{206}Pb/^{238}U$	Zircon	280 ± 15	Smith *et al.* (1994)
Swat	Gneiss	$^{206}Pb/^{238}U$	Zircon	276 ± 9	Anczkiewicz *et al.* (2001)
Swat	Gneiss	$^{206}Pb/^{238}U$	Zircon	267 ± 2	Anczkiewicz *et al.* (2001)
Swat	Gneiss	$^{206}Pb/^{238}U$	Zircon	265 ± 3	DiPietro & Isachsen (2001)
Swat	Gneiss	$^{206}Pb/^{238}U$	Zircon	278 ± 4	DiPietro & Isachsen (2001)
Swat	Diabase	$^{40}Ar/^{39}Ar$	Hornblende	284 ± 4	Baig (1990)
Manshera	Amphibolite	$^{40}Ar/^{39}Ar$	Hornblende	262 ± 1	Baig (1990)

At the moment, very few definitive statements can be made regarding the duration of Panjal magmatism but it is clear that magmatism was underway by the Early Permian. The initial volcanic rocks are considered to be the Agglomerate slate, which is interpreted to be an 'explosive volcanic' unit and consists of ash and possibly volcanic bombs, although non-volcanic fossiliferous material appears to predominate (Nakazawa & Kapoor 1973; Nakazawa *et al.* 1975). The Agglomerate slate is followed by the main effusive sequence. Gaetani *et al.* (1990) suggested that the eruption duration was 2–3 myr based on faunal markers from NW Lahul–SE Zanskar but the volcanic sequences are thinner than those around the Kashmir Valley and may be incomplete. The basalts at Guryal Ravine are capped by sedimentary rocks (novaculite and carbonate rocks) that are Early Permian (Artinskian) in age but that does not preclude the possibility that magmatism continued after the deposition of the Gangamopteris beds and the Zewan Formation (Nakazawa *et al.* 1975; Wopfner & Jin 2009; Tewari *et al.* 2015). It is possible that volcanism ended at different times in different localities and that the preserved 'continental' portion of the Panjal Traps is strictly Early–Middle Permian but as the rift developed into an ocean basin, and later the Neotethys Ocean, volcanism became strictly 'oceanic' in nature (Shellnutt *et al.* 2015).

Identifying Early Permian volcanic and plutonic rocks in the western Himalaya using high-precision radioisotope geochronology is a priority but is also problematic as many of the volcanic rocks are highly altered. As more Early Permian rocks are identified, their direct (magmatic differentiation) or indirect (crustal melting) relationship to the Panjal Traps can be evaluated with confidence and a greater understanding of the magmatic duration can be achieved.

Mafic Panjal Traps

Chemical characterization

The mafic Panjal Traps are described as massive aphyric tholeiitic to mildly alkalic continental flood basalts (Nakazawa & Kapoor 1973; Bhat & Zainuddin 1979; Bhat *et al.* 1981; Honegger *et al.* 1982; Papritz & Rey 1989; Chauvet *et al.* 2008; Shellnutt *et al.* 2014). The rocks from the Guryal Ravine near Srinagar are relatively unaltered as euhedral plagioclase laths and subhedral clinopyroxene are still preserved within a fine-grained groundmass (Fig. 5a, b). Rocks from other regions (Pir Panjal Range and Pahalgam) of the Traps have experienced greenschist-facies metamorphism and the primary mineralogy was largely replaced (Nakazawa & Kapoor 1973; Vannay & Spring 1993; Chauvet *et al.* 2008). Consequently, it is likely that some elements (e.g. large-ion lithophile elements (LILEs)) were mobilized during alteration.

Most of the Panjal Traps chemically classify as 'within-plate basalt', were probably derived from a relatively reducing mantle source (FMQ −1) and have light rare earth element (LREE)-enriched ($La/Yb_N \approx 4$: Sun & McDonough 1989)

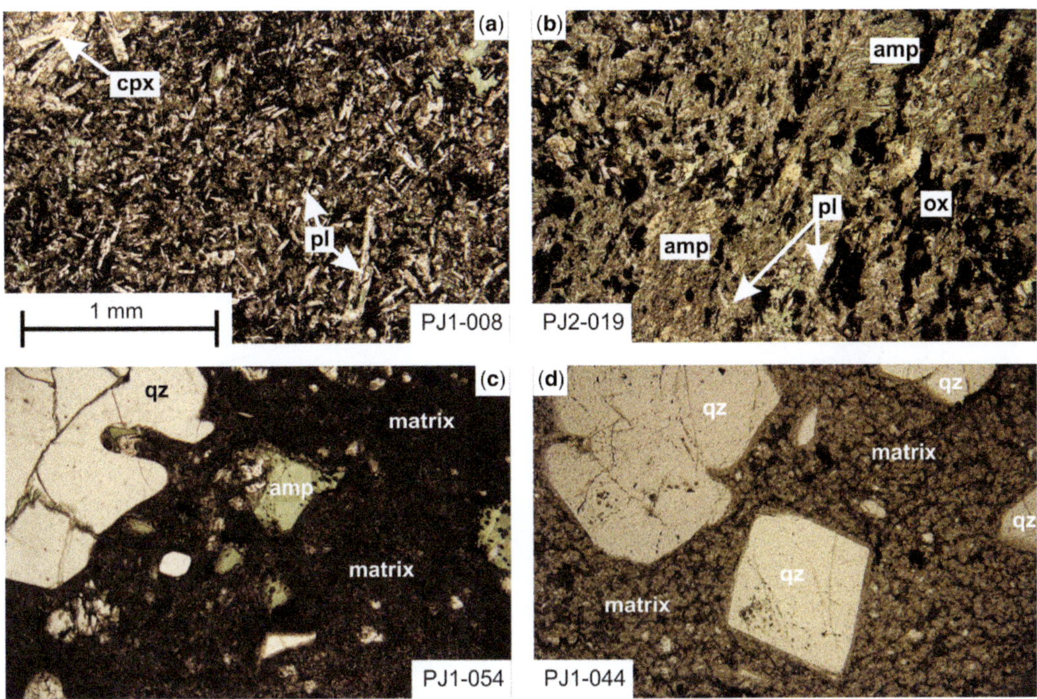

Fig. 5. Photomicrographs of the basaltic and silicic Panjal Traps. (**a**) Basalt from Guryal Ravine (sample PJ1-008) showing unaltered plagioclase (pl) and clinopyroxene (cpx). (**b**) Basalt from Pahalgam (sample PJ2-019) showing altered plagioclase (pl), amphibole (amp) and Fe–Ti oxide minerals (ox). (**c**) Dacite from Pampore (along the Guryal Ravine section) showing a quartz (qz) porphyry texture. (**d**) Rhyolite from Pampore (along the Guryal Ravine section) showing a quartz (qz) porphyry texture with microcrystalline texture. The scale in (a) is applicable to all the photomicrographs.

chondrite-normalized patterns (Figs 6–8). However, there are some rocks that are compositionally similar to ocean-floor basalt and have REE ($La/Yb_N < 2$; Sun & McDonough 1989) patterns similar to enriched mid-ocean ridge basalts (E-MORB). Some thick volcanic sequences around the Kashmir Valley show chemostratigraphic variations, specifically with respect to the TiO_2 concentration, as there appears to be groups of high-Ti ($TiO_2 > 1.7$ wt%) and low-Ti ($TiO_2 < 1.7$ wt%) basalts. There is distinct TiO_2 variability within specific sequences but considering all basaltic data there is a continuous spectrum of compositions (Fig. 9a). In broad terms, the Mg number (Mg#) and the Ti/Y ratio offer better discrimination of the basalts as there appears to be three groups: (1) high Mg# group (>57); (2) middle Mg# group (50–57); and (3) low Mg# group (<50). In combination with the Ti/Y ratio, it seems that samples with Ti/Y< 350 have the highest Mg# (Fig. 9b, c & d). Furthermore, the rocks with the highest Mg# values are likely to represent more 'primitive' lavas as they tend to have a higher Ni (>100 ppm) content than the rocks with lower Mg# values (Ni< 100 ppm). Within specific basalt sequences, the application of a Ti-based classification scheme is useful but its broad application to the Panjal Traps is somewhat cumbersome.

To date, there are only a few studies that report the Sr–Nd isotopes of the Panjal Traps (Spencer *et al.* 1995; Chauvet *et al.* 2008; Shellnutt *et al.* 2014, 2015). The Sr isotopes are quite variable ($^{87}Sr/^{86}Sr_i = 0.7043–0.7185$), which may, at least partially, be due to Rb or Sr mobility during greenschist-facies metamorphism (Fig. 10). The basalts with the lowest $^{87}Sr/^{86}Sr_i$ values are probably closer to their unmodified initial values as they are less altered ($^{87}Sr/^{86}Sr_i = 0.7043–0.7073$). The Nd isotopes are also variable ($\varepsilon_{Nd}(t) = -6.1$ to $+4.3$) but Sm and Nd are less susceptible to mobility and are probably indicative of their true magmatic value. There are six samples reported by Shellnutt *et al.* (2014) that may reflect the initial, uncontaminated basaltic magmas as they have $\varepsilon_{Nd}(t)$ values of between −1.4 and +1.3, $^{87}Sr/^{86}Sr_i$ values between 0.7043 and 0.7073, and do not have high Th/Nb_{PM} (i.e. >3) ratios or low Nb/U (i.e. <20) ratios

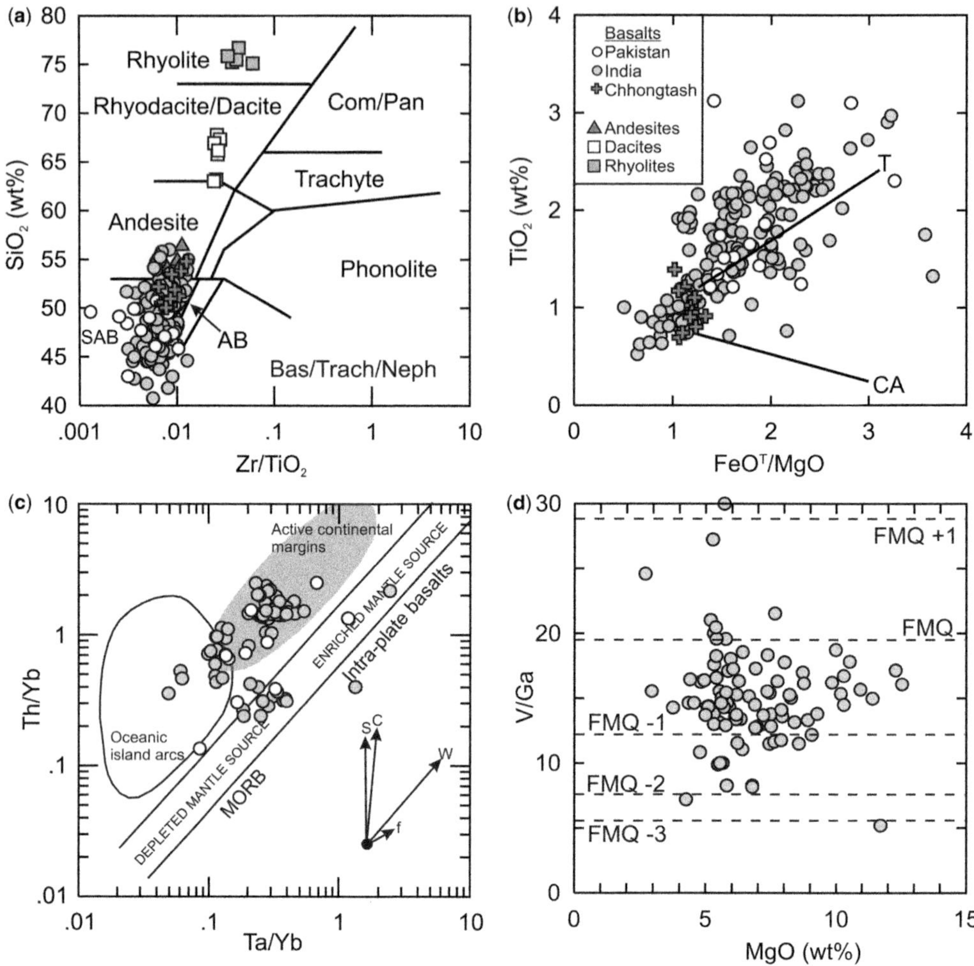

Fig. 6. (**a**) Rock classification of the Panjal Traps using immobile elements (Winchester & Floyd 1977). SAB, subalkaline basalts; AB, alkaline basalt; Com/Pan, comendite/pantellerite; Bas/Trach/Neph, basanite, trachybasanite and nephelinite. (**b**) Discrimination of the tholeiitic (T) basalts from the calc-alkaline (CA) basaltic rocks (Miyashiro 1974). (**c**) Th/Yb v. Ta/Yb basalt discrimination diagram of Wilson (1989) showing the differences between subduction and oceanic basalts derived from depleted and enriched sources. Vectors show the influence of each component: S, subduction component; C, crustal component; W, within-plate enrichment; f, fractional crystallization. (**d**) Binary diagram showing the use of bulk-rock V/Ga to indicate the redox condition of the mafic Panjal Traps. Reference lines at various f_{O_2} are after Mallmann & O'Neill (2009). Data from Pareek (1976), Bhat & Zainuddin (1978, 1979), Bhat *et al.* (1981), Honegger *et al.* (1982), Papritz & Rey (1989), Pogue *et al.* (1992), Vannay & Spring (1993), Rao & Rai (2007), Chauvet *et al.* (2008) and Shellnutt *et al.* (2012, 2014, 2015).

(Table 2). The remaining rocks, including the rocks with the highest $\varepsilon_{Nd}(t)$ values, appear to have been affected by crustal contamination.

Magma differentiation and crustal contamination

Most of the Panjal Traps have experienced varying degrees of either crystal fractionation or crustal contamination, or both, and therefore do not represent primary magma compositions. Shellnutt *et al.* (2014, 2015) suggested that the 'high-Ti' rocks from Pahalgam and some rocks from the Pir Panjal ('low-Ti') Range experienced crystal fractionation. The 'high-Ti' rocks are likely to have experienced olivine and plagioclase fractionation, whereas the 'low-Ti' rocks are likely to have experienced clinopyroxene and olivine fractionation. However, many

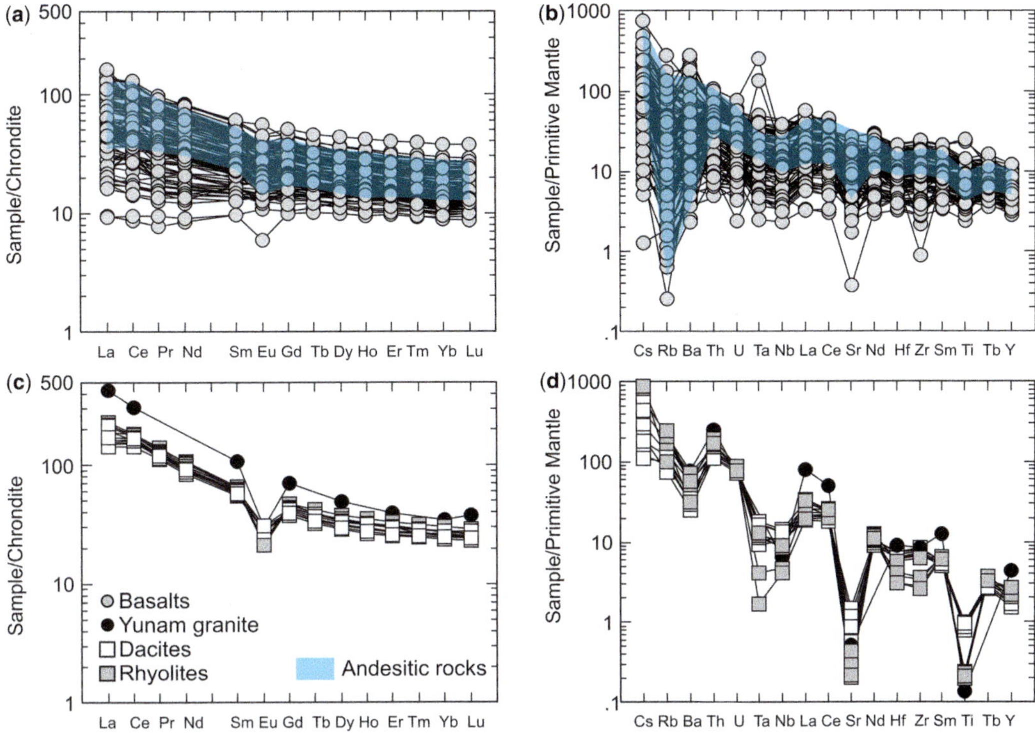

Fig. 7. (**a**) Chondrite-normalized REE and (**b**) primitive-mantle-normalized incompatible element plots of the plots of the mafic and andesitic rocks of the Panjal Traps. (**c**) Chondrite-normalized REE and (**d**) primitive-mantle-normalized incompatible element plots of the silicic volcanic and plutonic rocks of the Panjal Traps. Normalizing values of Sun & McDonough (1989). Data from Pogue *et al.* (1992), Spring *et al.* (1993), Rao & Rai (2007), Chauvet *et al.* (2008) and Shellnutt *et al.* (2012, 2014, 2015).

rocks do not demonstrate clear geochemical evidence of fractional crystallization and may represent a close approximation of the primary magma composition.

The range of trace element ratios (Nb/La, Th/Yb, Ta/Yb and Nb/U) sensitive to crustal contamination and the initial Nd isotopes suggest that many basalts experienced crustal contamination (Campbell 2002; Rudnick & Gao 2003). The Nb/La values of the Panjal Traps are generally <1.0, although a few samples have higher values, whereas the Nb/U (6–102) and Th/Nb_{PM} (0.5–6.9) ratios have a wide range of values (Fig. 11). The basalts with enriched Nd isotope signatures ($\varepsilon_{Nd}(t) < -4$) are likely to have experienced significant crustal contamination. The isotopic composition of the Himalayan crust has a wide range of $\varepsilon_{Nd}(t)$ values, from −10 to −15, and the $^{87}Sr/^{86}Sr_i$ values are generally >0.7200 (Spencer *et al.* 1995; Shellnutt *et al.* 2012). Isotopic modelling using an enriched end member similar to Himalayan crust and the least contaminated Panjal Traps (Table 2) indicates that the amount of crustal contamination is typically between 5 and 10% for specific units from the Guryal Ravine and Pahalgam sections, but there are some exceptions and larger amounts (>20%) of contamination probably occurred and produced the andesitic rocks (Shellnutt *et al.* 2014, 2015).

Primary magma composition

Most flood basalts do not represent primary magma compositions but rather derivative liquids that have experienced crystal fractionation or crustal contamination. The primary melt composition of basalt can be deduced from its bulk composition provided that it has only experienced olivine loss because the adiabatic temperature–pressure melting path is very similar to the olivine liquidus (Herzberg & Asimow 2008, 2015). Deducing the temperature and primary magma composition of ultramafic and mafic volcanic rocks can reveal important information regarding the possible thermal condition of flood basalt provinces (Herzberg & O'Hara 2002; Herzberg & Asimow 2008, 2015; Ali *et al.* 2010). For example, the

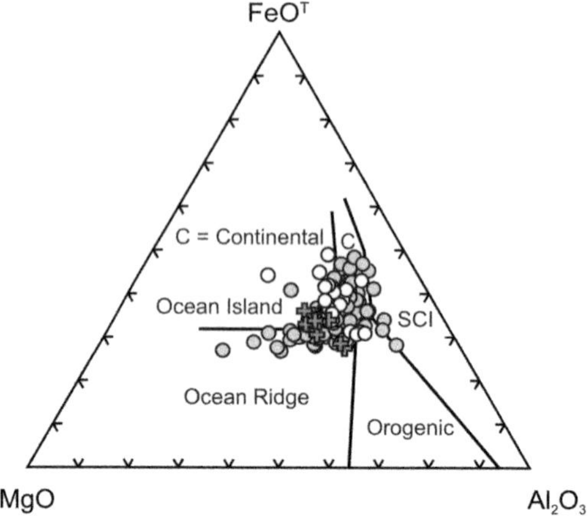

Fig. 8. Basalt tectonomagmatic discrimination diagram of Pearce *et al.* (1977) for the mafic Panjal Traps. C, continental basalts; SCI, spreading centre island (e.g. Iceland, Galapagos). Symbols as in Figure 2. Data from Pareek (1976), Bhat & Zainuddin (1978, 1979), Bhat *et al.* (1981), Honegger *et al.* (1982), Papritz & Rey (1989), Pogue *et al.* (1992), Vannay & Spring (1993), Rao & Rai (2007), Chauvet *et al.* (2008) and Shellnutt *et al.* (2012, 2014, 2015).

identification of anomalously hot mantle potential temperatures may be evidence for a mantle-plume origin (Ali *et al.* 2010; Hole 2015). The software PRIMELT3 was used to estimate the mantle potential temperature and the primary composition of the Panjal mafic rocks in order to determine if the volcanic system is indicative of an anomalously high thermal regime.

Only two 'low-Ti' basalts from the Kashmir Valley produce internally consistent results, whereas ultramafic rocks from Chhongtash did not produce meaningful results. The primary magma compositions and mantle potential temperature estimates are listed in Table 3. The calculations suggest the primary magmas were picritic (Le Bas 2000) and experienced approximately 10–20% olivine loss. The accumulated fraction melt (AFM) eruptive temperatures (T) and mantle potential temperatures (T_P) are estimated to be <1340°C and <1450°C, respectively. The results suggest that thermal regime was not anomalously hot, as the estimates are closer to ambient mantle (1300–1400°C).

Mantle source

The mantle source of the Panjal Traps was very likely a spinel peridotite (Shellnutt *et al.* 2014, 2015). The low Nb/Y (≤0.5) and Zr/Y (≤4.6) ratios of the Panjal Traps are more similar to basalts derived from a spinel–peridotite source than from a garnet–peridotite source (Fig. 12) (Fitton *et al.* 1997; Baksi 2001). Furthermore, the chondrite-normalized REE plots of the Panjal Traps have relatively flat chondrite-normalized heavy REE (HREE) patterns (Fig. 7a).

Trace element modelling suggests that low–moderate degrees of partial melting of a source, assuming a spinel lherzolite primitive mantle starting composition, can reproduce the range of the REE patterns seen in the Panjal rocks. The basalts from Guryal Ravine and Pahalgam that do not show clear evidence for crustal assimilation can be modelled by 3–7% batch melting of a spinel lherzolite (olivine = 57%, orthopyroxene = 26, clinopyroxene = 15% and spinel = 2%) (Fig. 13). The samples that require the largest amount of partial melting typically have >10 wt% MgO, Ni >100 ppm and fall within the ocean-floor basalt field of the tectonomagmatic discrimination diagram of Pearce *et al.* (1977) (Fig. 8). Samples with relatively flat REE patterns can be modelled with >10% partial melting. Moreover, moderate–small (≤10%) amounts of partial melting of a spinel peridotite source can explain the observed Nb–Y–Zr data trend in Figure 12 and implies a relatively shallow source (<75 km). Although it may not be possible to completely rule out the presence of garnet in the source, it would have to be a very minor (<1%) constituent.

It is likely that the Panjal Traps were derived from two isotopically distinct mantle sources: one similar to chondritic mantle (subcontinental lithospheric mantle) and another that was similar to asthenospheric mantle. Figure 14 shows three regression lines (Guryal Ravine line, Pahalgam line and PJ4 line) of the Panjal Traps using $\varepsilon_{Nd}(t)$ and the Th/Nb_{PM} ratio. The Th/Nb_{PM} ratio is an indicator for crustal contamination where average upper crust (UC) has a value of 7.3, depleted MORB mantle (DMM) has a value of 0.45 and primitive mantle (PM) is 1.1. Therefore, the 'uncontaminated' Traps should be closer to either the DMM or PM values, and the contaminated rocks should be between either DMM or PM and UC. It is clear that only a small cluster of samples (Table 2) have chondritic $\varepsilon_{Nd}(t)$ values ($\varepsilon_{Nd}(t) = 0 \pm 1$) with low Th/Nb_{PM} values (0.5 to 1.0). The uncontaminated samples can form two separate 'mixing lines': one with basalts from Guryal Ravine, whereas the other is with basalts from Pahalgam. When the 'mixing lines' are extended to higher Th/Nb_{PM} values, they passed through or very close to either the Panjal dacites (Guryal Ravine line) or the Panjal rhyolites (Pahalgam line) suggesting that specific silicic volcanic rocks may have acted as the enriched end member that contaminated specific basaltic sections (Fig. 14). A third 'mixing' line (PJ4 line) is observed involving rocks collected from the Pir Panjal Range near Gulmarg (PJ4 sampling site in Shellnutt *et al.* 2015). The PJ4 line when extended to higher

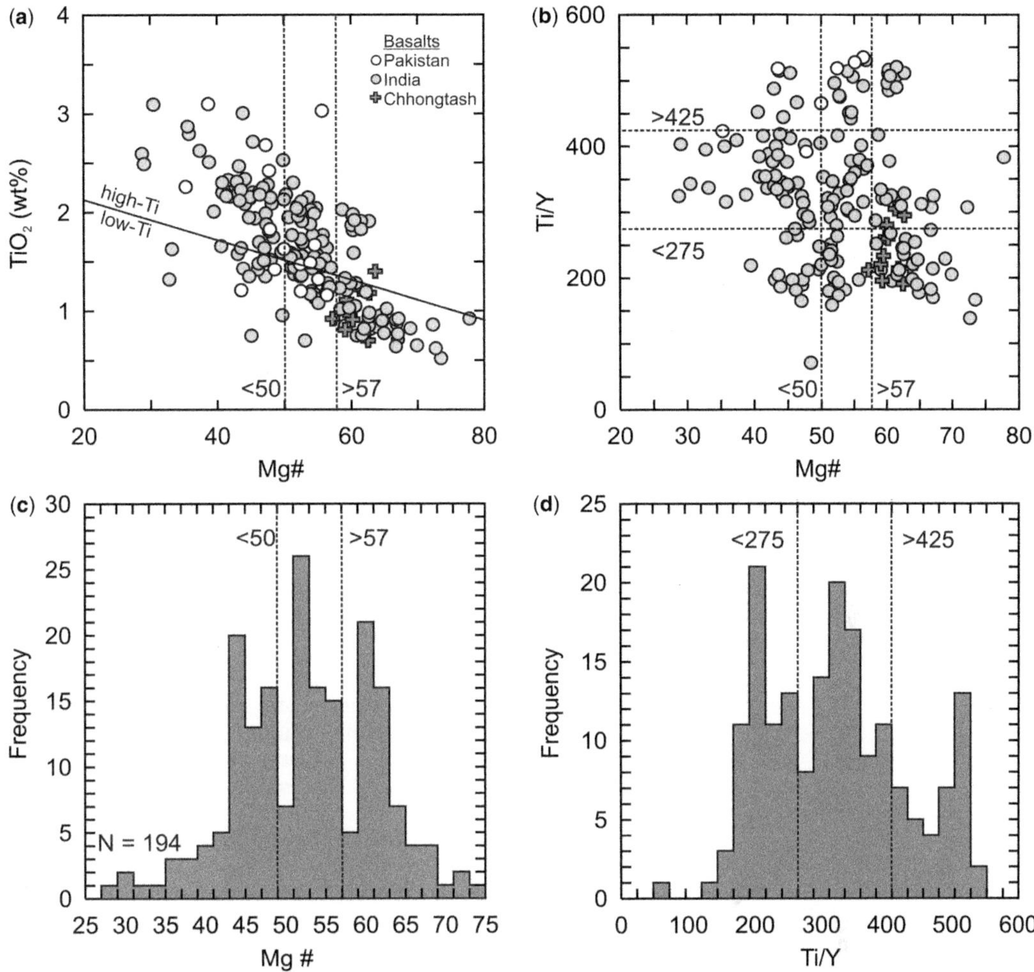

Fig. 9. Ti-classification of the Panjal Traps. (**a**) TiO_2 (wt%) v. Mg# (based on Lightfoot *et al.* 1993). (**b**) Ti/Y v. Mg#. (**c**) Histogram of the Mg# of the basaltic Panjal Traps. (**d**) Histogram of the Ti/Y of the basaltic Panjal Traps. Data from Pareek (1976), Bhat & Zainuddin (1978, 1979), Bhat *et al.* (1981), Honegger *et al.* (1982), Papritz & Rey (1989), Pogue *et al.* (1992), Vannay & Spring (1993), Rao & Rai (2007), Chauvet *et al.* (2008) and Shellnutt *et al.* (2014, 2015).

Th/Nb_{PM} values passes through the basalt data reported by Chauvet *et al.* (2008) from the Zanskar Valley, and crosses between the dacites and rhyolites. If the PJ4 regression line is extended to intercept either the DMM or PM line, then the corresponding $\varepsilon_{Nd}(t)$ value is between +8 and +9 and similar to depleted mantle. The samples that fall along the upper trend line appear to favour a contaminant similar to or something in-between the Panjal silicic rocks. Furthermore, the T_{DM} model (T_{DM} = 1240–2200 Ma) ages of the 'uncontaminated' rocks suggests that the source was probably indicative of a 're-worked or modified' subcontinental lithospheric mantle (SCLM). Thus, it is possible that the original mantle source that produced the uncontaminated chondritic basalts was the SCLM, whereas basalts from the Pir Panjal Range and some of those reported by Chauvet *et al.* (2008) were derived from a source that had a greater influence of asthenospheric mantle.

Andesitic Panjal Traps

The main volcanic sequences around the Kashmir Valley have horizons of andesitic rocks. Andesitic rocks are reported from Guryal Ravine, Pir Panjal and the Liddar Valley near Pahalgam (Bhat &

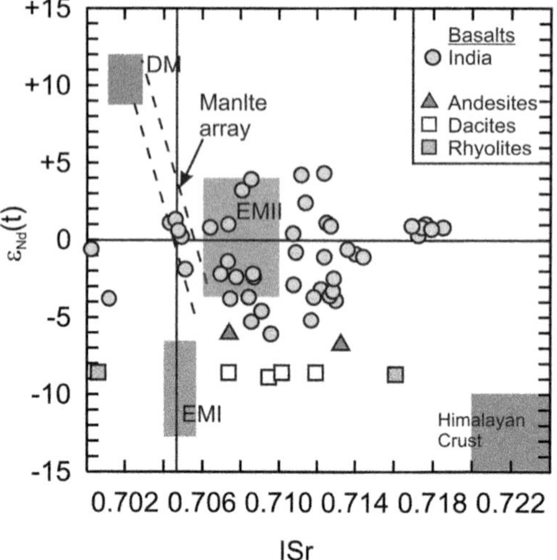

Fig. 10. Sr–Nd plot showing the mafic and silicic Panjal traps from the Kashmir Valley region. BA, basaltic andesite; DM, deplete mantle; EMI, enriched mantle I; EMII, enriched mantle II (Zindler & Hart 1986; Workman *et al.* 2004; Workman & Hart 2005). Isotopic range of the Himalayan crust from Spencer *et al.* (1995). Data from Chauvet *et al.* (2008) and Shellnutt *et al.* (2012, 2014, 2015).

Zainuddin 1978; Shellnutt *et al.* 2014, 2015). The whole-rock compositions are typically basaltic andesite as they have SiO_2 content between 54 and 56 wt%. Intermediate volcanic rocks, in general, are common within flood basalt provinces, and are interpreted as either differentiates of mafic magmas, hybrids between mantle melts and crustal melts, or predominantly crustal melts (Peate *et al.* 1992; Xu *et al.* 2001; Sensarma *et al.* 2002; Shimizu *et al.* 2005; Shellnutt & Jahn 2010).

The Panjal andesitic rocks are compositionally transitional between the mafic and silicic rocks, and are probably derived by mingling between mafic magmas and crustal melts. The whole-rock Sr–Nd isotopes ($\varepsilon_{Nd}(t) = -6.8$– to -6.1) are more enriched than the basalts but less than the silicic rocks (Fig. 14). Moreover, Th/Nb_{PM} (2.0–6.8) values and Nb/U (5.2–22.1) values are also between silicic and basaltic rocks (Figs 11 & 14). The silicic rocks have enriched $\varepsilon_{Nd}(t)$ values ($\varepsilon_{Nd}(t) = -8.6$ to -8.9) that are broadly similar to the $\varepsilon_{Nd}(t)$ values ($\varepsilon_{Nd}(t) = -10$ to -15) of the Himalayan crust, although the SiO_2 and TiO_2 contents are very different (dacites: $SiO_2 = c.$ 65 wt%, $TiO_2 = c.$ 1.1 wt%; rhyolites: $SiO_2 = c.$ 75 wt%, $TiO_2 = c.$ 0.4 wt%). Isotopic and trace element modelling suggests that between 20 and 30% mingling of a dacite or rhyolite and basalt could produce compositions similar to the andesitic rocks (Shellnutt *et al.* 2014). However, the type of contaminant will have a great effect on whether the bulk SiO_2 and TiO_2 are above or below 55 and 1 wt%, respectively. As discussed earlier, it seems that a specific contaminant (rhyolite or dacite) can be identified within some volcanic sequences (Fig. 14).

Although it is difficult to confirm at this moment, it is possible that the andesitic rocks throughout the Kashmir Valley are a marker horizon of a very specific eruptive episode that involved significant mingling between crustal melts and mafic magmas. The andesitic rocks in the Liddar Valley and Guryal Ravine section may be correlative as both sequences have two andesitic horizons that are separated by basalts with similar chemistry ($TiO_2 = c.$ 1.5 wt%). The significance of the andesitic rocks during the eruption of the Panjal Traps is uncertain but a detailed investigation on the reasons why there was an unusual amount of mingling, as well as the petrological consequences and possible environmental effects, is warranted.

Silicic Panjal Traps and Late Palaeozoic silicic plutons in the western Himalaya

Many continental mafic Large Igneous Provinces have a relatively small volume of silicic volcanic rocks typically, but not exclusively, in the upper portions of the volcanic sequences (cf. Lightfoot *et al.* 1987; Garland *et al.* 1995; Melluso *et al.* 2008; Xu *et al.* 2010). The Panjal Traps are no different in this respect, except that the silicic flows may be lower in the eruptive sequence. In many cases, the silicic volcanic rocks of other LIPs are interpreted to be derived by the differentiation of mafic magmas with or without crustal contamination (Lightfoot *et al.* 1987; Melluso *et al.* 2008; Xu *et al.* 2010). In contrast to their volcanic equivalents, silicic intrusive rocks are less commonly exposed within LIPs. The plutonic nature and/or the extent of erosion and exhumation are likely to be the primary reasons for the poor exposure of LIP silicic intrusive rocks.

The Panjal Traps have a volumetrically minor, but petrologically significant, portion of silicic volcanic rocks that appear to be restricted to the eastern part of the Kashmir Valley (Ganju 1944; Pareek 1976; Shellnutt *et al.* 2012). The volcanic rocks classify as dacite and rhyolite, and are quartz porphyrys with cryptocrystalline or crypto- to microcrystalline textures (Fig. 5c, d). The primary petrographical difference between the rocks is the amount of quartz phenocrysts. Thus far, silicic volcanic rocks have not been reported in the Zanskar, Lahul or Liddar valleys. Early investigations suggested they were derived by differentiation of mafic Panjal magmas but more recently studies indicate they are likely to be crustal melts (Shellnutt *et al.* 2012).

Table 2. *Summary of geochemical data of the mafic, intermediate and silicic Panjal Traps*

Sample	PJ1-033	PJ1-034	PJ2-007	PJ2-010	PJ2-019	PJ2-020	PJ2-022	PJ2-024
Location	Guryal	Guryal	Pahalgam	Pahalgam	Pahalgam	Pahalgam	Pahalgam	Pahalgam
Rock type	basalt	basalt	basalt	basalt	basalt	basalt	basalt	basalt
SiO_2 (wt%)	44.51	44.82	47.55	45.78	44.69	45.41	44.91	43.71
TiO_2	1.55	1.53	2.03	2.04	3.01	1.73	2.23	1.88
Al_2O_3	16.13	16.58	17.15	16.37	14.5	15.16	15.92	16.50
$Fe_2O_3^T$	12.81	12.83	12.17	11.78	15.83	13.33	17.65	14.87
MnO	0.20	0.19	0.14	0.16	0.25	0.27	0.30	0.24
MgO	7.75	8.27	8.72	7.20	6.26	7.45	6.89	8.76
CaO	10.11	9.08	2.18	8.41	6.73	9.62	3.15	6.12
Na_2O	2.19	2.84	2.80	3.25	3.54	3.03	4.73	3.32
K_2O	0.40	0.17	0.72	0.58	1.15	0.26	0.10	0.04
P_2O_5	0.22	0.22	0.32	0.28	0.49	0.23	0.29	0.32
LOI	3.44	3.83	5.88	4.05	2.74	3.08	3.67	4.21
Total	99.31	100.36	99.65	99.90	99.18	99.58	99.84	99.97
Mg#	54.5	56.1	58.7	54.8	43.9	52.5	43.6	53.9
Sc (ppm)	33	32	32	29	38	20	37	41
V	274	242	289	233	315	256	262	307
Cr	232	158	135	72	161	145	90	179
Co	55	54	51	44	54	56	52	61
Ni	182	152	133	93	97	138	101	103
Cu	52	199	20	51	113	59	168	108
Zn	100	77	85	71	156	96	138	130
Ga	16.8	16.1	17.0	17.0	19.3	16.6	20.6	19.0
Rb	11.9	3.6	31.0	17.2	30.8	2.4	2.1	0.7
Sr	185	158	37	236	162	346	201	403
Y	24.6	22.8	29.1	24.2	43.1	24.9	34.5	33.9
Zr	77	85	124	65	163	84	149	125
Nb	7.1	6.4	13.6	11.8	16.0	8.5	13.1	10.6
Cs	1.3	0.7	1.8	1.2	1.4	0.1	0.1	0.1
Ba	147	89	203	329	411	188	120	33
Hf	2.11	1.86	2.91	1.75	4.38	2.10	3.66	3.34
Ta	0.42	0.41	0.85	0.86	1.07	0.61	5.41	0.67
Th	0.63	0.54	0.94	0.73	1.30	0.59	1.66	1.38
U	0.14	0.13	0.15	0.14	0.16	0.09	0.26	0.22
$^{87}Sr/^{86}Sr_i$	0.7043	0.7046		0.7049	0.7047	0.7064	0.7073	0.7073
$\varepsilon_{Nd}(t)$	+1.1	+1.3	−0.6	+0.2	+0.6	+0.8	−1.4	+1.0
Eu/Eu*	1.1	1.1	0.9	1.1	1.0	1.1	0.9	1.1
$(La/Yb)_N$	2.5	2.1	2.9	3.6	3.0	2.7	2.4	2.8

Sample	PJ4-007	PJ4-008	PJ1-030	PJ4-003	PJ1-049	PJ1-054	PJ1-042	PJ1-040
Location	Pir Panjal	Pir Panjal	Guryal	Pir Panjal	Pampore	Pampore	Pampore	Pampore
Rock type	basalt	basalt	BA	BA	dacite	dacite	rhyolite	rhyolite
SiO_2 (wt%)	47.08	48.70	54.83	56.36	67.30	63.16	76.68	75.1
TiO_2	0.92	0.86	2.04	0.90	1.08	1.20	0.36	0.41
Al_2O_3	13.11	12.18	13.7.	14.48	12.87	14.55	10.82	10.41
$Fe_2O_3^T$	10.62	9.34	12.52	8.91	7.52	7.99	2.78	3.70
MnO	0.15	0.15	0.14	0.15	0.09	0.15	0.03	0.08
MgO	10.94	12.29	3.90	5.57	1.3	1.55	0.66	1.22
CaO	10.38	9.72	6.48	8.58	2.02	3.86	0.54	1.36
Na_2O	1.93	2.35	0	2.08	1.99	1.70	1.23	0.28
K_2O	0.33	0.17	1.59	0.17	4.17	3.10	4.73	5.34
P_2O_5	0.07	0.06	0.49	0.09	0.24	0.25	0.09	0.10
LOI	3.78	3.98	3.73	2.98	1.58	2.59	1.28	2.20
Total	99.31	99.8	99.42	100.27	100.16	100.10	99.20	100.20
Mg#	67.1	72.3	38.2	55.3	25.5	27.8	32.0	39.5

(*Continued*)

Table 2. (*Continued*)

Sample Location Rock type	PJ4-007 Pir Panjal basalt	PJ4-008 Pir Panjal basalt	PJ1-030 Guryal BA	PJ4-003 Pir Panjal BA	PJ1-049 Pampore dacite	PJ1-054 Pampore dacite	PJ1-042 Pampore rhyolite	PJ1-040 Pampore rhyolite
Sc (ppm)	38.6	36	37.3	31	18	20	9	9
V	235	206	280	228	81	113	24	17
Cr	1270	1040	94	470	59	75	47	237
Co	47	45	35	34	10	8	2	5
Ni	420	430	54	190	29	32	24	125
Cu	120	80	86	90	3	8	6	8
Zn	70	60	113	80	80	93	105	74
Ga	15	12	23	19	19	23	17	16
Rb	11	5	76	10	111	169	166	173
Sr	149	131	355	93	64	106	47	42
Y	17	17	46	22	38	51	43	50
Zr	45	42	206	101	302	07	156	247
Nb	2.5	2.2	18.2	6.7	27.6	29.0	19.5	20.4
Cs	0.5	0.3	5.1	0.7	1.5	4.3	8.9	6.8
Ba	70	38	377	25	619	493	802	964
Hf	1.3	1.2	5.2	2.7	7.7	7.8	4.6	6.4
Ta	0.19	0.19	1.2	0.56	1.9	2.0	1.5	1.4
Th	0.74	0.71	9.0	5.8	17.7	18.2	22	21
U	0.2	0.37	1.3	1.3	3.1	3.2	3.4	3.3
$^{87}Sr/^{86}Sr_i$	0.7112	−0.7124	0.7133	0.7074	0.7095	0.7074	0.7161	
$\varepsilon_{Nd}(t)$	+4.2	+4.3	−6.8	−6.1	−8.9	−8.6	−8.7	−8.6
Eu/Eu*	1.0	1.2	0.7	0.9	0.6	0.6	0.4	0.4
$(La/Yb)_N$	1.9	1.9	5.7	4.6	6.3	7.8	6.8	8.6

BA, basaltic andesite. Superscript ^{T} is the total Fe expressed as Fe_2O_3. LOI, loss on ignition. Mg# = $(Mg^{2+}/(Mg^{2+} + Fe^{2+})) \times 100$. Subscript $_N$ indicates normalized to the chondritic value of Sun & McDonough (1989). Data from Shellnutt *et al.* (2012, 2014, 2015).

The whole-rock chemistry shows the rocks are peraluminous, calcic to calc-alkalic and have enriched isotopic compositions ($\varepsilon_{Nd}(t) = -8.6$ to -8.9) that are more similar to average Himalayan crust ($\varepsilon_{Nd}(t) = -10$ to -14) than to the mafic Panjal rocks (Figs 10 & 15). Furthermore, the silicic rocks have very low Nb/U (<10) and high Th/Nb_{PM} (>3) values that are typical of crust-derived igneous rocks (Fig. 11). Petrological and geochemical modelling suggest that the rhyolites and dacites can each be derived by partial melting of rocks (different lithologies) typical of the middle crust (Shellnutt *et al.* 2012). However, it is also possible that the dacite could be representative of the rhyolite parental magma that experienced crystal fractionated as their whole-rock compositions fall along the same trend lines (decreasing TiO_2, Al_2O_3, $Fe_2O_3^T$, MgO and CaO with increasing SiO_2). Regardless of the relationship between the rhyolites and dacites, it is very likely that the injection of mafic Panjal magmas caused the crust to melt and produce at least the dacitic magma.

In addition to silicic volcanic rocks, there are number of silicic plutons throughout the western Himalaya that range in composition from granite and leucogranite to syenite and ijolite, and have Late Carboniferous–Middle Permian ages (Le Bas *et al.* 1987; Zeitler 1988; Baig 1990; Spring *et al.* 1993; Smith *et al.* 1994; Noble *et al.* 2001; Horton 2011; Ahmad *et al.* 2013). The Yunam granite from the Zanskar Valley is peraluminous, ferroan and calc-alkalic, and is similar to within-plate granites (A-type) derived by melting of the crust. Another granite reported by Horton (2011) from Zanskar has a mean zircon $\varepsilon_{Hf}(t)$ value of −5.1 that is typical of crust-derived granitic rocks. Late Carboniferous–Middle Permian granitic rocks and carbonatites from NW Pakistan are considered to be part of a larger and extension-related alkaline igneous province (Kempe & Jan 1970; Rafiq & Jan 1989; Jan & Karim 1990; Pogue *et al.* 1992; Ahmad *et al.* 2013). The rocks from the Ambela (315–280 Ma) complex are metaluminous to peraluminous with the syenites being alkalic and the granites being alkali-calcic to calc-alkalic (Rafiq & Jan 1989). Similar to the Zanskar granitic rocks, the Ambela rocks classify as within-plate (Fig. 15). Le Bas *et al.* (1987) and Tilton *et al.* (1998) examined the origin of the Koga and Jhambil carbonatites, and concluded they and the associated silicic intrusions are consistent with an intra-plate setting. The Nd isotopic compositions of the carbonatites ($\varepsilon_{Nd}(t) = +3.2$ to $+3.7$) overlap with the most depleted basaltic Panjal Traps ($\varepsilon_{Nd}(t) = +2.4$ to $+4.3$) from the Pir Panjal Range and therefore the rocks could be related to the same mantle source.

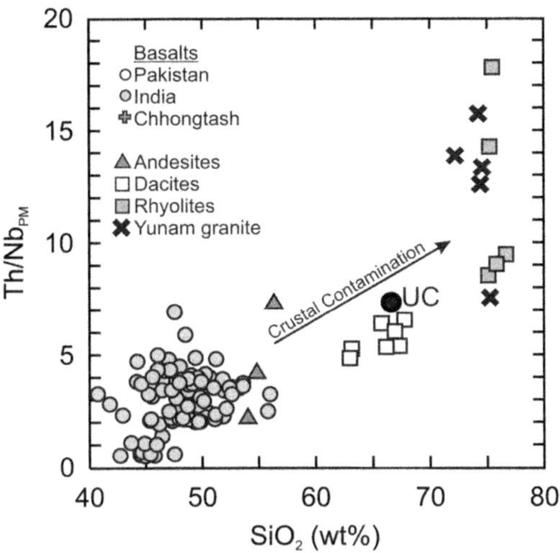

Fig. 11. Th/Nb_{PM} v. SiO_2 (wt%) of the Panjal mafic, intermediate and silicic rocks. The trend towards higher Th/Nb_{PM} and SiO_2 within the basaltic rocks is probably due to crustal contamination. UC, upper crust values from Rudnick & Gao (2003). Data from Spring *et al.* (1993), Chauvet *et al.* (2008) and Shellnutt *et al.* (2012, 2014, 2015).

The formation of the silicic rocks is expected within flood basalt provinces as they represent high-heat-flow regimes and are ideal settings for crustal melting, differentiation of mafic magmas and producing hybrid melts (Lightfoot *et al.* 1987; Melluso *et al.* 2008; Shellnutt *et al.* 2011*b*; Bryan & Ferrari 2013). The broad compositional spectrum of the Late Carboniferous–Middle Permian granitic rocks is therefore consistent with other LIPs but the range of ages (315–265 Ma) and the limited whole-rock and isotope chemistry makes a definitive link with the Panjal Traps difficult. In all likelihood, the Late Palaeozoic silicic plutons are related to the broader thermal and tectonic regime of the region that includes the Panjal Traps, and they merely represent the expression of high-temperature mafic magmas differentiating or partially melting the crust.

Ultramafic rocks in the Karakoram Range and their association with the Panjal Traps

Ultramafic volcanic rocks (picrites or komatiites) observed within LIPs are often considered to be evidence of anomalously high thermal conditions attributed to a mantle plume (Campbell 2007; Ali *et al.* 2010). Therefore, identifying ultramafic volcanic rocks is useful to surmise the possible thermal conditions of the mantle during the formation of a LIP. Rao & Rai (2007) reported ultramafic and mafic volcanic rocks associated with marine sedimentary rocks within the Permian Chhongtash Formation north of the Karakoram Batholith. The ultramafic rocks have a forsterite content ranging from 81 to 84, are chemically similar to alumina-undepleted komatiites, and are considered to be the parental magmas to the neighbouring mafic rocks.

Table 3. *Panjal Traps primary melt compositions and ultramafic rock compositions from Chhongtash*

Sample	PJ2-003	AFM	AFM	PJ4-006	AFM	AFM	KCV14	KCV16	KCV17	KCV18
SiO_2 (wt%)	51.23	50.62	50.90	52.14	51.97	52.33	43.70	44.18	43.61	43.73
TiO_2	0.76	0.64	0.66	0.98	0.87	0.90	0.40	0.41	0.35	0.36
Al_2O_3	14.70	12.30	12.63	12.37	10.90	11.25	8.22	7.86	7.82	7.97
Fe_2O_3		0.32	0.65		0.43	0.89				
FeO		9.01	8.62		8.19	7.67				
FeO^T	8.35			7.82			13.12	13.07	13.35	13.34
MnO	0.15	0.16	0.16	0.15	0.16	0.16	0.21	0.23	0.31	0.30
MgO	7.5	15.44	14.55	7.37	13.65	12.54	27.93	28.31	29.35	28.93
CaO	12.24	10.28	10.56	11.9	10.52	10.86	4.97	4.70	3.98	4.05
Na_2O	1.33	1.11	1.14	3.25	2.86	2.95	0.92	0.78	0.60	0.68
K_2O	0.07	0.06	0.06	0.43	0.38	0.39	0.50	0.41	0.58	0.58
P_2O_5	0.08	0.07	0.07	0.07	0.06	0.06	0.04	0.04	0.05	0.05
Pressure (bars)		1	1		1	1				
FeO (source)		8.54	8.53		8.47	8.43				
MgO (source)		38.12	38.12		38.12	38.12				
Fe_2O_3/TiO_2		0.5	1.0		0.5	1.0				
T (°C)		1340	1320		1330	1300				
T_P (°C)		1450	1420		1400	1370				
% ol addition		21.8	19.0		16.4	13.1				
Melt Fraction		0.30	0.29		0.28	0.27				

AFM, accumulated fractional primary melt. % ol = percent olivine. KCV14, KCV16, KCV17 and KCV18 from Rao & Rai (2007).

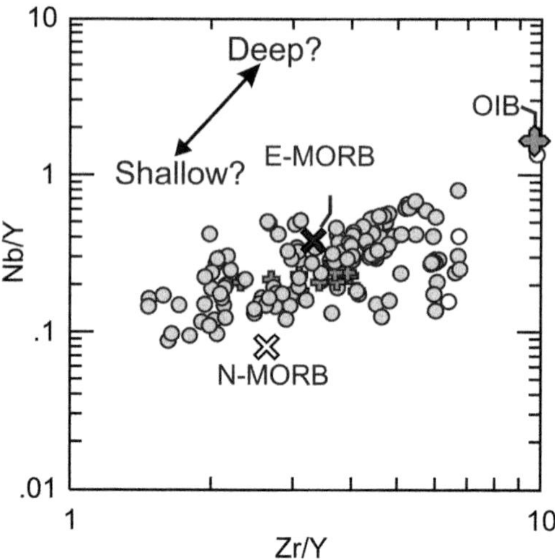

Fig. 12. Nb/Y v. Zr/Y ratios of the mafic Panjal Traps. The low ratio values are indicative of a shallow-mantle source (Baksi 2001). Normal mid-ocean ridge basalt (N-MORB), E-MORB and ocean island basalt (OIB: values from Sun & McDonough (1989). Data from Bhat & Zainuddin (1978, 1979), Bhat *et al.* (1981), Honegger *et al.* (1982), Papritz & Rey (1989), Rao & Rai (2007), Chauvet *et al.* (2008) and Shellnutt *et al.* (2014, 2015).

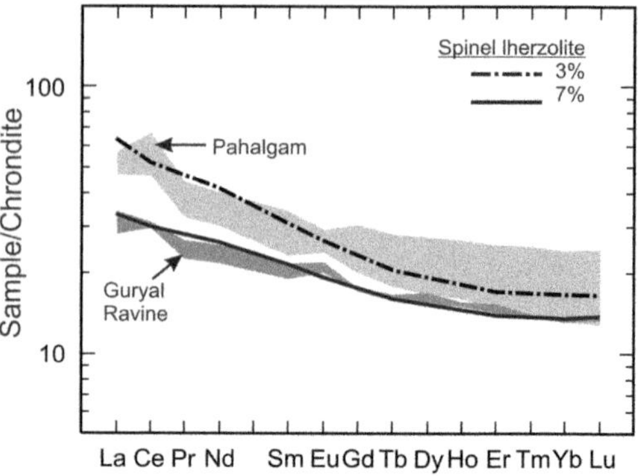

Fig. 13. Results of REE modelling with respect to the least-contaminated Panjal Traps from Guryal Ravine and Pahalgam. The models assume a primitive mantle starting composition (Sun & McDonough 1989). The composition of the spinel lherzolite is: olivine = 57%, orthopyroxene = 26%, clinopyroxene = 15% and spinel = 2%. Batch melting equation: $C_L/C_o = 1/D(1 - F) + F$, where C_L is the concentration in the liquid, C_o is the original rock composition, D is the bulk distribution coefficient and F is the weight fraction of melt produced. K_d values – olivine: La = 0.007, Ce = 0.006, Nd = 0.007, Sm = 0.007, Eu = 0.0068, Tb = 0.01, Dy = 0.014, Er = 0.011, Yb = 0.049 and Lu = 0.0454; orthopyroxene: La = 0.0003, Ce = 0.02, Nd = 0.03, Sm = 0.05, Eu = 0.03, Tb = 0.054, Dy = 0.09, Er = 0.152, Yb = 0.227 and Lu = 0.255; clinopyroxene: La = 0.056, Ce = 0.092, Nd = 0.15, Sm = 0.26, Eu = 0.39, Tb = 0.53, Dy = 0.38, Er = 0.54, Yb = 0.28 and Lu = 0.28; spinel: La = 0.01, Ce = 0.01, Nd = 0.01, Sm = 0.01, Eu = 0.01, Tb = 0.01, Dy = 0.01, Er = 0.01, Yb = 0.01 and Lu = 0.01. The K_d values are taken from Frey (1969), Schnetzler & Philpotts (1970), Paster *et al.* (1974), Irving & Frey (1978, 1984), Fujimaki *et al.* (1984), McKenzie & O'Nions (1991), Johnson (1998), Salters & Longhi (1999) and Green *et al.* (2000). Data from Rao & Rai (2007), Chauvet *et al.* (2008) and Shellnutt *et al.* (2012, 2014, 2015).

The term 'komatiite' to refer to the ultramafic rocks of the Chhongtash Formation is probably inappropriate considering the rather specific definition and the fact that olivine spinifex is not mentioned (Kerr & Arndt 2001; Arndt 2003; Dostal 2008). It is possible that the rocks are cumulates rather than liquids as they follow an olivine accumulation trend and do not plot close to the solidus of known Phanerozoic komatiites/picrites (Fig. 16). Furthermore, the neighbouring basalts have pillow textures, whereas the 'komatiites' do not. Shellnutt *et al.* (2015) reported a rock (PJ4-004) from the Pir Panjal Range that has >20 wt% MgO and is broadly similar to the Chhongtash 'komatiites' but identified it as porphyritic and therefore not a liquid composition. The results from PRIMELT3 suggest that the parental liquid of the low-Ti basalt is closer to a picrite than a komatiite (Table 3). At this point, it is difficult to link the 'komatiites' to the Panjal Traps as more textural and geochemical information is reuquired.

The basalts of the Chhongtash Formation are chemically similar to the mafic Panjal Traps but so are continental tholeiites in general (Figs 7a, b, 9 & 12). The lack of isotopic data, full suite of trace element and a radiometric age prevents a more definitive link with the Panjal Traps. Although it remains a possibility that the Chhongtash basalts and Panjal Traps are petrogenetically related, there are a number of other poorly constrained volcanic units (ophiolite, volcanic-arc) within the Karakoram Range that could be correlative.

Mantle plume or passive extension origin?

There is a tremendous debate regarding the existence of mantle plumes, let alone the association between LIPs and plumes (cf. King & Anderson 1995; Ernst & Buchan 2003; Ernst *et al.* 2005; Campbell 2007; Foulger 2007, 2010; Bryan & Ernst 2008; Ernst 2014). The mantle-plume model suggests that a relatively hot diapiric upwelling ascends from the deep mantle and impinges at the base of the lithosphere or crust and is followed by the injection of high-temperature mafic–ultramafic magmas into the crust.

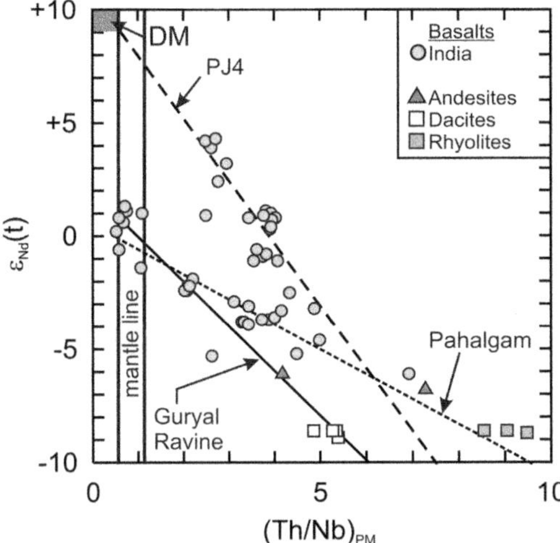

Fig. 14. $\varepsilon_{Nd}(t)$ v. Th/Nb_{PM} showing three mixing lines of the Panjal basaltic rocks and silicic rocks calculated by Shellnutt *et al.* (2015). Total mafic Panjal Traps data are superimposed on the calculated mixing lines. The mantle line is based on the depleted MORB mantle of Workman & Hart (2005). The calculated regression lines for the basalt data are extended to the *x*-axis and the mantle line. Th/Nb_{PM} is normalized to primitive mantle values of Sun & McDonough (1989). DM, depleted mantle of Workman & Hart (2005). Data from Chauvet *et al.* (2008) and Shellnutt *et al.* (2012, 2014, 2015).

Some of the magmas stall in the crust or form dyke networks, plutonic bodies and induce crustal melting, whereas the magmas that reach the surface erupt and form the spectacular basalt traps (Richards *et al.* 1989; Campbell 2007). The spatial association between some LIPs and volcanic rifted margins suggests that mantle plumes may exploit structural heterogeneities in the lithosphere that assist in continental break-up but some LIPs are unrelated to plate separation (Courtillot *et al.* 1999). In contrast, passive rifting and thermal convection models, known as the 'plate' model, is also put forth as an explanation of some LIPs (King & Anderson 1995; Anderson 2005; Foulger 2007, 2010). The premise of the 'plate' model is that 'causative processes of melting anomalies to the Earth's top thermal boundary layer' are a consequence of the lithosphere under tensional stress (Foulger 2010, p34). It is possible that the 'mantle-plume debate' is contentious because of the focus on a singular or restricted group of LIPs and that a 'one-size fits all' model is inappropriate for all LIPs given their chemical and tectonic differences (e.g. plate separation v. non-plate separation). In other words, the mantle-plume model may be applicable for some but not all LIPs (Sheth 2005, 2007; Ali *et al.* 2010; Ernst 2014; Shellnutt 2014).

Assessing the involvement of a mantle plume within an ancient LIP is based on a number of criteria but there are three critical factors. In addition to large volumes of magma (>100 000 km^3), a mantle-plume-derived LIP may exhibit: (1) a short duration of magmatism (≤1 myr); (2) a high thermal regime (the presence of ultramafic volcanic rocks); and (3) evidence of pre-volcanic uplift of the crust (Campbell 2007). In the best of circumstances, the criteria are difficult to assess but even more so if the LIP is dismembered or tectonized. The total magmatic duration of many mafic continental LIPs often exceeds 10 myr as the main effusive period may be preceded and followed by sporadic eruptions or intrusions. Consequently, the evaluation of rapid emplacement usually emphasizes peak effusion rates that represent a substantial (≥70%) portion of the volcanic system (Campbell 2007; Bryan & Ernst 2008). Identifying evidence for a high thermal regime is largely based on the presence of non-cumulate ultramafic volcanic rocks but assessing evidence for pre-volcanic uplift is difficult (cf. He *et al.* 2003; Sheth 2007; Ukstins Peate & Bryan 2008).

Evidence for the involvement of a mantle plume in the genesis of the Panjal Traps is limited (Chauvet *et al.* 2009; Zhai *et al.* 2013). Firstly, the total duration of volcanism is uncertain. Although Nakazawa & Kapoor (1973), Nakazawa *et al.* (1975), Gaetani *et al.* (1990) and Stojanovic *et al.* (2016) suggest that volcanism was probably short lived (<5 myr), there is a dearth of high-precision radioisotopic ages and it is possible that volcanism was continuous as the initial continental rift transitioned into an ocean basin (Shellnutt *et al.* 2015). Consequently, the only preserved remnants of the Panjal Traps are those that erupted within a continental setting, whereas the transitional to oceanic portions were probably subducted or highly deformed. Secondly, as discussed in the preceding section, the petrogenetic lineage of the Karakoram ultramafic rocks and the Panjal Traps is debatable but the calculated mantle potential temperatures of the primary magmas are not anomalously high compared to other mantle-plume-derived LIPs (Ali *et al.* 2010; Hole 2015). Thirdly, evidence of pre- and syn-volcanic uplift is documented by the progression from older marine sedimentation to younger continental sedimentation throughout the Kashmir and Zanskar valleys but the transition is attributed to rifting rather than thermal uplift (Gaetani *et al.* 1990; Garzanti *et al.* 1996*a*, *b*). Finally, the low La/Yb_N, Zr/Y and Nb/Y values, REE modelling, and the T_{DM} ages (1200–2200 Ma) of the early erupted basalts suggest that the initial magmas were likely to have been derived from an ancient lithospheric spinel–peridotite rather than a juvenile garnet–peridotite (Shellnutt *et al.* 2014, 2015).

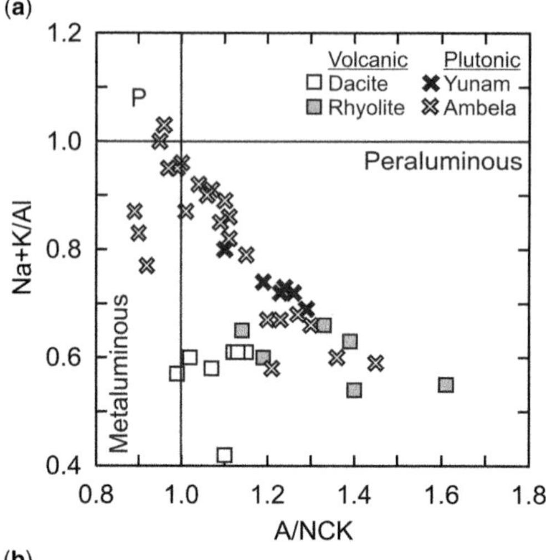

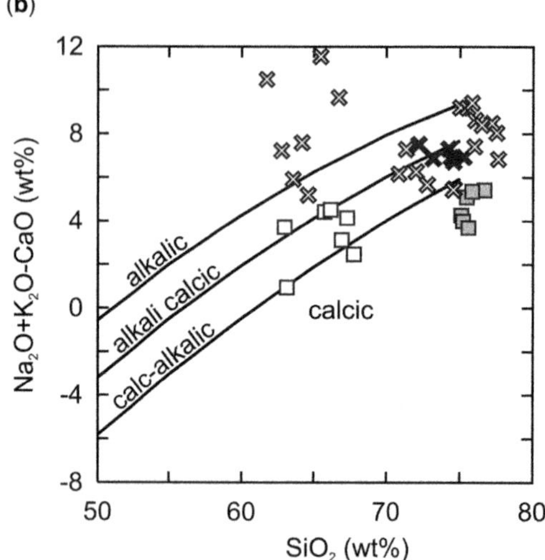

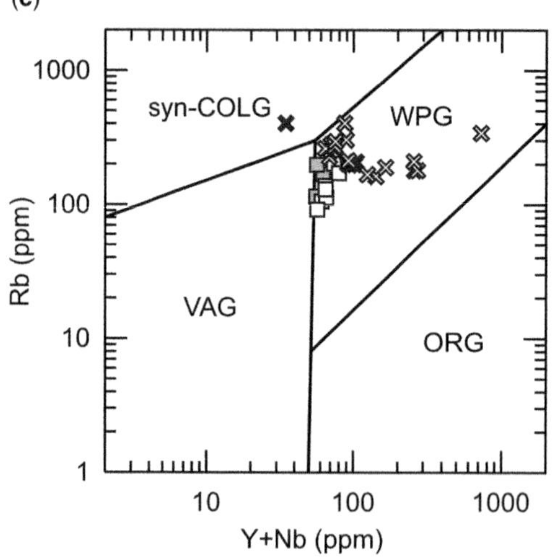

Fig. 15. (**a**) Alkali index (Na + K/Al) v. aluminum saturation index (ASI = Al/Ca + Na + K). (**b**) Classification of the Panjal silicic rocks according to the modified alkali-lime index ($Na_2O + K_2O$–CaO) of Frost *et al.* (2001). (**c**) Rb (ppm) v. Y + Nb (ppm) tectonomagmatic classification of granitic rocks of Pearce *et al.* (1984). Data of the silicic rocks from Rafiq & Jan (1989), Spring *et al.* (1993) and Shellnutt *et al.* (2012).

At this time, it appears that the Panjal Traps were not derived from a mantle plume but, rather, from a passive rift system that was influenced by the prevailing plate stress related to subduction on the northern margin of the Palaeotethys Ocean and possibly by the isostatic effects of deglaciation (Yeh & Shellnutt 2016). The low mantle potential temperature estimates and the variable Nd isotopes from chondritic to more depleted values is likely to be due to the transition from a continental rifting setting to a nascent ocean basin. In other words, the depleted mantle becomes more prominent within the source over time as the rift becomes a seafloor spreading centre.

Potential for mineral deposits with the Panjal magmatic system

Mafic continental LIPs (MLIPs) are ideal locations for magmatic ore deposit exploration. Many LIPs host massive sulphide or oxide deposits within layered mafic–ultramafic complexes or magma conduits that were derived from either primary or evolved magmas (Franklin *et al.* 2005; Galley *et al.* 2007; Ernst & Jowitt 2013). In some LIPs, the type of ore deposit (sulphide or oxide) may be related to the Ti-concentration of the flood basalts, with the 'low-Ti' type associated with sulphide deposits and the 'high-Ti' type associated with oxide deposits (Zhou *et al.* 2008). The compositional range of the Panjal Traps is similar to other flood basalt provinces (Emeishan LIP) that are known to be associated with sulphide and oxide deposits (Zhou *et al.* 2008; Shellnutt 2014). Unlike other Permian MLIPs, the Panjal Traps show a compositional transition from continental tholeiites to ocean-floor basalt. Moreover, columnar jointed basalt and pillow basalt are evidence indicating that both subaerial and subaqueous eruptions occurred. The identification of subaqueous basalts suggests that it is possible that exhalative volcanogenic massive sulphide (VMS) deposits may have formed. Therefore, it is possible that ore deposits associated with layered mafic–ultramafic intrusions, as well as VMS deposits, exist within the volcanic or plutonic sequences of the Panjal Traps.

The plutonic system of the Panjal Traps is poorly understood either because there are few exposures of moderately shallow to shallow intrusions or they have not been identified. Consequently, ore-deposit-bearing layered intrusions have not been reported. Mafic dykes and silicic intrusions are observed within the Zanskar Valley but the dykes are narrow (<10 m) and the granitic rocks are likely to have been derived by crustal melting (Gaetani *et al.* 1986; Fuchs 1987; Spring *et al.* 1993). The Peshawar Plain Alkaline Province in Pakistan contains Early–Middle Permian carbonatites and alkaline silicic rocks that could potentially host mineral deposits (rare metal deposits) but their direct association with the Panjal Traps is still uncertain (Ahmad *et al.* 2013). Thick exhalative VMS deposits between flows have not been identified within the Traps of the Pir Panjal Range or Zanskar Valley. Much like the layered intrusions, it could be due to the accessibility of the outcrop and/or limited detailed mapping, especially at higher altitudes of the Pir Panjal and Zanskar Ranges. The composition and tectonic setting of the Panjal Traps appears to be suitable for the formation of both ore-bearing layered intrusions and VMS deposits.

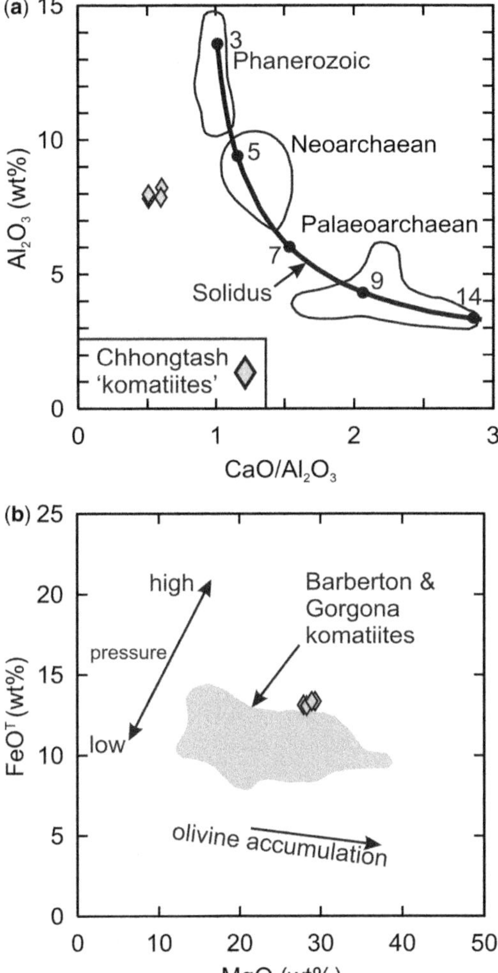

Fig. 16. (**a**) Al_2O_3 (wt%) v. CaO/Al_2O_3 diagram showing the distribution of komatiites and the Chhongtash 'komatiites' relative to the mantle solidus at a given pressure in GPa (1 GPa = *c.* 10 kbar). The komatiites are subdivided according to their age. Palaeoarchaean komatiites are mainly Al-depleted, Neoarchaean komatiites are predominantly Al-undepleted. The group of Phanerozoic rocks (<100 Ma) includes picrites (Dostal 2008). (**b**) FeO^T (wt%) v. MgO (wt%) of komatiitic rocks showing the effects of olivine accumulation and pressure (Arndt 2003). Data of the Chhongtash samples from Rao & Rai (2007).

Permian mafic rocks of the Himalaya and their association with the Panjal Traps

There are many occurrences of Permian rift-related volcanic rocks within the Tethyan domains of Oman, Pakistan, India and China (Chaudhry & Ashraf 1980; Bhat *et al.* 1981; Bhat 1984; Lapierre *et al.* 2004; Zhu *et al.* 2010; Ali *et al.* 2013; Shellnutt *et al.* 2014, 2015; Wang *et al.* 2014). The Panjal, Abor, Nar-Tsum, Bhote Kosi, Selong and Mojiang volcanic groups, Qiangtang mafic dykes, and the Garze Ophiolite are amongst the many Early–Middle Permian basaltic rocks that are attributed to rifting and formation of the Neotethys Ocean (Fig. 1) (Bhat *et al.* 1981; Sengör 1987; Yan *et al.* 2005; Fan *et al.* 2010; Zhu *et al.* 2010; Shellnutt *et al.* 2011*a*, 2012, 2014; Ali *et al.* 2012; Zhai *et al.* 2013; Wang *et al.* 2014; Xu *et al.* 2016). The rocks of east-central Himalaya (Abor and Nar-Tsum) are not as well studied as there are only a few published geochemical papers, none of which present the isotopic systematics; consequently, it is difficult to petrogenetically link all of the Permian rocks. In comparison to the Panjal Traps the Jilong and Selong basalts of Tibet are younger (<280 Ma), have a moderate TiO_2 content (1.8–2.0 wt%), a high MgO content (>10 wt%), high Mg# (>60), and high $^{87}Sr/^{86}Sr_i$ (0.7160–0.7185) and chondritic $\varepsilon_{Nd}(t)$ values (+0.7 to +1.2). The mafic dykes of the Qiangtang terrane range in age from 290–270 Ma and generally have higher $\varepsilon_{Nd}(t)$ (+2.3 to +7.6) values (Zhai *et al.* 2013; Xu *et al.* 2016). The Wusu basalts of the Mojiang volcanic group are 288 Ma and have high $\varepsilon_{Nd}(t)$ values (+4.0 to +5.5) but lower $^{87}Sr/^{86}Sr_i$ values (0.7038–0.7042) than the Panjal Traps (Fig. 17).

There are suggestions that the Panjal Traps are an axial offshoot from a mantle plume centred within

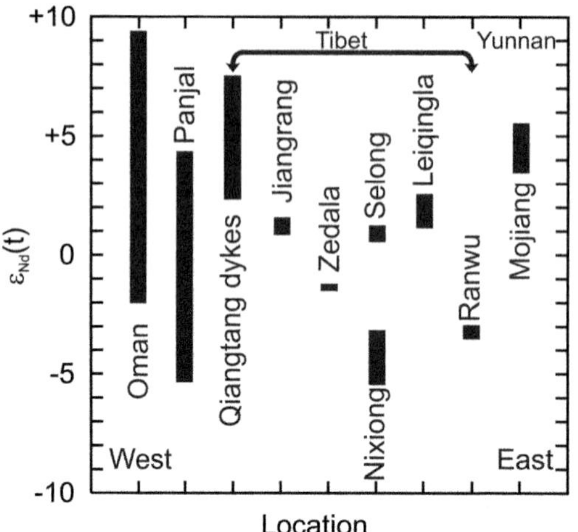

Fig. 17. The variability of $\varepsilon_{Nd}(t)$ of Early–Middle Permian mafic volcanic rocks from west to east across the Himalaya including the Permian volcanic rocks from Oman (Lapierre *et al.* 2004; Fan *et al.* 2010; Zhu *et al.* 2010; Zhai *et al.* 2013; Shellnutt *et al.* 2014, 2015; Xu *et al.* 2016).

either the Qiangtang block (east) or the Arabian Plate (west) but the calculated mantle potential temperatures of the Traps are not anomalously high, and they are the thickest and most continuous outcrops of basalts in the Himalaya, suggesting they may represent a central volcanic eruption location (Maury *et al.* 2003; Lapierre *et al.* 2004; Chauvet *et al.* 2008, 2009; Shellnutt *et al.* 2011*a*, 2014; Zhai *et al.* 2013). Furthermore, the work by Nakazawa & Kapoor (1973) and Nakazawa *et al.* (1975) suggest that volcanism was most intense within the Kashmir Valley and migrated to the south and SE. Although there is some compositional overlap between the different Permian Himalayan–Arabian basaltic groups, there are significant differences (>10 myr) in their age and Nd isotopes. It is likely that the Permian magmatic rocks in the Himalaya and Arabia are members of the same disjointed regional-scale rifting (HMP) that led to the formation of the Neotethys Ocean but that they represent separate magmatic systems. The precise reason (exploitation of a structural heterogeneity) or mechanism (mantle plume v. lithospheric extension) for the formation and propagation of the rift is not constrained

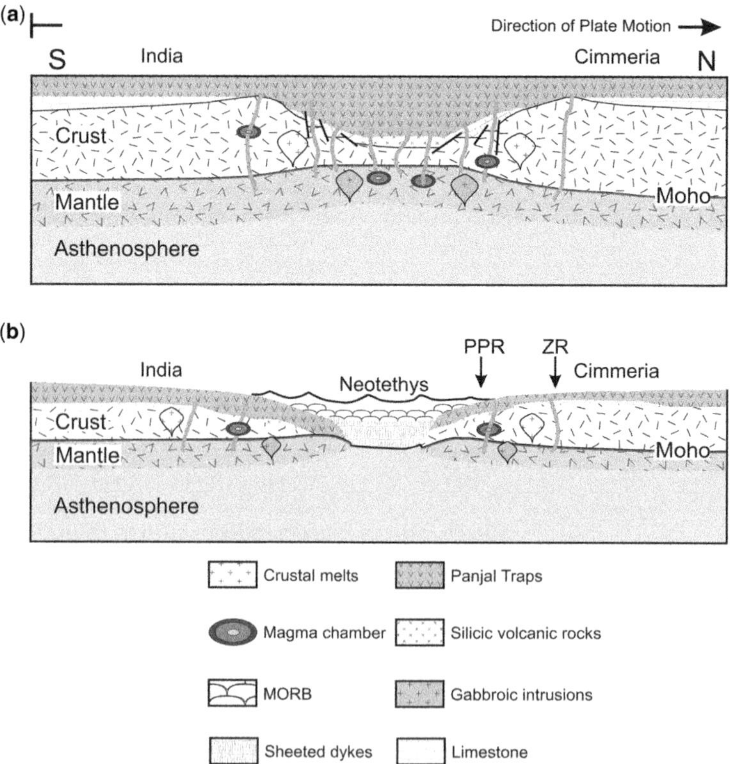

Fig. 18. Tectonic synthesis of the Panjal Traps. (**a**) The initial subaerial eruption followed by the (**b**) subaqueous eruptions and eventual opening of the Neotethys Ocean. The possible pre-India–Eurasia collision locations of the Pir Panjal (PPR) and Zanskar Range (ZR) basalts are shown.

but it could be that different regions of the rift had very specific tectonic conditions. For example, given the younger ages and isotopic differences between the Panjal Traps and the eastern Himalaya basalts, it could be possible that as eastern Cimmeria rifted from Gondwana it passed over an oceanic hotspot which produced the younger, more depleted Qiangtang dykes, whereas the Traps were produced within a passive extensional system during the initial rift.

Synthesis of the Panjal Traps

Neoproterozoic–Late Carboniferous continental to marine sediments were deposited on the passive

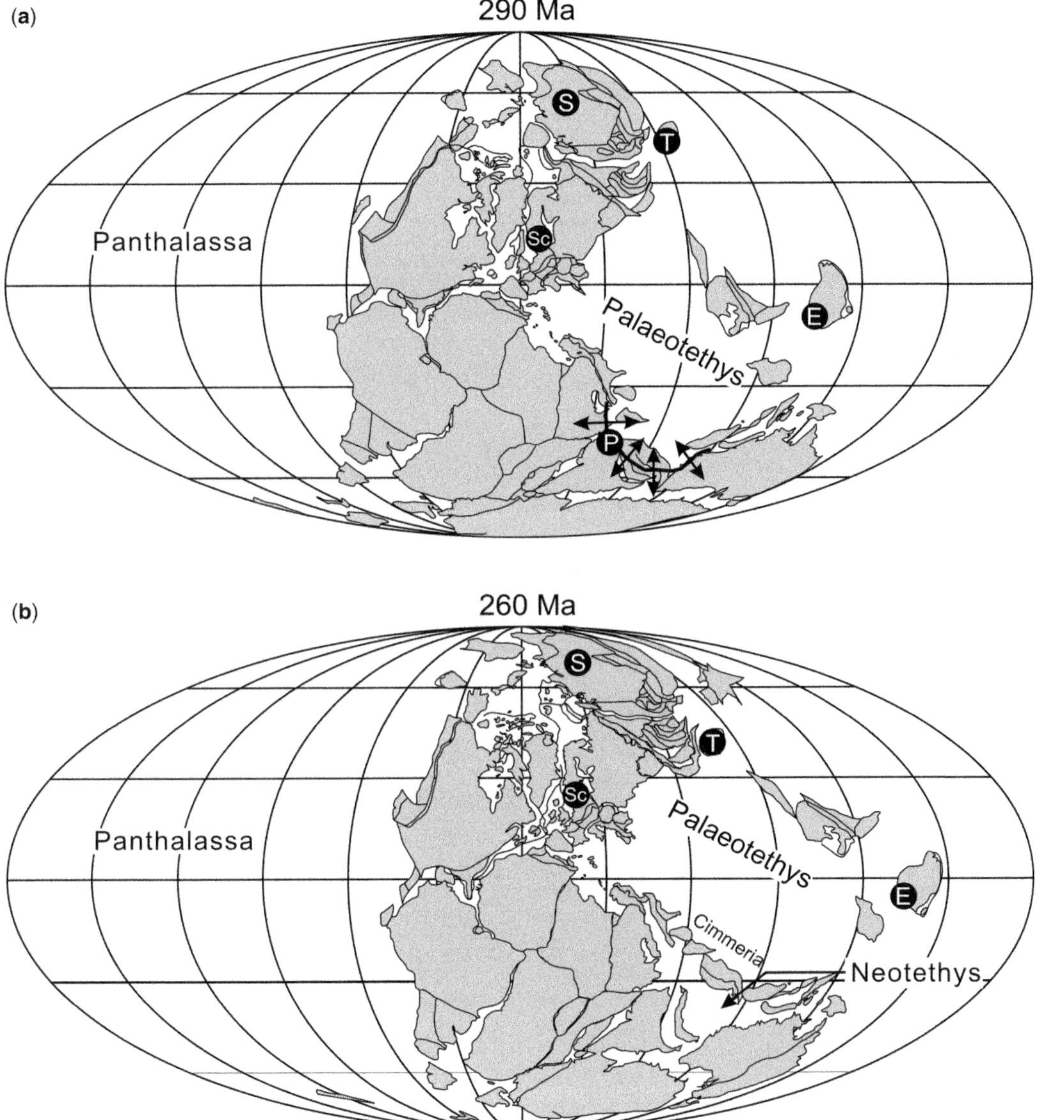

Fig. 19. Palaeogeographical reconstructions of Pangaea at (**a**) *c.* 290 Ma showing the location of the Panjal Traps (PT) and possible rift propagation, and the locations of the five major mafic continental LIPs of the Late Palaeozoic. (**b**) The Neotethys Ocean at *c.* 260 Ma after the rifting of Cimmeria from the northern margin of Gondwana. The reconstructions are based on Torsvik *et al.* (2014). E, Emeishan LIP (*c.* 260 Ma); P, Panjal Traps (*c.* 290 Ma); S, Siberian Traps (*c.* 250 Ma); Sc, Skagerrak-Centered LIP (*c.* 300 Ma); T, Tarim LIP (*c.* 280–270 Ma).

margin of Tethyan Gondwana at mid-southern latitudes (Stojanovic *et al.* 2016). The large Gondwanan ice sheet began to melt and the deposition of Middle Carboniferous fossiliferous (bryozoans, brachiopods and crinoids) Fenestella shale followed by the lower diamictite unit of the Upper Carboniferous Agglomerate slate occurred. The middle units of the Agglomeratic Slate appear to contain more tuffaceous material, possibly marking the first magmatism of the Panjal Traps, followed by the deposition of fossiliferous marine units marking the eustatic sea rise following deglaciation (Tripathi & Singh 1985, 1987; Wopfner & Jin 2009). Pyroclastic flows appear within the freshwater, plant-bearing siliciclastic Nishatbagh Beds and are followed by the main effusive sequence of the Panjal Traps.

The least contaminated lower basaltic flows have compositions similar to 'low-Ti' continental tholeiites and have chondritic Sr–Nd isotope compositions. The injection of mafic magmas into the crust likely induced melting that led to the formation of the silicic volcanic and plutonic rocks (Fig. 18a). Some of the mafic magmas mingled with the crustal melts to produce the andesitic rocks that erupted at distinct horizons, whereas other basalts experienced a smaller amount of contamination. It is possible that some of the mafic magmas or CO_2-rich fluids differentiated within the crust, and produced the carbonatites and syenites of the Peshawar Plain Alkaline Province. Younger flows appear to have erupted within a shallow-marine or lagoonal basin and developed pillow structures (Fig. 18b). Basalts from the Pir Panjal Range have more depleted Nd isotopic compositions and chemical characteristics of E-MORB. The geochemical change may be related to the tectonic setting transitioning from a predominantly continental setting to a predominantly oceanic setting. It is possible that during the continental–oceanic transition VMS deposits formed.

The final continental Traps were capped by marine sediments, whereas regions further from the initial volcanic epicentre were still likely to be under extension as seafloor spreading began and the first microcontinental blocks of Cimmera drifted away from mainland Gondwana (Fig. 19a). The Neotethys Ocean is born and the initial rift completely transitioned from a continental setting into a mid-ocean ridge setting (Fig. 19b). Cimmerian blocks drifted northwards until they accreted to the southern margin of Eurasia during the Middle Mesozoic (Metcalfe 2013; Torsvik *et al.* 2014). It is likely that only the continental portion and the earliest subaqueous Panjal Traps were preserved within the accreted Cimmerian blocks, whereas the younger oceanic equivalents were subducted during closure of the Neotethys Ocean. It is possible that the some of the subducted Panjal Traps were taken to a depth of 2–3 GPa and then brought back to the surface as the Stak Valley, Kaghan and Tso Morari eclogites (Spencer *et al.* 1995; Luais *et al.* 2001; Kouketsu *et al.* 2015; Rehman *et al.* 2016).

Future directions

A great deal of work is still needed to fully characterize the Panjal Traps and their plutonic equivalents. Much of the work to date is lacking high-precision geochronology and isotopic data, therefore correlating the volcanic and plutonic units is difficult. Basic geochemical (major and trace elemental and isotopic data) investigations are needed from volcanic sequences in Pakistan and the Kashmir Valley (Pir Panjal Range). Understanding the formation of the Late Permian silicic plutonic rocks is a major frontier as there are only rudimentary geochemical studies available. Furthermore, identifying the plumbing system is important for constraining crust–magma interactions, magmatic differentiation and the possible formation of orthomagmatic mineral deposits. Future studies should focus on linking the various disparate Early Permian igneous and metamorphic rocks throughout the western Himalaya so that Panjal magmatism and its effects can be fully understood. In order to link all of the Late Permian Himalayan rocks, geochemical, geochronological, structural, geophysical, palaeomagnetic and palaeontological studies remain priorities.

This paper benefited from the reviews of Andrea Marzoli and an anonymous reviewer, and editorial handling by Sarajit Sensarma. I am very appreciative of the field and laboratory support by Ghulam M. Bhat, Ghulam-ud-Din Bhat, G.M. Zaki, Kuo-Lung Wang, Bor-Ming Jahn, Sun-Lin Chung and Mike E. Brookfield, and the helpful discussions with Mary Yeh, Jaroslav Dostal and Kwan-Nang Pang. I would like to thank my mentor and friend Neil D. MacRae for his guidance and support all those years ago. This project was supported by Ministry of Science and Technology grant 102–2628-M-003-001-MY4 to JGS.

References

Ahmad, I., Khan, S., Lapen, T., Burke, K. & Jehan, N. 2013. Isotopic ages for alkaline igneous rocks, including a 26 Ma ignimbrite, from the Peshawar plain of northern Pakistan and their tectonic implications. *Journal of Asian Earth Sciences*, **62**, 414–424.

Ali, J.R., Fitton, J.G. & Herzberg, C. 2010. Emeishan large igneous province (SW China) and the mantle-plume up-doming hypothesis. *Journal of the Geological Society London*, **167**, 953–959, https://doi.org/10.1144/0016-76492009-129

Ali, J.R., Aitchison, J.C., Chik, S.Y.S., Baxter, A.T. & Bryan, S.E. 2012. Paleomagnetic data support Early Permian age for the Abor volcanics in the lower Siang Valley, NE India: significance for Gondwana break-up models. *Journal of Asian Earth Sciences*, **50**, 105–115.

Ali, J.R., Cheung, H.M.C., Aitchison, J.C. & Sun, Y. 2013. Palaeomagnetic re-investigation of Early Permian rift basalts from the Baoshan Block, SW China: constraints on the site-of-origin of the Gondwana-derived eastern Cimmerian terranes. *Geophysical Journal International*, **193**, 650–663.

Anczkiewicz, R., Oberli, F., Burg, J.P., Villa, I.M., Gunther, D. & Meir, M. 2001. Timing of normal faulting along the Indus suture in Pakistan Himalaya and a case of major $^{231}Pa/^{235}U$ initial disequilibrium in zircon. *Earth and Planetary Science Letters*, **191**, 101–114.

Anderson, D.L. 2005. Scoring hotspots: the plume and plate paradigms. *In*: Foulger, G.R., Natland, J.H., Presnall, D.C. & Anderson, D.L. (eds) *Plates, Plumes, and Paradigms*. Geological Society of America, Special Papers, **388**, 31–54.

Anderson, D.L. & Natland, J.H. 2005. A brief history of the plume hypothesis and its competitors: concepts and controversy. *In*: Foulger, G.R., Natland, J.H., Presnall, D.C. & Anderson, D.L. (eds) *Plates, Plumes, and Paradigms*. Geological Society of America, Special Papers, **388**, 119–145.

Arndt, N. 2003. Komatiites, kimberlites, and boninites. *Journal of Geophysical Research*, **108**, 2293, https://doi.org/10.1029/2002JB002157

Baig, M.S. 1990. *Structure and geochronology of pre-Himalayan and Himalayan orogenic events in the northwest Himalaya, Pakistan, with special reference to the Besham area*. PhD thesis, Oregon State University, Corvallis, OR.

Baksi, A.K. 2001. Search for a deep-mantle component in mafic lavas using a Nb–Y–Zr plot. *Canadian Journal of Earth Sciences*, **38**, 813–824.

Bhat, M.I. 1984. Abor Volcanics: further evidence for the birth of the Tethys Ocean in the Himalayan segment. *Journal of the Geological Society, London*, **141**, 763–775, https://doi.org/10.1144/gsjgs.141.4.0763

Bhat, M.I. & Zainuddin, S.M. 1978. Environment of eruption of the Panjal Traps. *Himalayan Geology*, **8**, 727–738.

Bhat, M.I. & Zainuddin, S.M. 1979. Origin and evolution of the Panjal volcanics. *Himalayan Geology*, **9**, 421–461.

Bhat, M.I., Zainuddin, S.M. & Rais, A. 1981. Panjal Trap chemistry and the birth of Tethys. *Geological Magazine*, **118**, 367–375.

Bond, D.P.G. & Wignall, P.B. 2014. Large igneous provinces and mass extinctions: an update. *In*: Keller, G. & Kerr, A.C. (eds) *Volcanism, Impacts, and Mass Extinctions: Causes and Effects*. Geological Society of America, Special Papers, **505**, 29–55.

Brookfield, M.E., Algeo, T.J., Hannigan, R., Williams, J. & Bhat, G.M. 2013. Shaken and stirred: seismites and tsunamites at the Permian–Triassic boundary, Guryal Ravine, Kashmir, India. *PALAIOS*, **28**, 568–582.

Bryan, S.E. & Ernst, R.E. 2008. Revised definition of large igneous provinces (LIPs). *Earth Science-Reviews*, **86**, 175–202.

Bryan, S.E. & Ferrari, L. 2013. Large igneous provinces and silicic large igneous provinces: progress in our understanding over the last 25 years. *Geological Society of America Bulletin*, **125**, 1053–1078.

Campbell, I.H. 2002. Implication of Nb/U, Th/U and Sm/Nd in plume magmas for the relationship between continental and oceanic crust formation and the development of the depleted mantle. *Geochimica et Cosmochimica Acta*, **66**, 1651–1661.

Campbell, I.H. 2007. Testing the plume theory. *Chemical Geology*, **241**, 153–176.

Campbell, I.H. & Griffiths, R.W. 1990. Implications of mantle plume structure for the evolution of flood basalts. *Earth and Planetary Science Letters*, **99**, 79–93.

Chaudhry, M.N. & Ashraf, M. 1980. The volcanic rocks of Poonch District, Azad Kashmir. *In*: *Proceedings of the International Committee of Geodynamics Group 6. Geological Bulletin of the University of Peshawar*, **13**, (Special Issue), 121–128.

Chauvet, F., Lapierre, K. *et al.* 2008. Geochemistry of the Panjal Traps basalts (NW Himalaya): records of the Pangea Permian break-up. *Bulletin de la Société Géologique de France*, **179**, 383–395.

Chauvet, F., Dumont, T. & Basile, C. 2009. Structures and timing of Permian rifting in the central Oman Mountains (Saih Hatat). *Tectonophysics*, **475**, 563–574.

Coffin, M.F. & Eldholm, O. 1992. Volcanism and continental break-up: a global complilation of large igneous provinces. *In*: Storey, B.C., Alabaster, T. & Pankhurst, R.J. (eds) *Magmatism and the Causes of Continental Break-up*. Geological Society, London, Special Publications, **68**, 17–30, https://doi.org/10.1144/GSL.SP.1992.068.01.02

Coffin, M.F. & Eldholm, O. 1994. Large igneous provinces: crustal structure, dimensions and external consequences. *Reviews in Geophysics*, **32**, 1–36.

Courtillot, V., Jaupart, C., Manighetti, I., Tapponier, P. & Besse, J. 1999. On the causal links between flood basalts and continental break-up. *Earth and Planetary Science Letters*, **166**, 177–195.

DiPietro, J.A. & Isachsen, C.E. 2001. U–Pb zircon ages from the Indian plate in northwest Pakistan and their significance to Himalayan and pre-Himalayan geologic history. *Tectonics*, **20**, 510–525.

Dostal, J. 2008. Komatiites. *Geoscience Canada*, **35**, 21–31.

Ernst, R.E. 2014. *Large Igneous Provinces*. Cambridge University Press, Cambridge.

Ernst, R.E. & Buchan, K.L. 2001. Large mafic magmatic events through time and links to mantle plume-heads. *In*: Ernst, R.E. & Buchan, K.L. (eds) *Mantle Plumes: Their Identification Through Time*. Geological Society of America, Special Papers, **352**, 483–457.

Ernst, R.E. & Buchan, K.L. 2003. Recognizing mantle plumes in the geological record. *Annual Review of Earth and Planetary Sciences*, **31**, 469–523.

Ernst, R.E. & Jowitt, S.M. 2013. Large igneous provinces (LIPs) and metallogeny. *In*: Colpron, M., Bissig, T., Rusk, B.G. & Thompson, J.F.H. (eds) *Tectonics, Metallogeny, and Discovery: The North American Cordillera and Similar Accretionary Settings*. Society of Economic Geologists, Special Publications, **17**, 17–51.

Ernst, R.E., Buchan, K.L. & Campbell, I.H. 2005. Frontiers in large igneous province research. *Lithos*, **79**, 271–297.

Fan, W., Wang, Y., Zhang, A., Zhang, F. & Zhang, Y. 2010. Permian arc–back-arc basin development along the Ailaoshan tectonic zone: geochemical, isotopic and geochronological evidence from the Mojiang volcanic rocks, southwest China. *Lithos*, **119**, 553–568.

Fitton, J.G., Saunders, A.D., Norry, M.J., Hardarson, B.S. & Taylor, R.N. 1997. Thermal and chemical structure of the Iceland plume. *Earth and Planetary Science Letters*, **153**, 197–208.

Foulger, G.R. 2007. Large mafic magmatic events through time and links to mantle plume-heads. *In*: Foulger, G.R. & Jurdy, D.M. (eds) *Plates, Plumes, and Planetary Processes*. Geological Society of America, Special Papers, **430**, 1–28.

Foulger, G.R. 2010. *Plates v. Plumes: A Geological Controversy*. Wiley-Blackwell, Chicester, UK.

Franklin, J.M., Gibson, H.L., Jonasson, I.R. & Galley, A.G. 2005. Volcanogenic massive sulphide deposits. *In*: Hedenquist, J.W., Thompson, J.F.H., Goldfarb, R. J. & Richards, J.P. (eds) *Economic Geology 100th Anniversary Volume, 1905–2005*. Society of Economic Geologists, Littleton, CO, 523–560.

Frey, F.A. 1969. Rare earth abundances in a high-temperature peridotite intrusion. *Geochimica et Cosmochimica Acta*, **33**, 1429–1447.

Frost, B.R., Barnes, C.G., Collins, W.J., Arculus, R.J., Ellis, D.J. & Frost, C.D. 2001. A geochemical classification for granitic rocks. *Journal of Petrology*, **42**, 2033–2048.

Fuchs, G. 1987. The geology of Southern Zanskar (Ladakh) – evidence for the autochthony of the Tethys zone of the Himalaya. *Jahrbuch der Geologischen Bundesanstalt*, **130**, 465–491.

Fujimaki, H., Tatsumoto, M. & Aoki, K.-i. 1984. Partition coefficients of Hf, Zr, and REE between phenocrysts and groundmasses. *Journal of Geophysical Research*, **89**, 662–672.

Gaetani, M., Casnedi, R., Fois, E., Garzanti, E., Jadoul, F., Nicora, A. & Tintori, A. 1986. Stratigraphy of the Tethys Himalaya in Zanskar, Ladakh initial report. *Rivisita Italiana di Paleontologia e Stratigrafia*, **91**, 443–478.

Gaetani, M., Garzanti, E. & Tintori, A. 1990. Permo-Carboniferous stratigraphy in SE Zanskar and NW Lahul (NW Himalaya, India). *Eclogae Geologicae Helvetiae*, **83**, 143–161.

Galley, A.G., Hannington, M.D. & Jonasson, I.R. 2007. Volcanogenic massive sulphide deposits. *In*: Goodfellow, W.D. (ed.) *Mineral Deposits of Canada: A Synthesis of Major Deposit-Types, District Metallogeny, the Evolution of Geological Provinces, and Exploration Methods*. Geological Association of Canada, Mineral Deposits Division, Special Publications, **5**, 141–161.

Ganju, P.N. 1944. The Panjal Traps: acid and basic volcanic rocks. *Proceedings of the Indian Academy of Sciences*, **18**, 125–131.

Garland, F., Hawkesworth, C.J. & Mantovani, S.M. 1995. Description and petrogenesis of the Parana rzhyolites, southern Brazil. *Journal of Petrology*, **36**, 1193–1227.

Garzanti, E., Nicora, A. & Tintori, A. 1992. Late Paleozoic to Early Mesozoic stratigraphy and sedimentary evolution of central Dolpo (Nepal Himalaya). *Rivisita Italiana di Paleontologia e Stratigrafia*, **98**, 271–298.

Garzanti, E., Nicora, A., Tintori, T., Sciunnach, D. & Angiolini, L. 1994. Late Paleozoic stratigraphy and petrography of the Thini Chu Group (Manang, Central Nepal): sedimentary record of Gondwana glaciation and rifting of Neotethys. *Rivisita Italiana di Paleontologia e Stratigrafia*, **100**, 155–194.

Garzanti, E., Angiolini, L. & Sciunnach, D. 1996*a*. The mid-Carboniferous to lowermost Permian succession of Spiti (Po Group and Ganmachidam Formations; Tethys Himalaya, northern India): Gondwana glaciation and rifting of Neo-Tethys. *Geodinamica Acta*, **9**, 78–100.

Garzanti, E., Angiolini, L. & Sciunnach, D. 1996*b*. The Permian Kulung Group (Spiti, Lahaul and Zanskar; NW Himalaya): sedimentary evolution during rift/drift transition and initial opening of Neo-Tethys. *Rivisita Italiana di Paleontologia e Stratigrafia*, **102**, 175–200.

Garzanti, E., Angiolini, L., Brunton, H., Sciunnach, D. & Balini, M. 1998. The Bashkirian "Fenestella Shales", the Moscovian 'Chaetetid Shales' tof the Tethys Himalaya (South Tibet, Nepal, India). *Journal ofAsian Earth Sciences*, **16**, 119–141.

Garzanti, E., Le Fort, P. & Sciunnach, D. 1999. First report of Lower Permian basalts in south Tibet: tholeiitic magmatism during break-up and incipient opening of Neotethys. *Journal of Asian Earth Sciences*, **17**, 533–546.

Green, T., Blundy, J., Adam, J. & Yaxley, G. 2000. SIMS determination of trace element partition coefficients between garnet, clinopyroxene and hydrous basaltic liquids at 2–7.5 GPa and 1080–1200°C. *Lithos*, **53**, 165–187.

Griffiths, R.W. & Campbell, I.H. 1990. Stirring and structure in mantle starting plumes. *Earth and Planetary Science Letters*, **99**, 66–78.

He, B., Xu, Y.G., Chung, S.L., Xiao, L. & Wang, Y.M. 2003. Sedimentary evidence for a rapid, kilometer scale crustal doming prior to the eruption of the Emeishan flood basalts. *Earth and Planetary Science Letters*, **213**, 391–405.

Herzberg, C. & Asimow, P.D. 2008. Petrology of some oceanic island basalts: PRIMELT2.XLS software for primary magma calculation. *Geochemistry, Geophysics, Geosystems*, **9**, Q09001, https://doi.org/10.1029/2008GC002057

Herzberg, C. & Asimow, P.D. 2015. PRIMELT3 MEGA. XLSM software for primary magma calculation: peridotite primary magma MgO contents from the liquidus to the solidus. *Geochemistry, Geophysics, Geosystems*, **16**, 563–578.

Herzberg, C. & O'Hara, M.J. 2002. Plume-associated ultramafic magmas of Phanerozoic age. *Journal of Petrology*, **43**, 1857–1883.

Hole, M.J. 2015. The generation of continental flood basalts by decompressional melting of internally heated mantle. *Geology*, **43**, 311–314.

Honegger, K., Dietrich, V., Frank, W., Gansser, A., Thoni, M. & Trommsdorff, V. 1982. Magmatism and metamorphism in the Ladakh Himalayas (the Indus-Tsangpo suture zone). *Earth and Planetary Science Letters*, **60**, 253–292.

Horton, F.M. 2011. *Geochronology and Zircon Geochemistry of Greater Himalaya Leucogranites in Zanskar,*

NW India. MSc thesis, San Francisco State University, San Francisco, CA.

Irving, A.J. & Frey, F.A. 1978. Distribution of trace-elements between garnet megacrysts and host volcanic liquids of kimberlitic to rhyolitic composition. *Geochimica et Cosmochimica Acta*, **42**, 771–787.

Irving, A.J. & Frey, F.A. 1984. Trace-element abundances in megacrysts and their host basalts – constraints on partition-coefficients and megacryst genesis. *Geochimica et Cosmochimica Acta*, **48**, 1201–1221.

Jan, M.Q. & Karim, A. 1990. Continental magmatism related to late Paleozoic–Early Mesozoic rifting in northern Pakistan and Kashmir. *Geological Bulletin of the University of Peshawar*, **23**, 1–25.

Jerram, D.A. & Widdowson, M. 2005. The anatomy of continental flood basalt provinces: geological constraints on the processes and products of flood volcanism. *Lithos*, **79**, 385–405.

Johnson, K.T.M. 1998. Experimental determination of partition coefficients for rare earth and high-field-strength elements between clinopyroxene, garnet, and basaltic melt at high pressures. *Contributions to Mineralogy and Petrology*, **133**, 60–68.

Kapoor, H.M. 1977. Pastannah section of Kashmir with special reference to 'Ophicreas' bed of Middlemiss. *Journal of the Palaeontological Society of India*, **20**, 339–347.

Kempe, D.R.C. & Jan, M.Q. 1970. An alkaline igneous province in the North-West Frontier Province, West Pakistan. *Geological Magazine*, **107**, 395–398.

Kerr, A.C. & Arndt, N.T. 2001. A note on the IUGS reclassification of the high-Mg and picritic volcanic rocks. *Journal of Petrology*, **42**, 2169–2171.

King, S.D. & Anderson, D.L. 1995. An alternative mechanism of flood basalt formation. *Earth and Planetary Science Letters*, **136**, 269–279.

Kouketsu, Y., Hattori, K. & Guillot, S. 2015. Protolith of the Stak eclogite in the northwestern Himalaya. *Italian Journal of Geosciences*, **134**, 64–72, https://doi.org/10.3301/IJG.2015.41

Lapierre, H., Samper, A. *et al.* 2004. The Tethyan plume: geochemical diversity of middle Permian basalts from the Oman rifted margin. *Lithos*, **74**, 167–198.

Le Bas, M.J. 2000. IUGS reclassification of the high-Mg and picritic volcanic rocks. *Journal of Petrology*, **41**, 1467–1470.

Le Bas, M.J., Mian, I. & Rex, D.C. 1987. Age and nature of carbonatite emplacement in north Pakistan. *Geologishe Rundschau*, **76**, 317–323.

Lightfoot, P.C., Hawkesworth, C.J. & Sethna, S.F. 1987. Petrogenesis of rhyolites and trachytes from the Deccan Trap: Sr, Nd and Pb isotope and trace element evidence. *Contributions to Mineralogy and Petrology*, **95**, 44–54.

Lightfoot, P.C., Hawkesworth, C.J., Hergt, J., Naldrett, A.J., Gorbachev, N.S., Fedorenko, V.A. & Doherty, W. 1993. Remobilisation of the continental lithosphere by a mantle plume: major-, trace-element, and Sr-, Nd-, and Pb-isotope evidence from picritic and tholeiitic laves of the Noril'sk district. *Contributions to Mineralogy and Petrology*, **114**, 171–188.

Luais, B., Duchene, S. & de Sigoyer, J. 2001. Sm–Nd disequilibrium in high-pressure, low-temperature Himalayan and Alpine rocks. *Tectonophysics*, **342**, 1–22.

Lydekker, R. 1883. *Geology of Kashmir and Chamba Territories and the British District of Khagan*. Geological Survey of India, Memoirs, **22**, 211–224.

Maury, R.C., Bechennec, F., Cotton, J., Caroff, J., Cordey, F. & Marcoux, J. 2003. Middle Permian plume-related magmatism of the Hawasina Nappes and the Arabian platform: implications on the evolution of the Neotethyan margin in Oman. *Tectonics*, **22**, 1073, https://doi.org/10.1029/2002TC001483

Mallmann, G. & O'Neill, H.St.C. 2009. Mantle melting as a function of oxygen fugacity compared with some other elements (Al, P, Ca, Sc, Ti, Cr, Fe, Ga, Y, Zr and Nb). *Journal of Petrology*, **50**, 1765–1794.

McKenzie, D. & O'Nions, R.K. 1991. Partial melt distributions from inversion of rare Earth element concentrations. *Journal of Petrology*, **32**, 1021–1091.

Melluso, L., Cucciniello, C., Petrone, C.M., Lustrino, M., Morra, V., Tiepolo, M. & Vasconcelos, L. 2008. Petrology of Karoo volcanic rocks in the southern Lembombo monocline, Mozambique. *Journal of African Earth Sciences*, **52**, 136–151.

Metcalfe, I. 2013. Gondwana dispersion and Asian accretion: tectonic and palaeogeographic evolution of eastern Tethys. *Journal of Asian Earth Sciences*, **66**, 1–33.

Middlemiss, C.S. 1910. Revision of Silurian–Trias Sequence in Kashmir. *Records of the Geological Survey of India*, **40**, 206–260.

Miyashiro, A. 1974. Volcanic rock series in island arcs and active continental margins. *American Journal of Science*, **274**, 321–355.

Myrow, P.M., Snell, K.E., Hughes, N.C., Paulsen, T.S., Heim, N.A. & Parcha, S.K. 2006. Cambrian depositional history of the Zanskar Valley region of the Indian Himalaya: tectonic implications. *Journal of Sedimentary Research*, **76**, 364–381.

Nakazawa, K. & Kapoor, H.M. 1973. Spilitic pillow lava in Panjal Trap of Kashmir, India. *Memoirs of the Faculty of Science, Kyoto University, Series of Geology and Mineralogy*, **39**, 83–98.

Nakazawa, K., Kapoor, H.M. *et al.* 1975. The upper Permian and the lower Triassic in Kashmir, India. *Memoirs of the Faculty of Science, Kyoto University, Series of Geology and Mineralogy*, **42**, 1–106.

Noble, S.R., Searle, M.P. & Walker, C.B. 2001. Age and significance of Permian Granites in western Zanskar, High Himalaya. *Journal of Geology*, **109**, 127–135.

Papritz, K. & Rey, R. 1989. Evidence for the occurrence of Permian Panjal Trap basalts in the lesser- and higher Himalayas of the western syntaxis area, NE Pakistan. *Eclogae Geologicae Helvetiae*, **82**, 603–627.

Pareek, H.S. 1976. On studies of the agglomerate slate and Panjal Trap in the Jhelum, Liddar, and Sind Valleys, Kashmir. *Records of the Geological Survey of India*, **107**, 12–37.

Paster, T.P., Schauwecker, D.S. & Haskin, L.A. 1974. The behavior of some trace elements during solidification of the Skaergaard layered series. *Geochimica et Cosmochimica Acta*, **38**, 1549–1577.

Pearce, J.A., Harris, N.B. & Tindle, A.G. 1984. Trace element discrimination diagrams for the tectonic interpretation of granitic rocks. *Journal of Petrology*, **25**, 956–983.

Pearce, T.H., Gorman, B.E. & Birkett, T.C. 1977. The relationship between major element chemistry and tectonic environment of basic and intermediate volcanic rocks. *Earth and Planetary Science Letters*, **36**, 121–132.

Peate, D.W., Hawkesworth, C.J. & Mantovani, M.S.M. 1992. Chemical stratigraphgy of magma types and their spatial distribution. *Bulletin of Volcanology*, **55**, 119–139.

Pogue, K.R., DiPeitro, J.A., Khan, S.R., Hughes, S.S., Dilles, J.H. & Lawrence, R.D. 1992. Late Paleozoic rifting in northern Pakistan. *Tectonics*, **11**, 871–883.

Rafiq, M. & Jan, M.Q. 1989. Geochemistry and petrogenesis of the Ambela granitc complex, NW Pakistan. *Geological Bulletin of the University of Peshawar*, **22**, 159–179.

Rampino, M.R. & Stothers, R.B. 1988. Flood basalt volcanism during the past 250 million years. *Science*, **241**, 663–668.

Rao, D.R. & Rai, H. 2007. Permian komatiites and associated basalts from the marine sediments of Chhongtash Formation, southeast Karakoram, Ladakh, India. *Mineralogy and Petrology*, **91**, 171–189.

Rehman, H.U., Lee, H.-Y., Chung, S.-L., Khan, T., O'Brien, P.J. & Yamamoto, H. 2016. Source and mode of the Permian Panjal Trap magmatism: evidence from zircon U–Pb and Hf isotopes and trace elemental data from the Himalayan ultrahigh-pressure rocks. *Lithos*, **260**, 286–299.

Richards, M.A., Duncan, R.A. & Courtillot, V.E. 1989. Flood basalts and hot-spot tracks: plume heads and tails. *Science*, **246**, 103–107.

Rudnick, R.L. & Gao, S. 2003. Composition of the continental crust. *In*: Rudnick, R.L. (ed.) *The Crust*. *In*: Holland, H.D. & Turekian, K.K. (eds) *The Crust*. Treatise on Geochemistry, **3**. Elsevier, Amsterdam, 1–64.

Salters, V. & Longhi, J. 1999. Trace element partitioning during the initial stages of melting beneath mid-ocean ridges. *Earth and Planetary Science Letters*, **166**, 15–30.

Saunders, A.D. & Reichow, M. 2009. The Siberian Traps and the end-Permian mass extinction: a critical review. *Chinese Science Bulletin*, **54**, 20–37.

Saunders, A.D., Jones, S.M., Morgan, L.A., Pierce, K.L., Widdowson, M. & Xu, Y.G. 2007. Regional uplift associated with continental large igneous provinces: the roles of mantle plumes and the lithosphere. *Chemical Geology*, **241**, 282–318.

Schnetzler, C.C. & Philpotts, J.A. 1970. Partition coefficients of rare-earth elements between igneous matrix material and rock-forming mineral phenocrysts; II. *Geochimica et Cosmochimica Acta*, **34**, 331–340.

Sengör, A.M.C. 1987. Tectonics of the Tethysides: orogenic collage development in a collisional setting. *Annual Review in Earth and Planetary Sciences*, **15**, 213–244.

Sensarma, S., Palme, H. & Mukhopadhyay, D. 2002. Crust–mantle interaction in the genesis of siliceous high magnesian basalts: evidence from the Early Proterozoic Dongargarh Supergroup, India. *Chemical Geology*, **187**, 21–37.

Shellnutt, J.G. 2014. The Emeishan large igneous province: a synthesis. *Geoscience Frontiers*, **5**, 369–394.

Shellnutt, J.G. & Jahn, B.-M. 2010. Formation of the Late Permian Panzhihua plutonic–hypabyssal–volcanic igneous complex: implications for the genesis of Fe–Ti oxide deposits and A-type granites of SW China. *Earth and Planetary Science Letters*, **289**, 509–519.

Shellnutt, J.G., Bhat, G.M., Brookfield, M.E. & Jahn, B.-M. 2011*a*. No link between the Panjal Traps (Kashmir) and the Late Permian mass extinctions. *Geophysical Research Letters*, **38**, L19308, https://doi.org/10.1029/2011GL049032

Shellnutt, J.G., Jahn, B.-M. & Zhou, M.-F. 2011*b*. Crustal-derived granites in the Panzhihua region, SW China: implications for felsic magmatism in the Emeishan Large Igneous province. *Lithos*, **123**, 145–157.

Shellnutt, J.G., Bhat, G.M., Wang, K.-L., Brookfield, M.E., Dostal, J. & Jahn, B.-M. 2012. Origin of the silicic volcanic rocks of the Early Permian Panjal Traps, Kashmir, India. *Chemical Geology*, **334**, 154–170.

Shellnutt, J.G., Bhat, G.M., Wang, K.-L., Brookfield, M.E., Jahn, B.-M. & Dostal, J. 2014. Petrogenesis of the flood basalts from the Early Permian Panjal Traps, Kashmir, India: geochemical evidence for shallow melting of the mantle. *Lithos*, **204**, 159–171.

Shellnutt, J.G., Bhat, G.M., Wang, K.-L., Yeh, M.-W., Brookfield, M.E. & Jahn, B.-M. 2015. Multiple mantle sources of the early Permian Panjal Traps, Kashmir, India. *American Journal of Science*, **315**, 589–619.

Sheth, H.C. 2005. From Deccan to Réunion: no trace of a mantle plume. *In*: Foulger, G.R., Natland, J.H., Presnall, D.C. & Anderson, D.L. (eds) *Plates, Plumes and Paradigms*. Geological Society of America, Special Papers, **388**, 477–501.

Sheth, H.C. 2007. Plume-related regional pre-volcanic uplift in the Deccan Traps: absence of evidence, evidence of absence: discussion. *In*: Foulger, G.R. & Jurdy, D.M. (eds) *Plates, Plumes, and Planetary Processes*. Geological Society of America, Special Papers, **430**, 803–813.

Shimizu, K., Nakamura, E. & Maruyama, S. 2005. The geochemistry of ultramafic to mafic volcanics from the Belingwe greenstone belt, Zimbabwe: magmatism in an Archean continental large igneous province. *Journal of Petrology*, **46**, 2367–2394.

Smith, H.A., Chamberlain, C.P. & Zeitler, P.K. 1994. Timing and Duration of Himalayan Metamorphism within the Indian Plate, Northwest Himalaya, Pakistan. *Journal of Geology*, **102**, 493–508.

Spencer, D.A., Tonarini, S. & Pognante, U. 1995. Geochemical and Sr–Nd isotopic characterization of Higher Himalayan ecologites (and associated metabasites). *European Journal of Mineralogy*, **7**, 89–102.

Spring, L., Bussy, F., Vannay, J.C., Hunon, S. & Cosca, M.A. 1993. Early Permian granitic dykes of alkaline affinity in the Indian High Himalaya of upper Lahul and SE Zanskar: geochemical characterization and geotectonic implications. *In*: Trelor, P.J. & Searle, M. (eds) *Himalayan Tectonics*. Geological Society, London, Special Publications, **74**, 251–264, https://doi.org/10.1144/GSL.SP.1993.074.01.18

Stojanovic, D., Aitchison, J.C., Ali, J.R., Ahmad, T. & Dar, R.A. 2016. Paleomagnetic investigation of the Early Permian Panjal Traps of NW India: regional tectonic implications. *Journal of Asian Earth Sciences*, **115**, 114–123.

Sun, S.-s. & McDonough, W.F. 1989. Chemical and isotopic systematics of oceanic basalts: implications for mantle composition and processes. *In*: Saunders, A.D. & Norry, M.J. (eds) *Magmatism in the Ocean Basins*. Geological Society, London, Special Publications, **42**, 313–345, https://doi.org/10.1144/GSL.SP.1989.042.01.19

Tewari, R., Awatarm, R. *et al.* 2015. The Permian–Triassic palynological transition in the Guryal Ravine section, Kashmir, India: implications for Tethyan–Gondwana correlations. *Earth-Science Reviews*, **149**, 53–66.

Tilton, G.R., Bryce, J.G. & Mateen, A. 1998. Pb–Sr–Nd isotope data from 30 and 300 Ma collision zone carbonatites in northwest Pakistan. *Journal of Petrology*, **39**, 1865–1874.

Timmerman, M.J., Heeremans, M., Kirstein, L.A., Larsen, B.T., Spencer-Dunworth, E.-A. & Sundvoll, B. 2009. Linking changes in tectonic style with magmatism in northern Europe during the late Carboniferous to latest Permian. *Tectonophysics*, **473**, 375–390.

Torsvik, T.H., Smethurst, M.A., Burke, K. & Steinberger, B. 2008. Long term stability in deep mantle structure: evidence from the *c.* 300 Ma Skagerrak-Centered large igneous province (the SCLIP). *Earth and Planetary Science Letters*, **267**, 444–452.

Torsvik, T.H., van der Voo, R. *et al.* 2014. Deep mantle structure as a reference frame for movements in and on the Earth. *Proceedings of the National Academy of Science of the United States of America*, **111**, 8735–8740.

Tripathi, C. & Singh, G. 1985. Carboniferous flora of India and its contemporaneity in the World. *Comptes Rendu X Congres International de Stratigraphie et de Geologie du Carbonifere, Madrid, Spain (1983)*, **4**, 295–306.

Tripathi, C. & Singh, G. 1987. Gondwana and associated rocks of the Himalaya and their significance. *In*: McKenzie, G.D. (ed.) *Gondwana Six: Stratigraphy, Sedimentology, and Paleontology*. American Geophysical Union, Geophysical Monograph Series, **41**, 195–205.

Ukstins Peate, I. & Bryan, S.E. 2008. Re-evaluating plume-induced uplift in the Emeishan large igneous province. *Nature Geosciences*, **1**, 625–629.

van de Schootbrugge, B. & Wignall, P.B. 2016. A tale of two extinctions: converging end-Permian and end-Triassic scenarios. *Geological Magazine*, **153**, 332–354.

Vannay, J.C. & Spring, L. 1993. Geochemistry of the continental basalts within the Tethyan Himalaya of Lahul-Spiti and SE Zanskar, northwest India. *In*: Trelor, P.J. & Searle, M. (eds) *Himalayan Tectonics*. Geological Society, London, Special Publications, **74**, 237–249, https://doi.org/10.1144/GSL.SP.1993.074.01.17

Veevers, J.J. & Tewari, R.C. 1995. Permian–Carboniferous and Permian–Triassic magmatism in the rift zone bording the Tethyan margin of southern Pangea. *Geology*, **23**, 467–470.

Wadia, D.N. 1934. The Cambrian–Trias sequence of North-Western Kashmir (Parts of Muzaffarabad and Baramular districts). *Records of the Geological Survey of India*, **68**, 121–176.

Wadia, D.N. 1961. *Geology of India*. Macmillan, London.

Wang, M., Li, C., Wu, Y.-W. & Xie, C.-M. 2014. Geochronology, geochemistry, Hf isotopic compositions and formation mechanism of radial mafic dykes in northern Tibet. *International Geology Review*, **56**, 187–205.

White, R.V. 2002. Earth's biggest 'whodunnit': unravelling the clues in the case of the end-Permian mass extinction. *Philosophical Transactions of the Royal Society, London*, **360**, 2963–2985.

White, R.V. & Saunders, A.D. 2005. Volcanism, impact and mass extinctions: incredible or credible coincidences? *Lithos*, **79**, 299–316.

Wignall, P.B. 2001. Large igneous provinces and mass extinctions. *Earth-Science Reviews*, **53**, 1–33.

Wilson, M. 1989. *Igneous Petrogenesis*. Unwin Hyman, London.

Winchester, J.A. & Floyd, P.A. 1977. Geochemical discrimination of different magma series and their differentiation products using immobile elements. *Chemical Geology*, **20**, 325–343.

Wopfner, H. & Jin, X.C. 2009. Pangea megasequences of Tethyan Gondwana-margin reflect global changes of climate and tectonism in Late Palaeozoic and Early Triassic times – a review. *Palaeoworld*, **18**, 169–192.

Workman, R.K. & Hart, S.R. 2005. Major and trace element composition of the depleted MORB mantle (DMM). *Earth and Planetary Science Letters*, **231**, 53–72.

Workman, R.K., Hart, S.R., Jackson, M., Regelous, M., Blusztajn, J., Kurz, M. & Staudigel, H. 2004. Recycled metasomatized lithosphere as the origin of the enriched mantle II (EM2) end-member: evidence from the Samoan volcanic chain. *Geochemistry, Geophysics, Geosystems*, **5**, Q04008, https://doi.org/10.1029/2003GC000623

Xu, W., Dong, Y., Zhang, X., Deng, M. & Zheng, L. 2016. Petrogenesis of high-Ti mafic dykes from southern Qiangtang, Tibet: implications for a ca. 290 Ma large igneous province related to the early Permian rifting of Gondwana. *Gondwana Research*, **36**, 410–422.

Xu, Y.-G., Chung, S.L., Jahn, B.M. & Wu, G.Y. 2001. Petrologic and geochemical constraints on the petrogenesis of Permian–Triassic Emeishan flood basalts in southwestern China. *Lithos*, **58**, 145–168.

Xu, Y.-G., Chung, S.-L., Chung Shao, H. & He, B. 2010. Silicic magmas form the Emeishan large igneous province, southwest China: petrogenesis and their link with the end-Guadalupian biological crisis. *Lithos*, **119**, 47–60.

Xu, Y.-G., Wei, X., Luo, Z.-Y., Liu, H.-Q. & Cao, J. 2014. The Early Permian Tarim large igneous province: main characteristics and a plume incubation model. *Lithos*, **204**, 20–35.

Yan, Q., Wang, Z. *et al.* 2005. Opening of the Tethys in southwest China and its significance to the breakup of East Gondwanaland in the late Paleozoic: evidence from SHRIMP U–Pb zircon analyses for the Garze ophiolite. *Chinese Science Bulletin*, **50**, 256–264.

Yang, S., Chen, H., Li, Z., Li, Y., Yu, X., Li, D. & Meng, L. 2013. Early Permian Tarim large igneous province in northwest China. *Science in China*, **56**, 2015– 2026.

Yeh, M.-W. & Shellnutt, J.G. 2016. The initial break-up of Pangaea elicited by Late Palaeozoic deglaciation.

Scientific Reports, **6**, 31442, https://doi.org/10.1038/srep31442
ZEITLER, P.K. 1988. Ion microprobe dating of zircon from the Malakand granite, NW Himalaya, Pakistan. A constraint on the timing of Tertiary metamorphism in the region. *Geological Society of America Abstracts with Programs*, **20**, 323.
ZHAI, Q.-G., JAHN, B.-M. ET AL. 2013. SHRIMP zircon U–Pb geochronology, geochemistry and Sr–Nd–Hf isotopic compositions of a mafic dyke swarm in the Qiangtang terrane, northern Tibet and geodynamic implications. *Lithos*, **174**, 28–43.
ZHOU, M.-F., ARNDT, N.T., MALPAS, J., WANG, C.Y. & KENNEDY, A.K. 2008. Two magma series and associated ore deposit types in the Permian Emeishan large igneous province, SW China. *Lithos*, **103**, 352–368.
ZHU, D.-C., MO, X.-X. ET AL. 2010. Presence of Permian extension- and arc-type magmatism in southern Tibet: paleogeographic implications. *Geological Society of America Bulletin*, **122**, 979–993.
ZINDLER, A. & HART, S.R. 1986. Chemical geodynamics. *Annual Review of Earth and Planetary Sciences*, **14**, 493–571.

Mantle source heterogeneity in continental mafic Large Igneous Provinces: insights from the Panjal, Rajmahal and Deccan basalts, India

K. VIJAYA KUMAR[1]*, MORE B. LAXMAN[1] & K. NAGARAJU[2]

[1]*School of Earth Sciences, SRTM University, Nanded 431606, Maharashtra, India*

[2]*Geological Survey of India, Raipur 493111, Chhattisgarh, India*

**Correspondence: vijay_kumar92@hotmail.com*

Abstract: In this paper we illustrate the geochemical diversity of the basalts of Panjal (*c.* 289 Ma), Rajmahal (*c.* 117 Ma) and Deccan (*c.* 65 Ma) continental large igneous provinces (continental LIPs) from India as controlled by multiple mantle sources. The Panjal primitive basalts are formed by 5–20%, Rajmahal basalts by 10–20% and Deccan picritic basalts by 1–20% melting of the mantle. The depth of melting beneath Deccan is greater than beneath Rajmahal, which greater than beneath Panjal. We have categorized the Panjal, Rajmahal and Deccan basalts into Type I and Type II. The Type I basalts with lower La/Nb, Th/Nb and Th/Yb ratios and higher ε_{Nd_i} values ($\varepsilon_{Nd_i} = -3$ to +8) are derived from the sublithospheric sources; some of the Type I basalts represent lithosphere-contaminated melts. The Type II basalts with higher La/Nb, Th/Nb and Th/Yb ratios and lower ε_{Nd_i} values ($\varepsilon_{Nd_i} = -20$ to -3) are produced from the subcontinental lithosphere. We estimate that lithosphere and sublithosphere (plume and/or E-MORB patches within asthenosphere) have contributed, respectively, approximately 67 and 33% to the Panjal, 52 and 48% to Rajmahal, and 28 and 72% to Deccan LIPs. Fertility of the Indian subcontinental lithosphere may have increased with decreasing age related to the break-up of Gondwana.

Supplementary material: References for the geochemical data, extracted from the GEOROC database, on the basaltic rocks from Panjal, Rajmahal and Deccan Large Igneous Provinces are available at: https://doi.org/10.6084/m9.figshare.c.3821881

Spatially extensive and temporally restricted magmatic activity in the Earth's crust, labelled as Large Igneous Provinces (LIPs: White & Saunders 2005; Bryan & Ernst 2008), is recorded both in oceanic and continental settings. Although the origin of the oceanic Large Igneous Provinces (oceanic LIPs) is better understood (Hofmann 2003), reasons for enormous geochemical diversity in the continental Large Igneous Provinces (continental LIPs) are not yet fully resolved in terms of relative contributions of shallow- and deep-mantle reservoirs. Magmatism in the continental LIPs is initiated by either lithosphere stretching or plume impinging (Turcotte & Emerman 1983; Zeyen *et al.* 1997). Both lithosphere stretching and plume impinging induces continental rifting, which, in some cases, may eventually lead to the formation of new oceanic crust. The igneous activity in the continental LIPs, although dominated by basaltic volcanism, produces a wide range of magmatic rocks on the Earth in a particular tectonic setting. The magmatic spectrum includes basalts with variable silica saturation, sodic and potassic alkaline rocks, carbonatites and A-type rhyolites, and their plutonic differentiates. The possible sources that can contribute to the geochemical variability in the continental LIP magmas (see Vijaya Kumar & Rathna 2008) (Fig. 1) include: (1) subcontinental lithospheric mantle SCLM (plus metasomes) (Hawkesworth *et al.* 1990; Goodenough *et al.* 2002; Li *et al.* 2015); (2) depleted asthenospheric mantle (Wedepohl *et al.* 1994; Aldanmaz *et al.* 2006; Fitton 2007; Heinonen *et al.* 2016); (3) thermal boundary layer that was metasomatized by mantle plume–asthenospheric melts (Wilson *et al.* 1995; Thompson *et al.* 2005; Hole 2015); (4) delaminated subcontinental lithosphere mantle that recycled into the asthenosphere (Zindler & Hart 1986); (5) mantle plumes (Nicholson & Shirey 1990; Rogers *et al.* 2000; Johnson *et al.* 2005; Campbell 2007; Colon *et al.* 2015); and (6) underplated and/or metasomatized lower continental crust (Shellnutt *et al.* 2015*b*. Among the possible mantle sources, the continental LIP basaltic magmas are dominantly derived from mantle plume (Mahoney *et al.* 1991; Xu *et al.* 2001) and subcontinental lithosphere (Hawkesworth *et al.* 1988; Gallagher & Hawkesworth 1992). For alternate views to plume hypothesis, the reader is referred to Hamilton (2011 and references therein). For the lithosphere to contribute significantly to continental LIP magmatism, it needs to fulfill at least one of the following conditions: (i) a favourable thermal state (Arndt & Christensen 1992); (ii) enough extension to cause decompression melting (McKenzie & Bickle 1988); (iii) the presence of volatiles (Ivanov

From: Sensarma, S. & Storey, B. C. (eds) 2018. *Large Igneous Provinces from Gondwana and Adjacent Regions*. Geological Society, London, Special Publications, **463**, 87–115.
First published online July 11, 2017, https://doi.org/10.1144/SP463.5

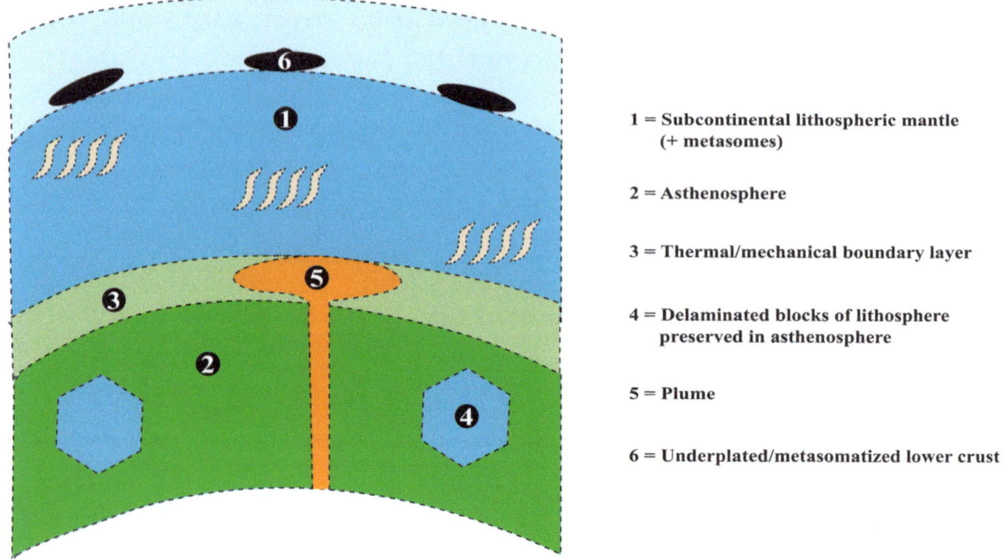

Fig. 1. A schematic model that summarizes the possible mantle and crustal sources that feed the magmas to build a continental LIP.

2015); and (iv) the presence of enriched pockets, which have lower solidii for melting (Foley 2008).

Although the sublithospheric mantle has distinct trace element characteristics, especially lower La/Nb, Th/La, Ba/Th and Th/Yb, and higher Nb/Yb, Zr/Sm, Nb/Y, Nb/U and $(La/Lu)_N$ ratios among others (Fitton & Dunlop 1985; Hofmann 1988, 2003; Sun & McDonough 1989 and references therein; Condie 2005; Johnson *et al.* 2005; Niu *et al.* 2012; Vijaya Kumar *et al.* 2015), interaction with the continental lithosphere makes the recognition of these discrete geochemical signatures in the sublithosphere-derived melts a complicated task (Wilson & Downes 1991; Baker *et al.* 1997; Macdonald *et al.* 2001; Späth *et al.* 2001; Bertrand *et al.* 2003; Shaw *et al.* 2003; Xu *et al.* 2005; Jung *et al.* 2006; Tang *et al.* 2006; Wei *et al.* 2014). The SCLM itself may have had an intricate geological history with many episodes of previous melt removal followed by later enrichment either by subduction-related fluids/melts or asthenospheric/plume-derived low-percentage partial melts (+fluids), or even both. The enrichment could be during melt-generation processes or an ancient event in a different tectonic setting. Foundering of the lower continental crust into the SCLM may further add to the geochemical heterogeneity of the continental lithosphere (Arndt & Goldstein 1989; Glazner 1994; Fan *et al.* 2004; Gao *et al.* 2004). Therefore, crustal signatures in the continental LIP basalts not only reflect assimilation/assimilation fractional crystallization of mantle-derived magmas but could also possibly reflect the presence of crustal material at mantle depths (Xu 2002). In addition to mantle source heterogeneity (e.g. Hawkesworth *et al.* 1988; Ellam & Cox 1991; Ellam *et al.* 1992), variations in the continental LIP magma geochemistry are attributed to various melting and magma chamber processes (Fodor 1987; Vijaya Kumar *et al.* 2010).

Identifying and quantifying the relative contributions of different mantle/crustal sources to the continental LIP magmatism is one of the fundamental problems in igneous petrogenesis. Parameterization of geochemical models on the continental LIP magmatism involves evaluating mantle source reservoirs and their spatiotemporal variation. In the present paper, we illustrate the geochemical diversity of the basalts of the Panjal (*c.* 289 Ma), Rajmahal (*c.* 117 Ma) and Deccan (*c.* 65 Ma) continental LIPs from India as controlled by distinct mantle reservoirs. For this purpose, we have extracted published geochemical data from the GEOROC database (see the Supplementary material for references) on the basalts from these three spatially and temporally distinct continental LIPs. We have considered basalts with MgO >4% in the present study. We have tested the data for post-magmatic elemental mobility and utilized only those elements that reflect magmatic lineages (see the subsection on 'Potential problems with elemental mobility' later in this paper). We have calculated age-corrected initial Nd isotopic (ε_{Nd_i}) compositions. As the oceanic lithosphere is geochemically homogeneous on a large scale and contributes very little to the oceanic LIPs, the relative contributions of different mantle reservoirs (plume, asthenosphere and lithosphere) to the oceanic LIPs are reasonably well known (see Hofmann 2003 for a comprehensive review). However, the same cannot be said for the continental LIPs due to the extreme geochemical variability and complexity of the subcontinental lithosphere, which also contributes substantially to the continental LIP magmatism. The objective of this study is, therefore, to quantitatively estimate the relative contributions of multiple mantle sources (depleted asthenosphere, plume and subcontinental lithosphere) in the generation of the Indian Phanerozoic continental LIPs, to evaluate the temporal variations in the Indian subcontinental lithosphere composition and structure, and to propose a comprehensive model for the origin of the spatially and temporally separated Indian Phanerozoic continental LIPs.

The Panjal, Rajmahal and Deccan basalts: a brief outline of the geology and models of origin

The Panjal, Rajmahal and Deccan LIPs are the predominant Phanerozoic crustal representatives of deep-mantle processes within the cratonic India. All three provinces are related to continental rifting during the initial to final stages of the Gondwana fragmentation (Hawkesworth *et al.* 1999; Segev 2002; Chauvet *et al.* 2008). Figure 2 illustrates the distribution of the Panjal, Rajmahal and Deccan traps in the northern, eastern and central portions of India. At the present exposure level, the areal extent of the Deccan basaltic province far outweighs the other two (Fig. 2; Table 1).

Panjal Traps

The Panjal Traps, mostly exposed along the Panjal and Zanskar mountain ranges of Kashmir Himalaya, represent the Permian LIP along the northern part of the Gondwana margin (Fig. 2a) (Wadia 1961; Veevers & Tewari 1995). The Panjal basalts are considered to be the igneous remnants of the Neo-Tethys opening, whose equivalents are found in Iran, Zagros and Oman (Lapierre *et al.* 2004 and references therein). The Panjal Traps are composed of basalts with subordinate amounts of felsic volcanic rocks (Ganju 1944; Bhat & Zainuddin 1979; Table 1). At the present level of exposure, the Panjal Traps occupy an area of 10^4 km^2 (Table 1); however, an area of 2×10^5 km^2 is estimated for the original extent of the Panjal eruption (Ernst & Buchan 2003). The thickness of the Traps decreases from approximately 3000 m in the west to $\leq$300 m in

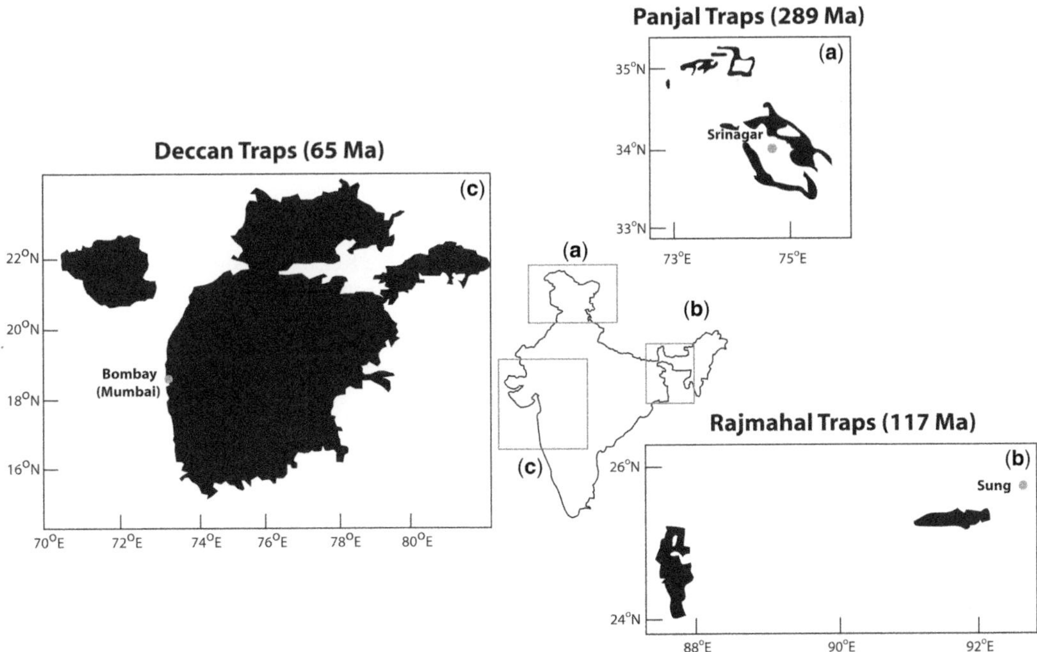

Fig. 2. Outline map of India with the location of the Panjal, Rajmahal and Deccan LIPs and a diagrammatic representation of the three LIPs.

Table 1. Some geological details of the Panjal, Rajmahal and Deccan continental LIPs, India

	Panjal	Rajmahal	Deccan	References
Area	10 000 km^2	200 000 km^2	500 000 km^2	Middlemiss (1910), Wadia (1934), Sengupta (1966), Nakazawa *et al.* (1975), Beane *et al.* (1986), Baksi (1994), Garzanti *et al.* (1999), Kent *et al.* (2002), Pande (2002), Ernst & Buchan (2003), Chauvet *et al.* (2008), Jay & Widdowson (2008), Valdiya (2010), Shellnutt *et al.* (2011)
Age	289 Ma	117 Ma	65 Ma	
Rock types	Mainly tholeiitic basalts with minor alkali basalt, basaltic andesite, rhyolite and dacite	Basalts with different Si-satuaration and subordinate amounts of trachyandesite, andesite, trachyte, rhyolite, aggolomerate and tuff, pyroxenite, lamproite, lamprophyre, nephelinite, ijolite, meteigite, syenite, and carbonatite	Mainly tholeiitic basalts with minor amounts of basanite, picrite, nephelinite, lamprophyre, carbonatites and rhyolite	
Thickness of lava flows (at the type area)	3 km	244 m	3 km	

the east (Middlemiss 1910; Wadia 1934; Fuchs 1987; Chauvet *et al.* 2008) (Table 1). The basalts have erupted both subaerially and in subaqueous proximal to shallow-marine environments (Wadia 1961; Nakazawa & Kapoor 1973; Nakazawa *et al.* 1975; Pareek 1976). Some of the Panjal basalts have undergone zeolite- to lower-greenschist-facies metamorphism (Shellnutt *et al.* 2014).

Many researchers have worked on the geochemical aspects of the Panjal Traps (Bhat & Zainuddin 1979; Bhat *et al.* 1981; Chauvet *et al.* 2008; Shellnutt *et al.* 2014). Shellnutt *et al.* (2014) have classified the Panjal basalts into high- and low-Ti basalts, with the former representing ocean island basalt (OIB)-type composition and the latter similar to continental tholeiities. The basalts are considered to have formed by melting in the spinel peridotite stability field of the SCLM (Shellnutt *et al.* 2014, 2015*a*).The high-Ti basalts are derived by low-degree partial melting, whereas low-Ti rocks are formed by a moderate degree of partial melting of spinel peridotite (Shellnutt *et al.* 2014). It has also been suggested that the rift opened into an oceanic margin, and basalts with mid-ocean ridge basalt (MORB) signatures eventually erupted, especially in the western segment of the province (Shellnutt *et al.* 2015*a*). A progressive changes in the minor and trace element geochemistry of the Panjal basalts from the OIB-type in the eastern margin to MORB-type in the western margin is also reflected in the isotopic composition, with ε_{Nd_i} varying from −5 in the east to +4 in the west (Shellnutt *et al.* 2015*a*). Some of the basalts have undergone as much as 30% crustal assimilation. Dacites, trachytes, rhyolites and acidic tuffs make up the felsic litho-package of the Panjal Traps, and it has been suggested that they possibly represent melts of Palaeoproterozoic mid-level crust (Shellnutt *et al.* 2012).

The continental rifting expressed by the eruption of Panjal Traps, Abor basalts and Nar-Tsum spilites, and the Bhote Kosi basalts of the eastern and central Himalaya eventually culminated in the formation of the Neo-Tethys Ocean and the ribbon continent Cimmeria (Nakazawa *et al.* 1975; Bhat *et al.* 1981; Bhat 1984; Sengör 1987; Garzanti *et al.* 1999; Ali *et al.* 2012; Shellnutt *et al.* 2011) at around 289 Ma (Shellnutt *et al.* 2011).

Rajmahal Traps

Outpouring of lavas, presumably from the Kerguelen hotspot, constructed the Rajmahal and Sylhet traps in NE India (Fig. 2b). The former were emplaced into the Gondwana Supergroup and the latter erupted over the Shillong Plateau. It is presumed that the Kerguelen plume erupted approximately 600-km-diameter-scale lavas over NE India (Kent *et al.* 2002). The Rajmahal–Sylhet Traps originally may have occupied an area of 2×10^5 km^2 (Baksi 1995; Kent *et al.* 2002) (Table 1) but, due to presence of sediment cover over the intermittent area, presently the Kerguelen-related traps crop out as two lobes at Rajmahal and Sylhet (Fig. 2b). The Rajmahal basalts, *sensu stricto*, occupy an area of approximately 4100 km^2 with a maximum thickness of 230 m (Kent *et al.* 2002). The Sylhet Traps, exposed in a linear 60–80 km-long and approximately 4 - km-wide strip along the southern edge of the Shillong Plateau (Baksi *et al.* 1987), have a maximum thickness of around 500–650 m (Sarkar *et al.* 1996). In addition to basalts of different silica saturation, many alkaline intrusives and carbonatites, and silicic volcanic rocks (Ghose *et al.* 2017), were other magmatic products of this igneous activity (Table 1).

Both the Rajmahal and Sylhet basalts show enriched characteristics with OIB-type signatures (Storey *et al.* 1992; Kent *et al.* 1997; Ghatak & Basu 2011). Both plume and non-plume hypotheses are proposed for the origin of the Rajmahal–Sylhet Traps. Storey *et al.* (1992) classified the Rajmahal Traps as Group I and Group II, with mantle plume and continental lithosphere as the possible sources, respectively. Recent studies by Ghatak and Basu (2011, 2013) re-affirm the existence of these two geochemically distinct groups. The Group I basalts have flat REE patterns with an average $\varepsilon_{Nd_i} = +2$, and Group II are slightly LREE-enriched and have $\varepsilon_{Nd_i} = -5$. The Group I basalts are considered to have formed by large-scale melting of the Kerguelen plume head (Kent *et al.* 1997; Ghatak & Basu 2011, 2013). Crustal contamination played a significant role in modifying the basalt geochemistry (Ghatak & Basu 2011, 2013); up to 20% of lower-crust granulite contamination may explain the geochemical characteristics of the Group II basalts (Ghatak & Basu 2011, 2013). In addition to the Rajmahal and Sylhet basaltic traps, many similar-age alkaline intrusives in NE India are considered to have sourced from the heterogeneous Kerguelen plume (Ghatak & Basu 2013). The eruption of the Rajmahal and Sylhet traps is considered to have been caused by the opening of the Indian Ocean at 117 ± 2 Ma (Baksi 1995; Kent *et al.* 2002).

Deccan Traps

The Deccan Traps were erupted predominantly on the Archaean crust of the Indian Shield (Chakrabarti & Basu 2006; Ray *et al.* 2008) (Fig. 2c). The Deccan Traps presently cover an area >5×10^5 km^2 (Table 1), but originally might have erupted over an area of 1.5×10^6 km^2 (Wadia 1975). The Deccan LIP volcanism is evidently associated with extensional tectonics caused by either lithospheric stretching (e.g. Sheth 2005*a*) or plume impinging (Courtillot *et al.* 2003; Jerram & Widdowson 2005). The western margin of the Deccan LIP is considered to be a major locus of eruptive centres (e.g. Raja Rao *et al.* 1978; Beane *et al.* 1986). The thickness of the Deccan Traps decreases from west (>3000 m) to east (<100 m), controlled by the location of eruptive centres and pre-Deccan topography (Mahoney *et al.* 2000; Jay & Widdowson 2008). The Deccan LIP contains basalts of different silica-saturation, and many alkaline intrusives, rhyolites and carbonatites (Table 1). The age-corrected ε_{Nd_i} in the Deccan basalts ranges from −18 to +7 (Table 2), indicating multiple mantle sources and large-scale crustal assimilation.

Both plume and non-plume hypotheses are considered for the origin of the Deccan basalts (Courtillot *et al.* 2003; Jerram & Widdowson 2005; Sheth 2005*a*). The geochemical diversity of the Deccan basalts is modelled by the mixing of the least-contaminated primary magmas and three types of assimilates: (1) continental lithosphere mantle; (2) amphibolitic, granulitic lower crust; and (3) granitic upper crust (Mahoney *et al.* 1982; Cox & Hawkesworth 1984; Lightfoot & Hawkesworth 1988; Lightfoot *et al.* 1990; Peng *et al.* 1994; Dessai *et al.* 2008). Additionally, assimilation of shale by the Deccan basaltic magma was suggested by Chandrasekharam *et al.* (2000). Further, Sheth (2005*b*) speculated that a part of the Deccan basaltic magmas might have been produced by the melting of eclogitized ancient oceanic crust preserved in the Indian subcontinental lithosphere. Dynamic partial melting of the mantle plume and shallow magma chamber processes are also considered to be two important mechanisms that created large trace element variation in the Deccan basalts (Vijaya Kumar *et al.* 2010). Progressive lithosphere thinning may have resulted in increasing degrees of partial melting from north to south at shallower depths (Peng & Mahoney 1995; Kumar 2003), which may explain the decrease in the incidence of high-Mg basalts with time during the Deccan volcanism (Krishnamurthy *et al.* 2000). The Deccan volcanism was initiated by continental rifting during the final phases of the Gondwana fragmentation at around 65 ± 1 Ma (see Pande 2002 for a review of Deccan ages).

Compositional diversity of the Panjal, Rajmahal and Deccan basalts

The geochemical diversity of the Panjal, Rajmahal and Deccan basaltic rocks is depicted through binary variation diagrams, histograms of element composition ranges, chondrite-normalized REE plots and isotope ratios. Average major and trace element compositions of the Panjal, Rajmahal and Deccan basalts are given in Table 2. Large amounts of data

Table 2. *Average major (wt%) and trace element (ppm) compositions* of the basaltic rocks from the Panjal, Rajmahal and Deccan continental LIPs, India*

	Panjal			Rajmahal			Deccan		
	Range			Range			Range		
	Min	Max	Average	Min	Max	Average	Min	Max	Average
SiO_2	40.74	56.36	47.69	47.2	58.11	51.73	40.03	55.65	49.61
TiO_2	0.75	3.78	1.8	0.5	3.01	1.92	0.20	5.66	2.30
Al_2O_3	12.18	18.88	15.21	12.15	16.23	14.36	7.46	19.45	13.74
FeO^T	7.29	18.28	12.17	7.69	15.45	11.14	7.88	18.50	12.91
MnO	0.11	0.35	0.2	0.12	0.29	0.18	0.01	1.71	0.20
MgO	4.35	12.54	6.91	4.49	10.1	5.99	4.00	18.60	6.48
CaO	1.75	13.46	7.38	5.95	13.17	10.12	6.02	18.34	10.36
Na_2O	0.94	5.9	3.2	2.10	3.50	2.60	0.12	5.20	2.43
K_2O	0.02	4.09	0.72	0.12	3.77	0.58	0.01	3.22	0.61
P_2O_5	0.02	0.83	0.2	0.07	1.00	0.23	0.03	3.02	0.28
Sc	19.9	56.7	38.7	14.5	62.0	37.6	0.97	210	34.6
V	145	504	272	130	428	268	3.00	919	377
Cr	15.4	2080	258	45.0	785	174	5.73	4880	206
Co	34.0	75.0	48.9	33.0	112	58.2	7.00	630	54.1
Ni	6.29	430	107	17.7	220	57.4	8.00	2200	98.3
Cu	7.89	210	70.3	15.0	298	111	10.0	3008	219
Zn	53.3	315	101	72.5	142	97.4	5.18	1784	129
Ga	6.00	26.0	17.6	19.0	29.0	22.7	4.00	120	22.2
Rb	0.16	174	24.9	0.30	44.0	14.0	0.17	204	17.11
Sr	36.0	667	196	135	1100	280	1.37	2456	269
Y	15.0	60.0	30.3	3.02	54.0	34.2	3.30	250	33.9
Zr	32.0	392	132	0.17	440	137	14.0	488	159
Nb	2.20	30.0	13.3	2.10	40.0	8.13	2.00	158	16.7
Ba	4.00	1992	358	44.4	767	208	0.02	7010	204
Hf	0.60	7.00	3.3	1.48	8.20	3.75	0.93	8.12	4.12
Ta	0.18	10.1	1.07	0.15	2.00	0.65	0.17	17.8	1.54
Pb				1.20	8.70	3.98	0.47	398	11.6
Th	9.13	9.13	3.74	0.39	6.00	1.88	0.32	48.0	3.37
U	1.61	1.61	0.64	0.08	0.71	0.36	0.02	25.1	0.79
La	3.88	35.0	17.1	3.90	54.6	15.5	3.10	124	19.8
Ce	8.85	80.4	40.1	4.09	111	30.3	6.81	225	43.8
Pr	1.28	9.25	4.72	2.10	13.9	4.35	0.95	40.2	5.28
Nd	2.50	42.0	19.4	5.50	59.8	18.2	4.56	373	23.1
Sm	1.10	10.3	4.85	2.50	13.2	5.06	1.37	23.8	5.78
Eu	0.55	3.50	1.52	0.60	4.30	1.75	0.65	19.2	1.99
Gd	2.00	11.3	5.33	2.60	9.90	6.01	0.83	17.9	6.29
Tb	0.44	1.51	0.86	0.56	1.40	1.00	0.28	16.4	1.13
Dy	2.83	10.3	5.46	2.68	8.54	6.20	0.85	14.4	5.56
Ho	0.58	1.97	1.15	0.83	1.70	1.30	0.28	2.23	1.08
Er	1.68	5.36	3.19	1.33	4.70	3.48	0.33	13.4	2.99
Tm	0.24	0.80	0.47	0.30	0.66	0.51	0.10	12.6	0.61
Yb	1.50	5.40	2.93	1.53	4.23	3.01	0.30	14.4	2.73
Lu	0.22	0.91	0.44	0.24	0.67	0.44	0.08	1.00	0.41
ε_{Nd_i}	−8.05	3.98	−2.88	−7.71	2.78	−2.39	−18.04	7.31	−1.46

*Data sources are given in Supplementary material and explanation to Figure 3.
Number of samples has a large range depending on the element considered.

are available for the Deccan basalts compared to the Panjal and Rajmahal basalts (Fig. 3). Average abundances of Si, Ti and Ca are lower, and Rb and Ba are higher, in the Panjal basalts compared to the Rajmahal and Deccan basalts (Table 2). V, Cu and Zn are distinctly higher in the Deccan basalts (Table 2). All three provinces contain positive and negative ε_{Nd_i} basalts, with the Deccan basalts showing a larger range (Table 2). Basalts from all the three provinces show comparable geochemical trends in the CaO–CaO/Al_2O_3 diagram, especially at lower CaO content. In the Panjal and Rajmahal provinces, CaO/Al_2O_3 and CaO show a well-defined positive relationship indicating plagioclase-dominated fractional crystallization (Fig. 3a, b). In the Deccan province, predominant mafic mineral fraction in the early stages of differentiation is depicted by a systematic decrease in the CaO/Al_2O_3 ratio at near-constant CaO content up until the CaO/Al_2O_3 ratio equals 1. Once plagioclase becomes a significant fractionating phase, CaO/Al_2O_3 shows a collinear decrease with CaO (Fig. 3c). In all three provinces, plagioclase becomes saturated when the CaO/Al_2O_3 ratio is between 0.8 and 1. Some of the high Al-basalts of the Deccan (Table 2) may represent plagioclase-laden melts (Vijaya Kumar *et al.* 2010).

Figure 4a–c illustrates the histogram of ΔNb (Fitton 1995: 1.74 + log(Nb/Y) − 1.92 log(Zr/Y)) variation in the Phanerozoic continental LIPs of India. All three provinces contain basalts with both positive and negative ΔNb values. The Panjal has 76% negative ΔNb basalts and 24% positive ΔNb basalts; the Rajmahal has 65% negative ΔNb basalts and 35% positive ΔNb basalts; and the Deccan basalts have 36% negative ΔNb basalts and 64% positive ΔNb basalts. In the oceanic magmatism, positive ΔNb values in the basalts are associated with OIB-type mantle sources, whereas negative ΔNb is related to N-MORB (normal-MORB)-type depleted mantle (Fitton 2007). However, the complex history of subcontinental lithosphere in terms of melt addition and depletion results in a wide range of ΔNb values from positive to negative. Further, plume-derived melts may produce negative ΔNb after crustal contamination. However, a wide range in the ΔNb values do suggest more than one mantle source for the basalts of the three continental LIPs.

Figure 4d–f demonstrates the range of $(La/Lu)_N$ values in the Panjal, Rajmahal and Deccan basalts. The $(La/Lu)_N$ ratios in the melts suggest a degree of partial melting of the source and the presence or absence of garnet in the melting regime. The Panjal basalts have $(La/Lu)_N$ ratios between 1 and 6, suggesting that the degree of melting is >5% and melts were derived from spinel peridotite (see Shellnutt *et al.* 2014, 2015*a*). The Deccan basalts have a wide range of $(La/Lu)_N$ ratios (from 1 to >70 × chondrite), suggesting the presence of garnet in the mantle sources and also, possibly, that the mantle sources were themselves enriched. The Rajmahal basalts also show higher $(La/Lu)_N$ ratios (up to 17 × chondrite) than those in Panjal but much less compared to those in Deccan basalts.

In the chondrite-normalized plot (Fig. 5), representative REE patterns covering the entire range of each province are illustrated. The Panjal basalts show moderately LREE-enriched (La = 100 × chondrite) and near-flat HREE patterns with minor negative Eu anomalies. There are crossings of the REE patterns, especially at the HREE segment. Significantly, some of the basalts show positive Ce anomalies; only one sample has a negative Ce anomaly. The Rajmahal basalts show similar moderately LREE-enriched REE patterns with flat to negative Eu anomalies. Some of the samples show much steeper patterns. The Rajmahal basalts also show crossing of the REE patterns. The Deccan basalts show the largest variation in the normalized REE plot, from LREE-enriched (La = 400 × chondrite) to near-flat patterns. The crossing of the REE patterns is prominently shown by the Deccan basalts.

The crossing of REE patterns is interpreted in terms of either the dynamic partial melting of homogeneous mantle or derivation from heterogeneous mantle sources (Vijaya Kumar & Rathna 2014 and references therein). In a rising mantle diapir, the initial stages of melting tend to be equilibrium batch melting, but once the mantle residue develops critical porosity, some amount of partial melt will be retained in the pores of residue and disequilibrium continuous melting will set in (Prinzhofer & Allegre 1985). The diapir is supposed to maintain its chemical homogeneity while liquids are extracted (Albarède 1996). A composite of the melts produced by both batch and continuous melting of a chemically homogenous source is what constitutes dynamic partial melts. The dynamic melting of the mantle (batch + continuous melting) produces LREE-enriched to LREE-depleted basaltic melts with crossing REE in the normalized plots. Alternatively, different degrees of partial melts derived from chemically distinct sources also result in the crossing REE patterns. Although dynamic melting is suggested for a small set of Deccan basalts drawn from a semi-continuous sequence (Vijaya Kumar *et al.* 2010), we consider that the complete geochemical diversity of the three continental LIPs need more than one mantle source feeding the magmas (see the Discussion).

Although the TiO_2 content in the oceanic basaltic melts has been used as a proxy for mantle sources (Chayes 1965), it is difficult to establish such a relationship for the basalts erupted on the continents. The OIB-type melts erupted onto the continents are supposed to contain a higher TiO_2 content than the lithosphere-derived melts (Xiao *et al.* 2004). But identification of such distinctions is rendered difficult by plume/asthenosphere–lithosphere interactions and crustal contamination (Wittke & Mack 1993; Macdonald *et al.* 2001; Späth *et al.* 2001; Xu *et al.* 2005; Tang *et al.* 2006). It is also suggested that high- and low-Ti basalts need not necessarily reflect two distinct sources (e.g. plume v. continental lithosphere). They can also be formed by higher and lower degree melting of a homogeneous source correspondingly at higher and lower mantle temperatures (MacGregor 1969; Fodor 1987; Xu *et al.*

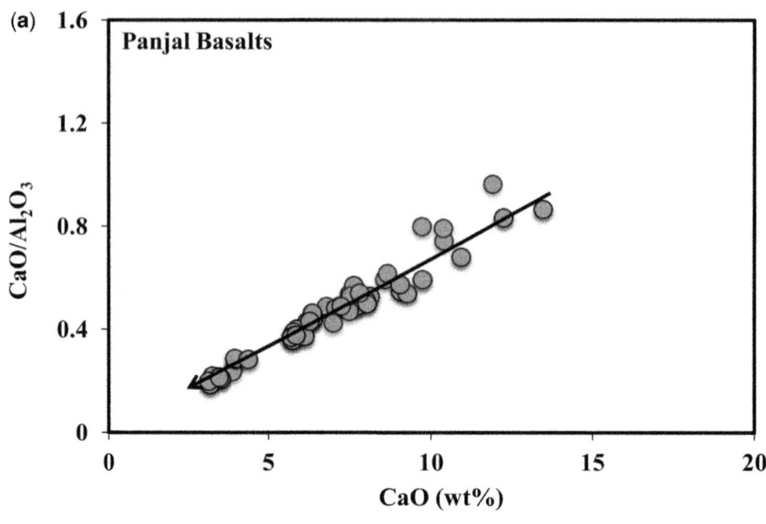

(a)
Panjal Basalts
CaO/Al_2O_3
CaO (wt%)
0
0.4
0.8
1.2
1.6
0
5
10
15
20

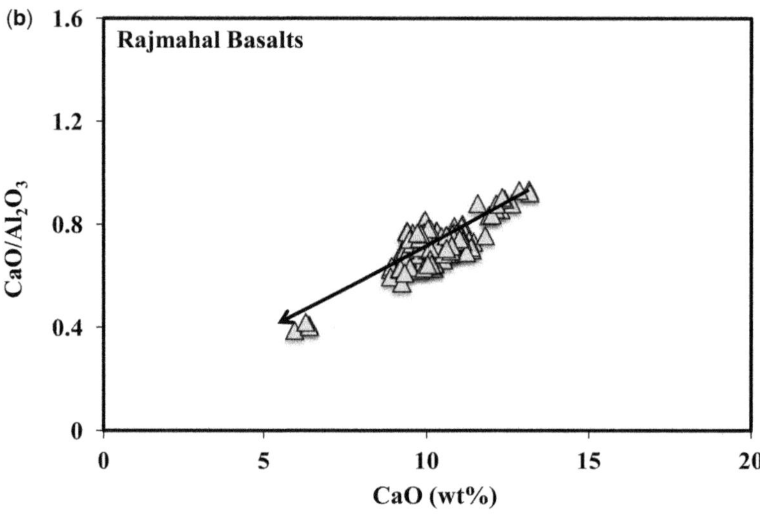

(b)
Rajmahal Basalts
CaO/Al_2O_3
CaO (wt%)
0
0.4
0.8
1.2
1.6
0
5
10
15
20

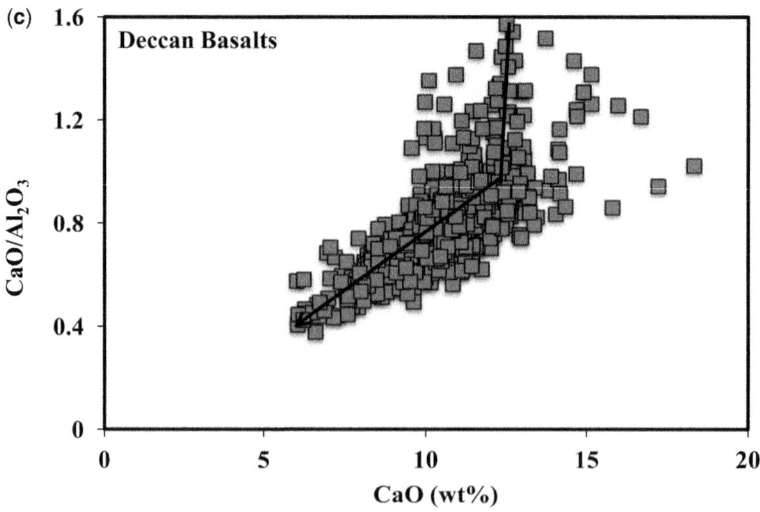

(c)
Deccan Basalts
CaO/Al_2O_3
CaO (wt%)
0
0.4
0.8
1.2
1.6
0
5
10
15
20

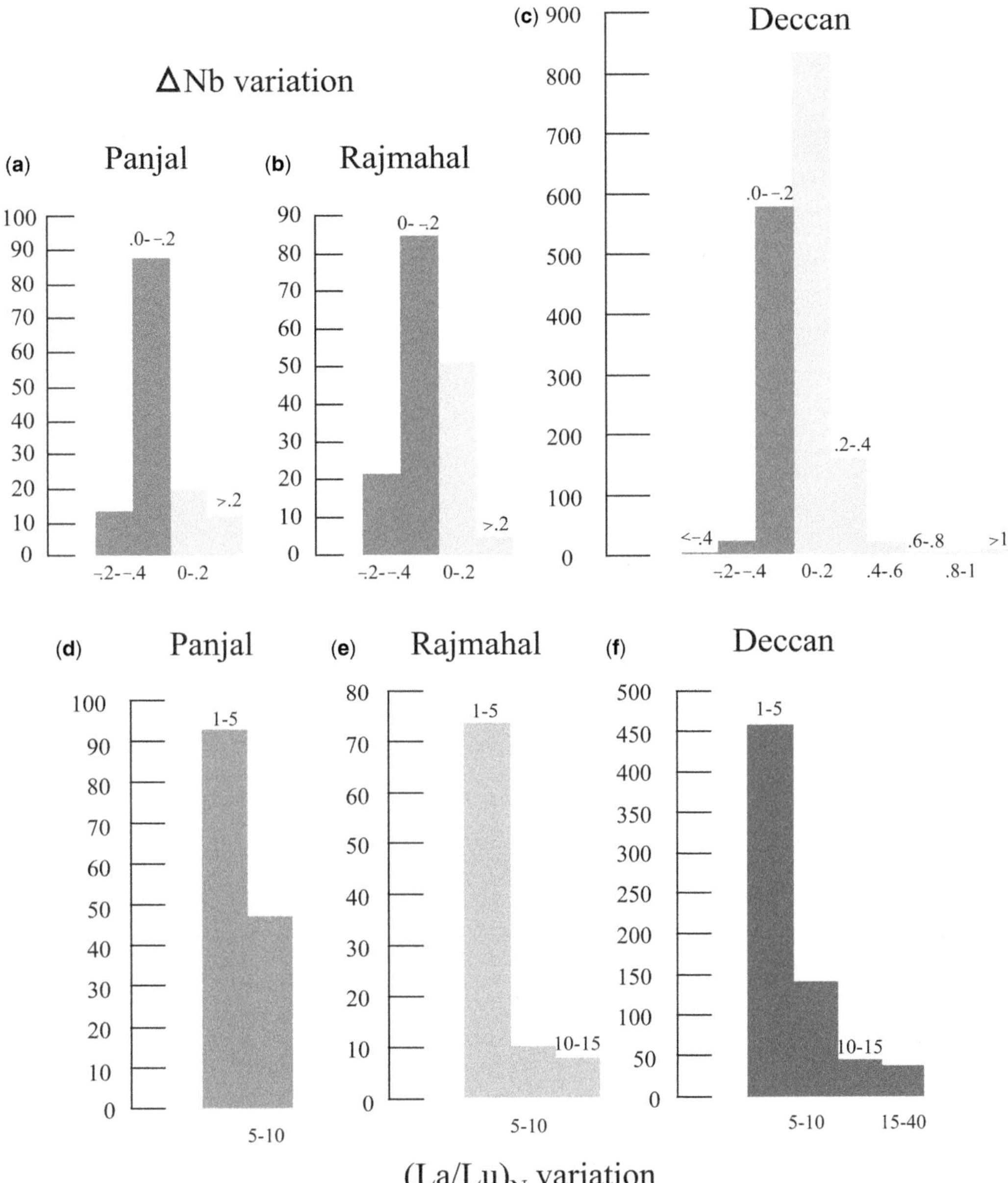

Fig. 4. The histograms showing the (**a**)–(**c**) ΔNb and (**d**)–(**f**) $(La/Lu)_N$ variations in the Panjal, Rajmahal and Deccan basalts. N represents normalization by chondrite.

Fig. 3. (**a**)–(**c**) CaO v. CaO/Al_2O_3 variations in the Panjal, Rajmahal and Deccan basalts. Geochemical data used in this and all other figures were mainly extracted from the GEOROC database maintained by the Max Planck Institute für Chemie in Mainz, Germany (http://georoc.mpch-mainz.gwdg.de/georoc/). Additional data were taken from Shellnutt *et al.* (2014) for Panjal, Ghatak and Basu (2013) for Rajmahal, and Vijaya Kumar *et al.* (2010) and Nagaraju (2012) for Deccan basalts. Samples with MgO >4% are only considered in the present study. Picritic basalts and picrites from the Deccan province (MgO >14%) are also included in the data analysis. The trends shown represent lines of best fit.

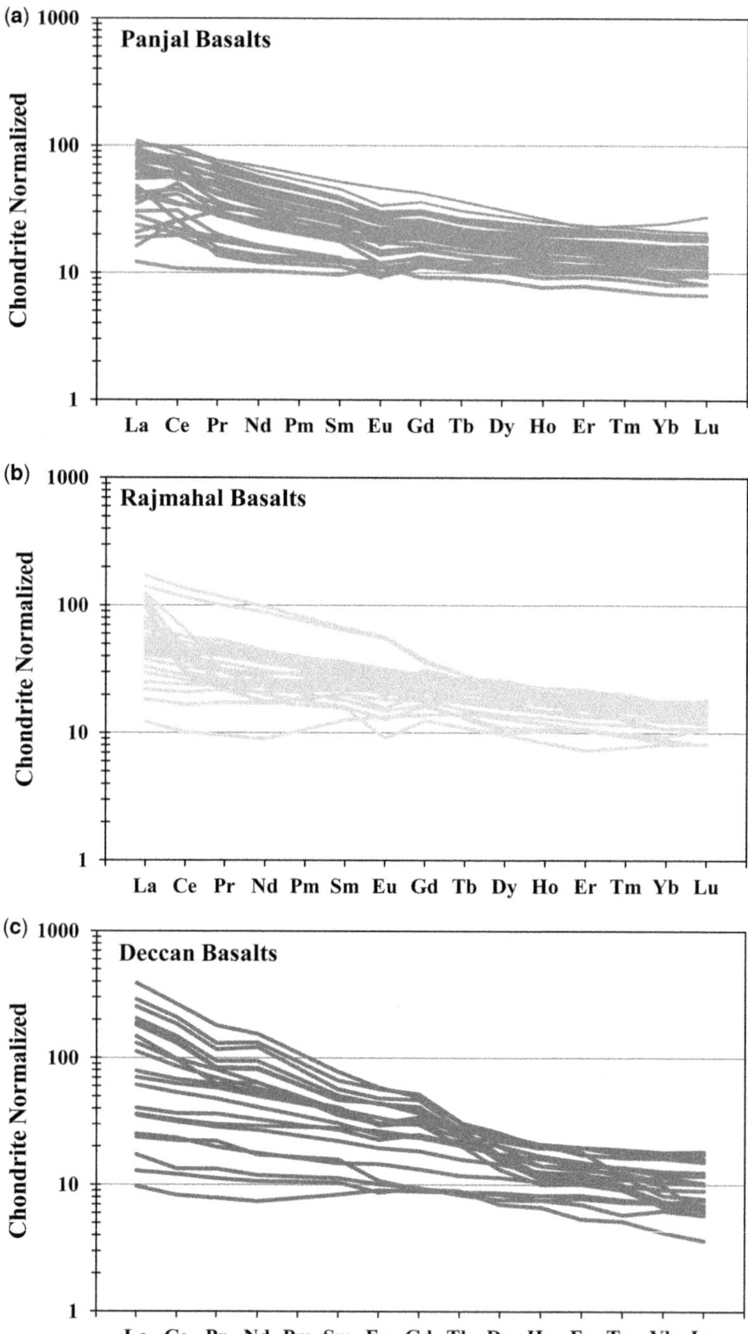

Fig. 5. Chondrite-normalized REE patterns for the Panjal, Rajmahal and Deccan basalts. Normalizing values are after Hanson (1980).

2001). Both high- and low-Ti basalts from the Phanerozoic continental LIPs of India do not show any perceptible relationship with ε_{Nd_i} (Fig. 6a). Both positive ε_{Nd_i} and negative ε_{Nd_i} basalts contain low- and high-Ti basalts (Fig. 6a). Using TiO_2 as a geochemical tool to decipher multiple mantle sources

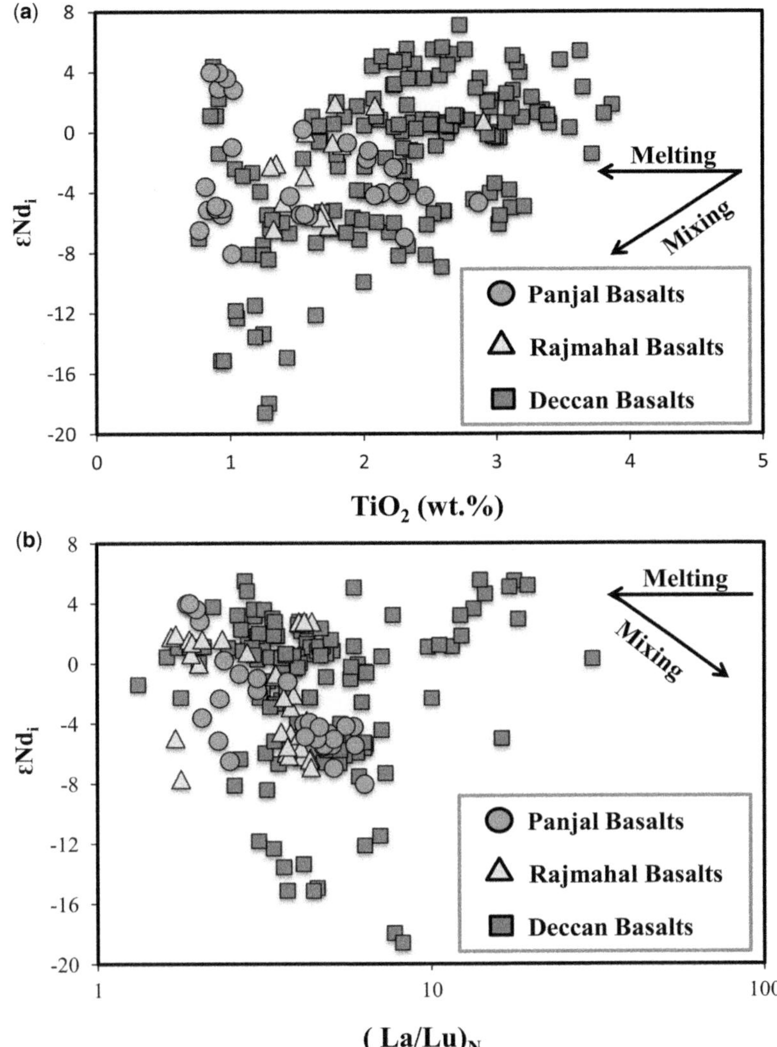

Fig. 6. (**a**) TiO_2 v. ε_{Nd_i} and (**b**) $(La/Lu)_N$ v. ε_{Nd_i} variation in the Panjal, Rajmahal and Deccan basalts. Vectors for melting and mixing are shown.

for continental basalts may not be straightforward (see MacGregor 1969; Fodor 1987). We suggest that the TiO_2 variation in the Indian continental LIPs may be an outcome of melting and fractionation mechanisms.

The $(La/Lu)_N$ ratio, an indicator of the degree of melting, also shows a complex relationship with ε_{Nd_i}, especially in the Deccan basalts (Fig. 6b). The Panjal and Rajmahal traps show an apparent negative relationship between $(La/Lu)_N$ and ε_{Nd_i} (Fig. 6b). The basalts with lower $(La/Lu)_N$ ratios, and, consequently, higher degree melts, show positive ε_{Nd_i} values, and the samples with higher $(La/Lu)_N$ ratios show negative ε_{Nd_i} values. This linear trend can be interpreted as mixing between melts derived from two distinct mantle sources or assimilation of continental crust during fractional crystallization. The variation in ε_{Nd_i} at constant $(La/Lu)_N$ ratios, shown by the Deccan basalts, cannot be explained by melting or crystallization mechanisms that do not change the ε_{Nd_i} values in the melts. This variation can only be achieved by the mixing of different mantle sources or mantle and continental crust. It looks as if higher $(La/Lu)_N$ ratios seem to be prominent in the basalts with positive ε_{Nd_i} values (Fig. 6b). This could be due to lower degrees of melting of the sublithospheric mantle within the garnet stability field (see the Discussion).

Discussion

In this section, we evaluate the large geochemical variation shown by the Panjal, Rajmahal and Deccan basalts in terms of variable degrees of partial melting of the mantle source, fractional crystallization and mantle source characteristics. Before espousing a quantitative assessment of the processes, we tested the geochemical data for post-magmatic mobility of the elements, if any.

Potential problems with elemental mobility

It is well known that the geochemical mobility of elements is commonly associated with alteration and metamorphism of the rocks of basaltic composition (Sun & Nesbitt 1978; Nyström 1984; Thompson 1991). To evaluate the post-magmatic mobility of the elements, if any, we used Zr as a monitor of the elemental mobility (see Pearce & Cann 1973; Weaver & Tarney 1981; Sheraton 1984; Vijaya Kumar & Rathna 2008) and plotted two high field strength elements (HFSEs), represented by Nb and La, and two large-ion lithophile elements (LILEs), represented by Sr and Rb, to assess the relative mobility of elements (Fig. 7). It is generally assumed that the elements that behave incompatibly during mantle melting and fractionation of basaltic magmas share positive relationships with Zr (Vijaya Kumar *et al.* 2006 and references therein). Both Nb and La share strong positive correlations with Zr (Fig. 7a–f), suggesting that they retained their original magmatic relationships; some of the Deccan basalts have a lower Nb content (Fig. 7c), which we interpret as due to the involvement of sediments in their genesis (see the discussion below) but not due to post-magmatic processes. Sr and Rb show dispersion, especially in the Panjal and Rajmahal basalts

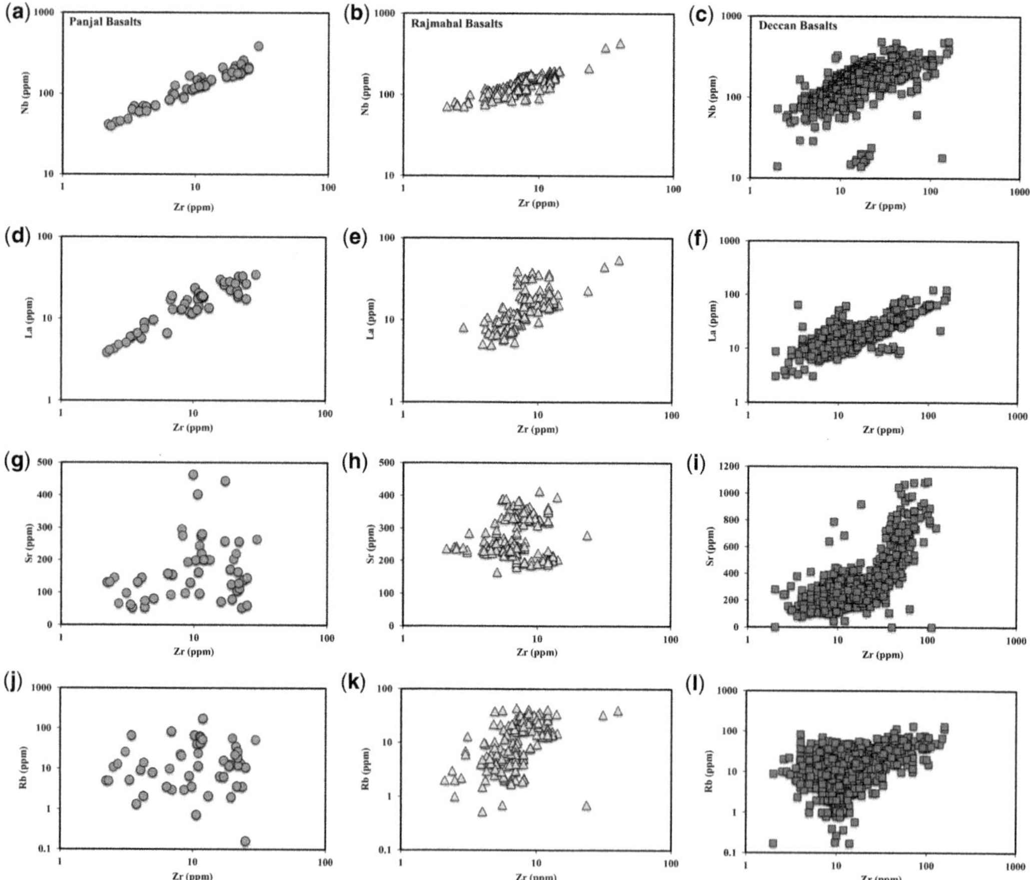

Fig. 7. Zr v. incompatible element variation in the Panjal, Rajmahal and Deccan basalts. Strong positive correlations for Nb and La suggest little mobility of HFS elements. In contrast, Sr and Rb (LIL elements) do show some scatter in the plots, which possibly reflects post-magmatic mobility of these elements. For an explanation, see the text.

(Fig. 7g–l). Based on the Zr trace element relationship, we presume that LILEs may have undergone different levels of post-magmatic mobilization. Therefore, we utilize only HFSE and Nd isotopic data in the petrogenetic modelling of the basalts of the three provinces. Due to the dispersion shown by Rb and Sr, we did not utilize Sr isotopic data to evaluate the petrogenesis.

Degree of partial melting and crystallization

To evaluate relative roles of partial melting and fractional crystallization on the geochemical variation of the Panjal, Rajmahal and Deccan basalts, we constructed a Ni v. Ce plot (Fig. 8). Nickel (Ni) preferentially partitions into residual olivine during partial melting of the mantle sources; as a result, the concentration of Ni in the melts would vary little for different extents of partial melting (Fig. 8). Conversely, the elements that preferentially partition into the liquid (e.g. Ce) would be very sensitive to different degrees of melting (Fig. 8). This contrasting behaviour of compatible and incompatible elements is utilized to construct the process identification diagrams (Allègre & Minster 1978), and is extensively used to model melting and crystallization processes. The partial melting curve was constructed assuming a mantle source with 1900 ppm Ni and 1.6 ppm Ce using the batch melting equation. We also constructed fractional (Neumann *et al.* 1954) and assimilation fractional crystallization (DePaolo 1981) curves. Details of calculations are given in the explanation to Figure 8. Melts that formed by different extents of melting plot parallel to the Ce axis (i.e. very high Ce variation), whereas melts that are related by different degrees of fractional crystallization plot parallel to Ni axis (i.e. very high Ni variation) (Fig. 8).

The Panjal primitive basalts may have formed by 5–20% melting of the mantle sources (Fig. 8a). Low-degree melts are not recorded in the Panjal basalts. The Rajmahal basalts show more evolved compositions and formed by 10–20% melting of the mantle sources (Fig. 8b). Only two basalts may have formed by lower degrees of melting (Fig. 8b). The Panjal and Rajmahal basalts show geochemical signatures of assimilation fractional crystallization of the parental liquids (Fig. 8a). The Deccan primary melts are formed by 1–20% of melting of the mantle sources (Fig. 8c). The geochemical variation in the Deccan basalts can be modelled by different degrees of melting of the mantle, followed by fractional crystallization (FC) and assimilation fractional crystallization (AFC) of the primary liquids (Fig. 8c). As has been established in many flood basalt provinces (Vijaya Kumar *et al.* 2010 and references therein), most of the erupted basalts in the Panjal, Rajmahal and Deccan provinces depict fractionated and assimilated evolved liquid compositions (Fig. 8).

Mantle source characteristics

To identify the mantle sources, we have filtered the GEOROC data on the Panjal, Rajmahal and Deccan basalts for which Nd isotopic data are available. When these data are plotted in the La/Nb v. Zr/Nb plot, two distinct trends are evident in all three basaltic provinces (Fig. 9). We have categorized them as Type I and Type II basalts. The Type I basalts show a steeper slope and higher Zr/Nb values at given La/Nb values compared to Type II basalts (Fig. 9). The Type II samples have a relatively more gentle slope and much higher La/Nb values compared to the Type I basalts. This pattern is apparent in basalts of all three provinces (Fig. 9). The Deccan Type II basalts also have higher Zr/Nb ratios (Fig. 9c). The bulk-rock La/Nb ratio in the Type I basalts is similar to that in oceanic basalts (0.6–1.2), and the average La/Nb in the Type II basalts is similar to that in the continental basalts (La/Nb = >2: Plank & Langmuir 1998; Barth *et al.* 2000). In the following discussion, we have combined the data of all three provinces to draw general conclusions on the mantle sources for the continental LIPs in general, and the Phanerozoic Panjal, Rajmahal and Deccan provinces in particular.

The Type I and Type II basalts categorized on the basis of Zr/Nb and La/Nb relationships show distinct Nd isotopic compositions. The Type I basalts have ε_{Nd_i} values between 8 and −3, and the Type II basalts between −20 and −3. This distinction between Type I and Type II basalts is clearly brought out in a plot between Th/Nb and ε_{Nd_i} (Fig. 10). The plotted parameters show a systematic negative correlation. If we assume a primitive sublithosphere mantle with ε_{Nd_i} = +6 and Th/Nb ratio of 0.07 (a typical enriched-MORB (E-MORB) composition), and GLOSS (global subducting sediment: Plank & Langmuir 1998) with $\varepsilon_{Nd_i} = -12$ and a Th/Nb ratio of 0.77, then the spread of the samples can be explained by mixing between the E-MORB-type mantle and GLOSS (Fig. 10). Most of the Type II basalts fall in two clusters (cluster 1: ε_{Nd_i} between −6 and −4; cluster 2: ε_{Nd_i} between −14 and −12) with a large variation in Th/Nb ratios (Fig. 10). Basalts with a restricted variation in ε_{Nd_i} and large variation in Th/Nb in each cluster indicate a variable degree melting of the mantle. Both mixing and melting trajectories are shown in Figure 10

The majority of Type I basalts fall in the field for sublithosphere mantle-derived basalts, with some of the samples falling above the field within the zone for active continental margin basalts in the plot defined by Th/Yb v. Ta/Yb variation (Fig. 11a) (Pearce 2003). This geochemical variation of progressive increase in the Th/Yb and Ta/Yb from E-MORB to OIB compositions can be explained by a systematic decrease in the degrees of melting

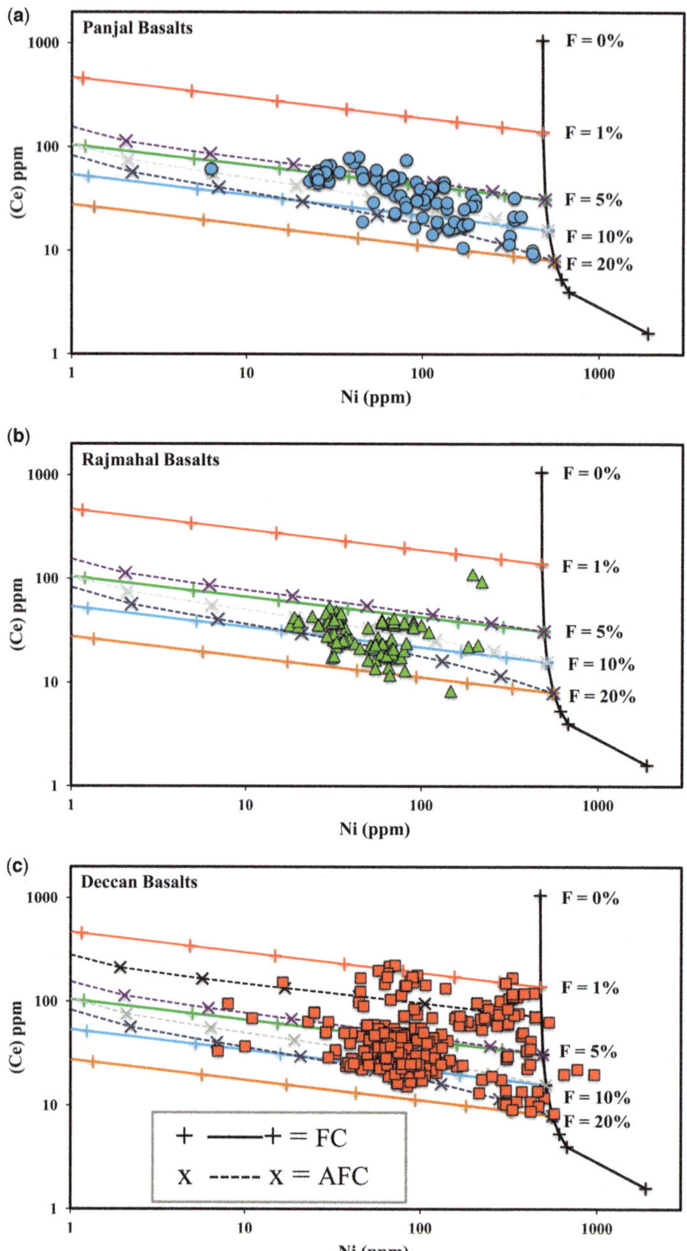

Fig. 8. Ni v. Ce variation in the Panjal, Rajmahal and Deccan basalts. Partial melting curve for different degrees of melting are calculated using modal batch melting equation. The mantle source is assumed to contain 1900 ppm Ni and 1.6 ppm Ce with bulk distribution coefficients (*D*s) of 4 and 0.0015 for Ni and Ce, respectively. Liquid evolution curves were also calculated for both fractional crystallization (FC: Neumann *et al.* 1954) and assimilation fractional crystallization (AFC: DePaolo 1981). Each tick on the melting and fractionation curves indicates 10% melting/fractionation. An assimilate with 20 ppm Ni and 100 ppm Ce is used in the AFC calculations. Ratio of the rate of assimilation to the rate of fractional crystallization (*r*) is considered as 0.2. Bulk distribution coefficients used in the FC and AFC calculations are 6 and 0.02 for Ni and Ce, respectively. Geochemical variation in the Panjal, Rajmahal and Deccan basalts is better explained by AFC. Equations for batch melting, fractional crystallization and assimilation fractional crystallization are after Schilling and Winchester (1967), Neumann *et al.* (1954) and DePaolo (1981) respectively. For an explanation, see the text.

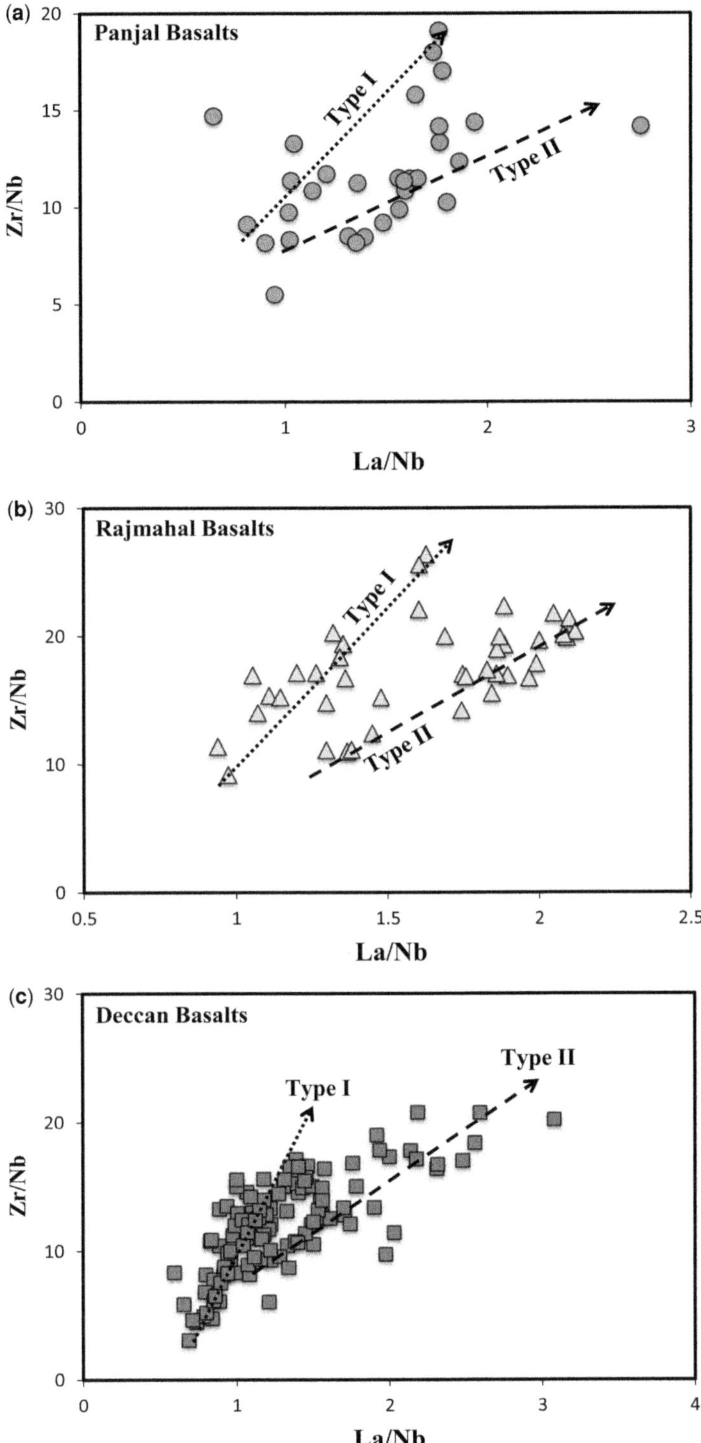

Fig. 9. La/Nb v. Zr/Nb variation in the (**a**) Panjal, (**b**) Rajmahal and (**c**) Deccan basalts. The basalts for which Nd isotopic data are available are only plotted in this diagram. Two distinct types (designated as Type I and Type II) can be categorized based on distinct variation trends. For an explanation, see the text.

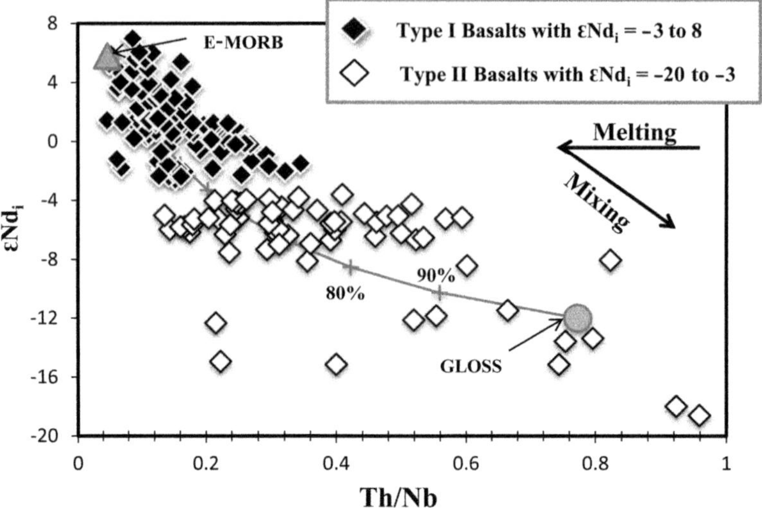

Fig. 10. Th/Nb v. ε_{Nd_i} variation in the Type I and Type II basalts from the Phanerozoic continental LIPs of India. Type I basalts have ε_{Nd_i} values between −3 and +8, and Type II basalts have ε_{Nd_i} values between −20 and −3. The binary mixing trajectory is calculated for assumed E-MORB (Sun & McDonough 1989) (Th = 1.8 ppm; Nb = 34 ppm; ε_{Nd_i} = 5.38) and GLOSS (Plank & Langmuir 1998) (Th = 6.91 ppm; Nb = 8.94; ε_{Nd_i} = −12) compositions. Each tick on the mixing curve indicates 10% mixing. Vectors for melting and mixing are shown.

and, possibly, an increase in the depths of melting. Some of the Type I basalts plot above the field for melts derived from sublithosphere mantle sources (Fig. 11a). Such a variation is interpreted in terms of a subduction overprinting on original sublithospheric mantle sources (Pearce 2003) or crustal contamination of deeper mantle-derived magmas (Arculus 1987). The Type II basalts plot exclusively above the field for melts derived from the sublithospheric mantle sources (Fig. 11b). The Type II basalts could represent either contaminated products of Type I basalts or they may have derived from the subcontinental lithospheric sources. Both Type I basalts with higher Th/Yb ratios and almost all the Type II basalts plot proximal to the GLOSS but not the UCC (upper continental crust: Rudnick & Gao 2003) (Fig. 11). It is clear that none of the basalts from the three large igneous provinces illustrate true N-MORB signatures (Fig. 11). We suggest that the depleted MORB mantle (DMM) source is ruled out for the basalts of the Phanerozoic LIPs of India.

The relative roles of asthenosphere/plume and lithosphere in the building of the Phanerozoic continental LIPs of India are depicted through the trace element signatures. A plot between Th/Nb and Zr/Nb demonstrates that the basalts with ε_{Nd_i} values of −3 to +8 have lower Th/Nb (<0.15) ratios (Fig. 12a) – geochemical characteristics similar to sublithospheric mantle sources, as indicated by the composition of OIB, E-MORB and PM (Fig. 12a). Typical depleted-mantle-derived N-MORB-type melts are absent in the primitive mantle (PM) Phanerozoic continental LIPs of India. We suggest that the Type II basalts with Th/Nb ratios >0.4 (Fig. 12a) represent melts exclusively derived from subcontinental lithospheric sources. Some of the lithospheric-derived basalts have chemical characteristic similar to GLOSS (Fig. 12a). Type I and Type II basalts with Th/Nb ratios between 0.15 and 0.4 overlap, which we interpret in terms of lithosphere–asthenosphere/plume interaction. Binary mixing trajectories between end-member compositions suggest that the upper continental crust did not contribute to the geochemical diversity of the Panjal, Rajmahal and Deccan LIPs (Fig. 12a).

Similar relationships are reinforced by the Zr/Nb v. Th/Ta plot (Fig. 12b). Type I basalts with Th/Ta ratios <2.4 are exclusively derived from the asthenosphere/plume sources (Fig. 12b). Once again, the absence of basalts with typical MORB signatures is reaffirmed by Zr/Nb and Th/Ta relationships. The Type II basalts with Th/Ta ratios of >4.6 display subcontinental lithosphere signatures (Fig. 12b); while the Type I and Type II basalts with Th/Ta ratios between 2.4 and 4.6 suggest asthenosphere/plume and lithosphere interaction. The role of ancient sediments in the geochemical diversity of continental LIP basalts is once again emphasized by the compositional similarity between Type II basalts and GLOSS (Fig. 12b). The geochemical diversity in the three continental LIPs seems to be

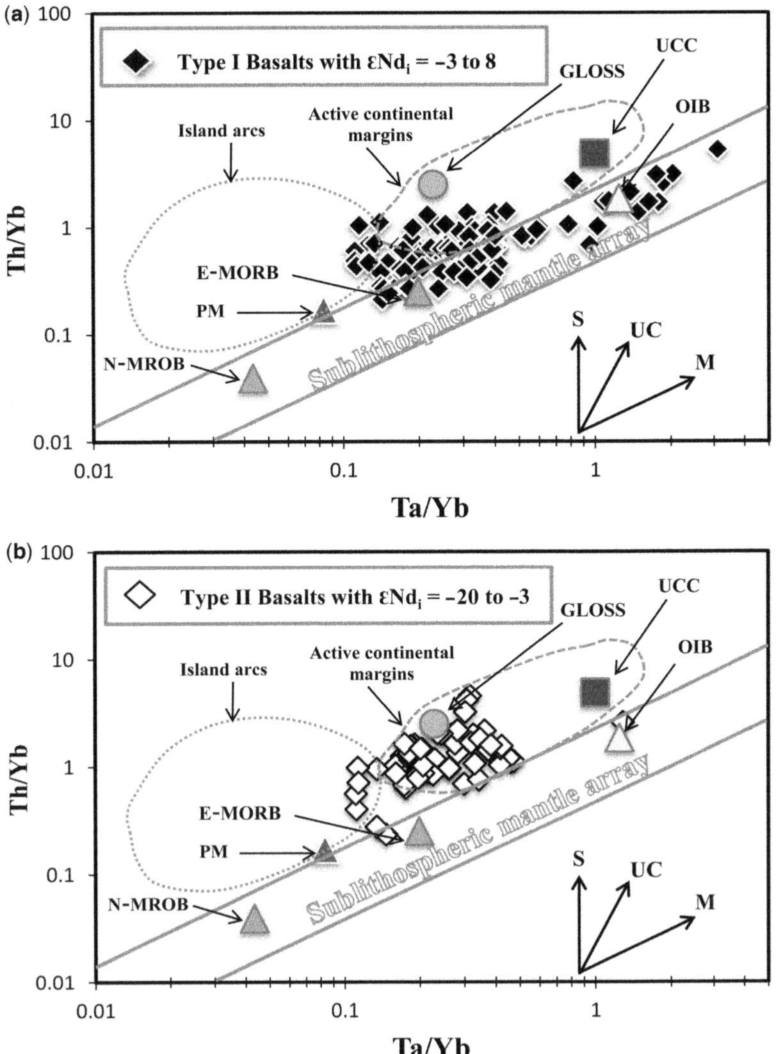

Fig. 11. Ta/Yb v. Th/Yb variation in the Type I and Type II basalts from the Phanerozoic continental LIPs of India. Fields for sublithospheric mantle, island arc and active continental margin are after Pearce (2003). The compositions for N-MORB, E-MORB, PM and OIB are after Sun & McDonough (1989); composition of GLOSS is after Plank & Langmuir (1998); composition of UCC is after Rudnick & Gao (2003). Vectors for the addition of subducting component (S), upper crustal contamination (UC) and partial melting (M) are shown.

an outcome of contributions from both sublithospheric (plume/asthenosphere) and SCLM sources. It is evident that the continental lithosphere has supplied significant amounts of material in building the continental LIP provinces. However, a fundamental question is whether this material is contributed through contamination of the asthenosphere/plume-derived melts or by partial melting of the lithosphere sources themselves. The mixing trajectories between OIB/E-MORB and GLOSS indicate that an unrealistic fraction of the latter is required to explain the geochemistry of some of the Type II basalts (Fig. 12b). The ε_{Nd_i} value of −20 to −3 for the Type II basalts, distinct trace element signatures (Figs 10–12) and their restriction to the active continental margin field (Fig. 11b) suggest that partial melting of the lithospheric sources is a greater possibility. Therefore, we suggest that the Type II basalts do not represent contaminated equivalents of Type I basalts but are, instead, the partial melts of subcontinental lithosphere. The extensive presence of lithosphere-derived basalts in the Phanerozoic continental LIPs

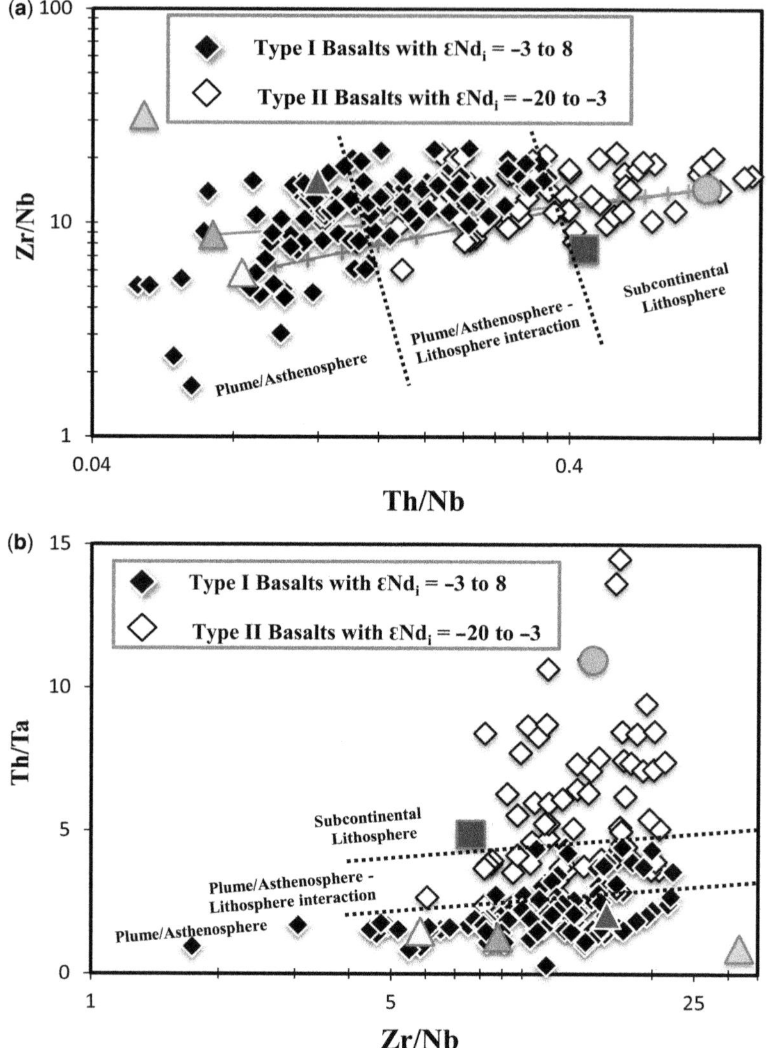

Fig. 12. (**a**) Th/Nb v. Zr/Nb and (**b**) Zr/Nb v. Th/Ta variations in the Type I and Type II basalts from the Phanerozoic continental LIPs of India. Binary mixing trajectories are calculated between E-MORB (Th = 0.6 ppm; Nb = 8.3 ppm; Zr = 73 ppm; Ta = 0.47 ppm), OIB (Th = 4 ppm; Nb = 48 ppm; Zr = 73 ppm; Ta = 2.7 ppm) and GLOSS (Th = 6.91 ppm; Nb = 8.94; Zr = 280 ppm; Ta = 0.63 ppm) compositions. Each tick on the mixing curve indicates 10% mixing. Symbols and references for mantle and crustal reservoirs are as in Figure 11.

suggests significant thermochemical erosion of the Indian subcontinental lithosphere.

Ghatak & Basu (2011, 2013) suggested that the isotope and trace element geochemistry of the Rajmahal–Sylhet Traps can be explained by the assimilation of Proterozoic meta-sediments of the Eastern Ghats Belt. The Deccan basalt province occupies the central portion of cratonic India, and the Panjal and Rajmahal volcanic provinces occur along the peripheral portions of the craton (Fig. 2). There is no reason why these spatially and temporally distinct provinces should be contaminated by the same assimilate. A more plausible mechanism could be presence of assimilate in the lithospheric source itself. It is pertinent to note that the Type II basalts show compositional similarities with GLOSS but not with UCC. The subcontinental lithosphere is capable of preserving the ancient subduction signatures (see Wang *et al.* 2006). The ancient sediments can be added to the SCLM either by subduction-related processes or by foundering lower-crustal meta-sediments into the lithosphere. The subduction-modified lithospheric mantle may provide a source for subsequent continental LIP magmatism either immediately after subduction or even after a prolonged time gap (Johnson *et al.* 1978; Goodenough *et al.* 2002; Vijaya Kumar & Rathna 2008). Understanding the long-term retention of subduction-related signatures by the continental lithosphere is imperative to evaluate the geochemical diversity of the continental LIP magmatism.

We have further used ΔNb to characterize the contribution of continental lithosphere and asthenosphere/plume in the continental LIP magmatism. The ΔNb value (Fitton 1995) is considered to be a fundamental source characteristic and is not affected by different extents of melting. Unless there is high proportion of continental crustal contamination or amphibole fractionation, ΔNb is also independent of magma chamber processes. Therefore, basaltic melts are expected to have values either greater than zero or less than zero depending on the nature of the source from which they are derived. The negative ΔNb anomalies characterize N-MORB and island arc basalt (IAB) (and, by implication, the continental lithosphere) mantle reservoirs, whereas OIB and E-MORB reservoirs have positive ΔNb values. If the melts are tapped from more than one source, then the basalts are expected to show a range of ΔNb values from positive to negative.

The dominance of negative ΔNb basalts with lower $(La/Lu)_N$ ratios (Fig. 4) suggest that the Panjal basalts were derived from shallow lithospheric mantle sources (see Shellnutt *et al.* 2014). Much larger range $(La/Lu)_N$ ratios and dominance of positive ΔNb basalts (Fig. 4) suggest that sublithospheric sources (plume/asthenosphere) played a significant role in building the Deccan volcanic province (Peng & Mahoney 1995; Courtillot *et al.* 2003; Jerram & Widdowson 2005). The geochemical variation in the Rajmahal basalts in terms of ΔNb values and $(La/Lu)_N$ ratios also suggest they may have been derived from more than one source. Plume/asthenosphere seem to have contributed a larger amount of material to the Deccan continental LIP than the lithospheric mantle, whereas the opposite is the case for Panjal where the contribution from the lithospheric sources far outweighs the sublithospheric sources (Shellnutt *et al.* 2014). In the Rajmahal, lithospheric sources also seem to have contributed significantly to the flood basalt volcanism.

The geochemical complexity of continental LIPs is illustrated by ΔNb and ε_{Nd_i} plots (Fig. 13). Both the positive and negative ΔNb basalts show a wide range in ε_{Nd_i} values. Pristine plume-derived melts are supposed to show positive ΔNb and $\varepsilon_{Nd_i} = 0 \pm 2$, whereas depleted N-MORB-type mantle-derived basalts have negative ΔNb but positive ε_{Nd_i} values from +8 to +10 (Fig. 13). However, continental lithosphere with its complex depletion and enrichment histories may show a wide range in ΔNb and ε_{Nd_i} values. Although ε_{Nd_i} values tend to be negative, especially for the Indian subcontinental lithosphere as it is thought to be of Archaean and Proterozoic ages

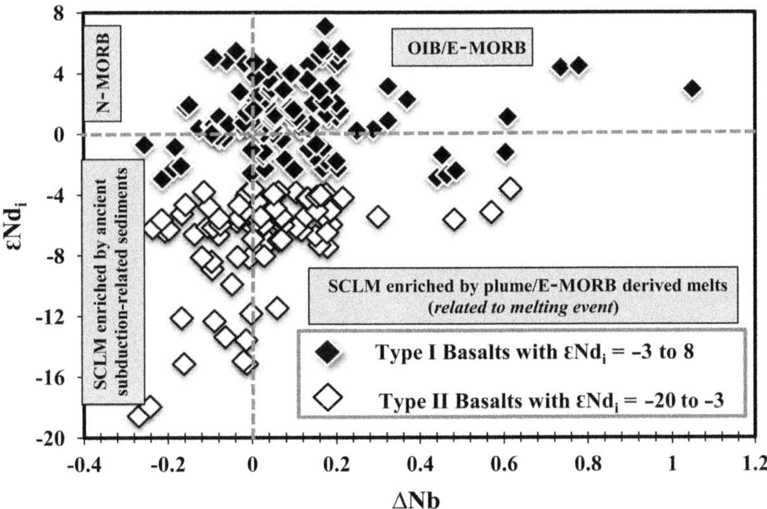

Fig. 13. ΔNb v. ε_{Nd_i} variation in the Type I and Type II basalts from the Phanerozoic continental LIPs of India. Vectors for melting and mixing are shown. Fields for different mantle reservoirs are demarcated with dashed lines. Newly formed IABs would fall in the field for N-MORB, and active continental margin basalts (ACMB) fall in the field for SCLM with subduction signatures. Abbreviations for mantle reservoirs are as in Figure 11. For an explanation, see the text.

(Mahadevan 2013), ΔNb values could be positive or negative depending on the process of enrichment. If the enrichment of continental lithosphere was affected by subduction-related mechanisms, then the lithosphere develops negative ΔNb values. However, if the enrichment is by melts/fluids leaked from the plume, during the melting event, then the lithosphere may retain the negative ε_{Nd_i} isotopic signature but develop positive ΔNb values as the plume-derived melts overprint the trace element signatures in the lithosphere but does not affect the isotopes as the enrichment is recent.

The geochemical diversity depicted by India's Phanerozoic continental LIP magmatism cannot be generated from a single mantle source even if it is heterogeneous. Let us consider three possible distinct mantle reservoirs: plume, depleted asthenosphere and subcontinental lithosphere. The heterogeneities in the plume sources induced by (1) oceanic crust injections (Hofmann 2003), (2) delamination of continental lithosphere (McKenzie & O'Nions 1995) or (3) injection of upper-crustal sediments (Hawkesworth & Schersten 2007 and references therein) do not produce the observed geochemical diversity. Melting of heterogeneous plumes produces melts with both positive and negative ΔNb and ε_{Nd_i} values (Revillon *et al.* 2002; Fitton 2007). The presence of typical N-MORB melts in the plume-derived suite is possibly related to the compositional and thermal zonation of the plumes (Campbell & Griffiths 1990). The Iceland (Fitton *et al.* 1997), Hawaiian (Huang *et al.* 2011) and Galapagos (Hoernle *et al.* 2016) plumes are zoned with central-zone plume material, which is enveloped by a thick layer of relatively hotter but compositionally N-MORB-type depleted upper mantle. The entrained depleted mantle within the plume produces melts with negative ΔNb and positive ε_{Nd_i} anomalies, whereas melting of plume material produces melts with positive ΔNb and positive ε_{Nd_i} anomalies. However, the extreme negative ΔNb and ε_{Nd_i} ratios shown by the Indian continental LIPs require long-term preservation of enriched characteristics, which can be expected only from the subcontinental lithosphere. Depleted asthenosphere with enriched streaks can produce melts with negative ΔNb and positive ε_{Nd_i} (N-MORB-type), and positive ΔNb and positive ε_{Nd_i} (E-MORB-type), signatures. The Indian Phanerozoic continental LIPs do not have typical N-MORB-type geochemical characteristics –ruling out depleted asthenosphere source. Additionally, extreme negative ε_{Nd_i} signatures shown by the Indian Phanerozoic continental LIPs cannot be produced in the asthenosphere. The chemical diversity cannot be generated by the melting of highly heterogeneous subcontinental lithosphere, even with ancient subduction-related and geologically recent asthenosphere/plume-derived low-degree melt additions. The former produces negative ΔNb and negative ε_{Nd_i} ratios in the lithospheric sources, and the latter imparts positive ΔNb signatures to the sources without affecting the ε_{Nd_i} values. Although, the Indian Phanerozoic continental LIPs with negative ΔNb and negative ε_{Nd_i}, and positive ΔNb and negative ε_{Nd_i}, values can be explained by the aforesaid mechanisms, melts with positive ΔNb and positive ε_{Nd_i} (OIB-type), and negative ΔNb and positive ε_{Nd_i} (N-MORB-type), values still require sublithospheric sources (Fitton 2007). Therefore, multiple mantle sources are prerequisite to explain the geochemical spread of the three continental LIPs. Based on La/Nb, Zr/Nb, ΔNb and ε_{Nd_i} values, we estimate that the lithosphere and sublithosphere (plume and/or E-MORB patches within the asthenosphere) contributions in building the Panjal continental LIP to be approximately 67 and 33%, Rajmahal continental LIP 52 and 48%, and Deccan continental LIP 28 and 72%, respectively. Considering all three provinces, we estimate that lithosphere and sublithosphere mantle sources contributed 38 and 62%, respectively, to the Indian Phanerozoic continental LIP magmatism.

Phanerozoic continental LIPs and fertility of the Indian subcontinental lithosphere

We have correlated La/Th and La/Nb ratios and ε_{Nd_i} values for the lithosphere-derived Type II basalts with age to evaluate the temporal variation in the fertility of the Indian subcontinental lithosphere, if any (Fig. 14). An apparent increase in La/Nb and La/Th ratios, and decrease in ε_{Nd_i} values, from the Panjal to the Deccan (Fig. 14) suggests that fertility of the Indian subcontinental lithosphere may have increased with decreasing age, possibly related to continual Gondwana fragmentation. Alternatively, the Indian SCLM was variably enriched in different segments, with the central portions underlying the Deccan province being strongly enriched.

One fundamental difference between the three Phanerozoic LIPs of India is the absence of alkaline rocks and carbonatites in the Panjal province, and their abundant presence in the Rajmahal and Deccan provinces. We believe that the absence of these rocks is controlled by the lithosphere thickness and fertility, and the degree of melting. In the case of the Panjal, rifting was initiated by lithosphere stretching that finally culminated in a new ocean (Shellnutt *et al.* 2015*a*). However, based on similar Nd isotopic characteristics, Shellnutt (2017) suggested that the most depleted Panjal Traps (ε_{Nd_i} = +2.4 to +4.3) and *c.* 300 Ma Koga and Jambil carbonatites (ε_{Nd_i} = +3.2 to +3.7: Le Bas *et al.* 1987; Tilton *et al.* 1998) of NW Pakistan (occurring *c.* 300 km west of the Panjal Traps) could be related to the same mantle source. We speculate that even if the carbonatites and Panjal

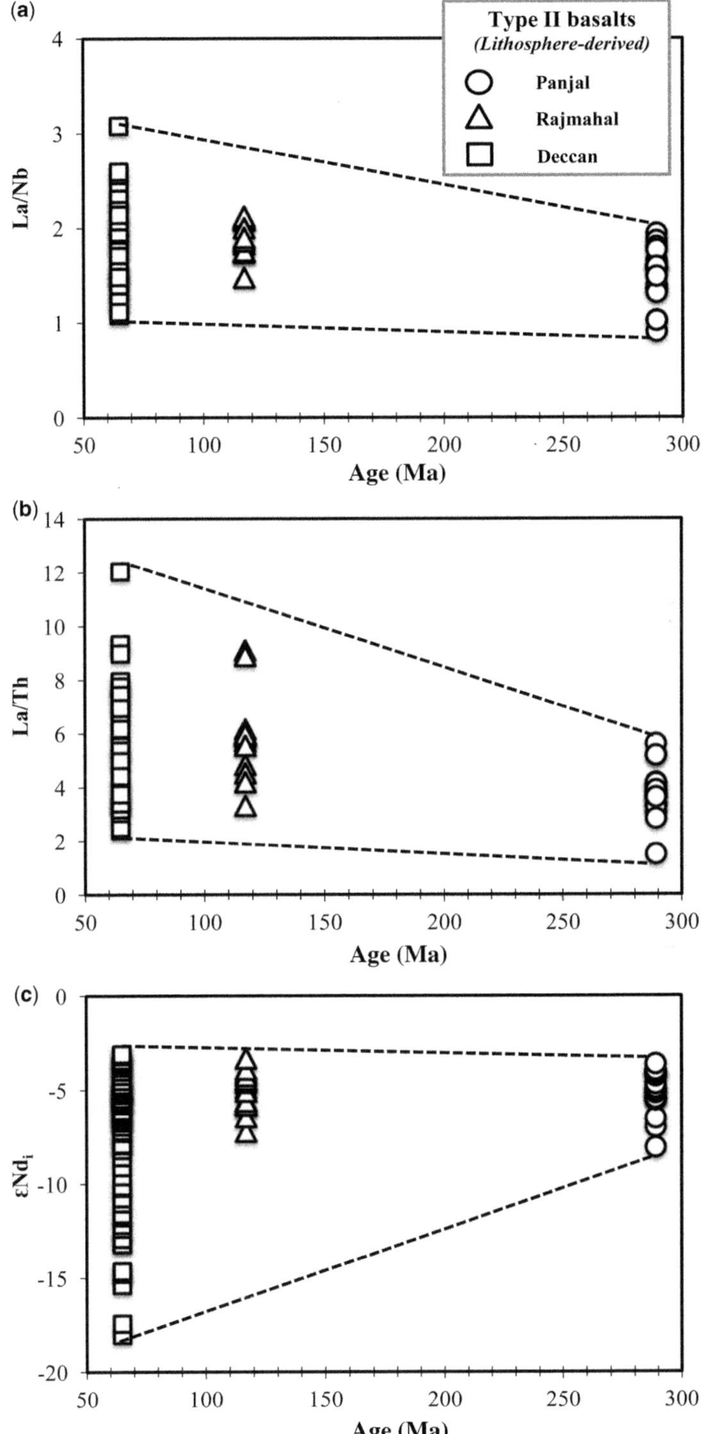

Fig. 14. (**a**) Age (Ma) v. La/Nb, (**b**) La/Th and (**c**) ε_{Nd_i} variations in the subcontinental lithosphere-derived basalts (Type II basalts) from the Phanerozoic continental LIPs of India. Fertility of the Indian subcontinental lithosphere seems to have increased with decreasing age, possibly related to advancing Gondwana fragmentation.

Traps are derived from comparable mantle sources, the carbonatites represent mantle-derived melts formed beneath a thick continental lithosphere far away from the rift, whereas the Panjal basalts are characteristically rift-related magmatism formed underneath a thinning continental lithosphere. Tilton *et al.* (1998) also suggested that the Koga and Jambil carbonatites were formed before the rift was established. In contrast, the Rajmahal and Deccan rifts were possibly induced by plumes impinging the continental lithosphere. This has a profound influence on the geodynamics of mantle melting (see Tappe *et al.* 2007; Vijaya Kumar & Rathna 2008). In the case of the Panjal, due to continuous rifting, the geothermal gradient and, consequently, the depth of melting would be shallower, resulting in the dominance of tholeiites with minor alkali basalts. The increasing percentages of partial melting of the mantle can be fuelled either by the increasing potential temperature of the mantle, if melting operated at the deeper depths, or the increasing segregation time of melt from the residue (i.e. melting at relatively shallower depths). At higher depths of melting, garnet would be retained in the residue, which would result in higher La/Lu ratios in the melts, but these are not shown by the Panjal basalts (Fig. 5). During partial melting of the mantle, the degree of melting increases with decreasing pressure (Asimow *et al.* 1997). Therefore, it is reasonable to presume that the Panjal basalts are derived by a higher degree of melting of mantle sources at shallower depths (spinel stability field). Melting has to be at much deeper depths and at a lower degree beneath the Rajmahal (spinel–garnet stability field), and more so with the Deccan (garnet stability field) in order to produce lamproites, ultramafic lamprophyres and carbonatites. Further, Deccan with lower- to higher-degree melts (Fig. 8) requires polybaric melt generation.

The Indian Phanerozoic continental LIP magmatism reflects how nature achieves such large-scale geochemical variation in a limited timescale. The present study illustrates how the fertility and past history of the subcontinental lithosphere play a significant role in the creation of enormous geochemical diversity that is characteristic of continental LIP magmatism. Although the terms 'fertility' (ability to produce magma on melting) and 'enrichment' (high content of incompatible elements) are apparently different, they both may be achieved by same process (see Anderson 2006). Fluxes (fluids or melts) internal or external to the mantle sources enrich the continental lithosphere. The former is by low-degree melting/degassing of sublithospheric mantle and the latter by subduction-related processes. Both these 'enrichment' processes essentially make the lithospheric mantle 'fertile'. The Phanerozoic continental LIP basalts of India show evidence of the operation of both these processes in modifying the subcontinental lithosphere mantle. Based on the geochemical signatures of the Type II basalts sampled from three spatially and temporally distinct LIPs, we suggest that the ancient subduction-related sediments were part of the SCLM, especially beneath the Deccan province.

Model for the Indian Phanerozoic continental LIPs

Our favoured model infers that the plume upwelling and partial melting that produced the melts with positive ΔNb and positive ε_{Nd_i} (labelled 1 in Fig. 15), and negative ΔNb and positive ε_{Nd_i} (labelled 2 in Fig. 15), signatures is likely to have heated the overlying lithospheric mantle, initiating melting of the metasomatized portions, which will melt readily due to their low solidii; these melts have negative ΔNb and negative ε_{Nd_i} signatures (labelled 3 in Fig. 15). We suggest that ancient subducted sediments preserved in the lithosphere are the possible metasomites. Melting of the continental lithosphere enriched by low-degree partial melts derived from plume sources, related to the melting event, produced melts with positive ΔNb and negative ε_{Nd_i} anomalies (labelled 4 in Fig. 15). Such a model explains the genesis of the basalts from the Rajmahal and Deccan LIPs. Based on isotopic evidence, many earlier workers suggested a mantle-plume origin for both the Rajmahal and Deccan basalts but the depth at which the plumes were melted seem to be different beneath the Rajmahal and Deccan provinces. It is possible that the depth of melting, and, consequently, the lithosphere thickness, is higher for the Deccan plume, whereas it seems that shallower melting of the plume, possibly beneath a thin lithosphere, produced the Rajmahal basalts (Fig. 15). The Panjal basalts are better explained by decompression melting of the subcontinental lithosphere (labelled 5 in Fig. 15). It is suggested that the Panjal rift has undergone asymmetrical extension, apparently more widespread in the west but limited in the east. Variations in the degree of lithosphere extension during the formation of continental rifts/ back-arc basins are also reported in the Rocas Verdes Basin, South America (Stern & de Wit 2003), central North Island, New Zealand (Davey *et al.* 1995), and the Proterozoic Eastern Ghats in India (Vijaya Kumar & Leelanandam 2008). Prolonged extension results in thinning of the lithosphere, and upwelling of asthenosphere diapirs finally generates N-MORB-type crust (i.e. the oceanization of continental crust). Such a mechanism is proposed for the western segment of the Panjal continental LIP (Shellnutt *et al.* 2015*a*). However, the present study speculates that lithosphere stretching was sufficient to melt E-MORB patches within the

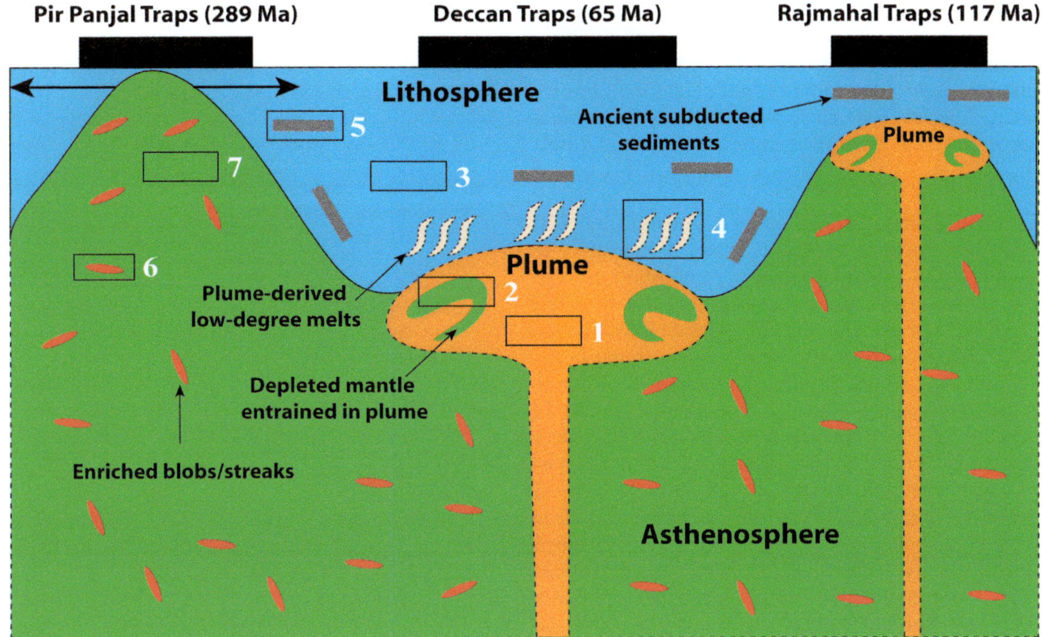

Fig. 15. A cartoon illustrating the possible mantle sources that feed the magmas to build the Panjal, Rajmahal and Deccan continental LIPs. 1, plume source (produces melts with positive ΔNb and positive ε_{Nd_i}); 2, entrained depleted mantle within the plume source (produces melts with negative ΔNb and positive ε_{Nd_i}); 3, hidden ancient subducted sediments (produces melts with negative ΔNb and highly negative ε_{Nd_i}); 4, enriched zones in the lithosphere by addition of low-degree melts derived from plume (produces melts with positive ΔNb and negative ε_{Nd_i}); 5, SCLM (produces melts with negative ΔNb and negative ε_{Nd_i}); 6, enriched blobs or streaks within the asthenosphere (produces melts with positive ΔNb and positive ε_{Nd_i}); 7, depleted asthenospheric mantle (produces melts with negative ΔNb and positive ε_{Nd_i}). The geochemical diversity in the Panjal, Rajmahal and Deccan basalts can be modelled by the melting and mixing of these sources. Typical N-MORB-type depleted melts are not observed in the Phanerozoic continental LIPs of India. Lithosphere doming due to plume impinging is not shown in the figure. For the discussion, see the text.

asthenosphere (labelled 6 in Fig. 15) but not the N--MORB-type depleted mantle (labelled 7 in Fig. 15). It is evident that both sublithospheric sources (possibly a mantle plume plus E-MORB patches within the asthenosphere) and subcontinental lithosphere played significant roles in building the Phanerozoic continental LIPs of India.

Conclusions

The present study illustrates the geochemical similarity and variability between the basalts from three spatially and temporally distinct Phanerozoic continental large igneous provinces (continental LIPs) of India, represented by Panjal (*c.* 289 Ma), Rajmahal (*c.* 117 Ma) and Deccan (*c.* 65 Ma), and evaluates the relative roles of asthenosphere (and/or plume) and subcontinental lithosphere mantle (SCLM) in the generation of the continental LIPs. Based on geochemical distinctions, we have classified the Panjal, Rajmahal and Deccan basalts into Type I and Type II. The Type I basalts with lower La/Nb, Th/Nb and Th/Yb ratios and distinctly higher ε_{Nd_i} values (−3 to +8) are derived from the sublithospheric sources – possibly a mantle plume plus E-MORB patches within the asthenosphere. Geochemistry of some of the Type I basalts can be explained by contamination of the plume-derived melts by continental lithosphere. The Type II basalts with higher La/Nb, Th/Nb and Th/Yb ratios and lower ε_{Nd_i} values (−20 to −3) are derived by partial melting of the subcontinental lithosphere. The Panjal primitive basalts are formed by 5–20%, the Rajmahal basalts by 10–20%, and the Deccan basaltic melts by 1–20% melting of the mantle sources. Assimilation fractional crystallization (AFC) is the prime differentiation process that amended the primitive magmas in the Phanerozoic continental LIPs of India.

In our favoured model, the Deccan and Rajmahal basalts represent the continental LIPs produced by a plume impinging beneath continental lithosphere, and the Panjal represents an LIP produced by lithosphere stretching in an extension regime. We suggest that the depth of melting beneath Deccan is greater than beneath Rajmahal, which greater than that beneath Panjal. Multiple mantle sources are involved in building the Phanerozoic continental LIPs of India. Plumes seem to have contributed a larger amount of material to the Deccan continental LIP than the lithospheric mantle, whereas the opposite is the case for Panjal where the contribution from the lithospheric sources far outweighs the sublithospheric sources. Also, in the Rajmahal continental LIP, lithospheric sources seem to have contributed significantly to the flood basalt volcanism. Plume–lithosphere interaction is impeccably recorded in the Deccan and Rajmahal basalts. The complete spectrum of the flood basalt magmatism in the Panjal, Rajmahal and Deccan is built by melts derived from E-MORB/OIB-like sublithospheric sources and continental lithosphere, and mixing between them. The N-MORB-like depleted mantle sources and upper continental crustal components did not play any significant role in the petrogenesis of the Phanerozoic continental LIPs of India. Our first-order estimates on relative contributions of the lithosphere and sublithosphere (plume and/or EMORB patches within asthenosphere) to the Panjal continental LIP are approximately 67 and 33%, to the Rajmahal continental LIP are 52 and 48%, and to the Deccan continental LIP are 28 and 72%, respectively. Considering all three provinces, we estimate that the lithosphere and sublithosphere mantle sources, respectively, contributed 38 and 62% in the generation of the Indian Phanerozoic continental LIP magmatism. A debate on plume v. non-plume sources for continental flood basalt magmatism should reconcile that highly heterogeneous subcontinental lithospheric mantle (SCLM) and its differential partial melting may contribute substantially to the geochemical diversity in the continental LIPs, as recorded in the Phanerozoic continental LIPs of India. The extensive presence of lithosphere-derived basalts in the Phanerozoic continental LIPs suggests significant thermochemical erosion of the Indian subcontinental lithosphere. The fertility of the Indian subcontinental lithosphere seems to have increased with decreasing age, possibly related to continual Gondwana fragmentation.

We are grateful to Profs Sensarma and Brain Storey for inviting us to contribute to this special volume on 'Large Igneous Provinces of Gondwana and Adjacent Regions'. We sincerely thank two anonymous journal reviewers for providing insightful suggestions on an earlier version of the manuscript. Geochemical data used in this paper were mainly extracted from the GEOROC database maintained by the Max Planck Institute für Chemie in Mainz, Germany (http://georoc.mpch-mainz.gwdg.de/georoc/.

References

Albarède, F. 1996. *Introduction to Geochemical Modeling*. Cambridge University Press, London.

Aldanmaz, E., Köprübaşı, N., Gürer, Ö.F., Kaymakçı, N. & Gourgaud, A. 2006. Geochemical constraints on the Cenozoic, OIB-type alkaline volcanic rocks of NW Turkey: Implications for mantle sources and melting processes. *Lithos*, **86**, 50–76.

Ali, J.R., Aitchison, J.C., Chik, S.Y.S., Baxter, A.T. & Bryan, S.E. 2012. Paleomagnetic data support Early Permian age for the Arbor volcanics in the lower Siang Valley, NE India: Significance for Gondwana break-up models. *Journal of Asian Earth Sciences*, **50**, 105–115.

Allègre, C.J. & Minster, J.F. 1978. Quantitative methods of trace element behaviour in magmatic processes. *Earth and Planetary Science Letters*, **38**, 1–25.

Anderson, D.L. 2006. Speculations on the nature and cause of mantle heterogeneity. *Tectonophysics*, **416**, 7–22.

Arculus, R.J. 1987. The significance of source v. process in the tectonic controls of magma genesis. *Journal of Volcanology and Geothermal Research*, **32**, 1–12.

Arndt, N.T. & Christensen, U. 1992. The role of lithospheric mantle in continental flood volcanism: thermal and geochemical constraints. *Journal of Geophysical Research*, **97**, 10 967–10 981.

Arndt, N.T. & Goldstein, S.L. 1989. An open boundary between lower continental crust and mantle: its role in crust formation and crustal recycling. *Tectonophysics*, **161**, 201–212.

Asimow, P.D., Firschmann, M.M. & Stolper, E.M. 1997. An analysis of variations in isotropic melt productivity. *Philosophical Transactions of the Royal Society of London*, **A355**, 255–281.

Baker, J.A., Menzies, M.A., Thirlwall, M.F. & Macpherson, C.G. 1997. Petrogenesis of Quaternary intraplate volcanism, Sana'a, Yemen: implication for plume–lithosphere interaction and polybaric melt hybridization. *Journal of Petrology*, **38**, 1359–1390.

Baksi, A.K. 1994. Revised plate motions relative to the hotspots from combined Atlantic and Indian Ocean hotspot tracks: comment and reply. *Geology*, **22**, 276–277.

Baksi, A.K. 1995. Petrogenesis and timing of volcanism in the Rajmahal flood basalt province, northeast India. *Chemical Geology*, **121**, 73–90.

Baksi, A.K., Barman, T.R., Paul, D.K. & Farrar, E. 1987. Widespread early Cretaceous flood basalt volcanism in Eastern India: Geochemical data from the Rajmahal–Bengal–Sylhet Traps. *Chemical Geology*, **63**, 133–141.

Barth, M.G., McDonough, W.F. & Rudnick, R.L. 2000. Tracking the budget of Nb and Ta in the continental crust. *Chemical Geology*, **165**, 197–213.

Beane, J.E., Turner, C.A., Hooper, P.R., Subbarao, K.V. & Walsh, J.N. 1986. Stratigraphy, composition and form of the Deccan basalts, Western Ghats, India. *Bulletin of Volcanology*, **48**, 61–83.

Bertrand, H., Chazot, G., Blichert-Toft, J. & Thoral, S. 2003. Implications of wide spread high-µ volcanism on the Arabian Plate for Afar mantle plume and lithosphere composition. *Chemical Geology*, **198**, 47–61.

Bhat, M.I. 1984. Abor Volcanics: further evidence for the birth of the Tethys Ocean in the Himalayan Segment.

Journal of the Geological Society, London, **141**, 763–775, https://doi.org/10.1144/gsjgs.141.4.0763

Bhat, M.I. & Zainuddin, S.M. 1979. Origin and evolution of the Panjal Volcanics. *Himalayan Geology*, **9**, 421–461.

Bhat, M.I., Zainuddin, S.M. & Rais, A. 1981. Panjal Trap chemistry and the birth of Tethys. *Geological Magazine*, **118**, 367–375.

Bryan, S.E. & Ernst, R.E. 2008. Revised definition of large igneous provinces (LIPs). *Earth Science Reviews*, **86**, 175–202.

Campbell, I.H. 2007. Testing the plume theory. *Chemical Geology*, **241**, 153–176.

Campbell, I.H. & Griffiths, R.W. 1990. Implications of mantle plume structure for the evolution of flood basalts. *Earth and Planetary Science Letters*, **99**, 79–93.

Chakrabarti, R. & Basu, A.R. 2006. Trace element and isotopic evidence for Archean basement in the Lonar crater impact breccias, Deccan Volcanic Province. *Earth and Planetary Science Letters*, **247**, 197–211.

Chandrasekharam, D., Vaselli, O., Sheth, H.C. & Keshav, S. 2000. Petrogenetic significance of ferro-enstatite orthopyroxene in basaltic dikes from the Tapi rift, Deccan flood basalt province, India. *Earth and Planetary Science Letters*, **179**, 469–476.

Chauvet, F., Lapierre, K. *et al.* 2008. Geochemistry of the Panjal Traps basalts (NW Himalaya): records of the Pangea Permian break-up. *Bulletin de la Société Géologique de France*, **179**, 383–395.

Chayes, F. 1965. Titania and alumina content of oceanic and circum-oceanic basalt. *Mineralogical Magazine*, **34**, 126–131.

Colon, D.P., Bindeman, I.N., Stern, R.A. & Fisher, C.M. 2015. Isotopically diverse rhyolites coeval with the Columbia river flood basalts: evidence for mantle plume interaction with the continental crust. *Terra Nova*, **27**, 270–276.

Condie, K.C. 2005. High field strength element ratios in Archean basalts: a window to evolving sources of mantle plumes? *Lithos*, **79**, 491–504.

Courtillot, V., Davaille, A., Besse, J. & Stock, J. 2003. Three distinct types of hotspots in the Earth's mantle. *Earth and Planetary Science Letters*, **205**, 295–308.

Cox, K.G. & Hawkesworth, C.J. 1984. Relative contribution of crust and mantle to flood basalt magmatism, Mahabaleshwar area, Deccan traps. *Philosophical Transactions of the Royal Society of London*, **A310**, 627–641.

Davey, F.J., Henrys, S.A. & Lodolo, E. 1995. Asymmetric rifting in a continental back-arc environment, North Island, New Zealand. *Journal of Volcanology and Geothermal Research*, **68**, 209–238.

DePaolo, D.J. 1981. Trace element and isotopic effects of combined wallrock assimilation and fractional crystallization. *Earth and Planetary Science Letters*, **53**, 189–202.

Dessai, A.G., Downes, H., López-Moro, F.-J. & López-Plaza, M. 2008. Lower crustal contamination of Deccan traps magmas: evidence from tholeiitic dykes and granulite xenoliths from western India. *Mineralogy and Petrology*, **93**, 243–272.

Ellam, R.M. & Cox, K.G. 1991. An interpretation of Karoo picrite basalts in terms of interaction between asthenospheric magmas and the mantle lithosphere. *Earth and Planetary Science Letters*, **105**, 330–342.

Ellam, R.M., Carlson, R.W. & Shirey, S.B. 1992. Evidence from Re–Os isotopes for plume–lithosphere mixing in Karoo flood basalt genesis. *Nature*, **359**, 718–721.

Ernst, R.E. & Buchan, K.L. 2003. Recognizing mantle plumes in the geological record. *Annual Reviews of Earth and Planetary Science Letters*, **31**, 469–523.

Fan, W.M., Guo, F., Wang, Y.J. & Zhang, M. 2004. Late Mesozoic volcanism in the northern Huaiyang tectonomagmatic belt, Central China: partial melts from a lithosphere mantle with subducted continental crustal relicts beneath Dabie orogen? *Chemical Geology*, **209**, 27–48.

Fitton, J.G. 1995. Coupled molybdenum and niobium depletion in continental basalts. *Earth and Planetary Science Letters*, **136**, 715–721.

Fitton, J.G. 2007. The OIB paradox. *In*: Foulger, G.R. & Jurdy, D.M. (eds) *Plates, Plumes, and Planetary Processes*. Geological Society of America, Special Papers, **430**, 387–412.

Fitton, J.G. & Dunlop, H.M. 1985. The Cameroon line, West Africa, and its bearing on the origin of oceanic and continental alkali basalt. *Earth and Planetary Science Letters*, **72**, 23–38.

Fitton, J.G., Saunders, A.D., Norry, M.J., Hardarson, B. S. & Tylor, R.N. 1997. Thermal and chemical structure of the Iceland plume. *Earth and Planetary Science Letters*, **153**, 197–208.

Fodor, R.V. 1987. Low- and high-TiO_2 flood basalts of southern Brazil: origin from picritic parentage and a common mantle source. *Earth and Planetary Science Letters*, **84**, 423–430.

Foley, S.F. 2008. Rejuvenation and erosion of the cratonic lithosphere. *Nature Geoscience*, **1**, 503–510.

Fuchs, G. 1987. The geology of southern Zanskar (Ladakh) – Evidence for the autochthony of the Tethys Zone of the Himalaya. *Jahrbuch der Geologischen Bundesanstal*, **130**, 465–491.

Gallagher, K. & Hawkesworth, C.J. 1992. Dehydration melting and the generation of continental flood basalts. *Nature*, **358**, 57–59.

Ganju, P.N. 1944. The Panjal Traps: acid and basic volcanic rocks. *Proceedings of the Indian Academy of Sciences*. **18**, 125–131.

Gao, S., Rudnick, R.L. *et al.* 2004. Recycling lower continental crust in the North China craton. *Nature*, **432**, 892–897.

Garzanti, E., Le Fort, P. & Sciunnach, D. 1999. First report of Lower Permian basalts in south Tibet: tholeiitic magmatism during break-up and incipient opening of Neotethys. *Journal of Asian Earth Sciences*, **17**, 533–546.

Ghatak, A. & Basu, A.R. 2011. The Sylhet Traps: Vestiges of the Kerguelen plume in northeastern India. *Earth and Planetary Science Letters*, **308**, 52–64.

Ghatak, A. & Basu, A.R. 2013. Isotopic and trace element geochemistry of alkali-mafic-ultramafic-carbonatitic complexes and flood basalts in NE India: origin in a heterogeneous Kerguelen plume. *Geochimica et Cosmochimica Acta*, **115**, 46–72.

Ghose, N.C., Chatterjee, N. & Windley, B.F. 2017. Subaqueous early eruptive phase of the Late Aptian Rajmahal volcanism, India: evidence from volcaniclastic rocks, bentonite, black shales, and oolite. *Geoscience Frontiers*, **8**, 809–822.

GLAZNER, A.F. 1994. Foundering of mafic plutons and density stratification of continental crust. *Geology*, **22**, 435–438.

GOODENOUGH, K.M., UPTON, B.G.J. & ELLAM, R.M. 2002. Long-term memory of subduction processes in the lithospheric mantle: evidence from the geochemistry of basic dykes in the Gardar Province of South Greenland. *Journal of the Geological Society, London*, **159**, 705–714, https://doi.org/10.1144/0016-764901-154

HAMILTON, W.B. 2011. Plate tectonics began in Neoproterozoic time, and plumes from deep mantle have never operated. *Lithos*, **123**, 1–20.

HANSON, G.N. 1980. Rare earth elements in petrogenetic studies of igneous systems. *Annual Review of Earth and Planetary Sciences*, **8**, 371–406.

HAWKESWORTH, C.J. & SCHERSTEN, A. 2007. Mantle plumes and geochemistry. *Chemical Geology*, **241**, 319–331.

HAWKESWORTH, C.J., MANTOVANI, M.S.M. & PEATE, D.W. 1988. Lithosphere remobilization during Parana magmatism. *In*: COX, K.G. & MENZIES, M. (eds) *Oceanic and Continental Lithosphere: Similarities and Differences. Journal of Petrology*, Special Volume 1, 205–223.

HAWKESWORTH, C.J., KEMPTON, P.D., ROGERS, N.W., ELLAM, R.M. & VAN CALSTEREN, P.W. 1990. Continental mantle lithosphere and shallow level enrichment processes in the Earth's mantle. *Earth and Planetary Science Letters*, **96**, 256–268.

HAWKESWORTH, C.J., KELLEY, S., TURNER, S., LE ROEX, A. & STOREY, B. 1999. Mantle processes during Gondwana break-up and dispersal. *Journal of African Earth Sciences*, **28**, 239–261.

HEINONEN, J.S., LUTTINEN, A.V. & BOHRSON, W.A. 2016. Enriched continental flood basalts from depleted mantle melts: modeling the lithospheric contamination of Karoo lavas from Antarctica. *Contributions to Mineralogy and Petrology*, **171**, 1–22.

HOERNLE, K., WERNER, R., MORGAN, J.P., GARBE-SCHONBERG, D., BRYCE, J. & MRAZEK, J. 2016. Existence of complex spatial zonation in the Galapagos plume. *Geology*, **28**, 435–438.

HOFMANN, A.W. 1988. Chemical differentiation of the Earth: the relationship between mantle, continental crust, and oceanic crust. *Earth and Planetary Science Letters*, **90**, 297–314.

HOFMANN, A.W. 2003. Sampling mantle heterogeneity through oceanic basalts: isotopes and trace elements. *In*: CARLSON, R.W. (ed.) *The Mantle and Core*. Treatise on Geochemistry, **2**. Elsevier, Amsterdam, 61–101.

HOLE, M.J. 2015. The generation of continental flood basalts by decompression melting of internally heated mantle. *Geology*, **43**, 311–314.

HUANG, S., HALL, P.S. & JACKSON, M.G. 2011. Geochemical zoning of volcanic chains associated with Pacific hotspots. *Nature Geoscience*, **4**, 874–878.

IVANOV, A.V. 2015. Why volatiles are required for cratonic flood basalt volcanism: two examples from the Siberian craton. *In*: FOULGER, G.R., LUSTRINO, M. & KING, S.D. (eds) *The Interdisciplinary Earth: A Volume in Honor of Don L. Anderson*. Geological Society of America, Special Papers, **514**, 325–338.

JAY, A.E. & WIDDOWSON, M. 2008. Stratigraphy, structure and volcanology of the SE Deccan continental flood basalt province: implications for eruptive extent and volumes. *Journal of the Geological Society, London*, **165**, 177–188, https://doi.org/10.1144/0016-76492006-062

JERRAM, D.W. & WIDDOWSON, M. 2005. The anatomy of Continental Flood Basalt Provinces: geological constraints on the processes and products of flood volcanism. *Lithos*, **79**, 385–405.

JOHNSON, J.S., GIBSON, S.A., THOMPSON, R.N. & NOWELL, G.M. 2005. Volcanism in the Vitim Volcanic Field, Siberia: geochemical evidence for a mantle plume beneath the Baikal rift zone. *Journal of Petrology*, **46**, 1309–1344.

JOHNSON, R.W., MACKENZIE, D.E. & SMITH, I.E.M. 1978. Delayed partial melting of subduction-modified mantle in Papua New Guinea. *Tectonophysics*, **46**, 197–216.

JUNG, C., JUNG, S., HOFFER, E. & BERNDT, J. 2006. Petrogenesis of Tertiary mafic alkaline magmas in the Hocheifel, Germany. *Journal of Petrology*, **47**, 1637–1671.

KENT, R.W., SAUNDERS, A.D., KEMPTON, P.D. & GHOSE, N. C. 1997. Rajmahal Basalts, eastern India: mantle sources and melt distribution at a volcanic rifted margin. *In*: MAHONEY, J. & COFFIN, M.F. (eds) *Large Igneous Provinces: Continental, Oceanic, and Planetary Flood Volcanism*. American Geophysical Union, Geophysical Monograph Series, **100**, 145–182.

KENT, R.W., PRINGLE, M.S., MULLER, R.D., SAUNDERS, A.D. & GHOSE, N.C. 2002. $^{40}Ar/^{39}Ar$ geochronology of the Rajmahal Basalts, India, and their relationship to the Kerguelen Plateau. *Journal of Petrology*, **43**, 1141–1153.

KRISHNAMURTHY, P., GOPALAN, K. & MACDOUGALL, J.D. 2000. Olivine compositions in picrite basalts and the Deccan volcanic cycle. *Journal of Petrology*, **41**, 1057–1069.

KUMAR, S. 2003. Variations in the thickness of the lithosphere underneath the Deccan volcanic province: evidence from rare earth elements. *In*: MOHAN, A. (ed.) *Milestones in Petrology and Future Perspectives*. Geological Society of India, Memoirs, **52**, 179–194.

LAPIERRE, H., SAMPER, A. *ET AL.* 2004. The Tethyan plume: geochemical diversity of middle Permian basalts from the Oman rifted margin. *Lithos*, **74**, 167–198.

LE BAS, M.J., MIAN, I. & REX, D.C. 1987. Age and nature of carbonatite emplacement in North Pakistan. *Geologische Rundschau*, **76**, 317–323.

LI, Y., PENG, P., WANG, X. & WANF, H. 2015. Nature of 1800–1600 Ma mafic dyke swarms in the North China craton: implications for the rejuvenation of the sub-continental lithospheric mantle. *Precambrian Research*, **257**, 114–123.

LIGHTFOOT, P.C. & HAWKESWORTH, C.J. 1988. Origin of Deccan Trap lavas: evidence from combined trace element and Sr-, Nd- and Pb-isotope studies. *Earth and Planetary Science Letters*, **91**, 89–104.

LIGHTFOOT, P.C., HAWKESWORTH, C.J., DEVEY, C.W., ROGERS, N.W. & VAN CALSTEREN, P.W.C. 1990. Source and differentiation of Deccan Trap lavas: implications of geochemical and mineral chemical variations. *Journal of Petrology*, **31**, 1165–1200.

MACDONALD, R., ROGERS, N.W., FITTON, J.G., BLACK, S. & SMITH, M. 2001. Plume–lithosphere interactions in the generation of the basalts of the Kenya rift, East Africa. *Journal of Petrology*, **42**, 877–900.

MacGregor, T.D. 1969. The system MgO–SiO_2–TiO_2 and its bearing on the distribution of TiO_2 in basalts. *American Journal of Science*, **267**, 342–363.

Mahadevan, T.M. 2013. Evolution of the Indian continental lithosphere: insights from episodes of crustal evolution and geophysical models. *Journal of the Indian Geophysical Union*, **17**, 9–38.

Mahoney, J., Macdougall, J.D., Lugmair, G.W., Murali, A.V., Sankar Das, M. & Gopalan, K. 1982. Origin of the Deccan Trap flows at Mahabaleshwar inferred from Nd and Sr isotopic and chemical evidence. *Earth and Planetary Science Letters*, **60**, 47–60.

Mahoney, J.J., Nicollet, C. & Dupuy, C. 1991. Madagascar basalts: tracking oceanic and continental sources. *Earth and Planetary Science Letters*, **18**, 350–363.

Mahoney, J.J., Sheth, H.C., Chandrasekharam, D. & Peng, Z.X. 2000. Geochemistry of flood basalts of the Toranmal section, northern Deccan Traps, India: implications for regional Deccan stratigraphy. *Journal of Petrology*, **41**, 1099–1120.

McKenzie, D. & O'Nions, R.K. 1995. The source regions of ocean island basalts. *Journal of Petrology*, **36**, 133–159.

McKenzie, D.P. & Bickle, M.J. 1988. The volume and composition of melt generated by extension of the lithosphere, *Journal of Petrology*, **29**, 625–679.

Middlemiss, C.S. 1910. Revision of the Silurian–Trias Sequence in Kashmir. *Records of the Geological Survey of India*, **40**, 206–260.

Nagaraju, K. 2012. *Geochemistry and petrogenesis of the basalts from the eastern Deccan volcanic province, India.* Unpublished PhD thesis, SRTM University, Nanded, Maharashtra, India.

Nakazawa, K. & Kapoor, H.M. 1973. Spilitic pillow lava in Panjal Trap of Kashmir, India. *Memoirs of the Faculty of Science, Kyoto University, Series of Geology and Mineralogy*, **39**, 83–98.

Nakazawa, K., Kapoor, H.M. et al. 1975. The upper Permian & the lower Triassic in Kashmir, India. *Memoirs of the Faculty of Science, Kyoto University, Series of Geology and Mineralogy*, **42**, 1–106.

Neumann, H., Mead, J. & Vitaliano, C.J. 1954. Trace element variation during crystallization as calculated from the distribution law. *Geochimica et Cosmochimica Acta*, **6**, 90–99.

Nicholson, S.W. & Shirey, S.B. 1990. Midcontinental rift volcanism in the Lake Superior region: Sr, Nd, and Pb isotopic evidence for a mantle plume origin. *Journal of Geophysical Research*, **95**, 10 851–10 868.

Niu, Y., Wilson, M., Humphreys, E.R. & O'hara, M.J. 2012. A trace element perspective on the source of ocean island basalts (OIB) and fate of subducted ocean crust (SOC) and mantle lithosphere (SML). *Episodes*, **35**, 310–327.

Nyström, J.O. 1984. Rare earth element mobility in vesicular lava during low-grade metamorphism. *Contributions to Mineralogy and Petrology*, **88**, 328–331.

Pande, K. 2002. Age & duration of the Deccan Traps, India: a review of radiometric & palaeomagnetic constraints. *Proceedings of Indian Academy of Science Earth and Planetary Sciences*, **111**, 115–123.

Pareek, H.S. 1976. On studies of the agglomerate slate & Panjal Trap in the Jhelum, Liddar & Sind Valleys, Kashmir. *Records of the Geological Survey of India*, **107**, 12–37.

Pearce, J.A. 2003. Supra-subduction zone ophiolites: the search for modern analogues. *In*: Dilek, Y. & Newcomb, S. (eds) *Ophiolite Concept and the Evolution of Geological Thought.* Geological Society of America, Special Paper, **373**, 269–293.

Pearce, J.A. & Cann, J.R. 1973. Tectonic setting of basic volcanic rocks determined using trace element analyses. *Earth and Planetary Science Letters*, **19**, 290–300.

Peng, Z.X. & Mahoney, J.J. 1995. Drillhole lavas from the northwestern Deccan Traps, and the evolution of Réunion hotspot mantle. *Earth and Planetary Science Letters*, **134**, 169–185.

Peng, Z.X., Mahoney, J.J., Hooper, P.R., Harris, C. & Beane, J.E. 1994. A role for lower continental crust in flood basalt genesis? Isotopic and incompatible element study of the lower six formations of the western Deccan Traps. *Geochimica et Cosmochimica Acta*, **58**, 267–288.

Plank, T. & Langmuir, C.H. 1998. The chemical composition of subducting sediment and its consequences for the crust and mantle. *Chemical Geology*, **145**, 325–394.

Prinzhofer, A. & Allegre, C.J. 1985. Residual peridotites and mechanisms of partial melting. *Earth and Planetary Science Letters*, **74**, 251–265.

Raja Rao, C.S., Sahasrabudhe, Y.S., Deshmukh, S.S. & Raman, R. 1978. *Distribution, Structure and Petrography of the Deccan Trap, India.* Report. Geological Survey of India, Kolkata, India.

Ray, R., Shukla, A.D. et al. 2008. Highly heterogeneous Precambrian basement under the central Deccan traps, India: direct evidence from xenoliths in dykes. *Gondwana Research*, **13**, 375–385.

Revillon, S., Chauvel, C., Arndt, N.T., Pik, R., Martineau, F., Fourcade, S. & Marty, B. 2002. Heterogeneity of the Caribbean plateau mantle source: Sr, O and He isotopic compositions of olivine and clinopyroxene from Gorgona island. *Earth and Planetary Science Letters*, **205**, 91–106.

Rogers, N.W., Macdonald, R., Fitton, J.G., George, R.W. W., Smith, M. & Barreiro, B.A. 2000. Two mantle plumes beneath the East African rift system: Sr, Nd and Pb isotope evidence from Kenya Rift basalts. *Earth and Planetary Science Letters*, **176**, 387–400.

Rudnick, R.L. & Gao, S. 2003. Composition of the continental crust. *In*: Holland, H.D. & Turekian, K.K. (eds) *The Crust.* Treatise on Geochemistry, **3**. Elsevier, Amsterdam, 1–64.

Sarkar, A., Datta, A.K., Poddar, B.C., Bhattacharyya, B.K., Kollapuri, V.K. & Sanwal, R. 1996. Geochronological studies of Mesozoic igneous rocks from eastern India. *Journal of Southeast Asian Earth Sciences*, **13**, 77–81.

Schilling, J.-G. & Winchester, J.W. 1967. Rare-earth fractionation and magmatic processes. *In*: Runcorn, S.K. (ed.) *Mantles for the Earth and Terrestrial Planets.* Wiley Interscience, New York, 267–283.

Segev, A. 2002. Flood basalts, continental break-up and the dispersal of Gondwana: evidence for periodic migration of upwelling mantle flows (plumes). *EGU Stephen Mueller Special Publication Series*, **2**, 171–191.

Sengör, A.M.C. 1987. Tectonics of the tethysides. Orogenic collage development in a collisional setting. *Annual Reviews of Earth and Planetary Science Letters*, **15**, 213–244.

Sengupta, S. 1966. Geological and geophysical studies in Western part of Bengal Basin, India. *Bulletin of the American Association of Petroleum Geologists*, **50**, 1001–1017.

Shaw, J.E., Baker, J.A., Menzies, M.A., Thirlwall, M.F. & Ibrahim, K.M. 2003. Petrogenesis of the largest intraplate volcanic field on the Arabian Plate (Jordan): a mixed lithosphere–asthenosphere source activated by lithospheric extension. *Journal of Petrology*, **44**, 1657–1679.

Shellnutt, J.G. 2017. The Panjal Traps. *In*: Sensarma, S. & Storey, B.C. (eds) *Large Igneous Provinces from Gondwana and Adjacent Regions*. Geological Society, London, Special Publications, **463**, https://doi.org/10.1144/SP463.4

Shellnutt, J.G., Bhat, G.M., Brookfield, M.E. & Jahn, B.-M. 2011. No link between the Panjal Traps (Kashmir) and the Late Permian mass extinctions. *Geophysical Research Letters: Solid Earth*, **38**, L19308.

Shellnutt, J.G., Bhat, G.M., Wang, K.-L., Brookfield, M.E., Dostal, J. & Jahn, B.-M. 2012. Origin of the silicic rocks of the Early Permian Panjal Traps, Kashmir, India. *Chemical Geology*, **334**, 154–170.

Shellnutt, J.G., Bhat, G.M., Wang, K.-L., Brookfield, M.E., Jahn, B.-M. & Dostal, J. 2014. Petrogenesis of the flood basalts from the Early Permian Panjal Traps, Kashmir, India: geochemical evidence for shallow melting of the mantle. *Lithos*, **204**, 159–171.

Shellnutt, J.G., Bhat, G.M., Wang, K.-L., Yeh, M.-W., Brookfield, M.E. & Jahn, B.-M. 2015*b*. Multiple mantle sources of the early Permian Panjal traps, Kashmir, India. *American Journal of Science*, **315**, 589–619.

Shellnutt, J.G., Usuki, T., Kennedy, A.K. & Chiu, H.-Y. 2015*a*. A lower crust origin of some flood basalts of the Emeishan large igneous province, SW China. *Journal of Asian Earth Sciences*, **109**, 74–85.

Sheraton, J.W. 1984. Chemical changes associated with high-grade metamorphism of mafic rocks in the east Antarctic shield. *Chemical Geology*, **47**, 135–157.

Sheth, H.C. 2005*a*. From Deccan to Réunion: no trace of a mantle plume. *In*: Foulger, G.R., Natland, J.H., Presnall, D.C. & Anderson, D.L. (eds) *Plates, Plumes, and Paradigms*. Geological Society of America, Special Papers, **388**, 477–501.

Sheth, H.C. 2005*b*. Were the Deccan flood basalts derived in part from ancient oceanic crust within the Indian continental lithosphere? *Gondwana Research*, **8**, 109–127.

Späth, A., Le Roex, A.P. & Opiyo-Akech, N. 2001. Plume–lithosphere interaction and the origin of continental rift-related alkaline volcanism: the Chyulu Hills volcanic province, southern Kenya. *Journal of Petrology*, **42**, 765–787.

Stern, R.J. & De Wit, M.J. 2003. Rocas Verdes ophiolites, southernmost South America: remnants of progressive stages of development of oceanic-type crust in a continental margin back-arc basin. *In*: Dilek, Y. & Robinson, E.T. (eds) *Ophiolites in Earth History*. Geological Society, London, Special Publications, **218**, 665–683, https://doi.org/10.1144/GSL.SP.2003.218.01.32

Storey, M., Kent, R.W. *et al.* 1992. Lower cretaceous volcanic rocks on continental margins and their relationship to the Kerguelen Plateaus. *In*: Wise, S.W. & Schlich, R. (eds) *Proceedings of the Ocean Drilling Program, Scientific Results, Volume 120*. Ocean Drilling Program, College Station, Texas, 33–53.

Sun, S.-s. & McDonough, W.F. 1989. Chemical and isotopic systematics of oceanic basalts: implications for mantle composition and processes. *In*: Saunders, A.D. & Norry, M.J. (eds) *Magmatism in the Ocean Basins*. Geological Society, London, Special Publications, **42**, 313–345, https://doi.org/10.1144/GSL.SP.1989.042.01.19

Sun, S.-s. & Nesbitt, R.W. 1978. Petrogenesis of Archaean ultrabasic and basic volcanics: Evidence from rare earth elements. *Contributions to Mineralogy and Petrology*, **65**, 301–325.

Tang, Y.J., Zhang, H.F. & Ying, J.F. 2006. Asthenosphere–lithospheric mantle interaction in an extensional regime: implication from the geochemistry of Cenozoic basalts from Taihang Mountains, North China Craton. *Chemical Geology*, **233**, 309–327.

Tappe, S., Foley, S.F., Stracke, A., Romer, R.F., Kjarsgaard, B.A., Heaman, L.M. & Joyce, N. 2007. Craton reactivation on the Labrador Sea margins: $^{40}Ar/^{39}Ar$ age and Sr–Nd–Hf–Pb isotope constraints from alkaline and carbonatite intrusives. *Earth and Planetary Science Letters*, **256**, 433–454.

Thompson, G. 1991. Metamorphic and hydrothermal processes: basalt–seawater interactions. *In*: Floyd, P.A. (ed.) *Oceanic Basalts*. Blackie, Glasgow, 148–173.

Thompson, R.N., Ottley, C.J. *et al.* 2005. Source of the Quaternary alkalic basalts, picrites and basanites of the Potrillo volcanic field, New Mexico, USA: lithosphere or convecting mantle? *Journal of Petrology*, **46**, 1603–1643.

Tilton, G.R., Bryce, J.G. & Mateen, A. 1998. Pb–Sr–Nd isotope data from 30 and 300 Ma collision zone carbonatites in Northwest Pakistan. *Journal of Petrology*, **39**, 1865–1874.

Turcotte, D.L. & Emerman, S.H. 1983. Mechanisms of active and passive rifting. *Tectonophysics*, **94**, 39–50.

Valdiya, K.S. 2010. *The Making of India: Geodynamic Evolution*. Macmillan, India.

Veevers, J.J. & Tewari, R.C. 1995. Permian–Carboniferous and Permian–Triassic magmatism in the rift zone bordering the Tethyan margin of southern Pangea. *Geology*, **23**, 467–470.

Vijaya Kumar, K. & Leelanandam, C. 2008. Evolution of the Eastern Ghats Belt, India: a plate tectonic perspective. *Journal of the Geological Society of India*, **72**, 720–749.

Vijaya Kumar, K. & Rathna, K. 2008. Geochemistry of the mafic dykes in the Prakasam alkaline province of Eastern Ghats Belt, India: implications for the genesis of continental rift-zone magmatism. *Lithos*, **104**, 306–326.

Vijaya Kumar, K. & Rathna, K. 2014. Chapter 3: Geochemical modeling of melting and cumulus processes: a theoretical approach. *In*: Kumar, S. & Singh, R.N. (eds) *Modelling of Magmatic and Allied Processes*. Springer, Cham, Switzerland, 47–73.

Vijaya Kumar, K., Narsimha Reddy, M. & Leelanandam, C. 2006. Dynamic melting of the Precambrian mantle: evidence from rare earth elements of the amphibolites from the Nellore–Khammam schist belt, south India. *Contributions to Mineralogy and Petrology*, **152**, 243–256.

Vijaya Kumar, K., Chavan, C. *et al.* 2010. Geochemical investigation of a semi-continuous extrusive basaltic section from the Deccan Volcanic Province, India: implications for the mantle and magma chamber processes. *Contributions to Mineralogy and Petrology*, **159**, 839–852.

VIJAYA KUMAR, K., RATHNA, K. & LEELANANDAM, C. 2015. Proterozoic subduction-related and continental rift-zone mafic magmas from the Eastern Ghats Belt, SE India: Geochemical characteristics and mantle sources. *Current Science*, **108**, 184–197.

WADIA, D.N. 1934. The Cambrian–Triassic sequence of north-western Kashmir (parts of Muzaffarabad and Baramular districts). *Records of the Geological Survey of India*, **68**, 121–176.

WADIA, D.N. 1961. *Geology of India*. Macmillan, London.

WADIA, D.N. 1975. *Geology of India*. 4th edn. Tata McGraw-Hill, New Delhi.

WANG, Y.J., FAN, W.M., ZHANG, H.F. & PENG, T.P. 2006. Early Cretaceous gabbroic rocks from the Taihang Mountains: implications for a paleosubduction-related lithospheric mantle beneath the central North China Craton. *Lithos*, **86**, 281–302.

WEAVER, B.L. & TARNEY, J. 1981. Chemical changes during dyke metamorphism in high-grade basement terrains. *Nature*, **289**, 47–49.

WEDEPOHL, K.H., GOHN, E. & HARTMANN, G. 1994. Cenozoic alkali basaltic magmas of western Germany and their products of differentiation. *Contributions to Mineralogy and Petrology*, **115**, 253–278.

WEI, X., XU, Y.G., FENG, Y.X. & ZHAO, J.X. 2014. Plume–lithosphere interaction in the generation of the Tarim large igneous province, NW China: geochronological and geochemical constraints. *American Journal of Science*, **314**, 314–356.

WHITE, R.V. & SAUNDERS, A.D. 2005. Volcanism, impact and mass extinctions: incredible or credible coincidences? *Lithos*, **79**, 299–316.

WILSON, M. & DOWNES, H.M. 1991. Tertiary–Quaternary extension-related alkaline magmatism in Western and Central Europe. *Journal of Petrology*, **32**, 811–849.

WILSON, M., ROSENBAUM, J.M. & DUNWORTH, E.A. 1995. Melilitites: partial melts of the thermal boundary layer? *Contributions to Mineralogy and Petrology*, **119**, 181–196.

WITTKE, J.H. & MACK, L.E. 1993. OIB-like mantle source for continental alkaline rocks of the Balcones province, Texas: trace-element and isotopic evidence. *Journal of Geology*, **101**, 333–344.

XIAO, L., XU, Y.G., MEI, H.J., ZHENG, Y.F., HE, B. & PIRAJNO, F. 2004. Distinct mantle sources of low-Ti and high-Ti basalts from the western Emeishan large igneous province, SW China: implications for plume–lithosphere interaction. *Earth and Planetary Science Letters*, **228**, 525–546.

XU, Y.G., CHUNG, S.L., JAHN, B.M. & WU, G.Y. 2001. Petrologic and geochemical constraints on the petrogenesis of Permian–Triassic Emeishan flood basalts in southwestern China. *Lithos*, **58**, 145–168.

XU, Y.G., MA, J.L., FREY, F.A., FEIGENSON, M.D. & LIU, J.F. 2005. Role of lithosphere-asthenosphere interaction in the genesis of Quaternary alkali and tholeiitic basalts from Datong, western North China Craton. *Chemical Geology*, **224**, 247–271.

XU, Y.J. 2002. Evidence for crustal components in the mantle and constraints for crustal recycling mechanisms: pyroxenite xenoliths from Hannuoba, North China. *Chemical Geology*, **182**, 301–322.

ZEYEN, H., VOLKER, F., WEHRLE, V., FUCHS, K., SOBOLEV, S. V. & ALTHERR, R. 1997. Styles of continental rifting: crust–mantle detachment and mantle plumes. *Tectonophysics*, **278**, 329–352.

ZINDLER, A. & HART, S. 1986. Chemical geodynamics. *Annual Review of Earth and Planetary Sciences*, **14**, 493–571.

Imprints of modal metasomatism in the post-Deccan subcontinental lithospheric mantle: petrological evidence from an ultramafic xenolith in an Eocene lamprophyre, NW India

ROHIT PANDEY, N. V. CHALAPATHI RAO*, DINESH PANDIT, SAMARENDRA SAHOO & PRASHANT DHOTE

EPMA Laboratory, Department of Geology, Centre of Advanced Study, Banaras Hindu University, Varanasi 221005, India

**Correspondence: nvcr100@gmail.com*

Abstract: We report here on the occurrence of an interesting mantle-derived ultramafic xenolith entrained in an Eocene (*c.* 55 Ma) lamprophyre dyke from the Dongargaon area of the Chhotaudepur alkaline subprovince located within the Narmada Rift Zone, NW India. The mineralogy of the xenolith comprises olivine, clinopyroxene and mica (phlogopite), with the latter occurring essentially as rims around the clinopyroxene. Inclusions of apatite, interstitial sulphide (pyrite) and micron-scale exsolved spinel are widespread. Olivine is forsteritic ($Fo_{85.34}$), displays little compositional variation and overlaps with that reported from worldwide mantle peridotite xenoliths. Clinopyroxene is a diopside with a compositional range of $Wo_{48.36}$ $En_{43.83}$, $Fs_{6.53}$ and $Ac_{1.27}$, and is conspicuous by its high CaO (up to 24.4 wt%) and TiO_2 (up to 1.6 wt%) content. Clinopyroxene is also compositionally similar to that reported from 'enriched' (metasomatized) peridotite xenoliths rather than those that occur in the 'normal' (depleted) peridotitic xenoliths. Phlogopites have a high concentration of fluorine (up to 1 wt%), whereas the apatites show an anomalous enrichment of F (up to 5 wt%), as well as enrichment in Sr (SrO up to 1.9 wt%). Our study provides the first direct petrographical evidence for the modal metasomatism in the post-Deccan subcontinental lithospheric mantle (SCLM) from this domain. From the textural and mineralogical assemblage of the xenolith, we infer that a possible olivine + garnet + orthopyroxene assemblage, in the presence of a metasomatic fluid, has given rise to clinopyroxene + phlogopite + spinel. The paragenesis of apatite essentially as inclusions suggests that it to be the earliest crystallized phase during the metasomatic event. Geothermobarometry of the clinopyroxene in the xenolith reveal temperatures of approximately 1200°C and pressures of approximately 12 kb, which are comparable with such data reported from other Deccan-related xenoliths. Preservation of phlogopite and apatite in the ultramafic xenolith imply that some of the readily fusible metasomatized portions in this domain escaped wholesale melting during the eruption of the Deccan Traps, possibly due to the variable thickness of the underlying SCLM.

Supplementary material: A table detailing natural and synthetic standards used for calibration is available at https://doi.org/10.6084/m9.figshare.c.3815296

Mantle-derived xenoliths entrained in kimberlites, lamproites, lamprophyres and alkali basalts provide the most direct information about the nature and composition of the lower portions of the Earth's continental plates. Our contemporary understanding of the geodynamic processes of petrogenetic significance, such as the micro- to macro-scale chemical heterogeneities and metasomatism in the Earth's subcontinental lithospheric mantle, is based primarily on studies of the mantle xenoliths (see Nixon *et al.* 1981; Harte & Hawkesworth 1989; Pearson *et al.* 2003; O'Reilly & Griffin 2013 and references therein). Two kinds of metasomatism have been recognized in the published literature, viz. modal and cryptic (Dawson 1984). In the former, the metasomatic process adds new minerals such as amphibole, mica, apatite, ilmenite and zircon to the parent rock; and the latter type involves a chemical change not accompanied by any change in modal mineralogy of the parent rock. Stealth (deceptive) metasomatism is a term that is also used for the processes that add minerals which are mineralogically indistinguishable from those present in common peridotites (O'Reilly & Griffin 2013). Evidence for cryptic, as well as modal, metasomatism from the mantle and lower-crustal xenoliths entrained in kimberlites, lamprophyres and alkali basalts from the on-cratons and off-cratons of the Indian Shield has been well documented (e.g. Ganguly & Bhattacharya 1987; Krishnamurthy *et al.* 1988, 1999;

From: Sensarma, S. & Storey, B. C. (eds) 2018. *Large Igneous Provinces from Gondwana and Adjacent Regions*. Geological Society, London, Special Publications, **463**, 117–136.
First published online July 5, 2017, https://doi.org/10.1144/SP463.6

Mukherjee & Biswas 1988; Karmalkar & Rege 2002; Dessai *et al.* 2004; Karmalkar *et al.* 2009 and references therein; Ray *et al.* 2016).

The purpose of this contribution is to report an unusual mantle-derived ultramafic nodule, which documents the imprints of modal metasomatism, occurring in a lamprophyre dyke in the Chhotaudepur alkaline subprovince located within the Narmada Rift of NW India (Fig. 1). As the host lamprophyre dyke has been dated to be of Eocene age (*c.* 55 Ma: Randive *et al.* 2012), the ultramafic nodule provides a rare opportunity to understand the nature of the sub-Deccan continental lithospheric mantle from this domain of the Deccan Large Igneous Province. We present the detailed petrological studies carried out on the various phases in the ultramafic nodule and explore their geodynamic significance in the context of the duration of the volcanism/magmatism in the Deccan Large Igneous Province.

Geological setting

The study area constitutes a part of the Deccan Large Igneous Province and comes under the Chhotaudepur alkaline subprovince (Fig. 1), where alkaline magmatism covering approximately 1200 km^2 and encompassing exotic rock types, such as carbonatites, nephelinites, lamprophyres, pseudoleucitites, phonolites, picrite basalts, tinguites, fenites and foid syenites, is concentrated (Sukeshwala & Sethna 1967; Viladkar 1981; Durgadmath 1984; Sethna 1989; Gwalani *et al.* 1993; Chawade 1996; Hari 1998; Chalapathi Rao *et al.* 2012; Hari *et al.* 2014). The Chhotaudepur alkaline subprovince is located within the Narmada Rift Zone, and the country rocks include the Deccan Traps, Precambrian basement (gneisses and schists) and Cretaceous sedimentary rocks (Bagh beds) represented by sandstones and limestones (Fig. 1). Gravity and magnetic studies indicate the presence of shallow, presently solidified, ancient magma chamber(s) beneath the Phenaimata layered igneous complex in this domain (Bijendra Singh *et al.* 2014).

The lamprophyre of this study intrudes the Deccan Traps (22° 23′ 31″ N, 74° 13′ 03″ E) in the vicinity of the Dongargaon village (Fig. 1). It strikes N40° W, can be traced to a length of up to 250 m with a width of about 1.5 m width, and is hard, compact and melanocratic in appearance. Rb–Sr mineral isochron dating of the Dongargaon lamprophyre gave a precise age of 55 ± 2.3 Ma (MSWD = 0.021: Randive *et al.* 2012). An oval-shaped ultramafic nodule, 3 mm in length and 2 mm in width (see Fig. 2), was encountered during the petrographical study of the Dongargaon lamprophyre. It should be mentioned here that the studied lamprophyre hosts a number of similar mantle-derived nodules but many of them are either completely altered and/or not well preserved, and hence are not amenable to microprobe studies. However, the nodule under study is fresh and unaltered, and thereby constitutes the subject matter of this paper.

Analytical techniques

Combined optical microscopy (Fig. 2) and backscattered electron (BSE) microscopy (Fig. 3: by electron probe microanalysis (EPMA)) was used to characterize the ultramafic nodule. Mineral chemistry of the various phases in the nodule was determined by a CAMECA-SXFive electron microprobe at the Department of Geology, Banaras Hindu University, Varanasi, India. Wavelength-dispersive spectrometry and a LaB_6 source were deployed for quantitative analyses. An accelerating voltage of 15 kV, a beam current of 10 nA and a beam diameter of 1 μm along with thallium acid pthalate (TAP), large pentaerythritol (LPET) and large lithium fluoride (LLIF) crystals were employed for measurement. A number of natural and synthetic standards were used for calibration, and their details (together with counting statistics) are provided in the Supplementary material. After repeated analyses, it was found that the error on major element concentrations is <1%, whereas the error on trace elements varied between 3 and 5%. X-ray elemental 'dot' mapping (Fig. 4) and line scans across the minerals (Fig. 5) were also conducted to infer the elemental distribution and compositional variation, respectively.

Petrography and mineral chemistry

The ultramafic nodule has a pronounced kelyphitic reaction rim, suggesting its xenolithic relationship with the host lamprophyre (Fig. 2). Textural aspects, commonly reported from mantle-derived ultramafic xenoliths, such as deformation, strained mineral assemblages and recrystallization are not observed. Olivine, clinopyroxene and mica are the major minerals in the xenoliths, with the latter occurring essentially as rims around the clinopyroxene (Fig. 2). BSE images reveal inclusions of hexagonal-shaped apatite and micro-scale exsolved spinels in the mica (Fig. 3). Olivine is subrounded in habit and constitutes a major portion of the xenolith. It is fractured and serpentinized along cracks, thereby implying the influence of fluids. Clinopyroxene is subhedral, feebly pleochroic (pinkish) and shows simple twinning. Mica is highly pleochroic (light pale brown to dark brown) and is essentially seen enveloping, discontinuously, the grain boundary of the clinopyroxene in the form of a rim. Apatite occurs essentially as inclusions in mica and very rarely in the

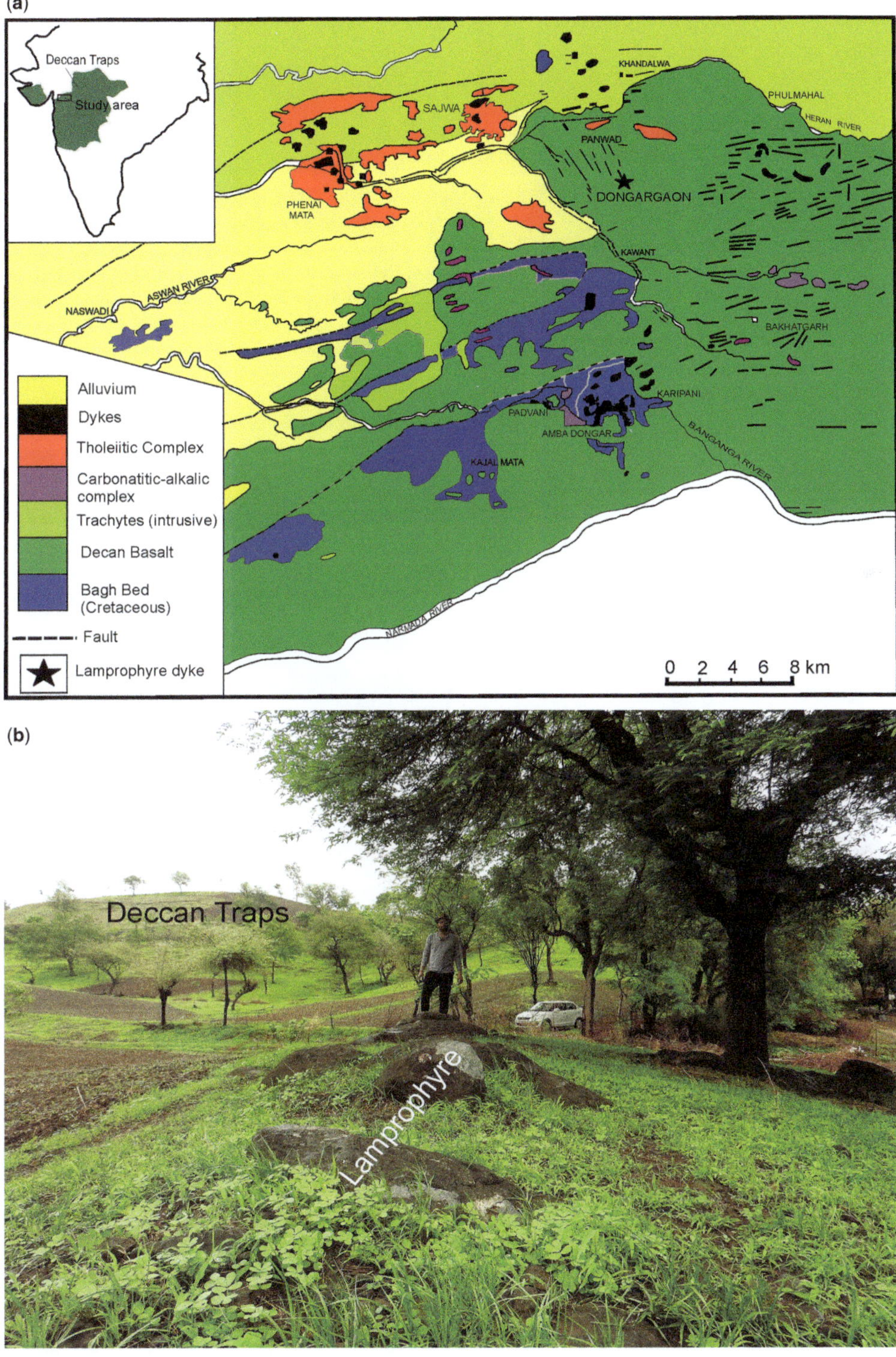

Fig. 1. (**a**) Generalized regional geological map of the Chhotaudepur alkaline province in NW India showing the location of the Dongargaon lamprophyre under study (after Gwalani *et al.* 1993). (**b**) The lamprophyre dyke of this study in the Dongargaon area intruding the Deccan Traps (outcropping in the background).

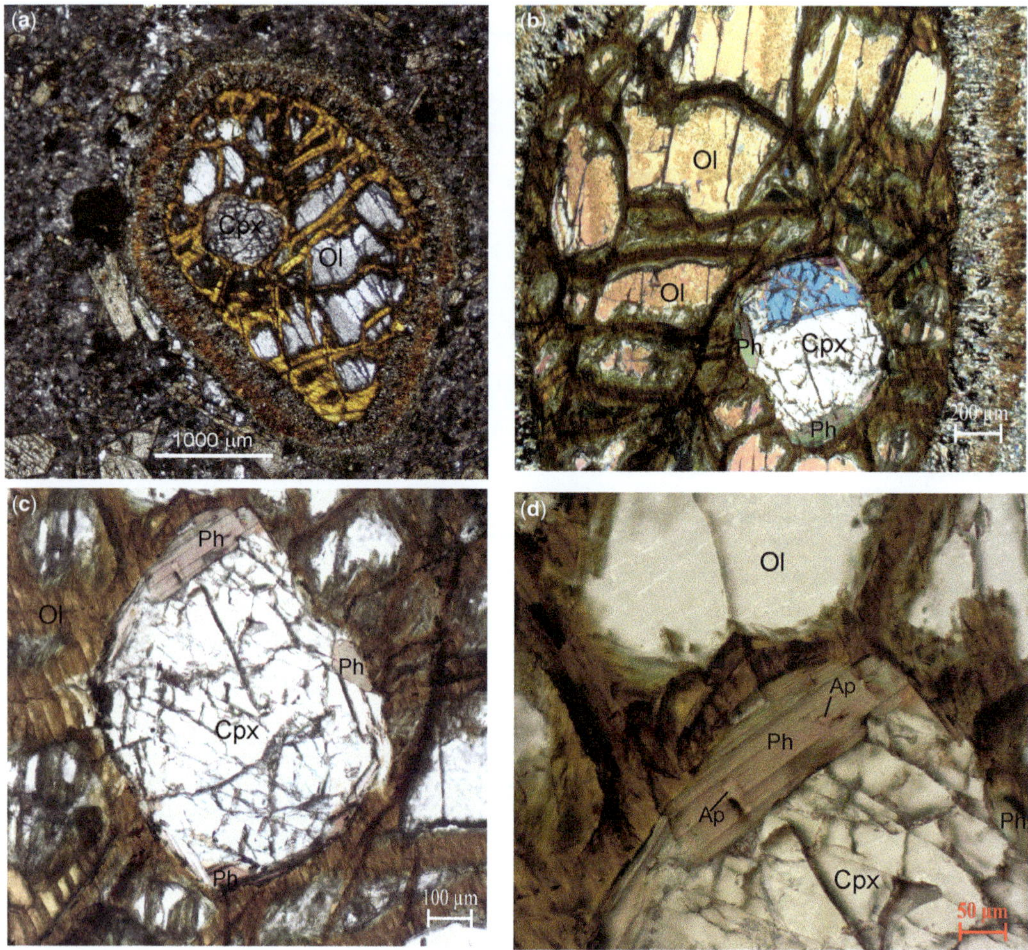

Fig. 2. Photomicrographs of the mantle xenolith encountered in the Dongargaon lamprophyre (sample N/D2/2). (**a**) Oval-shaped mantle xenolith showing olivine and clinopyroxene. Note the reaction rim surrounding the xenoliths (XPL). (**b**) & (**c**) Rimming of phlogopite around the clinopyroxene in the mantle xenoliths (PPL). (**d**) A high-magnification image showing the apatite inclusions within the phlogopite (PPL). Abbreviations: Ol, olivine; Cpx, clinopyroxene; Phl, phlogopite; Ap, apatite; XPL, crossed polars; PPL, plane polarized light.

clinopyroxene (Figs 2 & 3), with textural evidence suggesting apatite to be the earliest crystallized phase amongst these three minerals. Parallel to sub-parallel occurrence of spinel lamellae within the mica imply its formation due to exsolution (Fig. 3). Sulphide (pyrite) (Fig. 3) is a widespread as an interstitial phase in the xenolith. The representative mineral chemistry of the individual phases in the xenolith is provided in Table 1 and discussed below:

Olivine

Composition (average of 10 analyses) of the olivine has an Fo content of 85.34 (Table 1), implying its high magnesium nature. The CaO and NiO contents are 0.30 and 0.17 wt%, respectively (Table 1). X-ray elemental mapping involving Ca kα reveals that the Ca is channelized along the cracks/fractures in the olivine implying the mobility of this element (Fig. 4b). Elemental line scanning across the olivine reveals little variation in its Fo, CaO and MnO content (Fig. 5). Whereas the Fo content of the olivine of this study is similar to that recorded from garnet peridotite xenoliths, its Ni content is relatively lower than those reported from olivine occurring in: (i) spinel lherzolite xenoliths of *c.* 67 Ma alkali basalts from the Kutch, NW India; and (ii) garnet peridotites of 1.1 Ga Wajrakarur kimberlites, southern India (Fig. 6). Importantly, the Ni content in the olivine is substantially lower than those recorded in pyroxenites

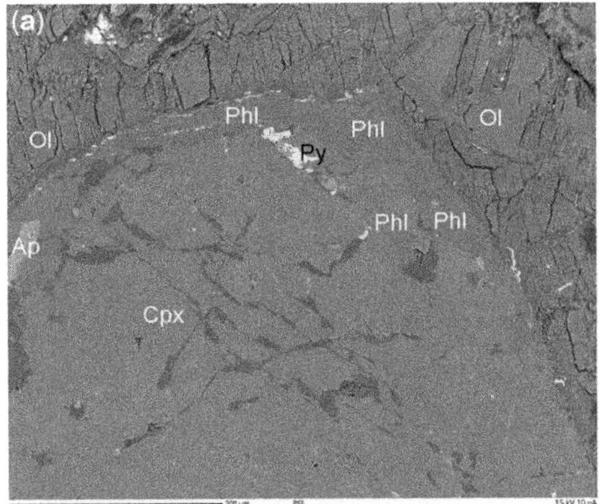

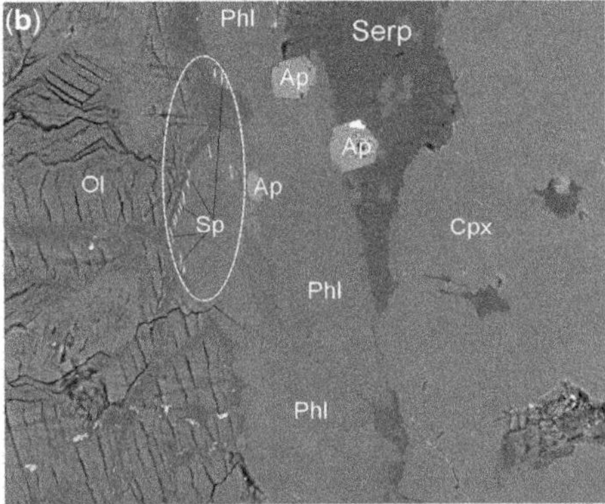

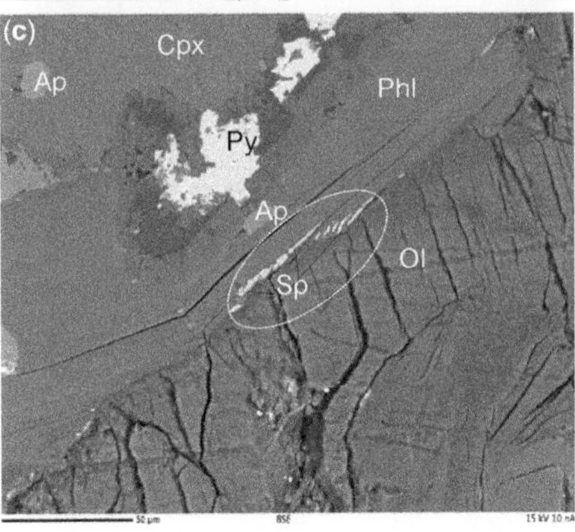

Fig. 3. Backscattered electron (BSE) images (by EPMA) of the various textural features in the mantle xenolith (sample N/D2/2). (**a**) Phlogopite rimming (with inclusions of apatite) along the grain boundary between clinopyroxene and olivine. Note the presence of pyrite along the crack/vein of clinopyroxene. (**b**) & (**c**) Parallel to sub-parallel exsolved spinel grains (domain highlighted by dotted white ellipse) in the phlogopite; note the inclusions of apatite in the phlogopite and clinopyroxene, as well as in the serpentinized olivine. Irregular patches of pyrite are also present. Abbreviations: Ol, olivine; Cpx, clinopyroxene; Phl, phlogopite; Ap, apatite; Serp, serpentinized olivine; Py, pyrite; Sp, spinel.

and rules out the possibility of the latter constituting the source rock of the xenoliths of this study.

Clinopyroxene

Chemistry of the clinopyroxene (average of 14 analyses: Table 1) reveals that it is a diopside with an average composition of $Wo_{48.36}$ $En_{43.83}$, $Fs_{6.53}$ and $Ac_{1.27}$. It is conspicuously high in CaO (up to 24.4 wt%) and TiO_2 (up to 1.6 wt%), and impoverished in Cr (Cr_2O_3 0.06 wt%) (see Table 1 and Fig. 4). Elemental line scanning across does not reveal any substantial deviation in the composition of its major elements (Fig. 5). The alumina content (3.09 wt%) of the clinopyroxene of this study is lower than that found reported from the sub-calcic pyroxene xenocrysts and Al-rich megacrysts from alkali basalts, and is closer in composition to the clinopyroxene occurring in peridotitic xenoliths in kimberlites (Fig. 7a). However, it should be pointed out here that the diopsides occurring in garnet peridotites and discrete nodules in kimberlites are essentially chrome diopsides. Likewise, the clinopyroxene reported from the spinel lherzolite xenoliths entrained in the Deccan-related alkali basalts of the Kutch, NW India (Krishnamurthy *et al.* 1988, 1999; Karmalkar & Rege 2002) and Murud Janjira lamprophyres, western India (Dessai *et al.* 2004) are also enriched in chromium. The Na_2O content of the clinopyroxene of this study is also distinct from that of the primary clinopyroxene of peridotite paragenesis (including the clinopyroxene reported from the xenoliths from Kutch and Murud Janjira) but is strikingly similar to the secondary (metasomatic) clinopyroxene formed by fluid–melt action in the mantle xenoliths (Fig. 7b). The high titania content of the clinopyroxene from this study also distinguishes it from the clinopyroxene occurring in the xenoliths from the Kutch, Murud Janjira and orangeites of the Bastar Craton, central India (Fig. 7c).

Mica

Mica in this study is compositionally a phlogopite (Fig. 8a), and is enriched in titania (TiO_2: up to 5.76 wt%), barium (up to 1.55 wt% BaO) and fluorine (up to 0.68 wt% F: Table 1). X-ray element mapping reveals the high potassium concentration in the phlogopite compared to the other phases in the xenoliths where potassium occurs as a widely dispersed element (Fig. 4). Whereas MgO and K_2O show little variation, a slight enrichment in titania is conspicuous from the core to the rim of the analysed phlogopite (Fig. 5; Table 1). The Al_2O_3 content of the phlogopite of this study is much higher (Fig. 8b) than those reported from the phlogopite in the mica–amphibole–rutile–ilmenite–diopside (MARID) suite of mantle xenoliths from the southern African

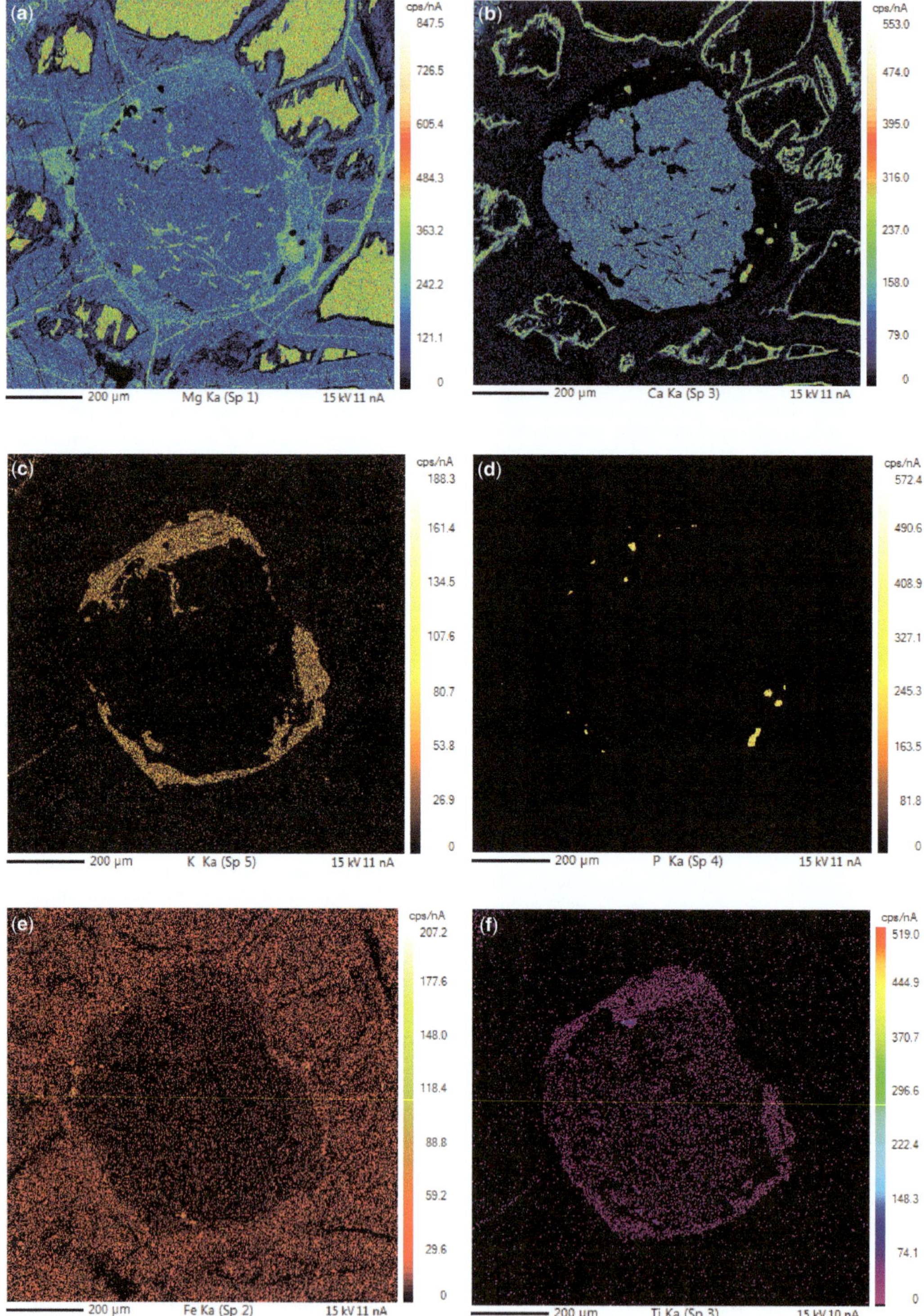

Fig. 4. X-ray elemental maps (generated by EPMA) of (**a**) Mg, (**b**) Ca, (**c**) K, (**d**) P, (**e**) Fe and (**f**) Ti in the olivine, clinopyroxene, phlogopite, apatite and spinel from the mantle xenolith characterized in Figures 2 and 3.

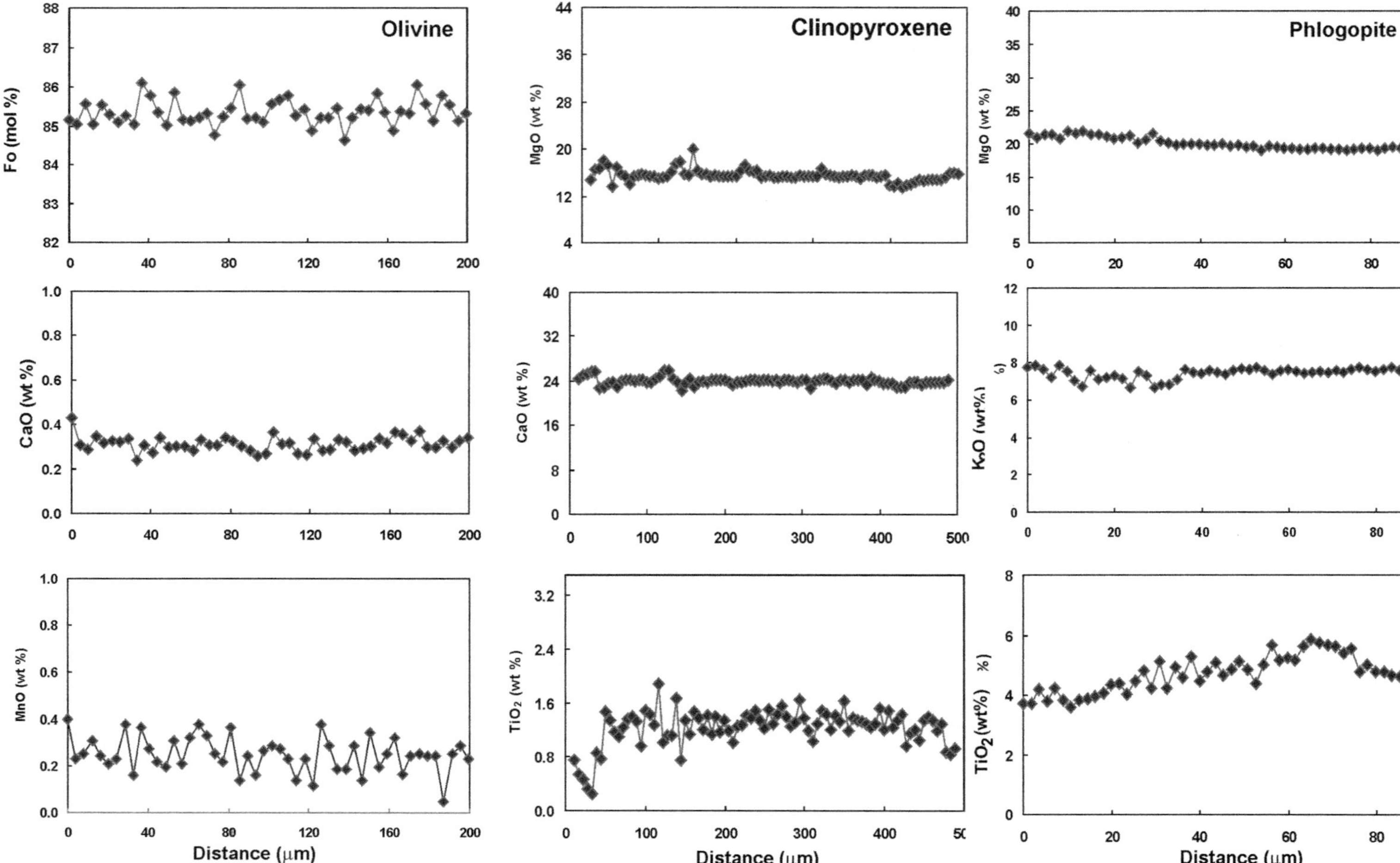

Fig. 5. Line profiles across olivine, clinopyroxene and phlogopite, which display an overall reduced variability in the Fo, CaO and MnO content.

Table 1. *Mineral chemistry (oxide wt%) of the olivine, diopside, phlogopite and apatite from the Mantle xenolith of this study*

Olivine	
Oxide	$n = 10$
SiO_2	39.54
TiO_2	0.05
Al_2O_3	0.00
Cr_2O_3	0.01
FeO	13.85
MnO	0.18
MgO	45.81
NiO	0.17
CaO	0.30
Total	99.92
Cations for 4 oxygen atoms	
Si	0.993
Ti	0.001
Al	0.000
Cr	0.000
Fe(II)	0.291
Mn	0.004
Mg	1.715
Ni	0.003
Ca	0.008
Total	3.01
Fo	85.34
Fa	14.47
Tp	0.19

Diopside	
Oxide	$n = 14$
SiO_2	50.80
TiO_2	1.10
Al_2O_3	3.09
Cr2O3	0.06
FeO	4.08
MnO	0.09
MgO	15.60
CaO	23.95
Na_2O	0.35
K_2O	0.02
Total	99.15
Cations for 6 oxygen atoms	
Si	1.89
Al	0.13
Fe(III)	0.081
Fe(II)	0.045
Cr	0.002
Ti	0.031
Mn	0.003
Mg	0.864
Ca	0.953
Na	0.025
K	0.001
Total	4.03
Wo	48.36
En	43.83
Fs	6.53
Ac	1.27
T (°C) (Grove & Juster 1989)	1221–1235
P (kb) (Putirka 2008)	8.6–11.3
T (°C) (Putirka 2008)	1188–1232
P (kb) (Ishii 1979)	8.5–12.4

Phlogopite															
	Rim	Rim	Core	Core	Core	Core	Core	Rim	Rim	Rim	Rim	Rim	core	core	core
SiO_2	36.76	36.15	36.17	36.24	36.61	36.59	36.58	36.15	36.83	36.35	36.19	36.26	36.82	37.20	37.03
TiO_2	3.73	4.82	5.28	4.67	5.12	5.66	5.62	5.76	5.65	5.03	4.61	4.81	2.95	3.16	4.83
Al_2O_3	14.85	15.47	15.01	15.22	15.05	15.19	15.20	15.06	14.92	15.22	15.28	14.67	15.73	15.81	14.66
FeO	7.55	7.88	7.95	7.31	7.48	7.89	8.58	8.36	7.67	7.89	8.24	8.10	8.23	7.53	7.92
MnO	0.06	0.06	0.02	0.05	0.04	0.01	0.06	0.04	0.13	0.02	0.02	0.10	0.05	0.00	0.00
MgO	21.02	20.61	20.04	20.00	19.77	19.75	19.17	19.31	19.22	19.39	19.50	19.39	19.42	19.49	19.47
CaO	0.00	0.00	0.00	0.00	0.00	0.00	0.00	0.00	0.00	0.00	0.00	0.00	0.00	0.00	0.17
BaO	0.25	1.13	1.55	1.55	0.53	0.28	0.67	0.35	0.42	0.64	0.67	0.64	1.02	2.25	0.18
Na_2O	1.13	1.25	1.42	1.19	1.30	1.34	1.25	1.29	1.29	1.35	1.34	1.38	0.89	1.13	1.39
K_2O	7.88	7.31	7.45	7.38	7.70	7.35	7.44	7.50	7.60	7.65	7.60	7.64	7.97	5.97	7.74
Cl	0.05	0.06	0.07	0.02	0.02	0.04	0.03	0.05	0.04	0.02	0.04	0.02	0.02	0.05	0.02
F	0.35	0.46	0.46	0.17	0.28	0.25	0.17	0.47	0.32	0.47	0.68	0.65	0.63	0.97	0.43
Total	93.63	95.20	95.42	93.80	93.90	94.36	94.77	94.33	94.10	94.02	94.18	93.63	93.71	93.57	93.84
Cations for 22 oxygen atoms															
Si	5.415	5.280	5.299	5.356	5.380	5.340	5.346	5.315	5.401	5.362	5.352	5.393	5.469	5.519	5.451
Ti	0.414	0.530	0.582	0.519	0.566	0.622	0.618	0.637	0.623	0.558	0.513	0.538	0.329	0.353	0.535
Al	2.579	2.663	2.591	2.650	2.608	2.613	2.618	2.610	2.579	2.646	2.663	2.571	2.754	2.765	2.544
Fe(II)	0.930	0.962	0.974	0.904	0.920	0.963	1.048	1.027	0.941	0.973	1.019	1.007	1.022	0.934	0.975
Mn	0.007	0.007	0.003	0.006	0.004	0.002	0.007	0.004	0.016	0.003	0.003	0.013	0.006	0.000	0.000
Mg	4.616	4.487	4.377	4.406	4.333	4.297	4.176	4.233	4.202	4.264	4.300	4.299	4.300	4.311	4.273
Ca	0.000	0.000	0.000	0.000	0.000	0.000	0.000	0.000	0.000	0.000	0.000	0.000	0.000	0.000	0.026
Ba	0.014	0.065	0.089	0.090	0.031	0.016	0.038	0.020	0.024	0.037	0.039	0.037	0.060	0.131	0.010
Na	0.323	0.354	0.404	0.342	0.370	0.379	0.353	0.367	0.368	0.386	0.383	0.397	0.255	0.325	0.395
K	1.481	1.362	1.393	1.392	1.443	1.368	1.387	1.407	1.422	1.439	1.434	1.450	1.510	1.130	1.452
Cl	0.012	0.014	0.017	0.005	0.004	0.011	0.008	0.012	0.011	0.005	0.011	0.005	0.005	0.011	0.005
F	0.165	0.212	0.213	0.080	0.132	0.114	0.080	0.218	0.150	0.217	0.318	0.304	0.294	0.457	0.201
Total	15.77	15.71	15.71	15.66	15.65	15.60	15.59	15.62	15.57	15.66	15.70	15.70	15.70	15.46	15.66
Mg#	0.83	0.82	0.82	0.83	0.82	0.82	0.80	0.80	0.82	0.81	0.81	0.81	0.81	0.82	0.81

Oxide (wt%)	Apatite									
	1	2	3	4	5	6	7	8	9	10
CaO	51.07	50.40	50.43	51.35	51.82	51.29	51.28	52.66	51.74	47.08
Na_2O	0.20	0.29	0.43	0.38	0.23	0.23	0.22	0.20	0.23	0.20
SrO	1.95	1.86	1.86	1.95	1.92	1.91	1.87	1.85	1.87	1.77
ZrO2	0.00	0.67	0.60	0.62	0.71	0.66	0.59	0.74	0.66	0.59
FeO	0.51	0.71	0.48	0.58	0.43	0.51	0.47	0.29	0.23	0.86
MnO	0.00	0.19	0.12	0.13	0.09	0.05	0.02	0.02	0.07	0.05
MgO	0.14	0.52	0.12	0.12	0.14	0.21	0.10	0.06	0.53	0.87
P_2O_5	40.32	38.33	37.40	39.56	39.46	40.85	41.15	40.10	39.40	40.68
SiO_2	0.76	1.61	1.52	0.96	0.50	0.55	0.53	0.53	0.96	1.82
Cl	0.75	0.65	0.76	0.75	0.64	0.70	0.71	0.69	0.68	0.69
F	4.74	4.24	4.42	5.42	3.13	4.73	3.58	3.58	3.94	4.91
Total	100.44	99.46	98.15	101.82	99.06	101.68	100.49	100.72	100.30	99.51
Cations per 25 (O, OH, F, Cl) atoms										
Ca	9.020	9.063	9.219	9.017	9.380	9.380	9.062	9.360	9.360	8.288
Na	0.062	0.094	0.143	0.121	0.074	0.074	0.071	0.064	0.064	0.062
Sr	0.186	0.181	0.184	0.186	0.188	0.188	0.179	0.178	0.178	0.168
Zr	0.000	0.109	0.100	0.100	0.117	0.117	0.095	0.119	0.119	0.095
Fe	0.070	0.100	0.069	0.079	0.061	0.061	0.065	0.040	0.040	0.118
Mn	0.000	0.026	0.017	0.018	0.013	0.013	0.003	0.003	0.003	0.007
Mg	0.035	0.130	0.031	0.029	0.035	0.035	0.026	0.015	0.015	0.212
P	5.627	5.447	5.402	5.489	5.645	5.645	5.745	5.632	5.632	5.659
Si	0.126	0.270	0.260	0.157	0.084	0.084	0.088	0.089	0.089	0.298
Cl	0.210	0.184	0.220	0.208	0.182	0.182	0.199	0.193	0.193	0.193
F	2.471	2.252	2.387	2.808	1.671	1.671	1.866	1.879	1.879	2.551
Total	16.12	16.43	16.44	16.20	16.61	16.61	16.33	16.51	16.51	15.91

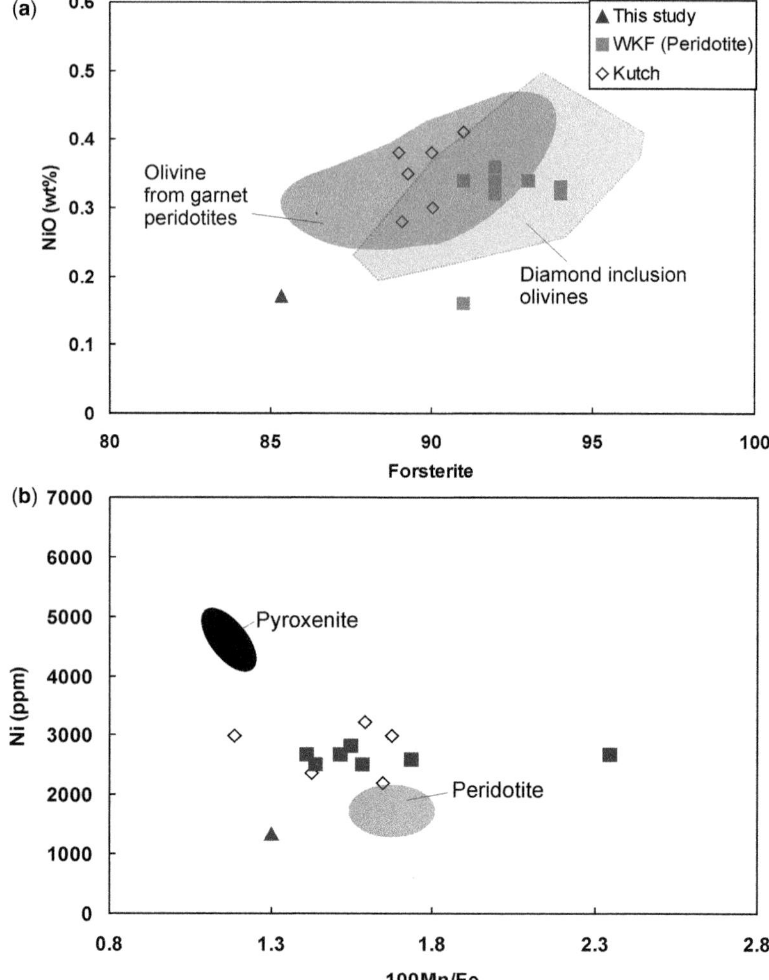

Fig. 6. (**a**) Variation in the forsterite (Fo) v. NiO (wt%) content in the olivine from the mantle xenoliths from this study. Fields of diamond inclusion olivines and olivine from garnet peridotites are taken from Rudnick *et al.* (1994). Also plotted are the data for olivine from the (i) garnet peridotite xenoliths (Ganguly & Bhattacharya 1987) from the 1.1 Ga kimberlites from the Wajrakarur kimberlite field (WKF), eastern Dharwar Craton, southern India and (ii) spinel lherzolite xenoliths (Krishnamurthy *et al.* 1988) from the 68 ± 2 Ma alkali basalts of Kutch, NW India. Note the lower Fo and NiO content of the olivine from this study. (**b**) 100 Mn/Fe v. Ni (ppm) bivariate plot showing the compositional fields of olivine from peridotite and pyroxenite (after Foley *et al.* 2013). Note the peridotitic affinity of the olivine from this study. Symbols and data sources are the same as in (a).

orangeites, whereas its Mg number (Mg#) is slightly lower than that found in micas from: (i) phlogopite–ilmenite–clinopyroxene ± rutile (PIC) xenoliths of southern African kimberlites; and (ii) garnet–peridotite xenoliths from the Wajrakarur kimberlites (Fig. 8b).

Apatite

The CaO and P_2O_5 content of the various apatites show little variation (Table 1). They are Sr-rich (SrO: up to 1.95 wt%) and have an anomalously high fluorine content (F: up to 5.42 wt%) compared to their chlorine content (Cl: up to 0.75 wt%). The composition of the apatite grains of this study excludes their magmatic origin and clearly confines them to the field of metasomatic apatite (Fig. 9).

Spinel

The spinel grains are too fine grained (1–2 µm in size) to get meaningful EPMA analyses. However,

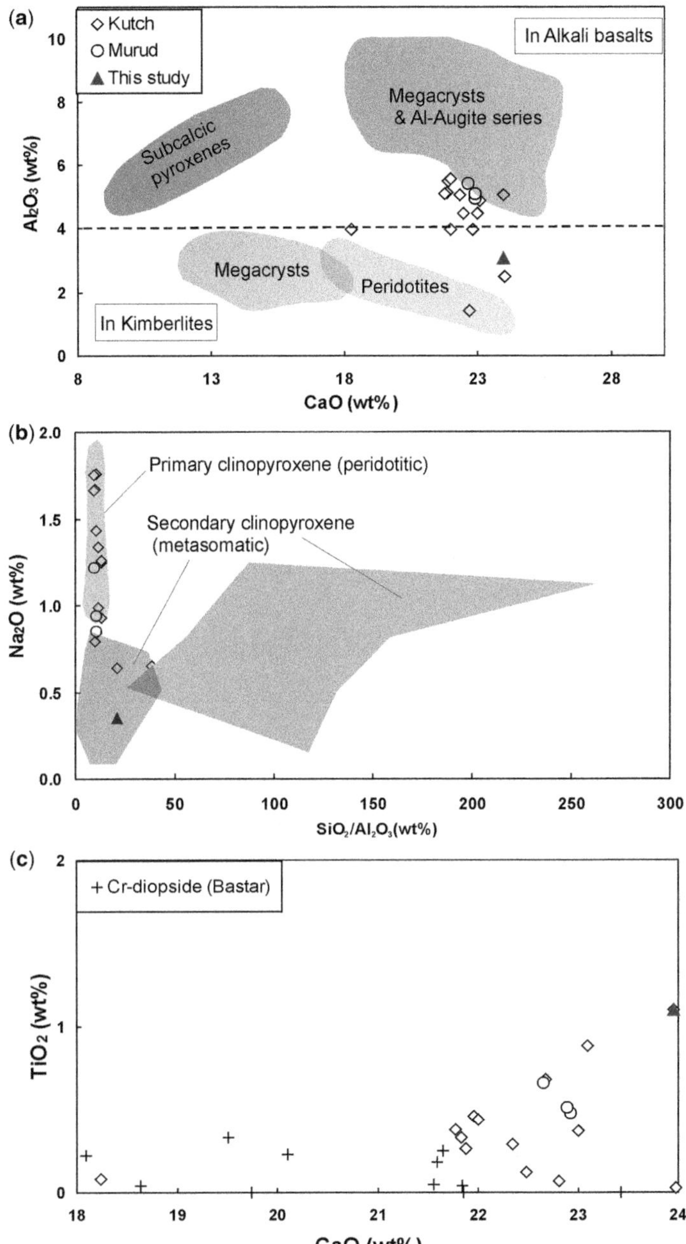

Fig. 7. (**a**) CaO (wt%) v. Al_2O_3 (wt%) bivariate plot for discriminating (dashed line) xenocrytic and megacrystic clinopyroxenes occurring in kimberlites and in alkali basalts (after Haggerty 1995). Note the peridotitic nature of the clinopyroxene of this study. The data for clinopyroxene in (i) spinel lherzolite xenoliths from Kutch (data from Karmalkar & Rege 2002) and (ii) clinopyroxenite and websterite xenoliths (data from Dessai *et al.* 2004) from the Murud Janjira lamprophyres, Mumbai, are also shown. (**b**) SiO_2/Al_2O_3 (wt%) v. Na_2O (wt%) plot distinguishing the primary (peridotitic) and secondary (metasomatic) clinopyroxene from the mantle xenoliths (after Bonadiman *et al.* 2005). Note the strongly metasomatic nature of the clinopyroxene of this study. The symbols and data sources remain the same as in (a). (**c**) CaO (wt%) v. TiO_2 (wt%) content of the clinopyroxene from this study. The data for Cr-diopside xenocrysts from the orangeites from the Bastar Craton (data after Chalapathi Rao *et al.* 2013) are also shown. The symbols and data sources are the same as in (a). Note the distinctly high titania content of the clinopyroxene of this study.

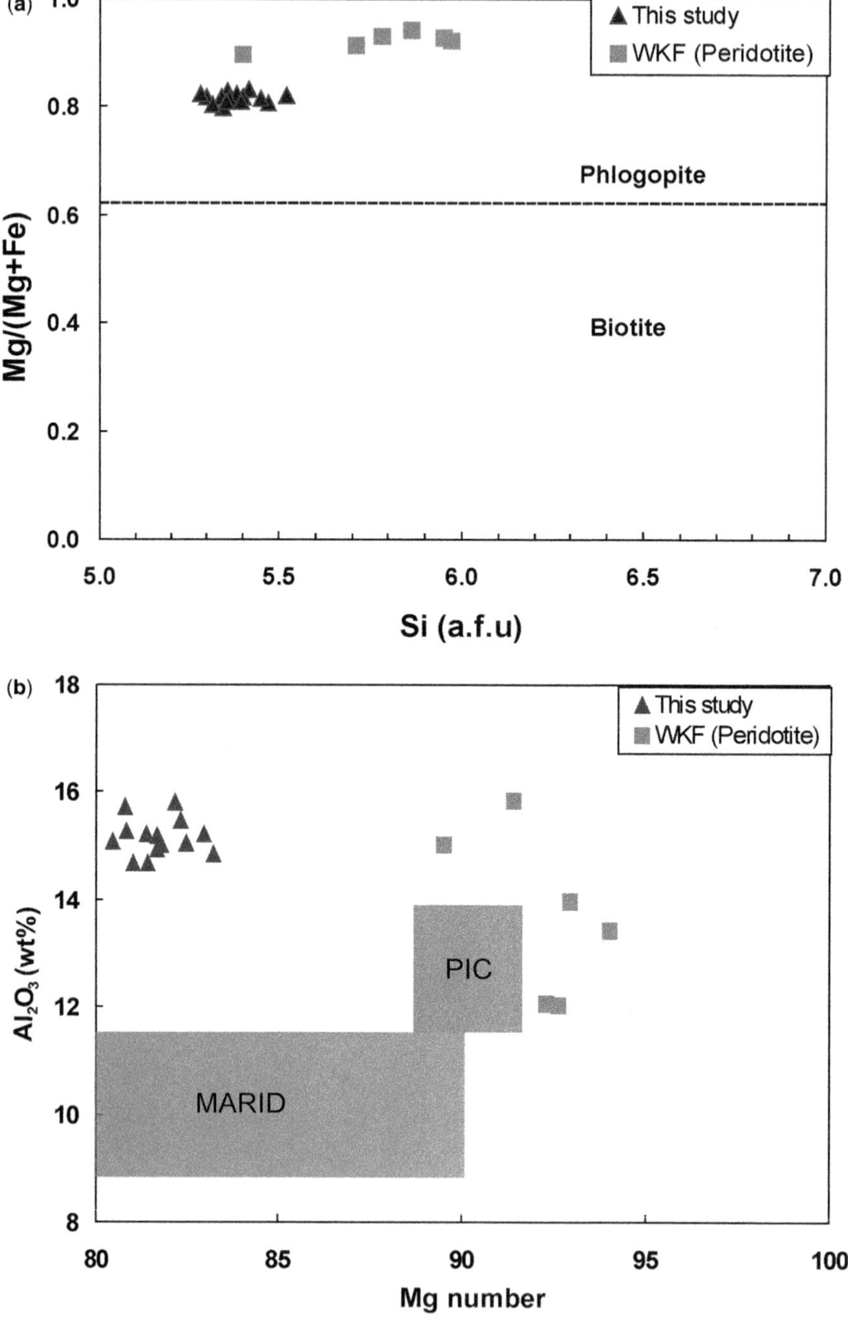

Fig. 8. (**a**) Si (atoms per formula unit) v. Mg/(Mg + Fe) mica classification diagram (after Reider *et al.* 1998) for micas (cores and rims) from the mantle xenolith of this study highlighting their phlogopitic nature. Data for phlogopites from the garnet peridotite xenoliths (Ganguly & Bhattacharya 1987) from the Wajrakarur kimberlite field (WKF), eastern Dharwar Craton, southern India, are also provided. (**b**) Mg number v. Al_2O_3 (wt%) plot of the phlogopites from this study compared with those from the garnet peridotite xenoliths (Ganguly & Bhattacharya 1987) from the WKF, MARID (mica–amphibole–rutile– ilmenite–diopside) and PIC (phlogopite–ilmenite–clinopyroxene ± rutile) suite of xenoliths from southern Africa (after Gregoire *et al.* 2002 and references therein). Note the distinctly high alumina content of the micas of this study.

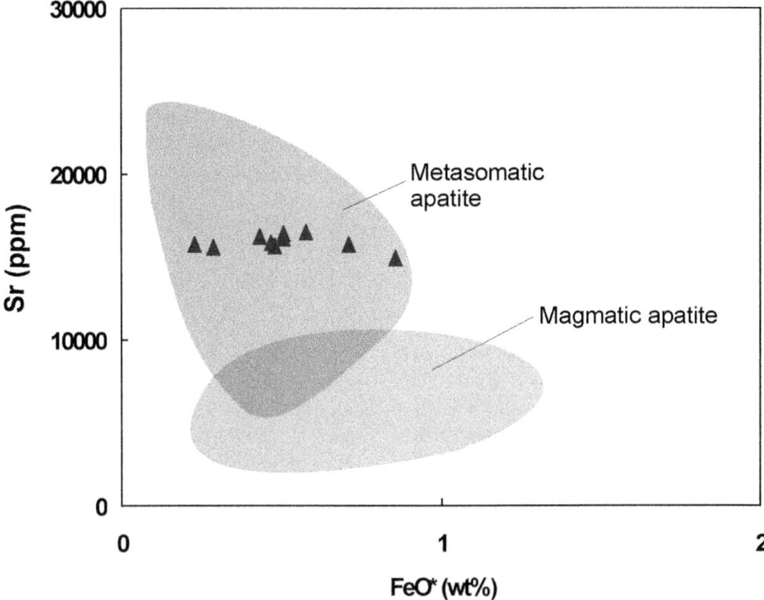

Fig. 9. FeO* (wt%) v. Sr (ppm) of the apatites from the mantle xenolith of this study. Fields of metasomatic apatite and magmatic apatite are also shown (after O'Reilly & Griffin 2000). Note the strongly metasomatic nature of the apatite of this study.

qualitative analyses reveal them to be magnetite with minor amounts of magnesium.

Discussion

Rounded appearance and disequilibrium reaction rim are generally considered to be indicative of mantle xenocrysts/xenoliths (Nixon 1987; Rock 1991). We excluded the possibility of the nodule being a cognate/cumulate, which is commonly known to occur from the crust–mantle boundary due to ponding of the magma. This is because the term 'cognate/cumulate' by definition should have same mineralogy as the host lamprophyre (see Rock 1991), which is not the case here. For example, olivine is entirely absent in the host lamprophyre; likewise, the mica present in the host lamprophyre is a biotite instead of phlogopite, and the clinopyroxene in the host lamprophyre is titania- and iron-enriched and Mg-depleted compared to that in the clinopyroxene from the nodule (unpublished data). The Mg# of olivine present in this ultramafic nodule is 85 (Table 1), which is lower (88–92) than that reported from the cratonic and off-cratonic Mg-rich, low-temperature mantle peridotite xenoliths (Type I: Pearson *et al.* 2003). Nevertheless, the Mg# of the olivine of this study is indistinguishable from that (83–89) reported from olivines found in the cratonic/cirum-cratonic coarse Fe-rich, low-temperature peridotites and pyroxenites (Type II: Pearson *et al.* 2003). The much lower Ni content in olivine compels us to exclude its pyroxenitic derivation (see above). It should be pointed out here that iron-rich olivines are also found in iron-rich lherzolite mantle xenoliths from the Hawaii, and are interpreted to be products resulting from mixing between iron-rich pyroxenites and normal peridotite, followed by subsequent re-equilibration (Sen & Leeman 1991). The possibility of a Fe-rich peridotitic mantle in the subcontinental lithospheric mantle of this domain, contributing to the Fe-rich nature of the olivine, also cannot be ruled out since olivines from the lamprophyres from the nearby Phenaimata and Mundwara alkaline complexes are known to contain an appreciable fayalitic content (Sahoo 2016). Interestingly, Sen & Chandrasekharam (2011) argued for the involvement of an Fe-rich peridotite source in the generation of some of the Deccan lavas. Likewise, the existence of a Fe-rich mantle underlying several cratonic domains of the Indian Shield has been documented (see Sensarma *et al.* 2013 and the references therein). Therefore, we conclude that the ultramafic nodule under study is a xenolith, very likely to have been derived from a Fe-rich peridotitic mantle, and entrained in the ascending lamprophyre magma.

The composition (Fo and Ni content) of the olivine implies the derivation of the xenolith from a peridotitic, rather than a pyroxenitic, mantle source. The presence of hydrous mafic silicate (phlogopite) and apatite provides direct and compelling evidence for the metasomatized nature of the xenolith (e.g. Lloyd & Bailey 1975; Haggerty 1989, 1995; Ionov *et al.* 1997; Ackerman *et al.* 2013). The mode of occurrence of the phlogopite as rims along the grain boundaries of the clinopyroxene highlights the role of infiltration of K-rich metasomatic fluids. The high CaO, and TiO_2, coupled with low Na_2O and paucity of Cr_2O_3, in the xenolithic clinopyroxene excludes its primary orgin and implies a secondary or metasomatic origin. In fact, several studies have shown that the clinopyroxenes encountered in the mantle xenoliths are of metasomatic origin (e.g. Pearson *et al.* 2003; Bonadiman *et al.* 2005; Sun *et al.* 2012). The Cr-poor nature of the clinopyroxene suggests that the metasomatic melt may itself have been depleted in chromium – possibly due to an earlier melt extraction experienced by its source. The occurrence of apatite essentially as inclusions implies it to be the earliest crystallized phase during the metasomatism; alternately, it may also be a product of an earlier metasomatic event that can only be confirmed by radiometric age determination. The presence of exsolved spinel, albeit in very minor proportions, in the phlogopite assumes petrogenetic significance and can be related to metasomatic reactions (below).

Extensive studies on the mantle-derived peridotite xenoliths derived from the kimberlites of the southern Africa have brought out a sequence of progressive metasomatism from garnet phlogopite peridotite (stage 1) through phlogopite-bearing peridotite (stage 2) to phlogopite- and amphibole (richterite)-bearing peridotite (stage 3) upon interaction with the metasomatic (enriching) fluids (see Erlank *et al.* 1987; Van Achterbergh *et al.* 2001). The mineralogy of the xenolith of this study points to it belonging to stage 2 metasomatism (above). The nature of the metasomatic fluid itself can also vary greatly from silicate (Kelemen *et al.* 1998; Simon *et al.* 2007) to carbonatitic (Green & Wallace 1988; Yaxley *et al.* 1998) in composition. From their experimental studies, Elkins-Tanton & Grove (2003) suggested that the presence of water-rich fluid reduces the stability of both orthopyroxene and garnet in a lherzolite, and renders phlogopite as the stable aluminous phase in the mantle. The reaction of garnet and orthopyroxene in a garnet lherzolite with the K-rich metasomatic fluid can, thus, transform it to a phlogopite + clinopyroxene + spinel-bearing peridotite, as shown by the following reaction (Erlank *et al.* 1987):

$$\text{Olivine} + \text{orthopyroxene} + \text{garnet} + K_2O\text{-rich fluid} = \text{phlogopite} + \text{diopside} + \text{spinel}$$

The absence of orthopyroxene (or its relict) and garnet coupled with the assemblage of phlogopite, diopside and spinel in the ultramafic xenolith of this study may be accounted for by the above metasomatic reaction. Consumption of orthopyroxene has also been recorded by Shaw *et al.* (2005) from the metasomatized xenoliths from the West Eifel volcanic field, Germany, wherein the precipitation of secondary pyroxene, phlogopite and olivine took place during the reaction in the presence of potassic melt. Formation of clinopyroxene (Ti-rich) and apatite (with Sr-enrichment) suggests that the metasomatic fluid is more of carbonatitic, rather than of silicic, composition (Griffin *et al.* 1999). In fact, the role of carbonatitic fluids leading to cryptic metasomatism in the spinel peridotites entrained in the alkali basalts from the nearby Kutch area (Karmalkar & Rege 2002) and the involvement of carbonated lherzolite in the genesis of alkali basalts of the Kutch (see Sen *et al.* 2009, 2016) are well documented. The occurrence in time and space of Amba Dongar carbonatite and several alkaline plugs in the Chhotaudepur subprovince (Fig. 1) provide field support for our proposal of involvement of a carbonatitic-type metasomatism. The occurrence of sulphide minerals as interstitial minerals at silicate grain boundaries is also suggestive of their metasomatic origin (see Lorand *et al.* 2003). Alternately, the xenolith of this study could also have been formed by high-pressure crystallization of a hydrous alkaline melt in a vein assemblage with a mantle plume (the Deccan plume in this case), which is likely to be the source of water – as inferred in the case of Hawaiian xenoliths (see Sen *et al.* 1996). However, the trace element content and chronology of various phases in the xenolith are required to evaluate this proposition.

To the best of our knowledge, garnet-bearing peridotite xenoliths were never reported from any of the litho-units of the Deccan Large Igneous Province. However, it may be pointed out that a solitary phase rich in TiO_2 and MgO and suspected to be an 'anomalous garnet' of the compositional range of pyrope–almandine–andradite(?) has been reported from the groundmass of a melanephelinite from the Kutch area (Krishnamurthy *et al.* 1999). In order to independently ascertain whether the xenolith of our study was sampled from a garnet stability field, we compared the REE pattern (unpublished data of the authors) of the host lamprophyre with that of the other Deecan-related lamprophyres from Murud Janjira, western India and from the Chhotaudepur – both of which are inferred to have been derived from melting of a metasomatized lithospheric mantle in the presence of a residual garnet (see Chalapathi Rao *et al.* 2012). The highly fractionated REE pattern (La/Yb: 65) of the Dongargaon lamprophyre is indistinguishable from that of the Murud Janjira (La/Yb: 66–80) and the Chhotaudepur (La/Yb: 66–80) lamprophyres (Fig. 10), implying an overall similarity in their petrogenesis. The presence of a residual garnet trace element signature in the alkali

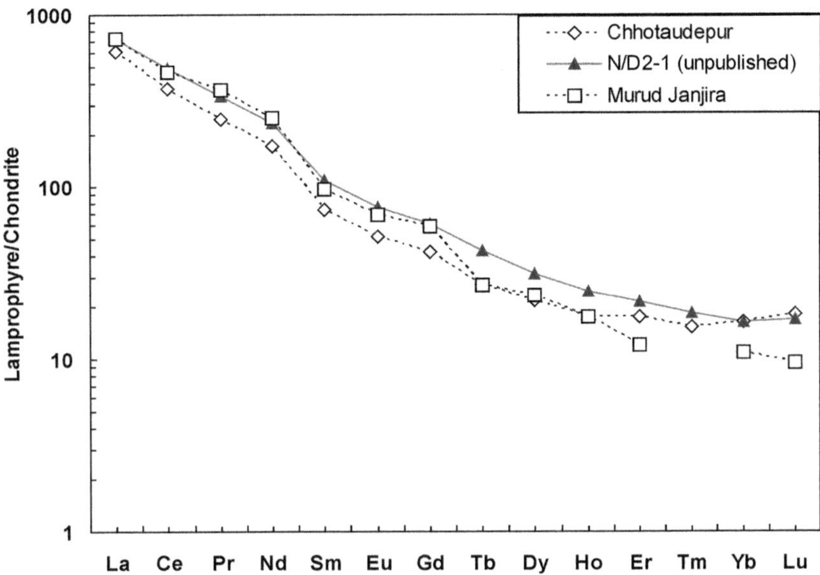

Fig. 10. Chondrite-normalized (after Evensen *et al.* 1978) rare earth element pattern of the Dongargaon lamprophyre of this study (ND2/2) compared with that of the other Deccan lamprophyres from Murud Janjira (after Dessai & Viegas 2010) and Chhotaudepur (after Chalapathi Rao *et al.* 2012).

basalts from the Kutch has also been recognized recently (Sen *et al.* 2009).

Pressure–temperature estimates

The mineralogical composition of mineral phases in the studied xenolith is a function of the temperature and pressure. Grove & Juster (1989) proposed a geothermometer to consider the sensitive parameters of Na, K, Mg# and TiO_2 to constrain the temperature condition of the clinopyroxene formation. The clinopyroxene composition obtained from the mantle xenolith of this study was used for temperature estimations, which ranged from 1221 to 1235°C. However, the concentrations of Na and K are very low in the clinopyroxene of this study, which provide a poor temperature estimation. Ishii (1979) suggested that the Fe and Mg content in clinopyroxene is directly proportional to the temperature of formation and can be a reliable geothermometer. Temperatures (1188–1232°C) obtained in this study using this method are also similar to those calculated from Grove & Juster (1989). However, given the fairly constant values of Fe and Mg and the paucity of alkali content, we opine that the estimations based on the Ishii (1979) method are far more realistic in our study. Textural and mineralogical evidence suggests that the Dongargaon xenolith was metasomatized in mantle conditions (above) where reliable pressure estimations will provide us with not only the depth of formation of the new phases but also the fluid-driven metasomatic activity. We applied the temperature-dependent clinopyroxene geobarometer of Putirka (2008) for our pressure estimation. The calculated pressure for clinopyroxene if this study ranges from 8.5 to 12.4 kb and from 8.6 to 11.3 kb, based on the input temperature from Ishii (1979) and Grove & Juster (1989), respectively. The *P–T* estimates (Fig. 11) indicate that clinopyroxene was most likely to have been formed at temperatures of around 1188–1232°C and pressures of 8.5–12.4 kb, which correspond to a depth of approximately 40 km.

Pressure–temperature estimates on spinel peridotite xenoliths from the alkaline basalts and plugs, which are regarded as the earliest manifestations of the Deccan Trap event at *c.* 68 Ma, from the Kutch area of NW India, gave equilibrium temperatures of 884–972°C and estimated pressures of around 12–15 kb (*c.* 40–45 km depth) (Krishnamurthy *et al.* 1988; Karmalkar *et al.* 2000). Whereas the pressure estimates obtained on the mantle nodule from this study are broadly similar to the above such estimates, their temperature estimates, on the other hand, differ by approximately 200°C. We attribute this to the difference in the geothermobarometers (together with their uncertainties) deployed, as well as the contrasting composition of the phases involved, and therefore regard the overall *P–T* estimates obtained from our study to be realistic approximates.

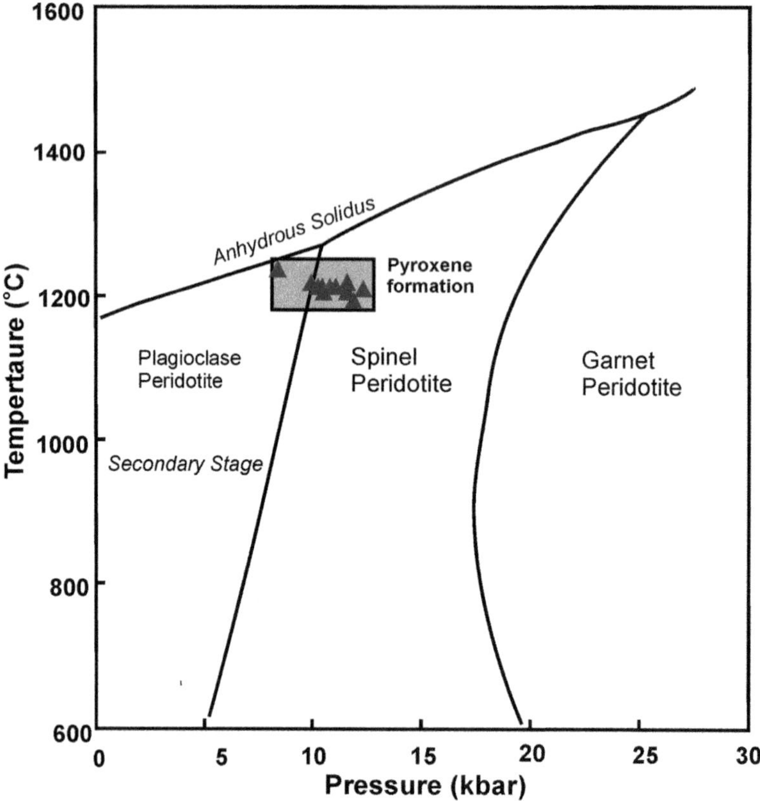

Fig. 11. The estimated pressure (after Green *et al.* 2001) and temperature conditions for metasomatized clinopyroxene (triangles) in the mantle xenolith of this study.

Geodynamic implications

In the Deccan Large Igneous Province, mantle- and lower-crustal-derived xenoliths are reported from the Kutch area of NW India, and also from the Murud Janjira area, near Mumbai, on the west coast. Cryptic metasomatism, involving enrichment of light REE Nb and Th in the spinel lherzolite xenoliths entrained in alkali basalts at *c.* 67–68 Ma (Venkatesan *et al.* 1986), has been recorded from the subcontinental lithospheric mantle (SCLM) below the Kutch (Karmalkar & Rege 2002). Even though, Dessai *et al.* (1999, 2004) noted the occurrence of kaersutite, phlogopite and apatite, and invoked metasomatism for their origin in the clinopyroxenite and granulite xenoliths occurring in the *c.* 65 Ma lamprophyres from the Murud Janjira, their proposal lacks petrographical or mineral chemistry support, such as that provided in this study. Present consensus favours the involvement of the Reunion plume for the Deccan event (see Chalapathi Rao *et al.* 2011; Sen & Chandrasekharam 2011 and references therein), which involved at least three main magmatic/volcanic events at (i) *c.* 68–67 Ma, (ii) *c.* 65 Ma and a terminal event at (iii) *c.* 62–61 Ma, with maximum flood basalt activity at the *c.* 65 Ma event (e.g. Hofmann *et al.* 2000; Widdowson *et al.* 2000; Chenet *et al.* 2007; Hooper *et al.* 2010; Sen *et al.* 2016). The arrival of the Reunion plume at the base of the Indian lithosphere at *c.* 70 Ma, well before the onset of the main phase of the Deccan basalt eruption, has also been proposed (e.g. Collier *et al.* 2008; Kerr *et al.* 2010).

The *c.* 55 Ma age of the Dongargaon lamprophyres provides only a minimum age constraint for the xenoliths and its maximum age remains unknown. However, the survival of phlogopite and apatite, which are considered to be the most fusible of the products of mantle metasomatism (see McKenzie 1989; Foley 1992; Konzett *et al.* 1998), along with the clinopyroxene in the studied xenoliths, demonstrate that wholesale melting of the metasomatized layer in the underlying SCLM in this domain did not occur during the Deccan Trap eruption. Given that the emplacement age of the Dongargaon lamprophyre succeeded the waning phase of the Deccan event by approximately 7 myr, it implies that at least some pockets and lenses of the metasomatized mantle in this domain escaped melting and remained intact during and after the Deccan event. As heterogeneity in the lithospheric thickness is known to be an important factor in controlling the extent and nature of mantle melting (Harry & Leeman 1995), we attribute the survival of the post-Deccan metasomatic layer to the non-uniform thickness of the underlying lithosphere in this domain (see also Gibson *et al.* 2006; Mohan *et al.* 2012). This situation is analogous to that found in several large igneous provinces such as the Parana-Etendeka and Siberia, where the small-volume alkaline magmatism (represented by kimberlites, lamproites, lamprophyres, etc.) succeeded the main flood basalt eruption by several millions of years (see Gibson *et al.* 1995, 2006; Griffin *et al.* 2005; Howarth *et al.* 2014; Le Roex & Class 2014). Further studies involving the trace element chemistry of the phases in the mantle xenolith and direct age determination of the metasomatic phases, such as phlogopite and apatite, are clearly required to evaluate the relative roles of the Deccan plume and the host lamprophyre as the possible agent(s) of metasomatism.

Conclusions

- This study documents the direct petrological evidence for modal metasomatism in the post-Deccan (Eocene) SCLM beneath the NW India.
- Garnet lherzolite has been inferred to be the protolith that, upon metasomatic reaction with a potassium-rich fluid, gave rise to the phlogopite + apatite + clinopyroxene + spinel assemblage in the xenolith.
- Pressure–temperature estimates of the clinopyroxene in the xenolith reveal temperatures of approximately 1200°C and pressures of approximately 12 kb, and correspond to a depth of approximately 40 km – all of which are broadly comparable with such data reported from other Deccan-related xenoliths.
- Survival of the phlogopite and apatite in the xenolith imply that some of the readily fusible portions of the SCLM escaped wholesale melting during the eruption of the Deccan Traps.

We thank the Head of the Department of Geology, BHU for extending all the facilities created by DST-FIST and DST PURSE. NVCR thanks DST-SERB for sanctioning major research projects (Project No. IR/S4/ESF-18/2011, dated 12 November 2013, and No. SR/S4/ES-599/2011, dated 27 May 2013) which made this study possible. DP thanks DST-SERB for a Research Scientist position and RP acknowledges financial support from CSIR for the award of a JRF (NET). Critical reviews by Gautam Sen (New York), Nitin Karmalkar (Pune) and Sujoy Ghosh (Kharagpur), together with the editorial suggestions by Sarajit Sensarma, improved the manuscript significantly and we extend our sincere thanks to all of them.

References

Ackerman, L., Spacek, P., Magna, T., Ulrych, J., Svojtka, M., Hegner, E. & Balogh, K. 2013. Alkaline and carbonate-rich melt metasomatism and melting of subcontinental lithospheric mantle: evidence from mantle xenoliths, NE Bavaria, Bohemian Massif. *Journal of Petrology*, **54**, 2597–2633.

Bijendra Singh, Prabhakara Rao, M.R.K., Prajapati, S.K. & Swarnapriya, C.H. 2014. Combined gravity and magnetic modeling over Pavagadh and Phenaimata igneous complexes, Gujarat, India: inference on emplacement history of Deccan volcanism. *Journal of Asian Earth Sciences*, **80**, 119–133.

Bonadiman, C., Beccaluva, L., Coltorti, M. & Siena, S. 2005. Kimberlite-like metasomatism and 'garnet signature' in spinel-peridotite xenoliths from Sal, Cape Verde Archipelago: relics of a subcontinental mantle domain within the Atlantic Ocean lithosphere? *Journal of Petrology*, **46**, 2465–2493.

Chalapathi Rao, N.V., Burgess, R., Lehmann, B., Mainkar, D., Pande, S.K., Hari, K.R. & Bodhankar, N. 2011. $^{40}Ar/^{39}Ar$ ages of mafic dykes from the Mesoproterozoic Chhattisgarh basin, Bastar craton, Central India: implication for the origin and spatial extent of the Deccan Large Igneous Province. *Lithos*, **125**, 994–1005.

Chalapathi Rao, N.V., Dharma Rao, C.V. & Das, S. 2012. Petrogenesis of lamprophyres from Chhota Udepur area, Narmada rift zone, and its relation to Deccan magmatism. *Journal of Asian Earth Sciences*, **45**, 24–39.

Chalapathi Rao, N.V., Lehmann, B., Belousova, E., Frei, D. & Mainkar, D. 2013. Petrology, bulk-rock geochemistry, indicator mineral composition, and zicron U–Pb geochronology of the end-Cretaceous diamondiferous Manipur orangeites, Bastar Craton, Central India. *In*: Pearson, D.G., Grütter, H.S., Harris, J.W., Kjarsgaard, B.A., O'Brien, H., Rao, N.V.C. & Sparks, S. (eds) *Proceedings of the 10th International Kimberlite Conference, Volume 1. Journal of the Geological Society of India*, Special Issue, 93–121.

Chawade, M.P. 1996. The petrology and geochemistry of dykes in Deccan basalts in parts of lower Narmada valley, around Chhaktala, Jhabua district, M.P. *In*: Deshmukh, S.S. & Nair, K.K.K. (eds) *Deccan Basalts*. Gondwana Geological Society, Special Volume, **2**, 185–200.

Chenet, A.L., Quidelleur, X., Fluteau, F., Courtillot, V. & Bajpai, S. 2007. $^{40}K–^{40}Ar$ dating of the main Deccan large igneous province: further evidence of KTB age and short duration. *Earth and Planetary Science Letters*, **263**, 1–15.

Collier, J.S., Sansom, V., Ishiuka, O., Taylor, R.N., Minshull, T.A. & Whitmarsh, R.B. 2008. Age of Seychelles–India break-up. *Earth and Planetary Science Letters*, **272**, 264–277.

Dawson, J.B. 1984. Contrasting types of upper mantle metasomatism. *In*: Kornprobst, J. (ed.) *Kimberlites*

II: The Mantle and Crust–Mantle Relationships. Elsevier, Amsterdam, 289–294.
DESSAI, A.G. & VIEGAS, A. 2010. Petrogenesis of alkaline rocks from Murud-Janjira, in the Deccan Traps, Western India. *Mineralogy and Petrology*, **98**, 297–311.
DESSAI, A.G., VASELLI, O. & KNIGHT, K. 1999. Thermal structure of the lithosphere beneath the Deccan Trap along the western Indian continental margin: evidence from xenolith data. *Journal of Geological Society of India*, **54**, 585–598.
DESSAI, A.G., MARKWICK, A., VASELLI, O. & DOWNES, H. 2004. Granulite and pyroxenite xenoliths from the Deccan Trap: insight into the nature and composition of the lower lithosphere beneath cratonic India. *Lithos*, **78**, 263–290.
DURGADMATH, M.B. 1984. Lamprophyre dykes from Phenai Mata area, Baroda District. *In*: Symposium on 'Three Decades of Developments on Petrology, Mineralogy and Geochemistry' 19–21 May, 1981, Jaipur, Geological Survey of India, Special Publication, **12**, 3–7.
ELKINS-TANTON, L.T. & GROVE, T.L. 2003. Evidence for deep melting of hydrous metasomatised mantle: Pliocene high-potassium magmas from the Sierra Nevadas. *Journal of Geophysical Research*, **108**, (B7), 2350, https://doi.org/10.1029/2002JB002168
ERLANK, A.J., WATERS, F.G., HAWKESWORTH, C.J., HAGGERTY, S.E., ALLSOPP, H.L., RICKARD, R.S. & MENZIES, M.A. 1987. Evidence for mantle metasomatism in peridotite nodules from the Kimberley pipes, South Africa. *In*: MENZIES, M.A. & HAWKESWORTH, C.J. (eds) *Mantle Metasomatism*. Academic Press, San Diego, CA, 221–311.
EVENSEN, N.M., HAMILTON, P.J. & O'NIONS, R.K. 1978. Rare earth abundances in chondritic meteorites. *Geochimica et Cosmochimica Acta*, **42**, 1199–1212.
FOLEY, S.F. 1992. Vein-plus-wall-rock melting mechanisms in the lithosphere and origin of potassic alkaline magmas. *Lithos*, **28**, 435–453.
FOLEY, S.F., PRELEVIC, D., REHFELDT, T. & JACOB, D. 2013. Minor and trace elements in olivines as probes into early igneous and mantle melting processes. *Earth and Planetary Science Letters*, **363**, 181–191.
GANGULY, J. & BHATTACHARYA, P.K. 1987. Xenoliths in Proterozoic kimberlites from southern India: petrology and geophysical implications. *In*: NIXON, P.H. (ed.) *Mantle Xenoliths*. John Wiley & Sons, Chichester, UK, 249–265.
GIBSON, S.A., THOMPSON, R.N., LEONARDOS, O.H., DICKIN, A.P. & MITCHELL, J.G. 1995. The Late Cretaceous impact of the Trindade mantle plume: evidence from large-volume, mafic, potassic magmatism in SE Brazil. *Journal of Petrology*, **36**, 189–229.
GIBSON, S.A., THOMPSON, R.N. & DAY, J.A. 2006. Timescales and mechanisms of plume–lithosphere interactions: $^{40}Ar/^{39}Ar$ geochronology and geochemistry of alkaline igneous rocks from the Parana-Etendeka large igneous province. *Earth and Planetary Science Letters*, **251**, 1–17.
GREEN, D.H. & WALLACE, M.E. 1988. Mantle metasomatism by ephemeral carbonatite melts. *Nature*, **336**, 459–462.
GREEN, D.H., FALLOON, T.J., EGGINS, S.M. & YAXLEY, G.M. 2001. Primary magmas and mantle temperatures. *European Journal of Mineralogy*, **13**, 437–451.
GREGOIRE, M., BELL, D.R. & LE ROEX, A.P. 2002. Trace element geochemistry of phlogopite-rich mafic mantle xenoliths: their classification and their relationship to phlogopite-bearing peridotties and kimberlites revisited. *Contributions to Mineralogy and Petrology*, **142**, 603–625.
GRIFFIN, W.L., SHEE, S.R., RYAN, C.G., WIN, T.T. & WYATT, B.A. 1999. Harzburgite to lherzolite and back again: metasomatic processes in ultramafic xenoliths from the Wesselton kimberlite, Kimberley, South Africa. *Contribution to Mineralogy and Petrology*, **134**, 232–250.
GRIFFIN, W.L., NATAPOV, L.M., O'REILLEY, S.Y., VAN ACHTERBERGH, E., CHERENKOVA, A.F. & CHERENKOV, V.G. 2005. The Kharamai kimberlite field, Siberia: modification of the lithospheric mantle by the Siberian Trap event. *Lithos*, **81**, 167–187.
GROVE, T.L. & JUSTER, T.C. 1989. Experimental investigations of low-Ca pyroxene stability and olivine–pyroxene- liquid equilibria at 1-atm in natural basaltic and andesitic liquids. *Contributions to Mineralogy and Petrology*, **103**, 287–305.
GWALANI, L.G., ROCK, N.M.S., CHANG, W-J., FERNANDEZ, S., ALLEGRE, C.J. & PRINZHOFER, A. 1993. Alkaline rocks and carbonatites of Amba Dongar and adjacent areas, Deccan Igneous Province, Gujarat, India. 1. Geology, petrography and petrochemistry. *Mineralogy and Petrology*, **47**, 219–253.
HAGGERTY, S.E. 1989. Mantle metasomes and the kinship between carbonatites and kimberlites. *In*: BELL, K. (ed.) *Carbonatites*. Unwin Hyman, London, 546–560.
HAGGERTY, S.E. 1995. A diamond trilogy: superplumes, supercontinents, and supernovae. *Science*, **285**, 851–860.
HARI, K.R. 1998. Mineralogical and petrochemical studies of the lamprophyres around Chhaktalao area, Madhya Pradesh. *Journal of the Geological Society of India*, **51**, 21–30.
HARI, K.R., CHALAPATHI RAO, N.V., SWARNKAR, V. & HOU, G. 2014. Alkali feldspar syenites with shoshonitic affinities from Chhotaudepur area: implication for mantle metasomatism in the Deccan large igneous province. *Geoscience Frontiers*, **5**, 261–276.
HARTE, B. & HAWKESWORTH, C.J. 1989. Mantle domains and mantle xenoliths. *In*: ROSS, J. (ed.) *Kimberlites and Related Rock, Volume 2*. Geological Society of Australia, Special Publications, **14**, 649–686.
HARRY, D.L. & LEEMAN, W.P. 1995. Partial melting of melt metasomatized subcontinental mantle and the magma source potential of the lower lithosphere. *Journal of Geophysical Research*, **100**, 10 255–10 269.
HOFMANN, C., FÉRAUD, G. & COURTILLOT, V. 2000. $^{40}Ar/Ar$ dating of mineral separates and whole rocks from the Western Ghats lava pipes: further constraints on duration and age of the Deccan Traps. *Earth and Planetary Science Letters*, **180**, 13–27.
HOOPER, P.R., WIDDOWSON, M. & KELLEY, S.P. 2010. Tectonic setting and timing of the final Deccan flood basalt eruptions. *Geology*, **38**, 839–842.
HOWARTH, G.H., BARRY, P.H. *ET AL.* 2014. Superplume metasomatism: evidence from Siberian mantle xenoliths. *Lithos*, **184–187**, 209–224.
IONOV, D.A., GRIFFIN, W.L. & O'REILLY, S.Y. 1997. Volatile-bearing minerals and lithophile trace

elements in the upper mantle. *Chemical Geology*, **141**, 153–184.

ISHII, T. 1979. Pyroxene geothermometry applied to a three-pyroxene achondrite from Allan Hills, Antractica and ordinary chondrites. *Mineralogical Journal*, **8**, 460–481.

KARMALKAR, N.R. & REGE, S. 2002. Cryptic metasomatism in the upper mantle beneath Kutch: evidence from spinel lherzolite xenoliths. *Current Science*, **82**, 1157–1165.

KARMALKAR, N.R., GRIFFIN, W.L. & O'REILLY, S.Y. 2000. Ultramafic xenoliths from Kutch (NW India): plume-related mantle samples? *International Geology Review*, **42**, 416–444.

KARMALKAR, N.R., DURAISWAMI, R.A., CHALAPATHI RAO, N. V. & PAUL, D.K. 2009. Mantle-derived mafic–ultramafic xenoliths and the nature of Indian sub-continental lithosphere. *Journal of Geological Society of India*, **73**, 657–679.

KELEMEN, P.B., HART, S.R. & BERNSTEIN, S. 1998. Silica enrichment in the continental lithosphere via melt/rock reaction. *Earth and Planetary Science Letters*, **164**, 387–406.

KERR, A.C., KHAN, M., MAHONEY, J.J., NICHOLSON, K.N. & HALL, C.M. 2010. Late Cretaceous alkaline sills of the south Tethyan suture zone, Pakistan: initial melts of the Réunion hotspot? *Lithos*, **117**, 161–171.

KONZETT, J., ARMSTRONG, R.A., SWEENEY, R.J. & COMPSTON, W. 1998. The timing of MARID metasomatism in the Kaapvaal mantle: an ion probe study of zircons from MARID xenoliths. *Earth and Planetary Science Letters*, **160**, 133–145.

KRISHNAMURTHY, P., PANDE, K., GOPALAN, K. & MACDOUGALL, J.D. 1988. Upper mantle xenoliths in alkali basalts related to Deccan Trap volcanism. *In*: SUBBARAO, K.V. (ed.) *Deccan Flood Basalts*. Geological Society of India, Memoirs, **10**, 53–67.

KRISHNAMURTHY, P., PANDE, K., GOPALAN, K. & MACDOUGALL, J.D. 1999. Mineralogical and chemical studies on alkaline basaltic rocks of Kutch, Gujarat, India. *In*: SUBBARAO, K.V. (ed.) *Deccan Volcanic Province*. Geological Society of India, Memoirs, **43**, 757–783.

LE ROEX, A.P. & CLASS, C. 2014. Metasomatism of the Pan African lithospheric mantle beneath the Damara Belt, Namibia, by the Tristan mantle plume: geochemical evidence from mantle xenoliths. *Contributions to Mineralogy and Petrology*, **168**, 1046.

LLOYD, F.E. & BAILEY, D.K. 1975. Light element metasomatism of the continental mantle: the evidence and consequences. *Physics and Chemistry of the Earth*, **9**, 389–416.

LORAND, J.P., ALARD, O., LUGUET, A. & KEAYS, R.R. 2003. Sulfur and selenium systematics of the subcontinental lithospheric mantle: inferences from the Massif Central xenolith suite (France). *Geochimica et Cosmochimica Acta*, **67**, 4137–4151.

MCKENZIE, D. 1989. Some remarks on the movement of small melt fractions in the Mantle. *Earth and Planetary Science Letters*, **95**, 53–72.

MOHAN, G., KUMAR, M.R., SAIKIA, D., KUMAR, K.A.P., TIWARI, P.K. & SURVE, G. 2012. Imprints of volcanism in the upper mantle beneath the NW Deccan volcanic Province. *Lithosphere*, **4**, 150–159, https://doi.org/10.1130/L178.1

MUKHERJEE, A.B. & BISWAS, S. 1988. Mantle-derived spinel lherzolite xenoliths from the Deccan volcanic province (India): implications for the thermal structure of the lithosphere underlying the Deccan Traps. *Journal of Volcanology and Geothermal Research*, **35**, 269–276.

NIXON, P.H. 1987. *Mantle Xenoliths*. John Wiley & Sons, Chichester, UK.

NIXON, P.H., ROGERS, N.W., GIBSON, I.L. & GREY, A. 1981. Depleted and fertile mantle xenoliths from southern African kimberlites. *Annual Reviews of the Earth and Planetary Sciences*, **9**, 285–309.

O'REILLY, S.Y. & GRIFFIN, W.L. 2000. Apatite in the mantle: implications for metasomatic processes and high heat production in Phanerozoic mantle. *Lithos*, **53**, 217–232.

O'REILLY, S.Y. & GRIFFIN, W.L. 2013. Mantle metasomatism. *In*: HARLOV, D.E. & AUSTRHEIM, H. (eds) *Metasomatism and the Chemical Transformation of Rock*. Lecture Notes in Earth System Sciences. Springer, Berlin, 471–533, https://doi.org/10.1007/978-3-642-28394-9_12

PEARSON, D.G., CANIL, D.G. & SHIREY, S.B. 2003. Mantle samples included in volcanic rocks: xenoliths and diamonds. *In*: HOLLAND, H.D. & TUREKIAN, K.K. (eds) *Treatise on Geochemistry*. Amsterdam, Elsevier, 171–275.

PUTIRKA, K.D. 2008. Thermometers and barometers for volcanic systems. *Reviews in Mineralogy and Geochemistry*, **69**, 61–120.

RANDIVE, K.R., SAHU, M.K., LANJEWAR, S., BELYATSKY, B. 2012. Eocene (*c.* 55 Ma) age for the lamprophyre dyke of Chota Udaipur carbonatite-alkaline subprovince, Lower Narmada Valley, Gujarat and Madhya Pradesh States, India. Extended Abstract presented at the *Tenth International Kimberlite Conference*, 6–11 February, 2012, Bangalore, India.

RAY, A., HATUI, K., PAUL, D., SEN, G., BISWAS, S.K. & DAS, B. 2016. Mantle xenolith–xenocryst-bearing monogenetic alkali basaltic lava field from Kutch Basin, Gujarat, Western India: estimation of magma ascent rate. *Journal of Volcanology and Geothermal Research*, **312**, 40–52.

REIDER, M., CAVAZZINI, D. ET AL. 1998. Nomenclature of micas. *Canadian Mineralogist*, **36**, 905–912.

ROCK, N.M.S. 1991. *Lamprophyres*. Blackie, London.

RUDNICK, R.L., MCDONOUGH, W.F. & ORPIN, A. 1994. Northern Tanzanian peridotite xenoliths: a comparison with Kaapvaal peridotites and inferences on metasomatic interactions. *In*: MEYER, H.O.A. & LEONARDOS, O.H. (eds) *Kimberlites. Related Rocks and Xenoliths. Proceedings of the 5th International Kimberlite Conference*. CPRM (Companhia de Pesquisa de Recursos Minerais), Special Publications, 336–354.

SAHOO, S. 2016. *Petrogenesis of lamprophyres associated with the igneous complexes from north western India*. Unpublished PhD thesis, Banaras Hindu University, Varanasi, India.

SEN, G. & CHANDRASEKHARAM, D. 2011. Deccan traps flood basalt province: an evaluation of the thermochemical plume model. *In*: RAY, J., SEN, G. & GHOSH, B. (eds) *Topics in Igneous Petrology*. Springer, Berlin, 29–53, https://doi.org/10.1007/978-90-481-9600-5_2

SEN, G. & LEEMAN, W.P. 1991. Iron-rich lherzolitic xenoliths from Oahu: origin and implication for Hawaiian magma sources. *Earth and Planetary Science Letters*, **102**, 45–57.

Sen, G., MacFarlane, A. & Srimal, N. 1996. Significance of rare hydrous alkaline melts in Hawaiian xenoliths. *Contributions to Mineralogy and Petrology*, **122**, 415–427.

Sen, G., Bizimis, M., Das, R., Paul, D.K. & Biswas, S. 2009. Deccan plume, lithosphere rifting, and volcanism in Kutch, India. *Earth and Planetary Science Letters*, **277**, 101–111.

Sen, G., Hames, W.E., Paul, D.K., Biswas, S.K., Ray, A. & Sen, I.S. 2016. Pre-Deccan and Deccan magmatism in Kutch, India: implications of new $^{40}Ar/^{39}Ar$ ages of intrusions. *In*: Thakkar, M.G. (ed.) *Recent Studies on the Geology of Kachchh*. Geological Society of India, Special Publications, **6**, 211–222.

Sensarma, S., Paul, D. & Chalapathi Rao, N.V. 2013. Large igneous provinces – global perspectives and prospects in India. *Current Science*, **105**, 182–192.

Sethna, S.F. 1989. Petrology and geochemistry of the acid, intermediate and alkaline rocks associated with the Deccan basalts in Gujarat and Maharashtra. *In*: Leelanandam, C. (ed.) *Alkaline Rocks*. Geological Society of India, Memoirs, **15**, 47–61.

Shaw, C.S.J., Eyzaguirre, J., Fryer, B. & Gagnon, J. 2005. Regional variations in the mineralogy of metasomatic assemblages in mantle xenoliths from the West Eifel Volcanic field, Germany. *Journal of Petrology*, **46**, 945–972.

Simon, N.S.C., Carlson, R.W., Pearson, D.G. & Davies, G.R. 2007. The origin and evolution of the Kaapvaal cratonic lithospheric mantle. *Journal of Petrology*, **48**, 589–625.

Sukeshwala, R.N. & Sethna, S.F. 1967. Giant pseudoleucites of Ghori, Chhota Udaipur. *American Mineralogist*, **52**, 1904–1910.

Sun, J., Liu, C.Z., Wu, F.Y., Yang, Y.H. & Chu, Z.Y. 2012. Metasomatic origin of clinopyroxene in Archean mantle xenoliths from Hebi, North China Craton: trace-element and Sr-isotope constraints. *Chemical Geology*, **328**, 123–136.

Van Achterbergh, E., Griffin, W.L. & Stiefenhofer, J. 2001. Metasomatism in mantle xenoliths from the Letlhakane kimberlites: estimation of element fluxes. *Contributions to Mineralogy and Petrology*, **141**, 397–414.

Venkatesan, T.R., Pande, K. & Gopalan, K. 1986. $^{40}Ar–^{39}Ar$ dating of Deccan basalts. *Journal of the Geological Society of India*, **27**, 102–109.

Viladkar, S.G. 1981. The carbonatites of Amba Dongar, Gujarat, India. *Bulletin of the Geological Society of Finland*, **53**, 17–28.

Widdowson, M., Pringle, M. & Fernandez, O.P. 2000. A post K–T boundary (Early Palaeocene) age for Deccan-type feeder dykes, Goa, India. *Journal of Petrology*, **41**, 1177–1194.

Yaxley, G.M., Green, D.H. & Kamenetsky, V.M. 1998. Carbonatite metasomatism in the Southeastern Australian lithosphere. *Journal of Petrology*, **39**, 1917–1930.

Origin of the Amba Dongar carbonatite complex, India and its possible linkage with the Deccan Large Igneous Province

JYOTI CHANDRA[1], DEBAJYOTI PAUL[1]*, SHRINIVAS G. VILADKAR[2] & SARAJIT SENSARMA[3]

[1]*Department of Earth Sciences, Indian Institute of Technology Kanpur, Kanpur 208 016, India*

[2]*Carbonatite Research Centre, Amba Dongar, Kadipani, District, Baroda 390 117, India*

[3]*Centre of Advanced Study in Geology, University of Lucknow, Lucknow 226 007, India*

**Correspondence: dpaul@iitk.ac.in*

Abstract: The genetic connection between Large Igneous Province (LIP) and carbonatite is controversial. Here, we present new major and trace element data for carbonatites, nephelinites and Deccan basalts from Amba Dongar in western India, and probe the linkage between carbonatite and the Deccan LIP. Carbonatites are classified into calciocarbonatite (CaO, 39.5–55.9 wt%; BaO, 0.02–3.41 wt%; ΣREE, 1025–12 317 ppm) and ferrocarbonatite (CaO, 15.6–31 wt%; BaO, 0.3–7 wt%; ΣREE, 6839–31 117 ppm). Primitive-mantle-normalized trace element patterns of carbonatites show distinct negative Ti, Zr–Hf, Pb, K and U anomalies, similar to that observed in carbonatites globally. Chondrite-normalized REE patterns reveal high LREE/HREE fractionation; average $(La/Yb)_N$ values of 175 in carbonatites and approximately 50 in nephelinites suggest very-low-degree melting of the source. Trace element modelling indicates the possibility of primary carbonatite melt generated from a subcontinental lithospheric mantle (SCLM) source, although it does not explain the entire range of trace element enrichment observed in the Amba Dongar carbonatites. We suggest that CO_2-rich fluids and heat from the Deccan plume contributed towards metasomatism of the SCLM source. Melting of this SCLM generated primary carbonated silicate magma that underwent liquid immiscibility at crustal depths, forming two compositionally distinct carbonatite and nephelinite magmas.

Supplementary material: Detailed sampling locations and descriptions, major and trace element composition of apatite in carbonatite and nephelinite, analytical reproducibility for major and trace elemental analyses, and details of trace element modelling are available at https://doi.org/10.6084/m9.figshare.c.3819457

Carbonatites contain more than 50 modal% carbonate minerals, such as calcite, dolomite, ankerite and siderite, that were crystallized from mantle-derived magma (Heinrich 1966; Streckeisen 1980; Woolley 1982; Le Maitre *et al.* 1989; Woolley & Kempe 1989). Although volumetrically insignificant compared to other mantle-derived magma, these have large economic significance – particularly contain abundant rare earth minerals, apatite and fluorite (Pirajno 1994; Rukhlov *et al.* 2015; Simandl 2015). The unusual geochemistry of these mantle-derived magmas provides an insight into the secular evolution of the source mantle, as well as petrogenetic aspects of carbonatite magma genesis and its interrelationship with the associated Large Igneous Province (LIP) (Ernst & Bell 2010).

Although carbonatites have been studied for the past four decades, no clear consensus has been reached with regards to its petrogenesis, particularly the nature of mantle source(s) involved (Bailey 1993; Lee & Wyllie 1994; Bell 1998; Bell & Tilton 2001; Bell & Simonetti 2010; Rukhlov *et al.* 2015). Nevertheless, extensive geochemical (including isotopic) work in the recent past aided by advancements in experimental petrology converges to three major hypotheses on the origin of carbonatites: (a) fractional crystallization of carbonated silicate magma (Lee & Wyllie 1994; Veksler *et al.* 1998); (b) liquid immiscibility of carbonated silicate (melilititic/nephelinitic) magmas (Von Eckermann 1961; Koster van Groos & Wyllie 1963; Kjarsgaard & Hamilton 1989; Lee & Wyllie 1994, 1996; Simonetti & Bell 1994; Bell 1998; Halama *et al.* 2005; Tappe *et al.* 2006; Brooker & Kjarsgaard 2011; Guzmics *et al.* 2011); and (c) partial melting of carbonated mantle peridotite (Wyllie & Huang 1975; Bell 1989; Gittins 1989; Harmer & Gittins 1998; Bell & Rukhlov 2004; Srivastava & Sinha 2004; Yaxley & Brey 2004; Gudfinnsson & Presnall 2005; Chakhmouradian 2006; Brey *et al.* 2009; Dasgupta *et al.* 2009; Ghosh *et al.* 2014). Nevertheless, several questions remain unanswered regarding carbonatite magma

From: Sensarma, S. & Storey, B. C. (eds) 2018. *Large Igneous Provinces from Gondwana and Adjacent Regions*. Geological Society, London, Special Publications, **463**, 137–169.
First published online July 10, 2017, https://doi.org/10.1144/SP463.3

generation by liquid immiscibility (Twyman & Gittins 1987; Gittins 1989). For example, liquid immiscibility does not explain the unusually high abundance of Nb, rare earth element (REE), P and Sr elements in carbonatites relative to the associated silicate rocks, and the apparent younger emplacement ages of several carbonatites compared to the associated alkaline silicates. Further, carbonatites of various ages ranging from 3 Ga to Recent are present in most of the continents (Woolley & Kjarsgaard 2008), which raises the question of whether the parental carbonate magma was generated from the subcontinental lithospheric mantle (SCLM) or deep mantle sources with a subsidiary but crucial role of lithosphere in the trapping of volatile-rich melts/fluids (Bell *et al.* 1999; Bell & Simonetti 2010; Ernst & Bell 2010). Additionally, the scientific community is still debating on the existence of primitive carbonatite melt within the mantle, stability of carbonate phases at different depths of the mantle, and the role of fluids in mantle metasomatism (Thibault *et al.* 1992; Bell & Simonetti 2010 and references therein).

Carbonatites are both spatially and temporally associated with several mantle-plume-derived LIPs (Simonetti *et al.* 1995; Le Roex & Lanyon 1998; Ray & Pande 1999; Bell 2001; Ernst & Bell 2010; Pirajno 2015). Notable examples of close associations of carbonatites with mantle-plume-generated LIPs are: (I) Amba Dongar (Gujarat: 65 ± 0.3 Ma), Sarnu-Dandali, Barmer (Rajasthan: 68.57 ± 0.08 Ma) and Mundwara carbonatites (Rajasthan: 68.53 ± 0.16 Ma) associated with the (70–63 Ma) Deccan LIP (Basu *et al.* 1993; Gwalani *et al.* 1993; Simonetti *et al.* 1995, 1998; Ray *et al.* 2000*a*); (II) Jacupiranga (Brazil: 131 ± 3 Ma) and Messum (Namibia: 126.8 ± 1.3 Ma) carbonatites associated with the 133 Ma Parana-Etendeka LIP (Pirajno 1994, 2015; Le Roex & Lanyon 1998; Bell 2001); (III) Maimecha Kotui province (250–220 Ma) with carbonatites associated with the 250 Ma Siberian LIP (Ernst & Bell 2010); (IV) carbonatite bodies associated with the Central Iapetus Magmatic Province (615–555 Ma) of eastern Laurentia (St-Honore: 642 Ma) and western Baltica (Alnö, Sweden: 589 Ma; and Fen, Norway: 583 ± 15 Ma) (Ernst & Buchan 2004); and (V) Phalaborwa (South Africa: 2060 ± 0.8 Ma) and Shield (2059 ± 35 Ma) carbonatites of 2055 Ma Bushveld LIP (Ernst & Bell 2010). Therefore, understanding the temporal relationship between carbonatite complexes and LIPs (Sensarma *et al.* 2013) will provide an insight into the origin of carbonatites.

The Deccan flood basalt (DFB) province in west-central India spans 500 000 km^2 (Subbarao & Sukheswala 1981), although the areal extent can be as high as 1.5 million km^2 taking into account its spread across the Seychelles Islands (Devey & Stephens 1992), whose origin has been linked to the mantle plume currently located beneath Réunion Island (Peng & Mahoney 1995). The major eruptive event occurred at *c.* 65–66 Ma (Allègre *et al.* 1999; Courtillot & Renne 2003; Renne *et al.* 2015; Richards *et al.* 2015; Schoene *et al.* 2015), with its epicentre located at the intersection of the Narmada-Tapti Rift and the west coast (Courtillot *et al.* 1999). Deep boreholes drilled to a depth of about 1.5 km in the Koyna-Warna region, DFB province (*c.* 520 km SSW of the Amba Dongar carbonatite complex) reveals a transition between massive basalt and the granitoid basement at a depth of approximately 930 m (Gupta *et al.* 2015).

The Amba Dongar carbonatite and associated alkaline silicate complex, located along the east–west-trending Narmada Rift, is contemporaneous with, or younger than, the main Deccan eruptive event (Basu *et al.* 1993; Gwalani *et al.* 1993, 1995; Simonetti *et al.* 1995; Ray & Pande 1999; Ray & Ramesh 1999). The reported emplacement ages of Amba Dongar carbonatites are 61 ± 2 and 76 ± 2 Ma (K–Ar dating: Deans *et al.* 1972), and 65 ± 0.3 Ma ($^{40}Ar/^{39}Ar$ dating: Ray & Pande 1999), which is slightly older or coeval with the *c.* 65 Ma main Deccan eruptive event. Basu *et al.* (1993) reported $^{40}Ar/^{39}Ar$ ages for the Phenai Mata alkali intrusion near Amba Dongar (64.96 ± 0.11 Ma, on biotite from gabbro), the Mundwara (68.53 ± 0.16 Ma, on biotite from an alkali olivine gabbro: 400 km NW of Amba Dongar) and Sarnu-Dandali/Barmer (68.57 ± 0.08 Ma, on biotite from an alkali pyroxenite: 150 km NW of Mundwara) carbonatites, and alkali complexes that are located north of the Narmada Rift. Using initial $^{87}Sr/^{86}Sr$, $^{143}Nd/^{144}Nd$, $^{206}Pb/^{204}Pb$, $^{207}Pb/^{204}Pb$ and $^{208}Pb/^{204}Pb$ isotope ratios of the Deccan alkaline complexes – for example, Bhuj (64.4–67.7 Ma, basanite-alkaline basalt), Mundwara (nephelinite, phonolite, tephrite and carbonatite), Barmer (alkali pyroxenite, melanephelinite, ijolite and melilitite) and Amba Dongar (basanite, nephelinites, phonolite plugs and one sample of carbonatite) alkaline complexes – Simonetti *et al.* (1998) concluded that these alkaline complexes were derived from a mixture of Indian mid-ocean ridge basalt (MORB), old continental lithosphere and Réunion mantle-plume sources. They also emphasized on the role of the Réunion mantle plume in the generation of parental carbonatite magma for the Amba Dongar complex. Ray & Ramesh (1999) used $\delta^{13}C$ and $\delta^{18}O$ compositions of carbonatites of Mundwara, Barmer and Amba Dongar to argue for the contribution of the Réunion plume in the parental carbonatite magma generation.

All these inferences hint towards some possible genetic linkage between carbonatite and associated alkaline silicate rocks with the Deccan LIP. Further, the Amba Dongar carbonatite complex is cross-cut by numerous doleritic and basaltic dykes (Sukheswala & Udas 1963; Viladkar 1996). To our knowledge, no geochemical data on doleritic and basaltic dykes have been reported in the published literature. Although major and trace element abundances in the Amba Dongar carbonatites are fairly well known, those in the associated alkaline rocks and Bagh sandstone and limestone are rather limited (e.g. Sethna 1971; Sukheswala & Viladkar 1978; Viladkar 1981, 2012; Viladkar & Dulski 1986; Viladkar & Wimmenauer 1992; Gwalani *et al.* 1993; Srivastava 1997; Ray *et al.* 2000*b*; Ray & Shukla 2004). Particularly, none of the previous studies reported the entire range of REE abundances. Further, the available geochemical data have been used to propose different hypotheses on the origin of the Amba Dongar carbonatites and associated rocks. For example, whereas Ray & Shukla (2004) hypothesized that fractional crystallization of two compositionally different parental melts produced the carbonate and alkaline silicate rocks of Amba Dongar, Viladkar (1984) argued that the genesis of the Amba Dongar carbonatite and associated nephelinite and phonolitic nephelinite are unrelated to the origin of the Deccan LIP. To elucidate the geochemical evolution of Amba Dongar carbonatites, we have undertaken systematic and extensive sampling of carbonatites and associated rocks from the area. We report new major and trace (including rare earth elements, REEs) compositions of Amba Dongar carbonatites, associated alkaline silicates, Deccan basalts, and Bagh sandstone and limestone. The major objectives of this study are to: (i) evaluate the genetic linkage between the Amba Dongar carbonatite complex and the Deccan LIP; and (ii) propose a plausible model for the generation of Amba Dongar carbonatites. Although earlier workers have suggested a possible link between Amba Dongar carbonatite and the Deccan LIP on the basis of temporal and spatial distribution and geochemical data (e.g. Simonetti *et al.* 1995; Ray *et al.* 2000*a*), we propose a petrogenetic model which shows that the Amba Dongar carbonatite and associated alkaline rocks are genetically linked with the Deccan LIP.

Geological setting

The Amba Dongar carbonatite ring complex (22°00′ N, 74°05′ E) covering an area of 30 km^2 is located at the NW periphery of the Deccan province, about 6 km north of the Narmada River in the Chhota Udaipur district of Gujarat state, India (Fig. 1). It was the first carbonatite complex in India identified by Sukheswala & Udas (1963), and has been mined mainly for fluorites. This complex was intruded into the Archaean basement gneiss, Cretaceous Bagh beds (*c.* 90 Ma sandstones and limestones: Jaitly & Ajane 2013) and Deccan basalts (Deans *et al.* 1972; Viladkar 1981). The domal nature of the DFB and Bagh beds, attested by the radial dips (30–60°) of the Bagh sandstone, near the Amba Dongar carbonatite complex has been ascribed to the upwards intrusion of carbonatite magma (Sukheswala & Udas 1963; Viladkar 1981; Gwalani *et al.* 1993).

The Amba Dongar carbonatite complex (65 ± 0.3 Ma, Ray & Pande 1999) comprises of predominantly calciocarbonatites and ferrocarbonatites, associated nephelinites and phonolites, Deccan basalts, and basaltic and doleritic dykes (Deans *et al.* 1972; Viladkar 1981). Compared to calciocarbonatites, volumetrically fewer ferrocarbonatites have intruded into the early formed carbonatite breccia and the main calciocarbonatite ring dyke; ferrocarbonatites occur as small plugs and permeated veins in calciocarbonatites in the SW and northern part of the ring dyke (Fig. 1). The associated nephelinites in the area are exposed in low grounds near the main carbonatite complex. Further, carbonatite breccia and the ring dyke have been intruded by calciocarbonatite dykes, which were termed alvikites by Viladkar (1981, 2012). Later, hydrothermal quartz and fluorite veins (or large cavity fillings) intruded into the pre-existing carbonatite rocks, which is a more prominent feature in the active mining area. The entire complex has been intruded by doleritic and basaltic dykes.

Sampling and analytical methods

In this study, a total of 71 rock samples were collected from all the different litho-units of the Amba Dongar carbonatite complex during 2013–14 (Fig. 1): calciocarbonatite (n = 34), ferrocarbonatite (n = 13), nephelinite (n = 5), Bagh sandstone (n = 4) and Bagh limestone (n = 3), DFB (n = 6), and doleritic (n = 2) and basaltic dykes (n = 4). Care was taken to collect unaltered samples from lithological exposures. Samples were collected mostly along the calciocarbonatite ring dyke; a few samples were also collected from late-stage calciocarbonatite cutting the main ring dyke. Samples were collected from five different ferrocarbonatite plugs and also from ferrocarbonatite veins cutting calciocarbonatite rocks.

For petrographical and textural studies, thin sections of rock samples (n = 38) were examined using a Leica microscope. Additionally, predominant minerals in carbonatites were also identified by X-ray diffraction (XRD) using a PANalytical X'Pert^3 Powder XRD, with a Cu target and operated at 30 kV and 10 mA at the Indian Institute of Technology Kanpur (IITK), India. Further, the composition of apatites occurring within both calciocarbonatites and

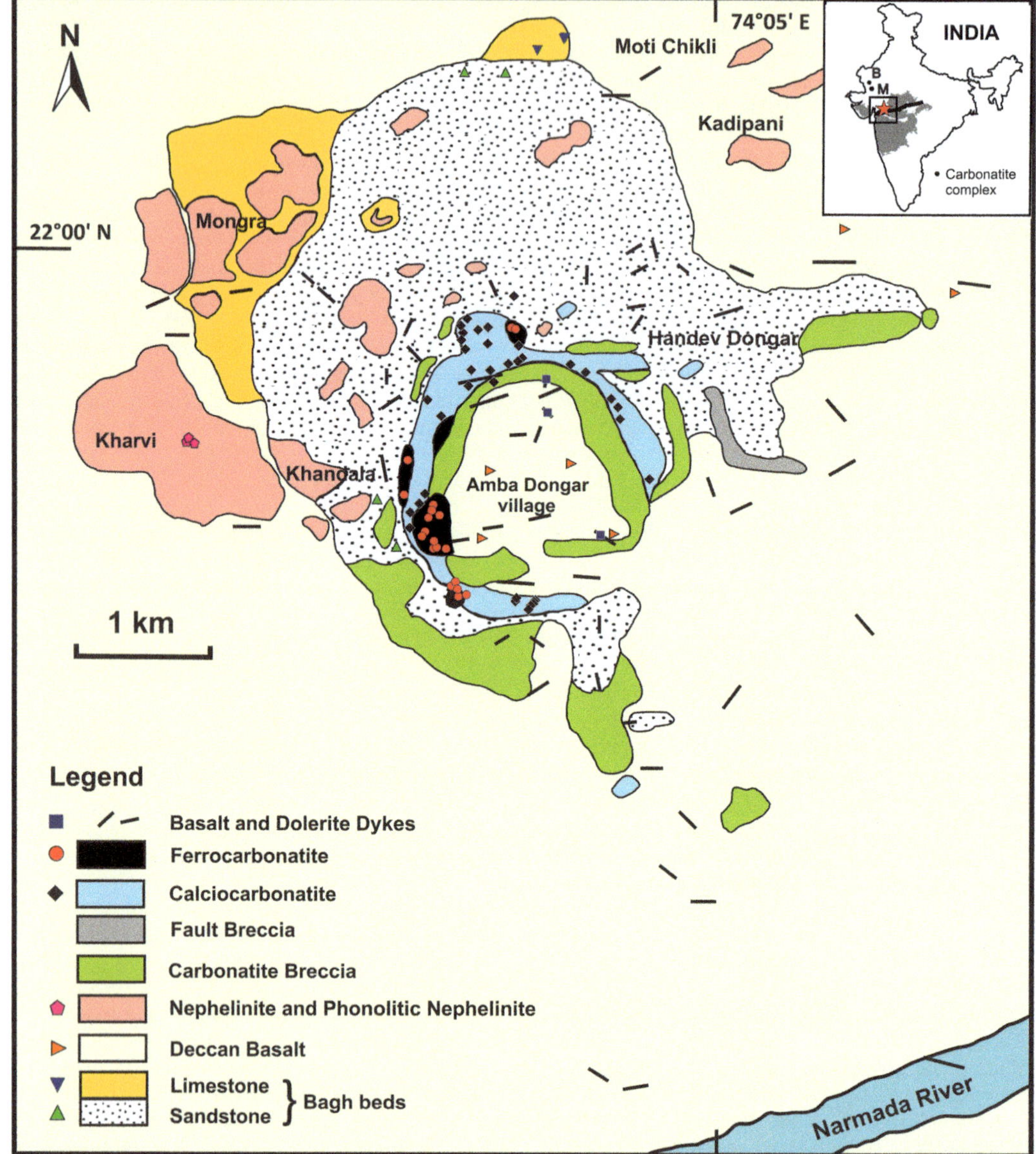

Fig. 1. Geological map of the Amba Dongar carbonatite complex, modified after Viladkar (1981). Rock types along with their sampling locations are shown. Inset shows the extent of the Deccan flood basalt province; the star is the location of Amba Dongar; B and M represent the locations of the Barmer and Mundwara carbonatite complexes; the solid line represents Narmada Rift.

nephelinites were measured using a CAMECA SXFive electron probe microanalyser (EPMA), with a lanthanum hexabromide (LaB_6) source, and operated at 15 kV and 20 nA with 1 µm diameter beam, at the Indian Institute of Technology Bombay (IITB), India. A total of 10 standards, including both naturally occurring minerals and synthetic elemental glass (apatite (F, P), albite (Na), NaCl (Cl), diopside (Ca), $SrSO_4$ (Sr), Th-glass (Si), La-glass (La), Ce-glass (Ce), Pr-glass (Pr) and Nd-glass (Nd)) were used for calibration.

For major element analyses, whole-rock samples were crushed and approximately 50 g of chips were handpicked and powdered to <75 µm size using a FRITSCH agate vibrating cup mill PULVERISETTE 9, while taking utmost care to minimize contact with metal. Fusion beads were prepared using a mixture of 1:10 ratio of sample powder to flux. The flux comprised a pure-grade mixture of lithium tetraborate (LiT), lithium metaborate (LiM) and lithium iodide (LiI); whereas a flux composition of 66.67–32.83–0.50 (LiT–LiM–LiI) was used for carbonate samples, it was 34.83–64.67–0.50 for the silicate rocks. The glass beads were analysed at IITK for major oxides and a few selected trace element abundances using a Rigaku ZSX Primus II wavelength dispersive X-ray fluorescence (WD-XRF) spectrometer with a 4 kW Rh target, and operated at 60 kV and 50 mA. About 30 international rock standards from the United States Geological Survey (USGS), Canadian Certified Reference Material Project (CCRMP) and International atomic energy agency (IAEA) were used for calibration. Geostandards carbonatite (COQ-1), basalt (BE-N), and granite (AC-E) were analysed repetitively with each batch of unknown samples to monitor the accuracy of analyses. Loss on ignition (LOI) of all samples was determined by igniting 1 g of bulk powder in Pt-crucibles to 1050°C for 4 h. The calibration conditions and detection limits of the elements analysed by EPMA, and the accepted (Govindraju 1994) and measured values of the three USGS rock standards used, are provided in the Supplementary material.

Concentrations of 30 trace elements in bulk-rock samples (n= 35) were determined with an inductively coupled plasma mass spectrometer (ICP-MS, Thermo Fisher Scientific iCAP Q) at IITK. Approximately 50 mg of carbonate sample powder was treated with 2.5 N HCl in pre-cleaned Savillex® vials to dissolve carbonates, and dried at 60°C. These dried samples containing silicate phases were digested using 1 ml of HF–HNO_3 (5:1) solution at 120°C for 24 h in sealed vials and dried. About 5 ml of 6 N HCl was added to the dried samples and sealed vials were heated at 120°C for 24 h, followed by drying at 120°C. The digested samples were stored in Suprapur grade 2% HNO_3. Six geostandards AGV-2, W-2a, BHVO-2, BE-N, BIR-1a and BCR-2 were used for data calibration. BHVO-2 was repetitively analysed with each batch of unknown samples to determine the accuracy of analyses, uncertainties associated with most trace elements was <9% (see the Supplementary material). Following sample-acid digestion using a mixture of HNO_3 + HF + $HClO_4$, a subset of the carbonatite samples (n = 17) were analysed with a Thermo Electron ICP-MS X Series II ICP-MS at the Laboratoire d'Hydrologie et de Géochimie de Strasbourg, France. Typical uncertainties associated with these analyses are 5% for trace elements up to 0.1 ppb. Three carbonatite samples analysed in France were re-digested and analysed at IITK to check for data reproducibility, which showed good agreement between the two datasets.

Results

Petrography

In hand specimen, milky white to reddish brown coloured calciocarbonatites show large variations in both grain size and mineralogy – ranging from fine to coarse grained, massive – and at times are intruded by veins of ferrocarbonatite, late-stage calciocarbonatite and (hydrothermal) quartz (Fig. 2). On the other hand, brown to dark brown coloured ferrocarbonatites are typically fine grained and heavier compared to calciocarbonatites. Calciocarbonatite is also intruded by late-stage calcicarbonatite veins (Fig. 2a, b). The intrusion of late-stage ferrocarbonatites into early formed calciocarbonatites can be observed both in the field and in hand specimens (Fig. 2c–f). Numerous quartz veins cutting through the ferrocarbonatites can be seen in hand specimens (Fig. 2e). Note that Viladkar (1981) reported the presence of calciocarbonatite xenoliths in ferrocarbonatites.

In thin sections, calciocarbonatites are fine to coarse grained (Fig. 3a–d). Calcite is the dominant carbonate phase. In coarse variety, the composition of calciocarbonatite varies from almost monomineralic (Fig. 3c) to those with accessory minerals such as apatite, pyrochlore, magnetite, barite and sometimes aegirine (Fig. 3d), forming interlocking and hypidiomorphic texture. Fine-grained calciocarbonatites are composed of tiny calcite crystals with flow bands, and are generally intruded by quartz and iron oxide veins (Fig. 3a). A medium-grained variety of calciocarbonatite shows alternate bands of calcite and magnetite forming a banded texture (Fig. 3b). Fine quartz veins and interstitial quartz grains with extended crystal margins are found in ferrocarbonatites. In a few ferrocarbonatite samples, brown coloured flow bands of ferrocarbonatite are intruded by thin veins of secondary cryptocrystalline calcite and minor fluorite (Fig. 3e). These thin veins of calcite are marked by meandering, tapering and necking in places (Fig. 3f). Mineralogically, ferrocarbonatites are essentially composed of ankerite and dolomite carbonate phases, which are also confirmed from XRD data (not shown here). XRD data also reveal the presence of barite, quartz, apatite, fluorite and REE-bearing minerals, such as bastnaesite, baddeleyite and britholite in ferrocarbonatites, which is consistent with that reported earlier (e.g. Viladkar 1981). Compared to calciocarbonatites, ferrocarbonatites contain less apatite and more barite, fluorite and REE minerals. Apatite in calciocarbonatites shows various crystal habits, from small subrounded to large tabular crystals. It occurs as stubby prisms, as well as cumulates, randomly scattered within the calcitic groundmass, and also occupies the interstitial space between calcites in calciocarbonatites (Fig. 4). Inclusions of apatite are

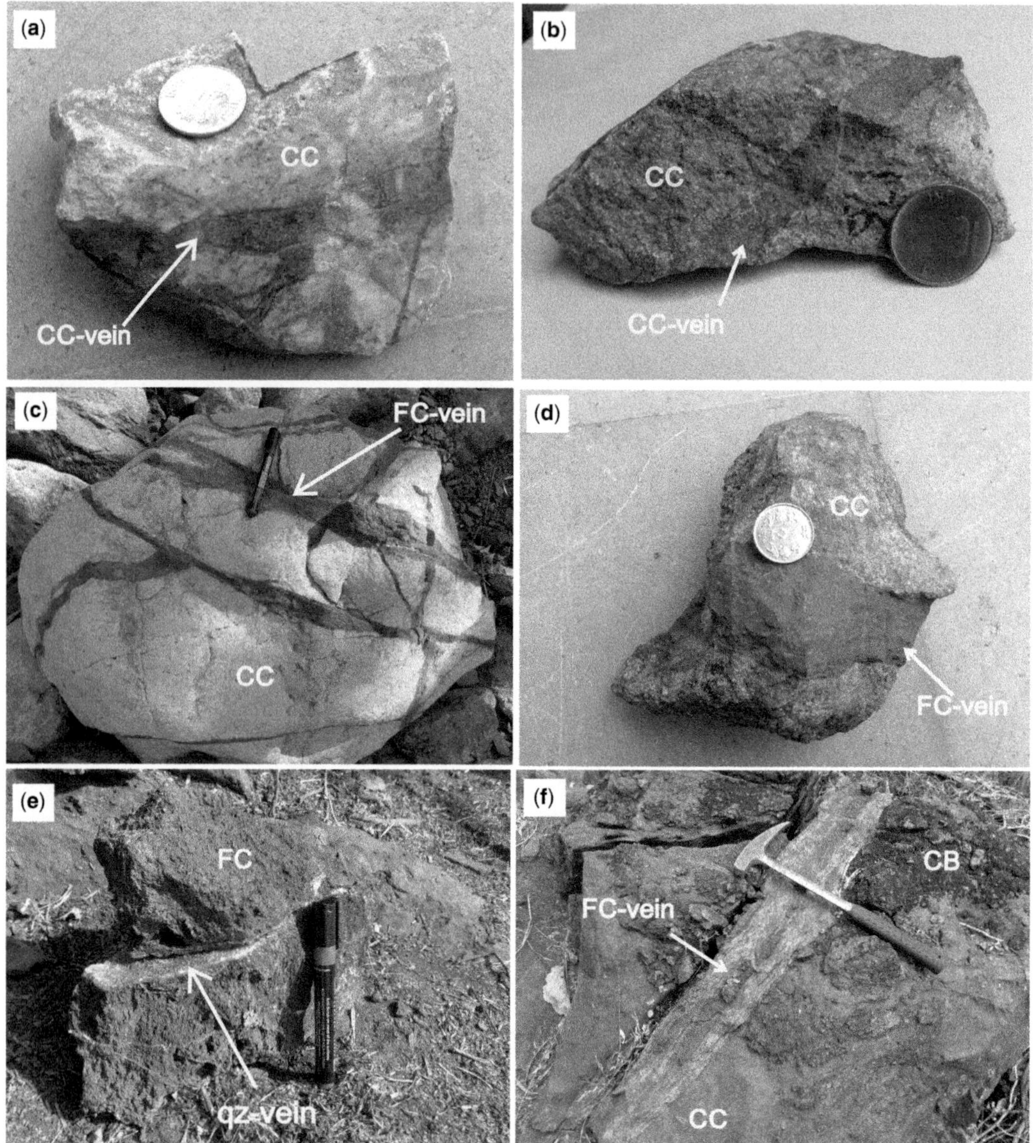

Fig. 2. Hand specimens of calciocarbonatites and ferrocarbonatites from the Amba Dongar carbonatite complex. (**a**) & (**b**) Late-stage calciocarbonatite vein in earlier calciocarbonatite. (**c**) & (**d**) Ferrocarbonatite vein in calciocarbonatite. (**e**) Quartz vein in ferrocarbonatite. (**f**) A ferrocarbonatite vein cutting through the early formed carbonatite breccia and calciocarbonatite. FC, ferrocarbonatite; CC, calciocarbonatite; CB, carbonatite breccia; qz, quartz.

found in calcite, and vice versa (Fig. 4a, b). Apatites in ferrocarbonatites occur as tiny randomly scattered crystals. Pyrochlore is generally honey brown in colour and occurs as randomly scattered small, euhedral/rhombohedral/hexagonal-shaped crystal with prominent zoning.

Nephelinite is composed mainly of phenocrysts of nepheline, aegirine, aegirine–augite and melanite in a fine-grained greyish brown groundmass forming a porphyritic texture (Fig. 3g, h). Large euhedral nephelines seem altered and partially fractured. Small euhedral nephelines are partially altered and randomly scattered throughout the groundmass. Rhombohedral, pentagonal to hexagonal crystals of melanite contain inclusions of aegirine–augite and apatite (Fig. 3h); similarly, several aegirine crystals contain tiny inclusions of melanite. Garnets are highly fractured and show prominent zoning, with darker rim and lighter core. Twinning and zoning are also observed in aegirine–augite. A sharp contact between melanite and aegirine–augite is visible in places. Randomly scattered apatite occurs as an accessory mineral in nephelinites. Porphyritic texture is common in Deccan basalts which show fresh euhedral plagioclase and subhedral augite embedded in a fine groundmass of plagioclase and pyroxene (Fig. 3i). Ophitic texture is common in dolerite dyke samples (Fig. 3j).

Finally, backscattered electron (BSE) images of apatites in three different samples of calciocarbonatites and one nephelinite are shown in Figure 4. Three forms of apatites are observed in calciocarbonatites, namely scattered/disseminated apatites within calcites (Fig. 4a, b), clusters near calcite boundaries (Fig. 4c, d) and apatite cumulates (Fig. 4e, f). Apatites of varying size and habit are seen in nephelinites (Fig. 4g, h).

Major oxide compositions

Bulk major oxide compositions of the Amba Dongar carbonatites are reported in Tables 1 and 2. Based on the chemical composition, these carbonatites are classified as calciocarbonatites and ferrocarbonatites (Fig. 5a). In the CMF (CaO–MgO–FeO + Fe_2O_3 + MnO) ternary diagram proposed after Woolley (1982) and Woolley & Kempe (1989), our calciocarbonatites show a restricted range compared to those reported by earlier workers. On the other hand, our ferrocarbonatites show more compositional variability compared to those reported by earlier workers (Fig. 5a). Only one calciocarbonatite sample (ADC-15) containing thin bands of magnetite crystals plot in the ferrocarbonatite field. Note that three samples (ADF-3, ADF-5 and ADF-9) plot in the magnesiocarbonatite field; XRD analyses showed the presence of a ferroan-dolomite phase. Some of the samples classified as ferrocarbonatites by previous workers also transition into the magnesiocarbonatite field (Fig. 5a). Our three magnesiocarbonatite samples have similar trace element patterns to those observed for the other ferrocarbonatites (shown later). Here, we group these three samples as ferrocarbonatite. The major oxide compositional variabilities in carbonatites (Fig. 6) reveal two distinct fields for our calciocarbonatites and ferrocarbonatites. Our calciocarbonatites define a narrow compositional range (CaO, 39.5–55.9 wt%) and plot within the field defined by previously published data (CaO, 34.5–54.1 wt%). Distinct negative correlations of SiO_2, $Fe_2O_3^T$, MgO, MnO and BaO v. CaO are observed in calciocarbonatites. On the other hand, our ferrocarbonatites exhibit large compositional variations compared to the already available data. In particular, the CaO content in our samples (15.6–31 wt%) is comparatively lower than that (CaO, 31.1–41.3 wt%) previously reported for Amba Dongar ferrocarbonatites. Compared to our calciocarbonatites (SiO_2, 0.55–9.67 wt%; BaO, 0.02–3.41 wt%), ferrocarbonatites have generally higher SiO_2 (1.83–34.52 wt%) and BaO (0.3–7 wt%). Notably, the SiO_2 content of our ferrocarbonatites ranges from 1.83 to 34.5 wt%; only five samples show a higher content >20 wt%. Note that approximately 40 wt% SiO_2 (corresponding to CaO of 22.5 wt%) has been reported for a ferrocarbonatite from Amba Dongar (Sukheswala & Viladkar 1978). The high SiO_2 content in a few of our ferrocarbonatites are attributed to the presence of thin bands of microcrystalline silica (confirmed from textural studies), which could not be separated out during sample preparation. Generally, the total alkali (K_2O + Na_2O) content in carbonatites is <0.4 wt%; in comparison, the total alkali content in the Oldoinyo Lengai natrocarbonatites, Tanzania, is approximately 40 wt% (GEOROC, Mainz database). Such an extremely low alkali content in the Amba Dongar carbonatites has been linked to a loss of alkalis (from the carbonatite magma) which has caused the widespread fenitization of country rocks in the area (Sukheswala & Viladkar 1981).

Major oxide, as well as selected trace element, compositions of apatites analysed in three calciocarbonatite samples and one nephelinite sample revealed strong compositional differences. In general, except for La, other REE concentrations, as well as F and FeO content, are similar between the two types of apatites. Similarity in the elemental abundance may indicate the same parental magma for both the types of apatites, also suggesting a genetic relationship between nephelinite and calciocarbonatite. Furthermore, although the apatites in carbonatites contain up to 0.4 wt% Na_2O, that in the bulk carbonatite is negligible. This is in agreement with much of the volatile content originally present in the parental carbonate magma having been lost to the surroundings, resulting in fenitization.

The major oxide composition of nephelinites, Bagh sandstone and limestone, and associated Deccan basalts, analysed in this study are reported in Table 3. A strong negative correlation between CaO (12.6–14.2 wt%) and SiO_2 (40.6–43.8 wt%) is observed in the nephelinites (not shown here). Generally, the Al_2O_3 (11.3–14.2 wt%), TiO_2 (0.83–1.59 wt%), P_2O_5 (0.29–0.88 wt%) and $Fe_2O_3^T$ (7.19–9.46 wt%) content in nephelinites increases with SiO_2. The chemical composition of nephelinites (except ADN-15) reported here are similar to the unfenitized nephelinites reported by Viladkar (2015). Evidence of fenitization was clearly seen both in hand specimens and in thin sections of ADN-15 (K_2O = 4.41 wt%: Table 3), which was not observed

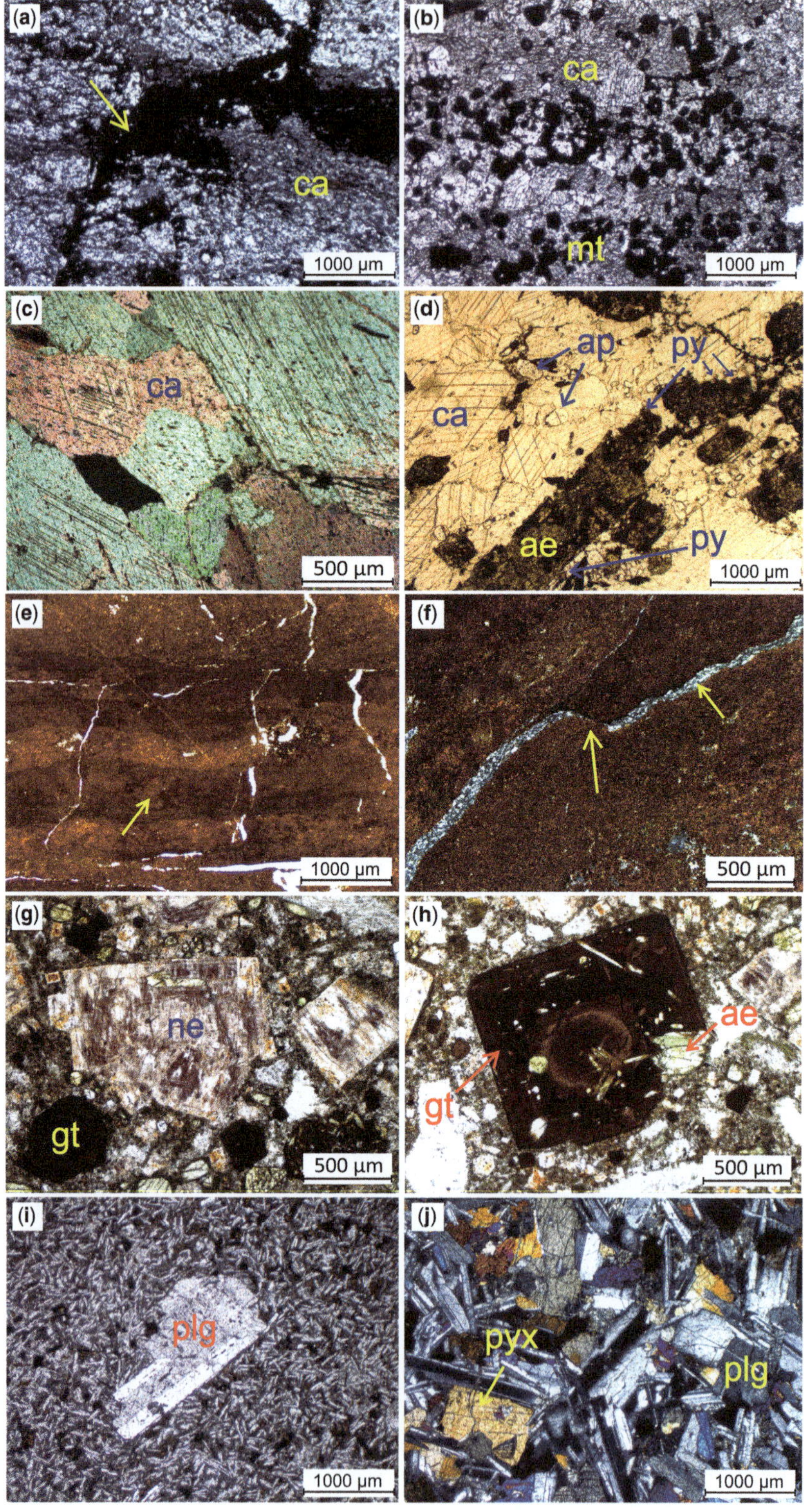
(a)
ca
1000 µm
(b)
ca
mt
1000 µm
(c)
ca
500 µm
(d)
ap
py
ca
ae
py
1000 µm
(e)
1000 µm
(f)
500 µm
(g)
ne
gt
500 µm
(h)
ae
gt
500 µm
(i)
plg
1000 µm
(j)
pyx
plg
1000 µm

in other nephelinites analysed in this study. The composition of Bagh limestones (Table 3) shows compositional diversity, particularly in SiO_2 (14.8–40 wt%), Al_2O_3 (4.61–10.8 wt%), CaO (20.7–42.7 wt%) and K_2O (0.46–4.22 wt%). One earlier reported sample of Bagh limestone (Gwalani *et al.* 1995) from the Amba Dongar is very similar in major oxide compositions, except for Al_2O_3 (0.34 wt%). A high content of CaO, and low SiO_2, Al_2O_3 and K_2O, in sample ADLs-2 suggest that it was least affected by fenitization. Out of the four Bagh sandstone samples, ADSt-2 and ADSt-3, with high $SiO_2 >$ 98 wt%, and low $Al_2O_3 < 0.6$ wt% and $K_2O <$ 0.1 wt%, seem to be least affected by fenitization. Note that the composition of ADSt-3 is very similar to the composition of the unfenitized Bagh sandstone reported earlier by Sukheswala & Viladkar (1981). The total alkali content (0.08–1.01 wt%) in Bagh sandstone is considerably less than that (0.85–5.17 wt%) in the Bagh limestone, which indicates that sandstone was relatively less affected by hydrothermal alteration/fenitization.

The major oxide compositions of other associated basaltic rocks analysed in this study (DFB and mafic dykes) are also reported in Table 3. The MgO content of these rocks ranges from 3.54 to 6.64 wt%, indicating the differentiated nature of these basalts. Similar values are also reported for the nearby Toranmal section (MgO, 2.98–7.4 wt%), Poladpur Formation (MgO, 3.74–12.01 wt%) and Ambenali Formation (MgO, 4.64–8.26 wt%) of the Deccan province (GEOROC database). In our samples, strong positive correlations in MgO v. Al_2O_3 (12–14.7 wt%), CaO (6.52–12.4 wt%) and Ni (20–72 ppm), and strong negative correlation between MgO and TiO_2 (1.3–3.36 wt%), suggest crystal fractionation effects of olivine, plagioclase and clinopyroxenes on the magma evolution. Generally, the CaO, P_2O_5, MnO and alkali content is higher in nephelinites than in basalts. The total alkali–silica (TAS) diagram (Le Bas *et al.* 1986) reveals the subalkaline (tholeiitic) nature of the analysed Deccan basalts (Fig. 5b).

Trace element compositions

Trace element concentrations of carbonatites, associated silicates and country rocks from Amba Dongar are reported in Tables 4 and 5. Similar to that observed for the major element contents, trace element concentrations and ratios show similar enrichment or depletion patterns to those observed in the literature. Note that, although trace element analyses of nearly 72 Amba Dongar carbonatite samples have been reported in the literature (Deans & Powell 1968; Sethna 1971; Sukheswala & Viladkar 1978; Viladkar 1981, 2012; Viladkar & Dulski 1986; Viladkar & Wimmenauer 1992; Gwalani *et al.* 1993; Srivastava 1997; Ray *et al.* 2000*b*; Ray & Shukla 2004), rarely any study has reported the full range of 30 elements (particularly HREEs) reported in this study. Usually, carbonatites show enrichment of REEs, Nb and Sr (Rukhlov *et al.* 2015; Simandl 2015). Our calciocarbonatites show a wide variability in trace element concentrations (Table 4), and are notably enriched in Sr (1421–13 146 ppm), Ba (149–30 500 ppm), Nb (10–2168 ppm) and Th (3.8–345 ppm). In comparison, ferrocarbonatites also show (Table 4) similar enrichment of Sr (2332–11 553 ppm) and Nb (76–2472 ppm), but are extremely enriched in Ba (2705–64 546 ppm) and Th (20–623 ppm). The variability in the incompatible element ratios v. the chondrite-normalized La/Sm ratio (La/Sm_N) in the carbonatite and nephelinite is shown in Figure 7. The La/Sm_N (5.1–24.6) ratio in the calciocarbonatites shows a narrow range, but in the ferrocarbonatite shows a wide range (La/Sm_N: 7.7–48.6). Widely variable Nb/Ta (842–29 278 in ferrocarbonatite v. 95–7177 in calciocarbonatite), Th/U (1.7–1454 v. 1–281) and Ba/Th (57–204 v. 3–361) ratios indicate the role of pyrochlore, thorite and barite crystal fractionation in the evolution of carbonatite magma.

A notable feature of these Amba Dongar carbonatites is the extreme enrichment in REE; the ΣREE of the calciocarbonatites and ferrocarbonatites varies by 1025–12 317 ppm and 6839–31 117 ppm, respectively. The LREE abundances, particularly La and Ce, are much higher in ferrocarbonatites than in calciocarbonatites. The average concentrations of La and Ce in ferrocarbonatites (5056 ppm and 8995 ppm, respectively) are higher by a factor of 4 than in calciocarbonatites. In comparison, even higher abundances (up to at least 3 times or more enrichment) in Sr, Ba, Th and U, and ΣREE have been reported in carbonatites in an intracontinental rift setting from South Nam Xe, NW Vietnam, one

Fig. 3. Petrographical features of different rocks from the Amba Dongar carbonatite complex. (**a**) Fine-grained calciocarbonatite with Fe-oxide veins indicated by an arrow. (**b**) Medium-grained calciocarbonatite with alternate bands of calcite and magnetite. (**c**) Coarse-grained monomineralic calciocarbonatite. (**d**) Coarse-grained calciocarbonatite with accessory phases. (**e**) & (**f**) Ferrocarbonatite with thin veins of secondary calcite and flow bands of dark material shown by the arrow. (**g**) Porphyritic texture in nephelinite. (**h**) A melanite garnet with inclusions of aegirine and apatite. (**i**) Deccan basalt with a phenocryst of plagioclase. (**j**) Ophitic texture in doleritic dyke. ap, apatite; ca, calcite; ae, aegirine; py, pyrochlore; mt, magnetite; ne, nepheline; gt, garnet; plg, plagioclase; pyx, pyroxene.

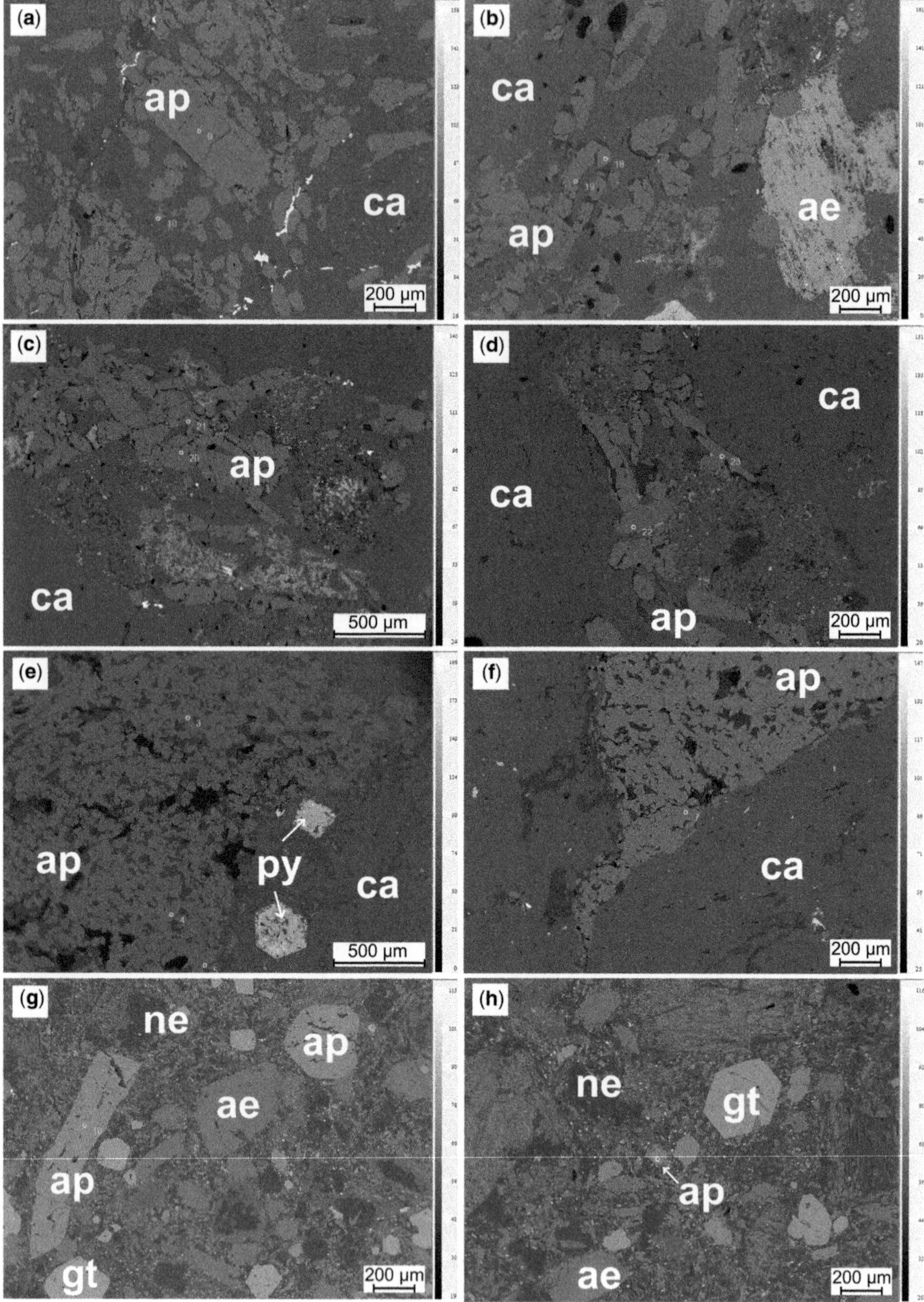

Fig. 4. Backscattered electron (BSE) images of apatite from three different calciocarbonatite samples (**a**)–(**f**) and one nephelinite sample (**g**) & (**h**). (a) Size variation in apatite intruded into calcitic groundmass (ADC-26). (b) Calcite inclusions in apatites (ADC-21). (c) & (d) Apatites located along the boundary of calcite crystals (ADC-21). (e) & (f) Apatite cumulate (ADC-29). (g) A large prism of apatite. (h) Tiny rounded crystal of apatite in nephelinite (ADN-5). The dots (green in the online version) represent the locations of microprobe spot analyses. Abbreviations are the same as in Figure 3.

Table 1. *Major oxide compositions (in wt%) of calciocarbonatites from the Amba Dongar carbonatite complex determined by WD-XRF of fusion beads*

	ADC-1	ADC-2	ADC-3	ADC-4	ADC-5	ADC-6	ADC-7	ADC-8	ADC-9	ADC-10	ADC-11	ADC-12	ADC-13	ADC-14	ADC-15	ADC-16	ADC-17
SiO_2	3.67	3.50	7.82	1.48	0.55	1.43	0.94	1.93	7.54	5.34	4.28	5.39	3.55	2.84	6.63	2.67	1.36
TiO_2	0.21	0.11	0.09	0.08	0.08	0.31	0.01	0.10	0.18	0.13	0.15	0.13	0.11	0.14	0.47	0.15	0.20
Al_2O_3	0.31	0.26	0.24	0.26	0.22	0.44	0.32	0.24	0.44	0.46	0.48	0.26	0.29	0.38	0.27	0.28	0.29
$Fe_2O_3^T$	2.10	0.78	2.92	0.10	0.08	6.24	0.32	0.78	4.90	1.96	1.86	3.17	1.55	1.05	14.2	2.67	1.32
MnO	0.45	0.33	0.37	0.37	0.14	0.21	0.22	0.27	0.40	0.51	0.54	0.40	0.29	0.14	0.48	0.25	0.40
MgO	0.13	0.09	0.34	1.22	0.28	0.38	0.31	0.99	0.18	0.12	0.16	0.20	0.16	0.42	1.56	0.38	0.17
CaO	51.0	54.3	50.8	53.3	55.8	50.5	55.9	52.7	49.7	48.4	48.5	49.1	51.5	51.7	41.7	52.3	52.8
Na_2O	0.03	0.06	bd	bd	bd	bd	0.12	0.02	bd	0.02	0.01	bd	0.07	bd	bd	bd	bd
K_2O	bd	bd	bd	bd	bd	0.06	bd	bd	0.02	0.07	0.08	bd	bd	0.03	bd	bd	0.00
P_2O_5	1.76	1.77	1.52	0.39	0.42	1.40	0.29	1.56	1.02	1.14	1.71	1.83	0.66	1.37	0.16	0.14	2.74
BaO	0.69	0.56	1.92	0.25	0.09	0.22	0.09	0.66	0.44	1.14	2.13	0.78	1.08	0.29	1.07	0.02	0.71
SrO	0.46	0.69	0.29	0.61	0.81	0.44	0.75	0.73	0.33	0.89	0.45	0.78	0.45	0.71	0.23	0.34	0.32
LOI	37.8	36.1	33.7	41.3	41.9	38.0	39.0	39.8	34.3	37.6	39.2	37.0	39.8	39.4	32.9	40.9	38.8
Sum	98.6	98.6	100.0	99.4	100.4	99.6	98.3	99.8	99.5	97.8	99.6	99.0	99.5	98.5	99.7	100.1	99.1

	ADC-18	ADC-19	ADC-20	ADC-21	ADC-22	ADC-23	ADC-24	ADC-25	ADC-26	ADC-27	ADC-28	ADC-29	ADC-30	ADC-31	ADC-32	ADC-33	ADC-34
SiO_2	9.67	9.27	0.89	2.41	8.99	3.81	1.51	9.29	8.31	4.83	6.02	4.59	0.95	6.09	4.97	8.40	3.74
TiO_2	0.11	0.78	0.12	0.08	0.10	0.12	0.08	0.15	0.18	0.08	0.15	0.10	0.07	0.15	0.08	0.11	0.12
Al_2O_3	0.29	0.36	0.39	0.70	0.29	0.25	0.27	0.81	0.38	0.11	0.34	0.49	0.21	0.27	0.28	0.40	0.70
$Fe_2O_3^T$	2.37	3.26	4.09	2.66	2.04	1.80	0.40	3.72	1.76	3.44	4.59	0.93	0.21	1.57	1.01	1.17	1.41
MnO	0.62	0.16	0.75	0.82	0.37	0.66	0.10	0.26	0.33	1.39	1.52	0.48	0.18	0.22	0.25	0.60	0.26
MgO	0.43	1.10	2.59	2.95	0.51	0.38	0.10	0.17	1.82	0.09	2.63	1.71	0.30	0.06	0.13	0.17	0.29
CaO	46.8	46.7	47.4	47.7	48.0	51.8	54.8	47.5	49.0	48.9	39.5	50.1	55.9	50.7	52.0	46.9	52.3
Na_2O	bd	0.33	bd	0.36	bd	bd	bd	bd	bd	0.00	bd	0.02	bd	0.02	bd	0.00	0.01
K_2O	0.00	0.02	0.01	0.01	0.01	bd	0.01	0.30	0.04	0.02	0.01	0.09	bd	bd	bd	0.08	0.18
P_2O_5	0.09	3.26	0.26	1.58	0.72	1.96	0.17	0.91	2.20	0.36	0.37	0.66	0.07	1.92	1.13	0.10	0.21
BaO	1.43	0.33	0.83	3.41	na	0.57	0.08	0.09	1.18	1.32	2.36	0.59	0.15	na	0.54	2.40	0.04
SrO	0.28	1.00	0.30	0.44	0.18	0.39	0.80	0.22	0.48	0.17	0.64	1.40	0.81	1.55	0.22	0.54	0.30
LOI	37.0	32.2	39.3	36.8	37.2	37.8	40.6	37.0	32.9	37.5	35.3	39.5	42.2	37.4	38.6	37.3	40.5
Sum	99.1	98.8	96.9	99.9	98.4	99.5	98.9	100.4	98.6	98.2	93.4	100.7	101.1	100.0	99.2	98.2	100.1

LOI, Loss on ignition; bd, below detection; na, not analysed.

Table 2. *Major oxide compositions (in wt%) of ferrocarbonatites from the Amba Dongar carbonatite complex determined by WD-XRF of fusion beads*

	ADF-1	ADF-2	ADF-3	ADF-4	ADF-5	ADF-6	ADF-7	ADF-8	ADF-9	ADF-10	ADF-11	ADF-12	ADF-13
SiO_2	11.3	3.28	19.8	32.7	9.78	12.1	1.83	na	9.14	3.04	29.6	34.5	24.6
TiO_2	0.33	0.03	0.02	0.09	0.10	0.29	1.50	0.06	0.10	0.40	0.12	0.21	0.24
Al_2O_3	1.04	0.35	0.31	0.36	0.26	0.16	0.27	0.09	0.31	0.27	0.42	1.19	0.63
$Fe_2O_3^T$	12.7	10.7	4.22	6.92	6.07	15.0	26.6	7.53	8.04	10.0	9.41	8.61	9.97
MnO	2.23	2.68	2.43	3.15	2.88	3.81	8.61	4.79	1.64	3.84	1.16	0.84	1.03
MgO	8.37	11.7	10.5	5.96	16.0	8.39	7.09	11.3	12.6	5.91	7.24	3.81	4.58
CaO	15.6	25.9	24.8	22.5	25.1	29.0	23.9	27.6	28.2	31.2	23.07	22.7	26.9
Na_2O	0.33	bd	0.11	0.08	bd	bd	bd	0.01	bd	bd	0.02	bd	0.32
K_2O	bd	bd	bd	bd	bd	bd	bd	0.01	bd	bd	bd	bd	bd
P_2O_5	0.82	0.14	0.75	1.79	0.37	1.11	0.99	0.20	0.11	0.33	2.30	5.96	3.48
BaO	na	3.98	na	3.22	1.51	na	na	7.21	0.87	4.38	0.92	0.30	1.83
SrO	0.59	0.95	0.49	0.45	0.56	0.90	1.31	0.97	0.28	1.37	0.28	0.31	0.30
LOI	25.6	31.1	28.0	20.3	36.5	16.0	19.9	29.9	39.1	21.9	26.8	17.7	19.2
Sum	78.9	90.8	91.4	97.5	99.1	86.8	92.0	89.7	100.4	82.6	101.3	96.1	93.1

LOI, Loss on ignition; bd, below detection; na, not analysed.

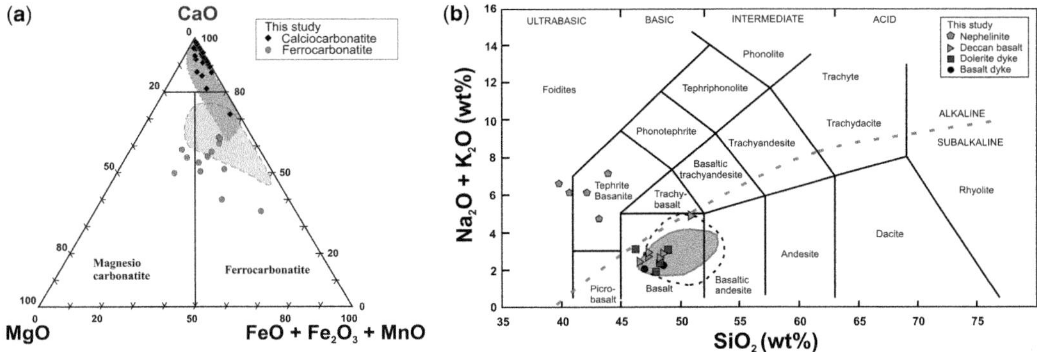

Fig. 5. (**a**) Chemical classification of the Amba Dongar carbonatites following the Woolley (1982) and Woolley & Kempe (1989) scheme. Grey and red (bounded by dashed lines) shaded fields encompass the composition of previously published calciocarbonatites ($n = 42$) and ferrocarbonatites ($n = 30$) from the study area, respectively. Data are from Sukheswala & Viladkar (1978), Viladkar (1981, 2012), Viladkar & Dulski (1986), Viladkar & Wimmenauer (1992), Gwalani *et al.* (1993) and Srivastava (1997). All ankeritic carbonatites from the previously published data are considered as ferrocarbonatites. (**b**) Total alkali–silica (TAS) diagram (Le Bas *et al.* 1986) showing the composition of associated silicate rocks from the study area. The grey field represents the previously published data (GEOROC, Mainz) for the basalts from the Ambenali Formation ($n = 110$) and the Poladpur Formation ($n = 78$), and the dashed field represents basalts of the Toranmal section ($n = 65$) in the northern part of DFB province.

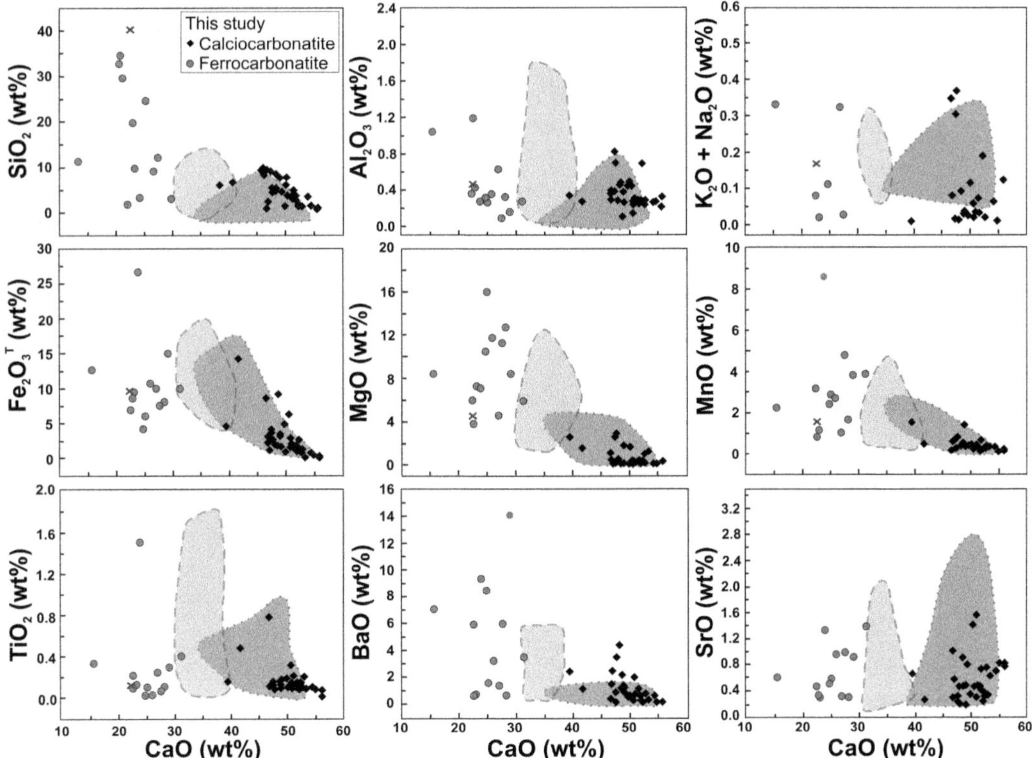

Fig. 6. Major oxide compositions (wt%) of calciocarbonatites and ferrocarbonatites from Amba Dongar. The grey and red (bounded by dashed lines) shaded fields represent previously published compositions whose sources are the same as in Figure 5. The cross symbol represents a sample of ankeritic carbonatite with quartz and barite from Amba Dongar reported by Sukheswala & Viladkar (1978).

Table 3. *Major oxide compositions (in wt%) of silicate rocks and sedimentary Bagh sandstone and Bagh limestone from Amba Dongar determined by WD-XRF of fusion beads*

	Nephelinite					Bagh sandstone				Bagh limestone		
	ADN-5	ADN-6	ADN-8	ADN-12	ADN-15	ADSt-1	ADSt-2	ADSt-3	ADSt-4	ADLs-1	ADLs-2	ADLs-3
SiO_2	42.0	43.0	40.6	43.8	39.7	94.4	98.1	98.8	96.7	23.8	14.8	40.0
TiO_2	1.59	1.31	0.95	0.83	0.76	0.40	0.10	0.06	0.08	0.47	0.39	0.72
Al_2O_3	11.3	13.9	13.5	14.2	12.5	3.30	0.59	0.39	2.04	7.16	4.61	10.8
$Fe_2O_3^T$	9.46	9.25	7.69	7.19	6.95	bd	bd	bd	bd	2.33	2.14	4.22
MnO	0.25	0.37	0.25	0.27	0.29	0.01	0.01	0.01	0.01	0.07	0.05	0.09
MgO	2.29	1.55	1.64	1.44	0.55	0.02	bd	bd	0.00	1.26	2.34	1.95
NiO	0.001	bd	bd	bd	bd	0.002	bd	bd	bd	0.002	0.002	0.003
CaO	13.4	13.1	14.2	12.5	14.9	0.52	0.44	0.47	0.47	36.5	42.7	20.7
Na_2O	4.78	3.84	4.16	5.64	2.28	bd	bd	bd	0.04	0.44	0.39	0.95
K_2O	1.42	0.94	2.00	1.58	4.41	1.01	0.10	0.08	0.45	1.93	0.46	4.22
P_2O_5	0.88	0.62	0.29	0.60	0.48	0.08	0.06	0.06	0.06	0.11	0.11	0.16
LOI	9.25	8.49	13.3	10.7	14.5	0.79	0.39	0.07	na	25.5	32.0	16.2
Sum	96.6	96.4	98.6	98.8	97.3	100.5	99.8	99.9	99.9	99.6	100.0	100.0
	Deccan basalt (DFB)						Dolerite dyke				Basalt dyke	
	ADB-1	ADB-5	ADB-6	ADB-8	ADB-17	ADB-20	ADDd-1	ADDd-2	ADDd-5	ADDd-6	ADBd-1	ADBd-4
SiO_2	50.8	48.5	46.5	48.1	47.2	47.3	48.9	47.9	48.2	46.1	46.9	48.4
TiO_2	3.13	2.45	2.54	2.27	2.66	3.16	3.36	1.30	1.84	3.28	2.10	1.43
Al_2O_3	12.8	13.5	14.7	14.0	13.8	13.1	12.5	13.9	12.6	12.0	13.5	14.1
$Fe_2O_3^T$	13.5	12.3	13.0	12.8	13.4	15.1	15.0	13.0	15.4	15.5	12.8	13.8
MnO	0.17	0.22	0.17	0.17	0.16	0.19	0.20	0.19	0.21	0.19	0.18	0.20
MgO	3.65	5.90	5.76	6.39	3.54	4.98	5.00	6.64	5.67	4.60	6.25	6.36
NiO	0.003	0.01	0.01	0.01	0.002	bd	0.01	0.01	bd	bd	0.01	0.01
CaO	6.52	10.7	10.5	10.8	9.76	9.58	8.87	12.4	9.62	8.85	11.7	11.9
Na_2O	2.80	2.53	2.08	2.12	2.04	2.33	2.42	1.71	1.89	2.54	1.78	2.03
K_2O	2.14	0.39	0.42	0.54	0.77	0.66	0.65	0.23	0.49	0.60	0.32	0.25
P_2O_5	0.37	0.26	0.28	0.25	0.42	0.36	0.35	0.11	0.20	0.31	0.21	0.15
LOI	1.34	1.05	3.60	1.74	5.14	3.62	1.80	3.12	2.73	3.66	2.25	2.09
Sum	97.2	97.8	99.6	99.2	98.9	100.4	99.1	100.5	98.9	97.6	98.0	100.7

LOI, Loss on ignition; bd, below detection; na, not analysed.

of the known high-trace-element-rich carbonatites globally (Thi *et al.* 2014).

Primitive-mantle-normalized (Palme & O'Neill 2014) trace element patterns of Amba Dongar rocks analysed in this study are presented in Figure 8. In general, carbonatites globally show characteristic large depletion in K, Zr, Hf and Ti, and moderate depletion in (fluid and mobile) U and Pb. These characteristics are also prominent in our carbonatites (Fig. 8a, b). Relatively more enrichment (with less variability) in element patterns is observed for our ferrocarbonatites compared to the calciocarbonatites. The global average compositions of calciocarbonatites and ferrocarbonatites reported by Woolley & Kempe (1989), as well as the average calciocarbonatite composition from Bizimis *et al.* (2003), are strikingly similar to the corresponding respective averages in our carbonatites, except for the more prominent K depletion in our samples. Spidergram patterns observed in carbonatites (Fig. 8a, b) do not display any similarity with that of average Ambenali Formation ($n = 30$; the least crustal-contaminated Deccan lavas). Primitive-mantle-normalized average trace element patterns of calciocarbonatite ($n = 300$), magnesiocarbonatite ($n = 26$), dolomitic-carbonatite ($n = 52$), silicocarbonatite ($n = 37$) and natrocarbonatite ($n = 128$) globally are all similar despite minor differences (compiled from the GEOROC database). Natrocarbonatites do not show U, K and Pb negative anomalies, but do show a positive Ba anomaly; similar features are also noticed for average plutonic East African carbonatites (Keller & Zaitsev 2012). Similarly, the plutonic carbonatites of the Kaapvaal Craton also do not show negative anomalies of U, Pb, Zr and Hf. The absence of these anomalies has been related to variable contamination of the carbonatite magma with the shallow-crustal material or surrounding country rocks. Post-magmatic processes can also play a role in either depletion or enrichment of fluid/mobile incompatible elements in carbonatites.

Primitive-mantle-normalized patterns of silicate rocks (DFB and mafic dykes) are similar to the average Ambenali Formation (Fig. 8c). However, nephelinites show significant enrichment in large-ion lithophile elements (LILEs) (Cs, 4.33–7 ppm; Rb, 30–56 ppm; Sr, 3056–3130 ppm; and Ba, 745–1990 ppm) and high field strength elements (HFSEs) (Nb, 48–344 ppm; Ta, 1.4–4.7 ppm; Zr, 503–565 ppm; Hf, 7–10 ppm) compared to the average Ambenali Formation. These patterns are similar to those observed for the earlier reported Amba Dongar nephelinites ($n = 12$); the majority of the early reported nephelinites may have experienced a variable extent of fenitization (Viladkar 2015). Interestingly, the pattern of nephelinites shows some resemblance (although not in absolute abundances) to that of carbonatites. The patterns defined by Bagh limestone show a similarity with the fenitized Bagh sandstone (ADSt-1), whereas the unfenitized sandstone sample (ADSt-3) is relatively depleted (except for Th, U and Zr) in the majority of the trace elements. Interestingly, the pattern of the unfenitized Bagh sandstone matches with that of nephelinites, except for the Rb, Ba and Sr patterns. Note that the abundances reported for the unfenitized sandstone (ADSt-3) are similar to those for the one unfenitized Bagh sandstone sample reported by Viladkar & Dulski (1986).

Chondrite-normalized (Palme & O'Neill 2014) REE patterns of rocks from Amba Dongar analysed in this study are presented in Figure 8d. Compared with the pattern of the least-contaminated Ambenali Formation lavas, Amba Dongar carbonatites exhibit much steeper REE patterns and a high degree of LREE/HREE fractionation; the average chondrite-normalized $(La/Yb)_N$ ratios in calciocarbonatites and ferrocarbonatites are 118 and 286, respectively. The Amba Dongar nephelinites also reveal steep REE patterns compared to the Ambenali Formation lavas ($La/Yb_N = 3$) and show high LREE/HREE fractionation ($La/Yb_N = 47$). Carbonatites are more highly enriched in LREE than nephelinites; the average La/Sm_N ratios in calciocarbonatites, ferrocarbonatites, nephelinites and the Ambenali Formation are 10, 22, 13 and 1, respectively. The average Gd/Yb_N ratios of the Ambenali Formation, nephelinite, calciocarbonatites and ferrocarbonatite are 2, 3, 7 and 8, respectively. In particular, the HREE content in the most-enriched carbonatites is about 20 times higher than the Ambenali Formation. Nevertheless, the somewhat flat HREE pattern (Gd–Lu) observed for the Ambenali Formation is strikingly similar to that of a few least-enriched calciocarbonatites (Fig. 8d). Further, the patterns for nephelinites are similar to those of the least-enriched carbonatites, as well as the unfenitized sandstone, irrespective of the abundance of elements.

Discussion

Low-temperature alteration

Petrographical evidence suggests that compared to the calciocarbonatites, some (ADF-3, ADF-4, ADF-11, ADF-12 and ADF-13) of the ferrocarbonatites were affected by low-temperature hydrothermal alteration. Unlike calciocarbonatites, these ferrocarbonatites mostly contain thin veins of secondary calcite (Fig. 3e), quartz and iron oxide, as well as fluorite and barite crystals. Evidence of hydrothermal alteration is more apparent in calciocarbonatite outcrops in the currently active mining area (mined mainly for fluorites) and less visible for carbonatite exposures away from the mining area. In the absence of other possible sources of silica in the study area, it

Table 4. *Trace element (including REE) compositions (in ppm) of calciocarbonatite and ferrocarbonatite from the Amba Dongar carbonatite complex determined by ICP-MS*

	Calciocarbonatite																			
ppm	ADC-1	ADC-2*	ADC-2†	ADC-3*	ADC-4	ADC-5	ADC-7*	ADC-9*	ADC-11	ADC-13*	ADC-14	ADC-15*	ADC-16	ADC-18	ADC-19	ADC-20	ADC-21*	ADC-22	ADC-23*	ADC-26*
Sc	0.4	1.4	1.5	13	1.8	1.6	0.8	3.3	1.7	7.1	2.2	13	7.7	7.7	1.9	3.8	16	0.4	5.2	16
V	113	33	31	81	23	7	11	492	216	88	50	634	103	109	435	226	310	127	126	202
Ni	0.4	na	5.7	3.0	5.4	6.1	na	155	5.3	na	5.7	4.0	5.2	4.1	5.2	5.2	na	5.3	3	na
Zn	300	139	117	84	54	na	18	142	379	197	8	375	55	162	26	585	457	495	48	76
Rb	1.2	0.8	0.8	0.3	0.3	0.1	bd	2.1	7.2	2.4	4.0	0.3	0.1	1.9	3.4	2.8	1.6	1.4	2.2	1.9
Sr	3915	5815	5754	2419	5194	6880	6350	2800	3773	3790	6007	1983	2874	2335	8419	2536	3760	1507	3340	4027
Y	146	186	206	453	96	109	48	121	211	179	140	88	na	122	73	na	363	131	189	132
Zr	20	20	na	61	11	14	11	21	21	50	187	180	30	29	389	31	77	17	23	16
Nb	33	74	68	900	28	75	24	203	350	417	2168	1457	579	632	1967	79	366	67	176	1323
Cs	0.04	0.1	0.1	0.1	bd	bd	0.1	0.2	0.1	bd	0.1	0.1	bd	bd	0.4	bd	bd	bd	bd	0.1
Ba	6150	5050	5096	17 200	2196	818	828	3979	19 075	9635	2605	9564	149	12 812	2923	7416	30 500	na	5137	10 600
La	1354	637	661	1500	263	253	243	684	2035	909	505	1650	1102	1706	284	2748	3410	1523	616	434
Ce	2279	1368	1395	2850	586	551	577	1200	2995	1870	1138	2600	2329	3409	635	4948	5780	2831	1210	826
Pr	181	98	107	329	59	57	42	128	254	235	116	264	248	363	65	513	648	281	150	86
Nd	555	290	307	1110	210	195	120	399	709	792	387	835	828	1230	206	1618	1920	875	511	277
Sm	90	34	37	150	31	30	15	45	77	101	63	92	100	165	24	190	222	113	65	36
Eu	4.3	7.7	9	39	10	8.6	3.8	11	20	23	18	19	23	36	8	49	53	26	17	9.3
Gd	31	26	28	102	26	30	10	28	74	66	52	47	70	101	23	150	159	96	48	26
Tb	5.4	3.9	3.4	14	2.9	3.1	1.3	3.4	7.0	7.0	6.0	4.7	6.6	6.9	2.5	15	15	7.9	6.1	3.6
Dy	13	21	25	67	14	15	6	17	33	29	28	18	29	24	12	61	61	31	30	18
Ho	3.6	4.2	4.4	12	2.5	2.7	1.1	3.1	5.9	4.6	4.6	2.7	5.0	3.6	2.1	9.3	10	4.5	5.2	3.4
Er	5.2	11	16	29	6.9	7.0	2.8	7.9	19	11	12	7	15	12	6.2	25	23	10	12	9
Tm	1.0	1.5	1.7	4	0.9	0.9	0.4	1.0	2.3	1.1	1.2	0.6	1.5	1.1	0.8	2.4	2.3	0.9	1.5	1.3
Yb	4.2	7.5	8	19	6.0	5.3	2.2	5.4	14	5.7	6.7	3.0	8.4	7.5	5.1	12	12	5.1	8.8	7.1
Lu	0.6	0.9	1.1	2.3	0.8	0.7	0.3	0.7	1.8	0.7	0.8	0.3	1.1	1.1	0.8	1.5	1.5	0.7	1.2	1.0
Hf	0.4	0.3	0.3	1	0.2	0.3	0.2	0.2	0.5	0.3	5	1	1	1	4	0.5	1	0.7	0.3	0.4
Ta	na	0.1	0.2	0.2	0.1	0.8	0.2	0.1	0.1	0.1	11	0.3	bd	0.1	12	bd	0.3	0.2	0.1	3.9
Pb	na	67	78	20	52	26	1.4	17	618	10	20	42	31	179	6.2	530	57	581	20	16
Th	161	90	101	115	65	75	6.0	104	111	158	345	219	49	71	53	27	169	24	81	77
U	na	14	12	1.2	0.7	2.2	1.2	1.1	63	3.1	12	3.2	1.2	1.0	48	1.2	0.6	3.3	1.3	7.2
∑REE	4527	2511	2604	6227	1219	1159	1025	2534	6247	4055	2338	5543	4767	7066	1275	10 342	12 317	5805	2682	1738

Table 4. *(continued)*

	Calciocarbonatite							Ferrocarbonatite													
ppm	ADC-27	ADC-27†	ADC-29*	ADC-30	ADC-31	ADC-33	ADC-34	ADF-1*	ADF-2	ADF-3*	ADF-4*	ADF-5*	ADF-6*	ADF-7*	ADF-8*	ADF-8†	ADF-9	ADF-10	ADF-11	ADF-12	ADF-13
Sc	na	0.4	bd	3	2.5	2.1	0.2	21	11	29	1.2	8.5	47	37	na	4	15	0.8	2.1	4.1	4.5
V	na	65	bd	23	273	120	69	940	134	447	178	271	319	403	na	124	499	192	278	341	515
Ni	7.1	6.2	10	7.6	15	4.4	6.1	6	4.8	5	2.3	4	7	6	3	2.9	3.7	3.3	11	7.9	6.7
Zn	1290	1142	370	bd	284	161	53	2148	1253	1010	1549	1195	1859	3564	1540	1556	2192	1998	487	222	753
Rb	0.8	1.1	2.45	0.05	1.02	5.83	12.54	0.40	0.43	2.00	0.54	1.90	bd	0.20	0.14	0.16	0.58	0.30	0.19	0.22	0.25
Sr	1421	1399	11 840	6865	13 146	4555	2498	4967	8009	4575	3764	4725	7637	11 100	8210	7994	2468	11 553	2332	2605	2564
Y	141	159	58	64	bd	236	116	168	318	595	na	126	286	173	365	375	160	106	169	267	389
Zr	22	20	10	9	bd	bd	24	66	15	127	24	11	82	88	na	6.7	3.3	16	bd	bd	bd
Nb	345	311	1307	10	bd	152	304	320	231	139	76	150	513	845	na	98	210	631	746	2472	913
Cs	0.1	0.1	0.08	bd	bd	0.04	0.95	0.16	0.00	bd	bd	bd	0.12	0.12	0.03	bd	bd	0.01	0.02	0.02	bd
Ba	11 854	12 421	5310	1361	na	21 503	382	na	35 631	na	28 824	13 500	na	na	64 546	63 501	7791	39 188	8238	2705	16 368
La	2921	2830	425	358	1331	2542	986	6560	8160	9710	6969	2340	9100	1670	6915	7190	3001	898	3116	2987	4300
Ce	5402	5510	803	593	2394	3448	1853	10 600	8053	10 800	7070	3270	16 900	18 200	13 301	13 200	4424	7865	4091	5591	6764
Pr	436	501	56	58	235	284	187	966	515	899	467	342	1440	456	1050	1105	375	167	332	549	599
Nd	1487	1471	254	180	751	809	436	2370	1265	2010	1028	637	3250	1320	2580	2611	1065	477	880	1729	1711
Sm	146	155	32	22	117	99	75	155	171	181	91	102	183	138	209	212	123	57	91	215	199
Eu	32	31	8.2	6.2	25	23	18	26	46	43	27	23	31	28	45	39	28	19	17	39	38
Gd	100	107	27	19	118	93	62	80	156	160	115	66	130	90	160	144	105	96	85	168	165
Tb	8.4	8.7	2.5	2.1	11	7.7	5.3	6.0	13	17	9.0	6.6	7.9	9.0	14	13	8.8	6.4	5.4	12	12
Dy	35	32	12	10	54	34	21	24	55	85	39	27	34	41	68	63	34	24	24	49	58
Ho	5.3	5.1	2.0	1.8	10	5.7	3.4	4.1	9.1	16	7.1	4.6	6.0	7.1	12	10	5.5	4.4	4.3	7.7	11
Er	16	14	6	5	25	17	9	14	26	42	23	12	21	19	38	35	17	14	13	22	30
Tm	1.4	1.3	0.7	0.6	3.0	1.8	0.9	1.4	2.9	4.9	2.8	1.3	2.2	2.4	3.3	3.1	1.7	1.8	1.3	2.2	3
Yb	7.8	7.3	4.2	3.6	16	11	4.6	7.9	16	24	16	6.6	11	13	17	15	10	10	6	12	16
Lu	0.9	0.8	0.6	0.5	1.9	1.3	0.6	1.1	1.9	2.8	1.9	0.8	1.3	1.6	1.9	2	1.3	1.3	0.8	1.4	2
Hf	0.5	0.3	bd	0.2	0.9	0.6	0.6	0.6	1.1	2.0	0.5	0.2	0.7	0.8	1.0	1.2	0.5	0.6	0.4	1.2	1.8
Ta	0.1	0.1	bd	0.03	bd	0.4	0.3	0.4	0.1	0.1	bd	bd	0.1	0.1	0.1	0.1	0.2	0.1	0.2	0.1	0.1
Pb	344	336	276	14	236	217	18	388	524	520	88	187	54	316	344	378	606	145	24	149	220
Th	38	37	42	3.8	bd	132	24	376	623	266	142	139	306	536	509	522	128	266	56	20	130
U	11	10	12	2.7	bd	16	3.5	2.1	2.4	36	0.4	1.4	0.4	3.4	0.4	0.3	44	4.1	3.6	12	7.2
ΣREE	10 599	10 674	1633	1260	5092	7377	3662	20 816	18 490	23 995	15 866	6839	31 117	21 995	24 414	24 642	9199	9641	8667	11 384	13 908

*Sample analysed at the Laboratoire d'Hydrologie et de Géochimie de Strasbourg, France.
†Duplicate analysis done at IITK, India.

Table 5. *Trace element compositions (in ppm) of nephelinite, sedimentary Bagh sandstone, Bagh limestone, DFB and a dolerite from the Amba Dongar carbonatite complex determined by ICP-MS*

	Nephelinite			Bagh sandstone		Bagh limestone			DFB					Dolerite dyke
ppm	ADN-5	ADN-6	ADN-12	ADSt-1	ADSt-3	ADLs-1	ADLs-2	ADLs-3	ADB-1	ADB-5	ADB-6	ADB-8	ADB-17	ADDd-1
Sc	4.4	2.2	2.2	6.3	0.1	7.5	7.1	11	21	30	26	28	22	24
V	372	337	276	63	7.8	57	58	79	355	401	317	309	375	426
Ni	4.7	2.6	1.8	16	bd	14	15	22	25	52	64	72	20	52
Zn	118	125	109	na	na	na	na	na	77	78	69	59	71	93
Rb	42	30	56	21	2.3	94	22	175	59	9.1	4.0	11	11	15
Sr	3082	3056	3130	682	10	763	719	682	437	258	363	565	651	252
Y	48	40	37	11	3.2	13	13	24	41	35	32	32	40	44
Zr	545	565	503	693	101	84	42	164	221	120	177	161	237	191
Nb	48	344	327	7	6	5	6	12	41	16	25	23	35	22
Cs	4.9	4.3	6.9	0.20	0.06	0.62	0.23	2.96	0.6	0.2	0.9	9.5	5.7	0.7
Ba	1990	745	1850	27	10	247	227	615	536	79	178	230	na	136
La	158	176	168	15	7.8	17	15	39	46	16	27	24	39	23
Ce	233	241	258	26	13	37	30	76	101	37	61	54	81	56
Pr	16	15	16	3.2	1.8	3.9	3.4	8.8	9	4	6	5	10	6
Nd	63	62	70	12	5.5	15	13	33	50	22	32	29	30	33
Sm	8.7	7.2	8.0	2.4	0.5	3.5	2.9	6.7	8.4	5.0	6.1	5.7	8.5	7.0
Eu	2.5	1.9	2.2	0.6	0.1	0.7	0.6	1.5	2.4	1.6	1.9	1.7	2.3	2.1
Gd	9.3	7.8	8.2	2.4	0.5	3.4	2.8	6.5	8.1	5.4	6.1	5.8	8.3	7.5
Tb	1.1	0.8	0.9	0.3	0.1	0.5	0.4	0.9	1.1	0.8	0.9	0.8	1.1	1.1
Dy	6.2	4.6	4.7	1.9	0.3	2.3	2.1	4.6	6.3	5.0	5.1	4.8	6.0	6.6
Ho	1.2	0.9	0.9	0.4	0.1	0.5	0.5	1.0	1.2	1.0	1.0	0.9	1.1	1.3
Er	3.5	2.8	2.6	1.1	0.2	1.3	1.2	2.5	3.3	2.8	2.9	2.6	3.3	3.6
Tm	0.5	0.4	0.3	0.2	0.04	0.2	0.2	0.4	0.4	0.4	0.4	0.3	0.4	0.5
Yb	3.0	2.4	2.2	1.0	0.2	1.1	1.1	2.3	2.6	2.3	2.3	2.2	2.7	2.9
Lu	0.4	0.4	0.3	0.2	0.03	0.2	0.2	0.3	0.4	0.3	0.3	0.3	0.4	0.4
Hf	8.5	10	7.2	0.4	0.8	1.0	1.3	2.9	4.7	2.7	4.0	3.6	5.1	3.6
Ta	1.4	4.7	4.0	bd	0.03	bd	0.4	1.0	2.5	1.0	1.5	1.5	2.4	1.3
Pb	30	40	32	6	0.7	4	5	18	8.4	1.8	4	3.3	6.8	2.7
Th	14	16	18	5.1	3.1	7.3	6.2	13	9	2.8	5.3	4.4	9.8	2.7
U	7.2	8.9	7.8	0.8	1.7	0.9	1	1.9	1.9	0.6	1.2	1.0	2.1	0.5
∑REE	506	523	542	67	30	87	73	184	240	104	153	137	194	151

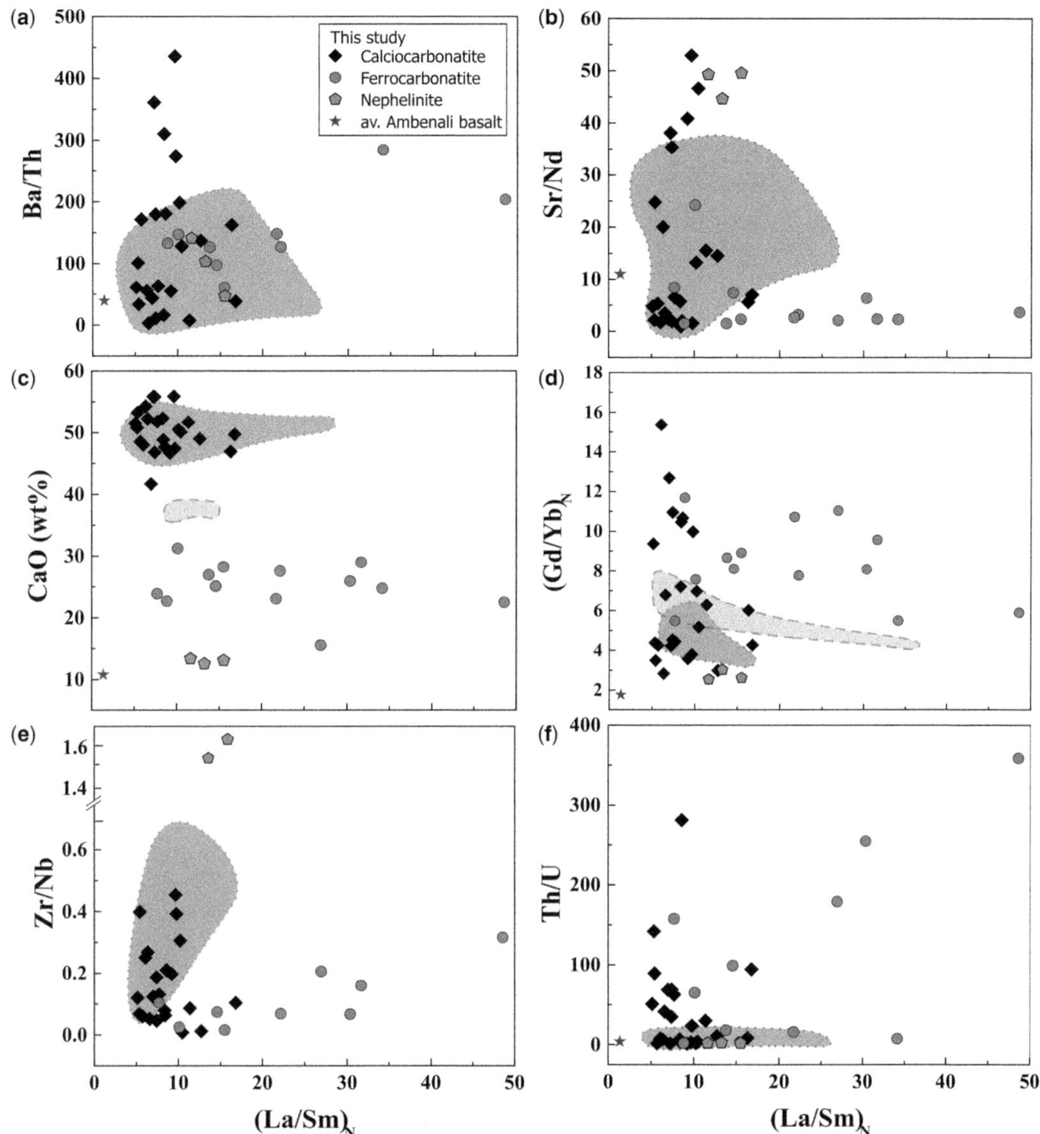

Fig. 7. Incompatible trace element ratios and CaO v. chondrite-normalized $(La/Sm)_N$ ratios in Amba Dongar rocks. Grey and red (bounded by dashed lines) shaded fields represent published data of calciocarbonatites ($n = 30$) and ferrocarbonatites ($n = 13$); data sources are the same as in Figure 5.

is possible that the excess of SiO_2, in the form of hydrothermally deposited quartz veins, may have been sourced from the Bagh sandstone. According to Sukheswala & Udas (1967), large fluorite deposits were formed by the reaction between Bagh sediments and carbonatite/associated alkaline magma. Both hand-specimen and thin-section studies suggest that most of our calciocarbonatites are fresh, whereas a few ferrocarbonatites have been partially affected by hydrothermal process.

Nephelinites are partially altered by fenitization, revealed from the presence of hydromuscovite formed due to the alteration of nepheline (Fig. 3g). According to Viladkar (2015), fenitization of Amba Dongar nephelinites was caused by the reaction of nepheline and pyroxene with fluids (containing Na, K, Mg, H_2O, CO_2, F and Cl) released from the carbonatite magma, at a temperature of 500–600°C; the fenitized/altered nephelinites are relatively more enriched in LILE (Ba, Rb, Th and Nb) and

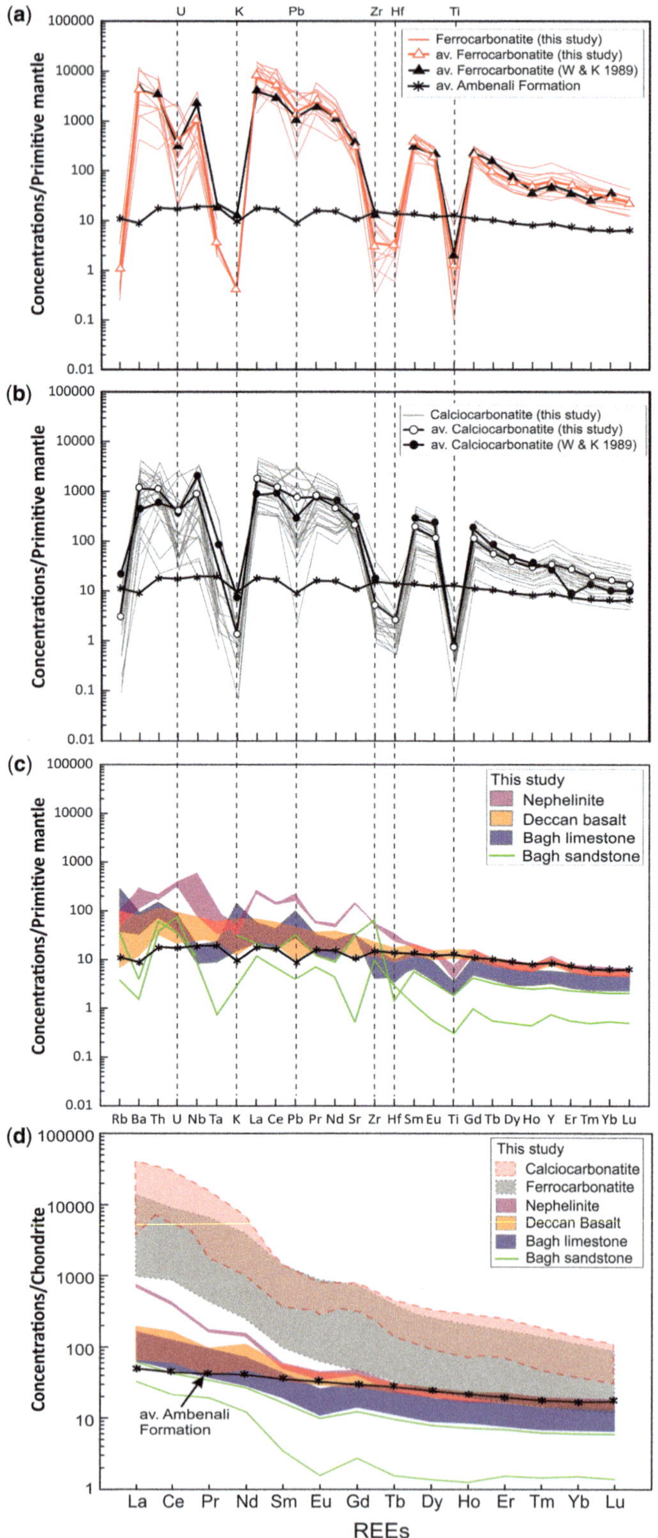
(a)
U K Pb Zr Hf Ti
Ferrocarbonatite (this study)
av. Ferrocarbonatite (this study)
av. Ferrocarbonatite (W & K 1989)
av. Ambenali Formation
Concentrations/Primitive mantle
(b)
Calciocarbonatite (this study)
av. Calciocarbonatite (this study)
av. Calciocarbonatite (W & K 1989)
Concentrations/Primitive mantle
(c)
This study
Nephelinite
Deccan basalt
Bagh limestone
Bagh sandstone
Concentrations/Primitive mantle
Rb Ba Th U Nb Ta K La Ce Pb Pr Nd Sr Zr Hf Sm Eu Ti Gd Tb Dy Ho Y Er Tm Yb Lu
(d)
This study
Calciocarbonatite
Ferrocarbonatite
Nephelinite
Deccan Basalt
Bagh limestone
Bagh sandstone
Concentrations/Chondrite
av. Ambenali Formation
La Ce Pr Nd Sm Eu Gd Tb Dy Ho Er Tm Yb Lu
REEs

LREE relative to the unfenitized ones. Others have shown that the host Bagh sandstone has experienced intense fenitization close to the main carbonatite body (Sukheswala & Viladkar 1981). As discussed before, the Bagh sandstone (ADSt-3) analysed in this study has been least affected by fenitization. However, the Bagh limestone samples analysed in this study may have been partially affected by fenitization. Chemical evidence for the alteration of our Deccan basalts is somewhat limited. Except for one sample (ADB-17), the LOI content in these samples is <3.6 wt%, which suggests relatively minor post-magmatic alteration. Further, thin-section studies did not reveal any significant alteration in either the basaltic or the doleritic dyke samples.

Crystal fractionation

Petrography of calciocarbonatites suggested that the fractionation of calcite was followed by fractionation of apatite cumulate (Figs 3c, d & 4e, f). Apatite is a common accessory phase in both carbonatites and nephelinites (Fig. 4). In calciocarbonatites, major oxide compositions of apatites indicate a higher CaO content for apatite cumulate compared to other forms of apatite. Apatite shows various textures, such as apatite cumulate and scattered apatite crystals in calcitic groundmass (Fig. 4), which reflect continuous apatite fractionation in carbonatite melt with concomitant physicochemical changes during calciocarbonatite formation. Similar to the abundances of apatite, pyrochlore is also an abundant accessory phase in calciocarbonatite but is less abundant in ferrocarbonatite. The presence of numerous inclusions of apatite and calcite in magnetite (Fig. 3b) suggests that magnetite crystallization occurred after calcite and apatite fractionation. Interestingly, compared to apatites in nephelinite (SiO_2, 0.69–1.10 wt%; SrO, 1.32–2.6 wt%; P_2O_5, 38.6–39.6 wt%; and Na_2O, 0.06–0.13 wt%), apatites of calciocarbonatite have distinctly lower SiO_2 (0.01–0.34 wt%) and SrO (0.58–0.95 wt%), and distinctly higher P_2O_5 (39.9–41.3 wt%) and Na_2O (0.1–0.38 wt%). This compositional difference suggests that apatite crystallization may have continued following liquid immiscibility of the carbonate and the silicate magma. Ferrocarbonatites show the presence of REE-bearing minerals, such as bastnaesite, baddeleyite and monazite. In nephelinites, large euhedral crystals of nepheline with subhedral melanite garnet and aegirine are present (Fig. 3g, h). Bizimis *et al.* (2003) showed that Zr, Hf and Ti mostly reside in the non-carbonate fraction, whereas Nd and Sr are compatible in carbonate fractions of carbonatites. It is possible that fractionation of early non-carbonate or silicate phases from the parental carbonatite magma may have generated the notable depletion in Zr, Hf and Ti in carbonatites.

Inferences on carbonatite magma genesis by trace element modelling

Carbonatite magmas are thought to be generated from melting of the metasomatized mantle. Primary carbonatite magmas are believed to represent very-low-degree partial melts (<0.1%) of carbonated peridotite or carbonated eclogite sources (Wyllie & Huang 1975; Eggler 1978; McKenzie 1985; Falloon & Green 1989; Dalton & Presnall 1998; Dasgupta & Hirschmann 2006; Ray *et al.* 2013). To draw inferences on aspects of carbonatite magma genesis, we performed trace element modelling (see the Supplementary material for details), taking into account trace element partitioning behaviour during melting of different mantle sources. Our goal was to test if it is possible to generate the trace element patterns (Fig. 8a, b) observed in the Amba Dongar carbonatites, specifically by melting three different types of mantle sources, viz.: depleted MORB mantle (DMM: Salters & Stracke 2004), primitive mantle (PM; Palme & O'Neill 2014) and SCLM (McDonough 1990). Following Dasgupta *et al.* (2009), we also assume that the initial (and primary) carbonatite melt can be generated at a depth of 300–400 km (12–14 GPa) where the plume adiabat (1500–1600°C mantle potential temperature) intersects the solidus of a carbonated (100–1000 ppm CO_2) garnet lherzolite source. The modal composition of the source (61% olivine, 20% clinopyroxene and 19% garnet) and the peridotite-melt bulk distribution coefficients for trace elements are adopted from Dasgupta *et al.* (2009). Assuming a very low degree of melting (0.03%), we performed calculations for both equilibrium melting and fractional melting scenarios (Shaw 1970).

The trace element patterns of primary carbonatite melts derived from a carbonated garnet lherzolite source under the aforementioned conditions are shown in Figure 9. Note that the difference in trace element abundances between batch and fractional melting calculations is negligible because of the extremely small extent of melting ($F = 0.03\%$)

Fig. 8. Primitive mantle-normalized (Palme & O'Neill 2014) spider diagram for the Amba Dongar rocks: (**a**) calciocarbonatite; (**b**) ferrocarbonatite; (**c**) nephelinite, basalt, Bagh limestone and Bagh sandstone; and (**d**) chondrite-normalized (Palme & O'Neill 2014) REE concentrations. Also shown for comparison are the average compositions of calciocarbonatites and ferrocarbonatite from Woolley & Kempe (1989), and the average Ambenali Formation (GEOROC database).

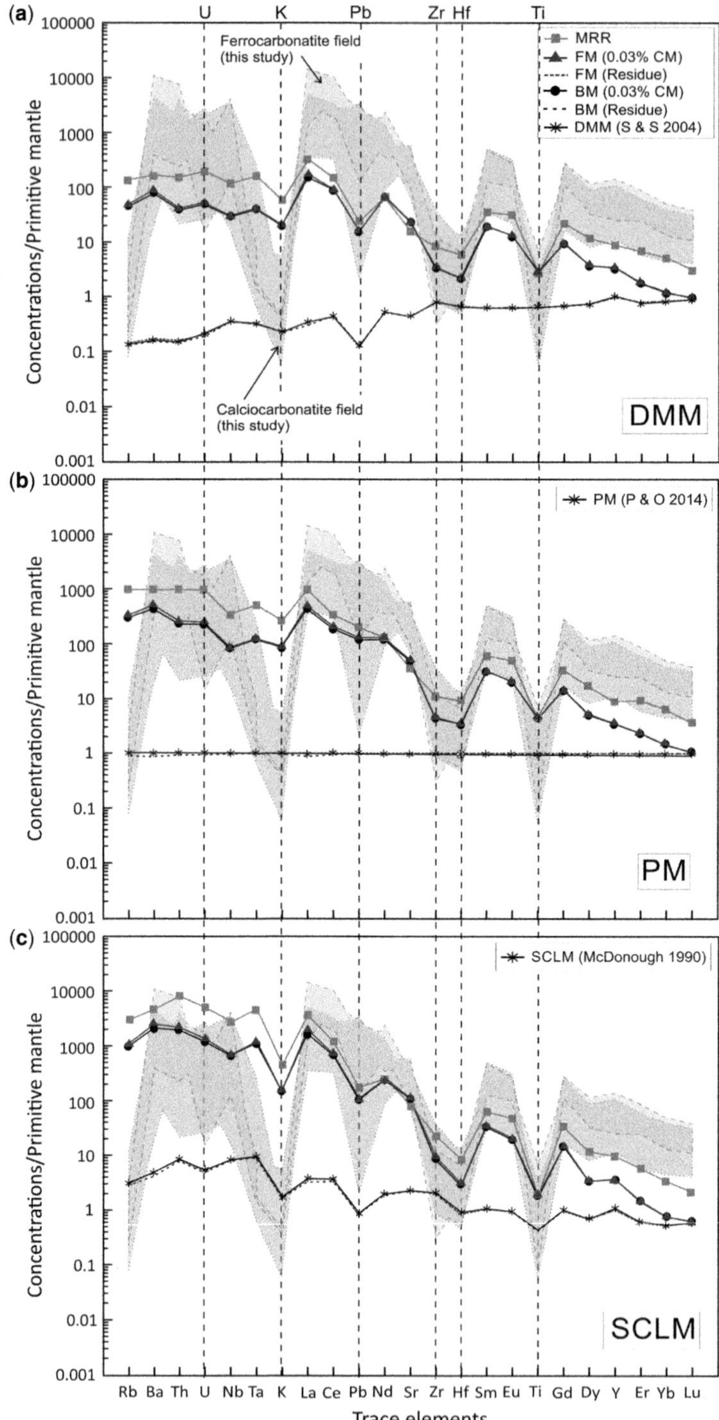

Fig. 9. Primitive mantle-normalized (Palme & O'Neill 2014) compositions of near-solidus carbonatite melt (CM) for the Amba Dongar carbonatites generated by batch melting (BM), fractional melting (FM) and melt–rock reaction (MRR) from: (**a**) depleted mantle (DMM) source; (**b**) primitive mantle (PM) source; and (**c**) subcontinental lithospheric mantle (SCLM). S & S (2004), Salters & Stracke (2004); P & O (2014), Palme & O'Neill (2014).

considered for carbonatite melt generation. Our results show that (batch/fractional) melts generated from both DMM and PM sources do not reproduce the entire pattern observed for the Amba Dongar carbonatites, although some signatures are retained (Fig. 9a, b). Although the DMM-generated melt retains depletion signatures for Ti, Pb, K and Zr–Hf, abundances in the melt approximates the least-enriched carbonatites only; these melts do not reproduce Rb–Nb abundances of carbonatites. Similar results are also produced for carbonatite melt generated from a PM source, except that these melts do not have the characteristic Pb depletion and weak K depletion, and higher LILE concentrations than the melts generated from the DMM source. Nevertheless, none of these melting models reproduces the HREE abundances of the Amba Dongar carbonatites, a result of our assumption that the carbonatite melt was generated in the garnet stability depths of approximately 300 km. On the other hand, batch carbonatite melts generated from a SCLM source (McDonough 1990) produced comparatively better trace element patterns that approximate the Amba Dongar carbonatite compositions, particularly the distinct Ti, Zr–Hf, Pb and K negative anomalies, the high LREE/HREE fractionation, and the 2000–3000 times higher than PM abundances in Rb–Ta (Fig. 9c). Note that by assuming a melt fraction of 3%, the calculated trace element patterns for near-solidus batch carbonatite melts become enriched by only 200 times higher than PM in Rb–Ta. Like the previous two cases, simple batch/fractional melting of a SCLM source also fails to generate (Fig. 9c) the flat HREE patterns observed for Amba Dongar carbonatites (Fig. 8a, b); in particular, the HREE abundances in the calculated melts are 10 times less enriched than the carbonatites.

It is possible that the primary carbonatite melt evolves to a wide range of compositions due to subsequent (metasomatic) reaction of this melt with a shallower mantle lherzolite source (with minor garnet). Dasgupta *et al.* (2009) argued that the extensive reaction of carbonatite melt with shallow-mantle peridotite (at *c.* 3 GPa) can elevate HREE abundances in carbonatites while also producing natural patterns of carbonatites from oceanic settings, particularly the enrichment in LREE, Ba, Th, U and Nb. To obtain elevated HREE in our primary carbonatite melts, we performed calculations using relevant equations for the zone refinement process, and assumed a rock:melt ratio of 3000:1, bulk distribution coefficient (D_i) values for elements and 3% garnet in the metasomatized shallow-mantle peridotite source (depth of *c.* 90 km). The melt–rock reaction or the simple zone refinement model resulted in elevated HREE abundances in the calculated melts, a feature that clearly could not be generated by simple batch/fractional melting. However, the melt–rock reaction process still could not reproduce the much higher HREE abundances observed in the Amba Dongar carbonatites. Although the melt–rock reaction process involving a SCLM-type source is able to reproduce trace element patterns (and abundances) that bear some similarity with the patterns of Amba Dongar carbonatites, we could not reproduce a single melt mimicking the average pattern shown by the entire range of trace element abundances. Therefore, it highly likely that the primary carbonatitic melt generated in equilibrium with garnet lherzolite at mantle depths of approximately 300 km must have evolved by other processes to obtain trace element characteristics of the parental carbonatite magma for the Amba Dongar carbonatites. The other processes, such as carbonate–silicate liquid immiscibility and crystal fractionation of apatite, pyrochlore or other REE-bearing minerals, impart distinct elemental fractionations characteristic of carbonatites. In addition, the composition of primary carbonatite magma during its ascent through the lithosphere can be modified by crustal assimilation and loss of volatiles (e.g. CO_2, Cl, F, H_2O) and alkalis causing widespread fenitization of the host rocks. In this regard, finding or establishing the primary magma composition of the Amba Dongar carbonatites will be the first step towards quantifying melting conditions and the subsequent processes leading to compositional diversification of these carbonatites.

Link between Amba Dongar carbonatites and Deccan flood basalts

The coexistence of carbonatite and alkaline silicate rocks in several LIPs globally– both field relationships and contemporaneous emplacement ages– has been used to establish a genetic linkage between the carbonatite and plume-origin magma (Veena *et al.* 1998; Vladykin 2009; Ernst & Bell 2010). Several authors have previously argued for a link between the Amba Dongar carbonatites and Deccan LIP (Simonetti *et al.* 1995, 1998; Ray & Pande 1999; Ray & Ramesh 1999; Ray *et al.* 2000*a*, 2003) because of the close spatial and temporal association between the two, particularly the similar or contemporaneous emplacement ages of both. Mantle-plume activity around 65 Ma initiated continental rifting between India and Seychelles, and subsequently lead to large-volume magmatism (plume head eruption) within a short time span that resulted in the formation of the Deccan LIP and contemporaneous alkaline magmatism along the Narmada Rift (Simonetti *et al.* 1998; Ray *et al.* 2000*a*, 2003). The emplacement ages of the Amba Dongar carbonatite complex (*c.* 65 ± 0.3 Ma: Ray & Pande 1999) and associated nephelinites (64.8 ± 0.6 Ma: Ray & Pande 1999) along the Narmada Rift is contemporaneous with the 70–63 Ma main eruptive pulse of Deccan volcanism (see the age compilation in Table 1 of Richards *et al.* 2015). Simonetti *et al.* (1995) argued that the Pb isotope ratios in picritic and basaltic lavas in drillholes from the NW Deccan Traps (Peng & Mahoney 1995) are similar to those from the Amba Dongar carbonatites. The initial $^{87}Sr/^{86}Sr$, $^{143}Nd/^{144}Nd$, $^{206}Pb/^{204}Pb$, $^{207}Pb/^{204}Pb$ and $^{208}Pb/^{204}Pb$ isotope ratios of Deccan alkaline complexes have led others (Basu *et al.* 1993; Simonetti *et al.* 1998) to believe that alkaline magmatism (with or without associated carbonatite rocks) along the northern part of the Deccan province is also genetically linked to the volumetrically more significant tholeiitic volcanism; complexes include the Phenai Mata alkali intrusion (64.96 Ma: gabbro), Mundwara (68.5 Ma: nephelinite, phonolite, tephrite and carbonatite), Barmer (68.57 Ma: alkali pyroxenite, mela-nephelinite, ijolite and melilitite), Bhuj (64.4–67.7 Ma: basanite-alkaline basalt) and Amba Dongar (65 Ma: basanite, nephelinites and phonolite plugs and carbonatite). Further, Ray & Ramesh (1999) reported $\delta^{13}C$ and $\delta^{18}O$ compositions of Mundwara, Barmer and Amba Dongar carbonatites, and argued for the contribution of the Réunion plume in the generation of these carbonatites.

Others (Mahoney *et al.* 1985; Sen *et al.* 2009) have argued that alkaline complexes in the northern part of the Deccan province were derived from a metasomatized mantle source and the tholeiitic magma was derived from non-metasomatized regions of enriched mantle-type source. It is becoming clear that mantle metasomatism is an essential requirement for the generation of CO_2-rich and alkali-rich magma (Green & Wallace 1988; Ionov *et al.* 1993; Veksler & Keppler 2000). Further, radiogenic, stable and noble gas isotopic signatures in carbonatites globally are consistent with a sublithospheric source for the generation of parental carbonatite magma, associated with either asthenospheric upwelling or plume-related LIP magmatism (Bell & Simonetti 2010). In this regard, our trace element modelling results (Fig. 9c) favour carbonatite magma generation from a subcontinental lithospheric mantle source metasomatized by fluids and heat released from the plume.

Principal component analysis (PCA)

We performed factor analysis on a subset of variables (major and selected trace element concentrations of carbonatites) to determine the number of latent factors responsible for variance in the chemical composition of carbonatites and the host nephelinites. The latent factors may be related to source heterogeneity, partial melting, fractional crystallization, country rock assimilation, etc. (White & Duncan 1996). The factor analysis model assumes that the original variables are linear combinations of the factors and uses principal component analysis (PCA) to extract components to maximize the proportion of variability explained by each component subject to the orthogonality constraint. It simplifies the interpretation obtained by PCA by rotating the system of Cartesian coordinates around the origin by using a transform matrix. Mathematical details on factor analysis model can be found in Davis (2002).

A 41×24 matrix was used for factor analysis where the rows are the number of observations (calciocarbonatites and ferrocarbonatite, and nephelinite samples) and the columns are represented by variables (major oxides, REE, and Ba and Sr concentrations). Factors were extracted using the PCA method, and the number of factors to retain was determined by plotting the eigen values (>1) of the correlation matrix against the components where Cattell's Scree can be observed. Four significant factors that together account for 86% variance in the dataset were extracted. Factor loadings for the four significant factors are illustrated in Figure 10. In order to view a simpler factor loading pattern that is easier to interpret, an orthogonal Kaiser Varimax rotation method was also employed (that converged in eight iterations) to adjust the factor loadings such that they are either near ± 1 or near 0. The results of a component plot in rotated space are also shown in Figure 10, in which data are projected into the plane containing the two principle eigenvectors (factors). The first factor, which explains 54% of the variance, has high positive loadings for REE and Ba, and negative loadings for SiO_2, CaO and Al_2O_3, which may be attributed to source characteristics and degree of melting (i.e. low-degree melting of a carbonated lherzolite source). The second factor explains 16% of variance and has moderate–high positive loadings for Fe_2O_3, MgO, La, Ce, Pr and Nd, and Ba, and a high negative loading for CaO. This factor may be associated with the fractionation of calciocarbonatites to ferrocarbonatites, consistent with a higher LREE/HREE fractionation in ferrocarbonatites (Fig. 8d). Because this factor also shows positive loadings of SiO_2, Sr and Ba, we also associate late-stage hydrothermal alteration that has notably affected a few of our ferrocarbonatites to this factor. The third factor accounts for 10% variance in the dataset and has high positive loadings for SiO_2, and negative loadings for Fe_2O_3, MnO, MgO, CaO, Sr and Ba. It is likely that this factor is associated with liquid immiscibility of the carbonate and silicate magma. The fourth factor explains 6% of variance, and has mostly negative loadings for LREE and positive loadings for HREE. We attribute this factor to melt generation in the garnet stability field. Note that any single factor may represent the effects of multiple processes. The component plot in rotated space shows that compositional variability

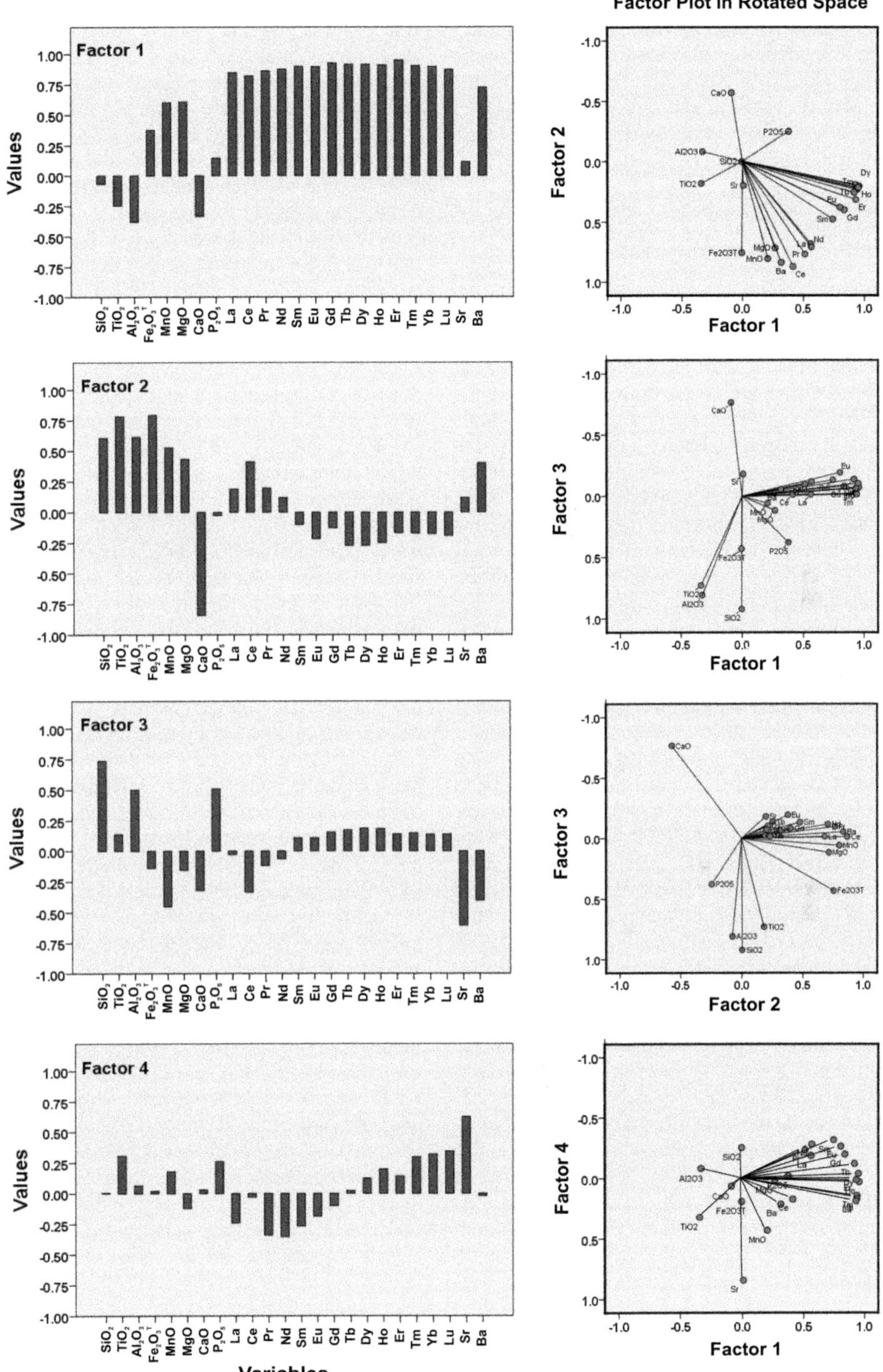

Fig. 10. Factor analysis of dataset (major oxides, REEs, Ba and Sr) of nephelinites, calciocarbonatites and ferrocarbonatites. Subplots in the left-hand side column show factor loadings for 24 variables from 41 observations. The four factors account for 86% of variance in the dataset. Subplots in the right-hand side column show factors in rotated space where data are projected into a plane containing two principal eigenvectors.

is largely affected by both factors 1 and 2. Factor analysis performed considering only carbonatites (excluding nephelinites) also resulted in four factors accounting for 86% variance. This suggests the involvement of similar factors in the generation of carbonatite and associated nephelinite of Amba Dongar.

Petrogenetic model for Amba Dongar carbonatite magma generation

Even though carbonatites have caught the attention of experimental petrologists and geochemists over the past few decades, petrogenesis of primary carbonatite melts are still debated (e.g. Bell & Simonetti 2010). Based on inferences drawn by others, the Amba Dongar carbonatite complex seems to have undergone a complicated petrogenetic evolutionary history (Simonetti *et al.* 1995, 1998; Ray 1998; Ray & Pande 1999; Viladkar & Schidlowski 2000; Ray *et al.* 2000*a*). Combining results from experimental petrology and geochemical observations, we propose a petrogenetic model for the generation of Amba Dongar carbonatites and associated nephelinites in this subsection.

Experimental studies have shown that a tiny amount of carbon in the mantle has a significant effect on the melting behaviour of mantle rocks (Presnall & Gudfinnsson 2005). Carbon is generally present in the mantle in various forms, namely graphite, diamond, CO_2 fluid, C^{4+} substitution for Si^{4+} in volatile-free mantle minerals, magnesite in the lower mantle, and dolomite in the shallower mantle (Keppler *et al.* 2003; Luth 2003; Isshiki *et al.* 2004). Although a few earlier experimental studies have suggested that dolomite is stable up to a pressure of 3–4.5 GPa (Dalton & Presnall 1998; Lee *et al.* 2000), Dasgupta *et al.* (2004) extended its stability to 5–8.5 GPa. On the other hand, magnesite is stable at >8.5 GPa and 1245± 35°C, but transforms to an unknown phase at approximately 115 GPa (*c.* 2600 km) and 1827–27°C (Biellmann *et al.* 1993; Dasgupta *et al.* 2004; Isshiki *et al.* 2004). Experimental studies have shown that a minor amount of carbonate (200 ppm CO_2) can reduce the solidus of volatile-free peridotite by approximately 300°C at ≥1.9 GPa; at shallower depths (<1.9 GPa) CO_2 has a negligible effect on the solidus (Falloon & Green 1989; Presnall & Gudfinnsson 2005). The source of carbon in the mantle may either be primitive and stored non-uniformly within the less accessible parts of the mantle or added by subduction of carbonated eclogites or sedimentary carbonate (Bell & Simonetti 2010).

Our preferred petrogenetic model for the origin of Amba Dongar carbonatite complex, depicting the role of the Réunion mantle plume, is shown in Figure 11. It is hypothesized that the lithologically heterogeneous Réunion plume consists of low-solidus eclogitic/pyroxenitic plums (recycled oceanic crust) that are randomly distributed in a high-solidus peridotite matrix. Further, these plums are incompatible trace element enriched, whereas the matrix is depleted in composition. Being low-solidus material, the trace element-enriched plums start to melt at greater mantle depths. On the other hand, the high-solidus peridotitic matrix will melt only at a much shallower depth as the plume continues to rise; extensive melting of the matrix produced tholeiites. It is hypothesized that CO_2-rich fluids comprising low-degree partial melts of eclogitic plums and heat released from the mantle-plume head (that generated DFB) at deeper depth rose through the lithosphere and metasomatized a part of the SCLM (garnet lherzolite) source. Trace element modelling results suggest that low-degree (≤1%) partial melting of this carbonated garnet lherzolite at >3 GPa (>100 km) gave rise to primitive/primary carbonated silicate melt. Note that our trace element modelling could not reproduce a single primary carbonatitic melt (in equilibrium with garnet lherzolite at mantle depths) mimicking the average pattern shown by the entire range of trace element abundances. Therefore, we hypothesize the generation of a primary carbonated silicate melt that must have evolved by other complex geochemical processes, during its ascent through the continental lithosphere, to obtain trace element characteristics of the parental carbonatite magma for Amba Dongar carbonatites. Owing to the low-viscosity and low-density nature of carbonatite melts (5×10^{-2} Pa s and 2.2 g cm^{-3}: Treiman & Schedl 1983), these melts rose rapidly through the lithosphere and became enriched in the highly incompatible trace elements (particularly the HREE) due to the melt–rock reaction process. The primary carbonated silicate melt fractionated into two compositionally distinct carbonate and silicate liquids at shallower crustal depths (about ≤2 GPa and 1000–1200°C: Brooker & Kjarsgaard 2011; Kamenetsky & Yaxley 2015), both of which subsequently underwent separate evolutionary paths and erupted coevally. Kjarsgaard & Hamilton (1989) suggested that the liquid immiscibility of the carbonate and alkaline silicate melts is initiated as a result of fractional crystallization of silicate minerals from the primary carbonated silicate magma and the decreasing CO_2 solubility at lower pressure. It is also likely that the primary carbonated silicate magma composition may be modified by wall-rock assimilation, in addition to simultaneous crystal fractionation and separation of immiscible carbonate and alkaline melts, which subsequently undergo separate evolutionary paths. This complicated evolution was modelled by Ray (1998), who proposed a modified version of the AFC (assimilation-fractional crystallization) model of DePaolo (1981) that additionally

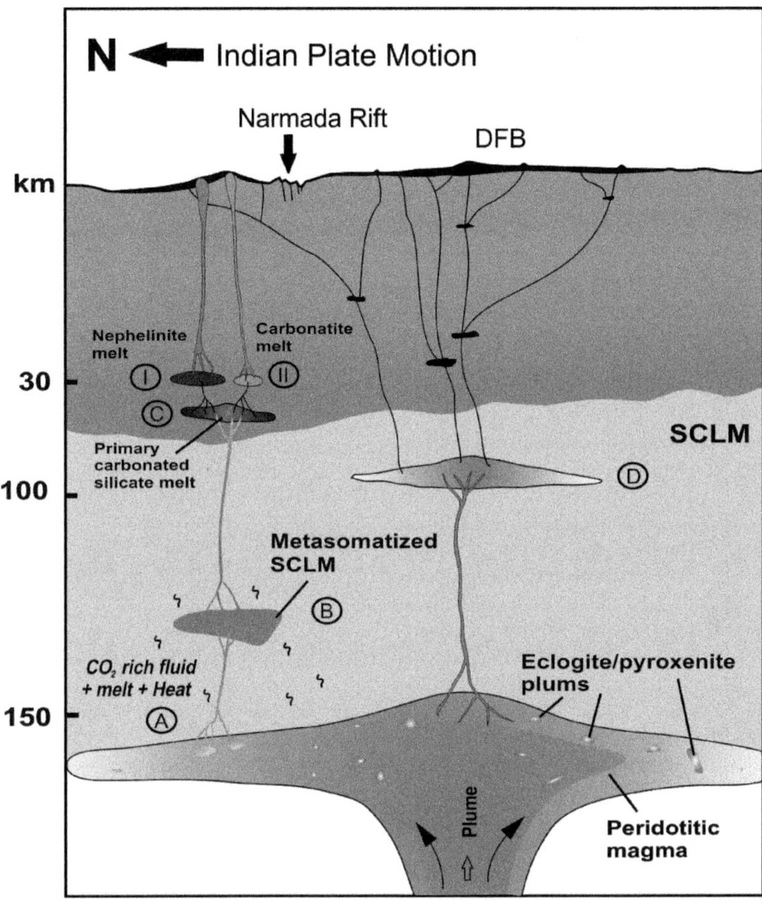

Fig. 11. Petrogenetic model (not drawn to scale) showing the origin of the Amba Dongar carbonatite complex. The plume comprises low-solidus eclogite/pyroxenite plums distributed within the high-solidus peridotite matrix. (A) CO_2-rich fluid, melt (low-degree melts from plums) and heat released from the Deccan plume at a depth of approximately 150 km, metasomatizes are part of a subcontinental lithospheric mantle (SCLM). The carbonate-metasomatized garnet lherzolite source (B) undergoes low-degree melting ($F < 1\%$) at mantle depths of >100 km (>3 GPa) to form primary carbonated silicate melt. (C) This primary carbonated silicate melt rises rapidly and separates into two compositionally distinct nephelinitic (I) and carbonatitic (II) magmas via liquid immiscibility at crustal depths, following which respective parental magmas evolve via crystal fractionation or country-rock assimilation. Simultaneously, large-scale (decompression) melting of the rising mantle plume at depths shallower than 100 km generates extensive tholeiitic magma (D) that give rise to the Deccan flood basalts.

incorporated the effects of simultaneous liquid immiscibility – the so-called AFCLI model. Ray (1998) only used the Sr isotope systematics ($^{87}Sr/^{86}Sr$ ratio v. Sr concentration) of both carbonatites and associated alkaline silicates of the Deccan province to strongly argue in favour of the AFCLI process for the evolution of carbonatites and associated silicates from a primary carbonated silicate magma that was contaminated by the addition of up to 5% of the lower-crustal material (Dharwar gneissic/granulitic basement rocks). Although testing this AFCLI model to explain the compositional variation in our dataset (entire range of trace elements) seems tempting, its applicability requires an accurate estimation of the mineral–melt and silicate melt–carbonate melt partition coefficients for each trace element. Because of the lack of such data, we have refrained from testing the AFCLI model. However, such a model cannot be ruled out and definitely needs to be tested rigorously once partitioning coefficients are accurately determined.

The temperatures of the carbonatite and nephelinite magmas are estimated to be 900–1050°C (Le Bas 1987; Wallace & Green 1988; Kjarsgaard 1998). It is possible that (calcite, apatite and pyrochlore) crystal fractionation and assimilation of country rock (basement gneiss, Bagh sandstone and limestone) played a role in the evolution of the primary carbonate and silicate magmas, which can be quantified using isotopic mixing models. Nevertheless, our proposed petrogenetic model highlights the importance of the mantle plume in supplying the necessary CO_2-rich fluids, melt and heat required for the metasomatism of the subcontinental lithospheric mantle source in the generation of the Amba Dongar carbonatites and associated nephelinites.

Conclusions

Many carbonatite complexes worldwide are spatially and temporally associated with mantle-plume-originated Large Igneous Provinces (LIPs). The Amba Dongar carbonatite complex in the Chhota Udaipur District, Gujarat is the youngest Indian carbonatite complex, whose emplacement age coincides with the age of main pulse of Deccan flood basalts at *c.* 65 Ma. In this study, we present extensive geochemical data on carbonatites and associated silicate rocks of Amba Dongar, and probe the role of mantle plume that formed the Deccan LIP in the generation of spatially and temporally associated carbonatites. Compositionally, Amba Dongar carbonatites can be classified into two classes: calciocarbonatite and ferrocarbonatite. Major oxide composition reveals a narrow range for calciocarbonatites (CaO, 39.5–55.9 wt%; SiO_2, 0.55–9.67 wt %), whereas ferrocarbonatites (CaO, 15.6–31.0 wt %; SiO_2, 1.83–34.52 wt%) exhibit large compositional variations compared to the already available data. Further, the concentration of BaO (0.02–3.41 wt% in calciocarbonatites v. 0.3–7 wt% in ferrocarbonatites), Nb (10–2168 v. 76–2472 ppm), Sr (1421–13 146 v. 2332–11 553 ppm) and Th (3.8–345 v. 20–623 ppm) are unusually higher in the carbonatites relative to the associated nephelinites and Deccan basalts. A notable feature of Amba Dongar carbonatites is the extreme enrichment in REE (ΣREE 1025–31 117 ppm), typically higher in ferrocarbonatites relative to the calciocarbonatites. Trace element modelling shows that the primary carbonatite melt generated from a carbonated peridotite source does not reproduce the trace element characteristics observed in the Amba Dongar carbonatites. However, the melt–rock reaction or zone refinement process involving a metasomatized SCLM source is able to reproduce trace element patterns (and abundances) bearing some similarity with the patterns observed for Amba Dongar carbonatites, although it still fails to explain the Rb, Ba, Th, U, Nb and HREE patterns. Using experimental petrology constraints on carbonate magma generation, we propose a model in which the contribution of CO_2-rich fluids and heat from a mantle plume plays a significant role in the metasomatism of SCLM. In this model, low-degree partial melts of a carbonated garnet lherzolite (SCLM source) produced at a depth of >100 km represents parental carbonated silicate magma. The parental carbonated silicate magma underwent liquid immiscibility at crustal depths to give rise to two compositionally distinct carbonatite and nephelinite magmas that further evolved due to possible crystal fractionation and host-rock assimilation. Our proposed petrogenetic model highlights the importance of the Deccan mantle plume in the generation of the Amba Dongar carbonatites and associated nephelinites.

This paper is part of the PhD thesis of JC. JC is grateful to the Council of Scientific and Industrial Research for a Senior Research Fellowship for his PhD dissertation. We thank Dr François Chabaux and Dr Gopal Joshi (Thermo Fisher Scientific, India) for analysis of part of the trace element data. Thanks are due to Gujarat Mineral Development Corporation for kind assistance during fieldwork in the Amba Dongar mine area. We are grateful to two anonymous reviewers and Prof. Bryan Storey (editor) for their constructive comments, which significantly improved clarity of the paper. SS thanks the Head, Department of Geology, University of Lucknow for support.

References

Allègre, C.J., Birck, J.L., Capmas, F. & Courtillot, V. 1999. Age of the Deccan traps using ^{187}Re–^{187}Os systematics. *Earth and Planetary Science Letters*, **170**, 197–204, https://doi.org/10.1016/S0012-821X(99)00110-7

Bailey, D. 1993. Carbonate magmas. *Journal of the Geological Society, London*, **150**, 637–665, https://doi.org/10.1144/gsjgs.150.4.0637

Basu, A.R., Renne, P.R., Dasgupta, D.K., Teichmann, F. & Poreda, R.J. 1993. Early and late alkali igneous pulses and a high-^{3}He plume origin for the Deccan flood basalts. *Science*, **261**, 902–906, https://doi.org/10.1126/science.261.5123.902

Bell, K. (ed.) 1989. *Carbonatites: Genesis and Evolution*. Unwin Hyman, London.

Bell, K. 1998. Radiogenic isotope constraints on relationships between carbonatites and associated silicate rocks – a brief review. *Journal of Petrology*, **39**, 1987–1996, https://doi.org/10.1093/petroj/39.11-12.1987

Bell, K. 2001. Carbonatites: relationships to mantle-plume activity. *In*: Ernst, R.E. & Buchan, K.L. (eds) *Mantle Plumes: Their Identification through Time*. Geological Society of America, Special Papers, **352**, 267–290, https://doi.org/10.1130/0-8137-2352-3

Bell, K. & Rukhlov, A.S. 2004. Carbonatites from the Kola Alkaline Province: origin, evolution and source characteristics. *In*: Zaitsev, A. & Wall, F. (eds) *Phoscorites and Carbonatites from Mantle to Mine: The Key Example of the Kola Alkaline Province*. Mineralogical Society, London, Book Series, **10**, 421–455.

BELL, K. & SIMONETTI, A. 2010. Source of parental melts to carbonatites – critical isotopic constraints. *Mineralogy and Petrology*, **98**, 77–89, https://doi.org/10.1007/s00710-009-0059-0

BELL, K. & TILTON, G.R. 2001. Nd, Pb and Sr isotopic compositions of East African carbonatites: evidence for mantle mixing and plume inhomogeneity. *Journal of Petrology*, **42**, 1927–1945, https://doi.org/10.1093/petrology/42.10.1927

BELL, K., KJARSGAARD, B.A. & SIMONETTI, A. 1999. Carbonatites – into the twenty-first century. *Journal of Petrology*, **39**, 1839–1845, https://doi.org/10.1093/petroj/39.11-12.1839

BIELLMANN, C., GILLET, P., GUYOT, F., PEYRONNEAU, J. & REYNARD, B. 1993. Experimental evidence for carbonate stability in the Earth's lower mantle. *Earth and Planetary Science Letters*, **118**, 31–41, https://doi.org/10.1016/0012-821X(93)90157-5

BIZIMIS, M., SALTERS, V.J.M. & DAWSON, J.B. 2003. The brevity of carbonatite sources in the mantle: evidence from Hf isotopes. *Contributions to Mineralogy and Petrology*, **145**, 281–300, https://doi.org/10.1007/s00410-003-0452-3

BREY, G.P., BULATOV, V.K. & GIRNIS, A.V. 2009. The influence of water and fluorine on the melting of carbonated peridotite at 6 and 10 GPa. *Lithos*, **112S**, 249–259, https://doi.org/10.1016/j.lithos.2009.04.037

BROOKER, R.A. & KJARSGAARD, B.A. 2011. Silicate–carbonate liquid immiscibility and phase relations in the system SiO_2–Na_2O–Al_2O_3–CaO–CO_2 at 0.1–2.5 GPa with applications to carbonatite genesis. *Journal of Petrology*, **52**, 1281–1305.

CHAKHMOURADIAN, A.R. 2006. High-field-strength elements in carbonatitic rocks: geochemistry, crystal chemistry and significance for constraining the sources of carbonatites. *Chemical Geology*, **235**, 138–160, https://doi.org/10.1016/j.chemgeo.2006.06.008

COURTILLOT, V., JAUPART, C., MANIGHETTI, I., TAPPONNIER, P. & BESSE, J. 1999. On causal links between flood basalts and continental breakup. *Earth and Planetary Science Letters*, **166**, 177–195, https://doi.org/10.1016/S0012-821X(98)00282-9

COURTILLOT, V.E. & RENNE, P.R. 2003. On the ages of flood basalt events. *Comptes Rendus Geoscience*, **335**, 113–140, https://doi.org/10.1016/S1631-0713(03)00006-3

DALTON, J.A. & PRESNALL, D.C. 1998. Carbonatitic melts along the solidus of model lherzolite in the system CaO–MgO–Al_2O_3–SiO_2–CO_2 from 3 to 7 GPa. *Contributions to Mineralogy and Petrology*, **131**, 123–135, https://doi.org/10.1007/s004100050383

DASGUPTA, R. & HIRSCHMANN, M.M. 2006. Melting in the Earth's deep upper mantle caused by carbon dioxide. *Nature*, **440**, 659–662.

DASGUPTA, R., HIRSCHMANN, M.M. & WITHERS, A.C. 2004. Deep global cycling of carbon constrained by the solidus of anhydrous, carbonated eclogite under upper mantle conditions. *Earth and Planetary Science Letters*, **227**, 73–85, https://doi.org/10.1016/j.epsl.2004.08.004

DASGUPTA, R., HIRSCHMANN, M.M., MCDONOUGH, W.F., SPIEGELMAN, M. & WITHERS, A.C. 2009. Trace element partitioning between garnet lherzolite and carbonatite at 6.6 and 8.6 GPa with applications to the geochemistry of the mantle and of mantle-derived melts. *Chemical Geology*, **262**, 57–77, https://doi.org/10.1016/j.chemgeo.2009.02.004

DAVIS, J.C. 2002. *Statistics and Data Analysis in Geology*. 3rd edn. John Wiley, New York.

DEANS, T. & POWELL, J.L. 1968. Trace element and Sr isotope Pakistan. *Nature*, **218**, 750–752.

DEANS, T., SUKHESWALA, R.N., SETHNA, S.F. & VILADKAR, S. G. 1972. Metasomatic feldspar rocks (potash fenites) associated with the fluorite deposits and carbonatites of Amba Dongar, Gujarat, India. *Transactions of the Institution of Mining and Metallurgical, Section B: Mining Technology*, **81**, B1–B9.

DEPAOLO, D.J. 1981. Trace element and isotopic effects of combined wallrock assimilation and fractional crystallization. *Earth and Planetary Science Letters*, **53**, 189–202, https://doi.org/10.1016/0012-821X(81)90153-9

DEVEY, C.W. & STEPHENS, W.E. 1992. Deccan-related magmatism west of the Seychelles–India rift. *In*: STOREY, B.C., ALABASTER, T. & PANKHURST, R.J. (eds) *Magmatism and the Causes of Continental Break-up*. Geological Society, London, Special Publications, **68**, 271–291, https://doi.org/10.1144/GSL.SP.1992.068.01.17

EGGLER, D.H. 1978. The effect of CO_2 upon partial melting of peridotite in the system (Na_2O–CaO–Al_2O_3–MgO–SiO_2–CO_2) to 35 kb, with an analysis of melting in a peridotite–H_2O–CO_2 system. *American Journal of Science*, **278**, 305–343, https://doi.org/10.2475/ajs.278.3.305

ERNST, R.E. & BELL, K. 2010. Large igneous provinces (LIPs) and carbonatites. *Mineralogy and Petrology*, **98**, 55–76, https://doi.org/10.1007/s00710-009-0074-1

ERNST, R.E. & BUCHAN, K.L. 2004. Igneous rock associations in Canada 3. Large Igneous Provinces (LIPs) in Canada and adjacent regions: 3 Ga to present. *Geoscience Canada*, **31**, 103–126.

FALLOON, T.J. & GREEN, D.H. 1989. The solidus of carbonated, fertile peridotite. *Earth and Planetary Science Letters*, **94**, 364–370, https://doi.org/10.1016/0012-821X(89)90153-2

GHOSH, S., LITASOV, K. & OHTANI, E. 2014. Phase relations and melting of carbonated peridotite between 10 and 20 GPa: a proxy for alkali- and CO_2-rich silicate melts in the deep mantle. *Contributions to Mineralogy and Petrology*, **167**, 964, https://doi.org/10.1007/s00410-014-0964-z

GITTINS, J. 1989. The origin and evolution of carbonatite magmas. *In*: BELL, K. (ed.) *Carbonatites: Genesis and Evolution*. Unwin Hyman, London, 580–600.

GOVINDRAJU, K. 1994. 1994 compilation of working values and sample description for 383 geostandards. *Geostandards Newsletter*, **18**, 1–158.

GREEN, D.H. & WALLACE, M.E. 1988. Mantle metasomatism by ephemeral carbonatite melts. *Nature*, **336**, 459–462.

GUDFINNSSON, G.H. & PRESNALL, D.C. 2005. Continuous gradations among primary carbonatitic, kimberlitic, melilititic, basaltic, picritic, and komatiitic melts in equilibrium with garnet lherzolite at 3–8 GPa. *Journal of Petrology*, **46**, 1645–1659.

GUPTA, H., RAO, N.P. *ET AL.* 2015. Investigations related to scientific deep drilling to study reservoir-triggered earthquakes at Koyna, India. *International Journal of Earth Sciences*, **104**, 1511–1522.

Guzmics, T., Mitchell, R.H., Szabó, C., Berkesi, M., Milke, R. & Abart, R. 2011. Carbonatite melt inclusions in coexisting magnetite, apatite and monticellite in Kerimasi calciocarbonatite, Tanzania: melt evolution and petrogenesis. *Contributions to Mineralogy and Petrology*, **161**, 177–196, https://doi.org/10.1007/s00410-010-0525-z

Gwalani, L.G., Rock, N.M.S., Chang, W.J., Fernandez, S., Allégre, C.J. & Prinzhofer, A. 1993. Alkaline rocks and carbonatites of Amba Dongar and adjacent areas, Deccan Igneous Province, Gujarat, India: 1. Geology, petrography and petrochemistry. *Mineralogy and Petrology*, **47**, 219–253, https://doi.org/10.1007/BF01161569

Gwalani, L.G., Fernandez, S., Karanth, R.V., Demeny, A., Chang, W.J. & Avasia, R.K. 1995. Alkaline and tholeiitic dyke swarm associated with Amba Dongar and Phenai Mata complexes, Chhota Udaipur alkaline sub-province, Western India. *In*: Devaraju, T.C. (ed.) *Dyke Swarms of Peninsular India*. Geological Society of India, Memoirs, **33**, 391–423.

Halama, R., Vennemann, T., Siebel, W. & Markl, G. 2005. The Grønnedal-Ika carbonatite–syenite complex, South Greenland: carbonatite formation by liquid immiscibility. *Journal of Petrology*, **46**, 191–217, https://doi.org/10.1093/petrology/egh069

Harmer, R.E. & Gittins, J. 1998. The case for primary, mantle-derived carbonatite magma. *Journal of Petrology*, **39**, 1895–1903, https://doi.org/10.1093/petroj/39.11-12.1895

Heinrich, E.W. 1966. *The Geology of Carbonatites*. Rand McNally and Company, Chicago, IL.

Ionov, D.A., Dupuy, C., O'Reilly, S.Y., Kopylova, M.G. & Genshaft, Y.S. 1993. Carbonated peridotite xenoliths from Spitsbergen: implications for trace element signature of mantle carbonate metasomatism. *Earth and Planetary Science Letters*, **119**, 283–297, https://doi.org/10.1016/0012-821X(93)90139-Z

Isshiki, M., Irifune, T. *et al.* 2004. Stability of magnesite and its high-pressure form in the lowermost mantle. *Nature*, **427**, 60–63, https://doi.org/10.1038/nature02181

Jaitly, A.K. & Ajane, R. 2013. Comments on Placenticeras mintoi (Vredenburg, 1906) from the Bagh Beds (Late Cretaceous), Central India with special reference to turonian nodular limestone horizon. *Journal of the Geological Society of India*, **81**, 565–574.

Kamenetsky, V.S. & Yaxley, G.M. 2015. Carbonate–silicate liquid immiscibility in the mantle propels kimberlite magma ascent. *Geochimica et Cosmochimica Acta*, **158**, 48–56.

Keller, J. & Zaitsev, A.N. 2012. Reprint of 'Geochemistry and petrogenetic significance of natrocarbonatites at Oldoinyo Lengai, Tanzania: Composition of lavas from 1988 to 2007'. *Lithos*, **152**, 47–55, https://doi.org/10.1016/j.lithos.2012.08.012

Keppler, H., Wiedenbeck, M. & Shcheka, S.S. 2003. Carbon solubility in olivine and the mode of carbon storage in the Earth's mantle. *Nature*, **424**, 414–416, https://doi.org/10.1038/nature01828

Kjarsgaard, B.A. 1998. Phase relations of a carbonated high-CaO nephelinite at 0.2 and 0.5 GPa. *Journal of Petrology*, **39**, 2061–2075, https://doi.org/10.1093/petroj/39.11-12.2061

Kjarsgaard, B.A. & Hamilton, D.L. 1989. The genesis of carbonatites by immiscibility. *In*: Bell, K. (ed.) *Carbonatites: Genesis and Evolution*. Unwin Hyman, London, 388–409.

Koster van Groos, A.F. & Wyllie, P.J. 1963. Experimental data bearing on the role of liquid immiscibility in the genesis of the carbonatites. *Nature*, **199**, 801–802, https://doi.org/10.1038/199801a0

Le Bas, M.J. 1987. Nephelinites and carbonatites. *In*: Fitton, J.G. & Upton, B.G.J. (eds) *Alkaline Igneous Rocks*. Geological Society, London, Special Publications, **30**, 53–83, https://doi.org/10.1144/GSL.SP.1987.030.01.05

Le Bas, M.J.L., Maitre, R.W.L., Streckeisen, A., Zanettin, B. & Rocks, I.S. 1986. A chemical classification of volcanic rocks based on the total alkali–silica diagram. *Journal of Petrology*, **27**, 745–750, https://doi.org/10.1093/petrology/27.3.745

Lee, W.J. & Wyllie, P.J. 1994. Experimental data bearing on liquid immiscibility, crystal fractionation, and the origin of calciocarbonatites and natrocarbonatites. *International Geology Review*, **36**, 797–819.

Lee, W.J. & Wyllie, P.J. 1996. Liquid immiscibility in the join $NaAlSi_3O_8$–$CaCO_3$ to 2.5 GPa and the origin of calciocarbonatite magmas. *Journal of Petrology*, **37**, 1125–1152, https://doi.org/10.1093/petrology/37.5.1125

Lee, W.J., Huang, W.L. & Wyllie, P. 2000. Melts in the mantle modeled in the system CaO–MgO–SiO_2–CO_2 at 2.7 GPa. *Contributions to Mineralogy and Petrology*, **138**, 199–213, https://doi.org/10.1007/s004100050557

Le Maitre, R.W.B., Dudek, P. *et al.* 1989. *A Classification of Igneous Rocks and Glossary of Terms: Recommendations of the International Union of Geological Sciences, Subcommission on the Systematics of Igneous Rocks*. Blackwell Scientific, Oxford.

Le Roex, A.P. & Lanyon, R. 1998. Isotope and trace element geochemistry of cretaceous damaraland lamprophyres and carbonatites, northwestern namibia: evidence for plume–lithosphere interactions. *Journal of Petrology*, **39**, 1117–1146.

Luth, R.W. 2003. Mantle volatiles – distribution and consequences. *In*: Carlson, R.L. (ed.) *The Mantle and Core*. Treatise on Geochemistry, **2**. Elsevier, Amsterdam, 319–361.

Mahoney, J.J., MacDougall, J.D., Lugmair, G.W., Gopalan, K. & Krishnamurthy, P. 1985. Origin of contemporaneous tholeiitic and K-rich alkalic lavas: a case study from the northern Deccan Plateau, India. *Earth and Planetary Science Letters*, **72**, 39–53, https://doi.org/10.1016/0012-821X(85)90115-3

McDonough, W.F. 1990. Constraints on the composition of the continental lithospheric mantle. *Earth and Planetary Science Letters*, **101**, 1–18, https://doi.org/10.1016/0012-821X(90)90119-I

McKenzie, D. 1985. The extraction of magma from the crust and mantle. *Earth and Planetary Science Letters*, **74**, 81–91.

Palme, H. & O'Neill, H.S.C. 2014. Cosmochemical estimates of mantle composition. *In*: Carlson, R.W. (ed.)

The Crust. Treatise on Geochemistry, **3**. Elsevier, Amsterdam, 1–39, https://doi.org/10.1016/B978-0-08-095975-7.00201-1

Peng, Z.X. & Mahoney, J.J. 1995. Drillhole lavas from the northwestern Deccan Traps, and the evolution of Réunion hotspot mantle. *Earth and Planetary Science Letters*, **134**, 169–185, https://doi.org/10.1016/0012-821X(95)00110-X

Pirajno, F. 1994. Mineral resources of anorogenic alkaline complexes in Namibia: a review. *Australian Journal of Earth Sciences*, **41**, 157–168, https://doi.org/10.1080/08120099408728123

Pirajno, F. 2015. Intracontinental anorogenic alkaline magmatism and carbonatites, associated mineral systems and the mantle plume connection. *Gondwana Research*, **27**, 1181–1216, https://doi.org/10.1016/j.gr.2014.09.008

Presnall, D.C. & Gudfinnsson, G.H. 2005. Carbonate-rich melts in the oceanic low-velocity zone and deep mantle. *In*: Foulger, G.R., Natland, J.H., Presnall, D.C. & Anderson, D.L. (eds) *Plates, Plumes and Paradigms*. Geological Society of America, Special Papers, **388**, 207–216, https://doi.org/10.1130/0-8137-2388-4.207

Ray, J.S. 1998. Trace element and isotope evolution during concurrent assimilation, fractional crystallization, and liquid immiscibility of a carbonated silicate magma. *Geochimica et Cosmochimica Acta*, **62**, 3301–3306, https://doi.org/10.1016/S0016-7037(98)00237-3

Ray, J.S. & Pande, K. 1999. Carbonatite alkaline magmatism associated with continental flood basalts at stratigraphic boundaries: cause for mass extinctions. *Geophysical Research Letters*, **26**, 1917–1920, https://doi.org/10.1029/1999GL900390

Ray, J.S., Pande, K., Bhutani, R., Shukla, A.D., Rai, V.K., Kumar, A., Awasthi, N., Smitha, R.S. & Panda, D.K. 2013. Age and geochemistry of the Newania dolomite carbonatites, India: Implications for the source of primary carbonatite magma. *Contributions to Mineralogy and Petrology*, **166**, 1613–1632, https://doi.org/10.1007/s00410-013-0945-7

Ray, J.S. & Ramesh, R. 1999. Evolution of carbonatite complexes of the Deccan flood basalt province: stable carbon and oxygen isotopic constraints. *Journal of Geophysical Research: Solid Earth*, **104**, 29 471–29 483, https://doi.org/10.1029/1999JB900262

Ray, J.S. & Shukla, P.N. 2004. Trace element geochemistry of Amba Dongar carbonatite complex, India: Evidence for fractional crystallization and silicate-carbonate melt immiscibility. *Proceedings of the Indian Academy of Sciences, Earth and Planetary Sciences*, **113**, 519–531, https://doi.org/10.1007/BF02704020

Ray, J.S., Pande, K. & Venkatesan, T.R. 2000*a*. Emplacement of Amba Dongar carbonatite–alkaline complex at Cretaceous/Tertiary boundary: evidence from ^{40}Ar–^{39}Ar chronology. *Journal of Earth System Science*, **109**, 39–47, https://doi.org/10.1007/BF02719147

Ray, J.S., Ramesh, R., Pande, K., Trivedi, J.R., Shukla, P.N. & Patel, P.P. 2000*b*. Isotope and rare earth element chemistry of carbonatite–alkaline complexes of Deccan volcanic province: implications to magmatic and alteration processes. *Journal of Asian Earth Sciences*, **18**, 177–194, https://doi.org/10.1016/S1367-9120(99)00030-9

Ray, J.S., Pande, K. & Pattanayak, S.K. 2003. Evolution of Amba Dongar carbonatite complex: constrains from ^{40}Ar–^{39}Ar chronologies of the inner basalt and alkaline plug. *International Geological Review*, **45**, 857–862.

Renne, P.R., Sprain, C.J., Richards, M.A., Self, S., Vanderkluysen, L. & Pande, K. 2015. State shift in Deccan volcanism at the Cretaceous–Paleogene boundary, possibly induced by impact. *Science*, **350**, 76–78, https://doi.org/10.1126/science.aac7549

Richards, M.A., Alvarez, W. *et al.* 2015. Triggering of the largest Deccan eruptions by the Chicxulub impact. *Geological Society of America Bulletin*, **127**, 1507–1520.

Rukhlov, A.S., Bell, K. & Amelin, Y. 2015. Carbonatites, isotopes and evolution of the subcontinental mantle: an overview. *In*: Simandl, G.J. & Neetz, M. (eds) *Symposium on Strategic and Critical Materials Proceedings*, 13–14 November 2015, Victoria, British Columbia. British Columbia Ministry of Energy and Mines, British Columbia Geological Survey Paper, **3**, 39–64.

Salters, V.J. & Stracke, A. 2004. Composition of the depleted mantle. *Geochemistry, Geophysics, Geosystems*, **5**, 1–27.

Schoene, B., Samperton, K.M. *et al.* 2015. U–Pb geochronology of the Deccan Traps and relation to the end-Cretaceous mass extinction. *Science*, **347**, 182–184, https://doi.org/10.1126/science.aaa0118

Sen, G., Bizimis, M., Das, R., Paul, D.K., Ray, A. & Biswas, S. 2009. Deccan plume, lithosphere rifting, and volcanism in Kutch, India. *Earth and Planetary Science Letters*, **277**, 101–111, https://doi.org/10.1016/j.epsl.2008.10.002

Sensarma, S., Paul, D. & Rao, N.V.C. 2013. Large igneous provinces–global perspectives and prospects in India. *Current Science*, **105**, 182–192.

Sethna, S.F. 1971. A note on the trace element contents of carbonatites of Amba Dongar and surrounding areas, Chhota Udaipur. *Journal of the Geological Society of India*, **12**, 311–317.

Shaw, D.M. 1970. Trace element fractionation during anatexis. *Geochimica et Cosmochimica Acta*, **34**, 237–243.

Simandl, G.J. 2015. Carbonatites and related exploration targets. *In*: Simandl, G.J. & Neetz, M. (eds) *Symposium on Strategic and Critical Materials Proceedings*, 13–14 November 2015, Victoria, British Columbia. British Columbia Ministry of Energy and Mines, British Columbia Geological Survey Paper, **3**, 31–37.

Simonetti, A. & Bell, K. 1994. Isotopic and geochemical investigation of the Chilwa Island carbonatite complex, Malawi: evidence for a depleted mantle source region, liquid immiscibility, and open-system behaviour. *Journal of Petrology*, **35**, 1597–1621, https://doi.org/10.1093/petrology/35.6.1597

Simonetti, A., Bell, K. & Viladkar, S.G. 1995. Isotopic data from the Amba Dongar carbonatite complex, west-central India: evidence for an enriched mantle source. *Chemical Geology*, **122**, 185–198, https://doi.org/10.1016/0009-2541(95)00004-6

Simonetti, A., Goldstein, S.L., Schmidberger, S.S. & Viladkar, S.G. 1998. Geochemical and Nd, Pb, and Sr isotope data from Deccan alkaline complexes – inferences for mantle sources and plume–lithosphere interaction.

Journal of Petrology, **39**, 1847–1864, https://doi.org/10.1093/petroj/39.11-12.1847

Srivastava, R.K. 1997. Petrology, geochemistry and genesis of rift-related carbonatites of Ambadungar, India. *Mineralogy and Petrology*, **61**, 47–66.

Srivastava, R.K. & Sinha, A.K. 2004. Geochemistry of Early Cretaceous alkaline ultamafic–mafic complex from Jasra, Karbi Anglong, Shillong Plateau, Northeastern india. *Gondwana Research*, **7**, 549–561.

Streckeisen, A. 1980. Classification and nomenclature of volcanic rocks, lamprophyres, carbonatites and melilitic rocks. IUGS Subcommission on the systematics of igneous rocks. *Geologische Rundschau*, **69**, 194–207.

Subbarao, K.V. & Sukheswala, R.N. (eds) 1981. *Deccan Volcanism and Related Basalt Provinces in Other Parts of the World.* Geological Society of India, Memoirs, **3**.

Sukheswala, R.N. & Udas, G.R. 1963. Note on the carbonatite of Ambadongar (Gujarat State) and its economic potentialities. *Science and Culture*, **29**, 563–568.

Sukheswala, R.N. & Udas, G.R. 1967. Fluorspar mineralization related to carbonatite-alkalic complex at Amba Dongar, Gujarat State. *Current Science*, **36**, 14–16.

Sukheswala, R.N. & Viladkar, S.G. 1978. Carbonatite of India. *In*: *Proceedings of the First International Symposium on Carbonatites*, Poços de Caldas Brazil. Ministério das Minas e Energia, Departamento Nacional da Produção Mineral, Rio de Janeiro, Brazil, 277–293.

Sukheswala, R.N. & Viladkar, S.G. 1981. Fenitized sandstones in Amba Dongar carbonatites, Gujarat, India. *Journal of the Geological Society of India*, **22**, 368–374.

Tappe, S., Foley, S.F. *et al.* 2006. Genesis of ultramafic lamprophyres and carbonatites at Aillik Bay, Labrador: a consequence of incipient lithospheric thinning beneath the North Atlantic Craton. *Journal of Petrology*, **47**, 1261–1315, https://doi.org/10.1093/petrology/egl008

Thi, T.N., Wada, H., Ishikawa, T. & Shimano, T. 2014. Geochemistry and petrogenesis of carbonatites from South Nam Xe, Lai Chau area, northwest Vietnam. *Mineralogy and Petrology*, **108**, 371–390, https://doi.org/10.1007/s00710-013-0301-7

Thibault, Y., Edgar, A.D. & Lloyd, F.E. 1992. Experimental investigation of melts from a carbonated phlogopite lherzolite: implications for metasomatism in the continental lithospheric mantle. *American Mineralogist*, **77**, 784–794.

Treiman, A.H. & Schedl, A. 1983. Properties of carbonatite magma and processes in carbonatite magma chambers. *Journal of Geology*, **91**, 437–447.

Twyman, J.D. & Gittins, J. 1987. Alkalic carbonatite magmas: parental or derivative? *In*: Fitton, J.G. & Upton, B.G.J. (eds) *Alkaline Igneous Rocks*. Geological Society, London, Special Publications, **30**, 85–94, https://doi.org/10.1144/GSL.SP.1987.030.01.06

Veena, K., Pandey, B.K., Krishnamurthy, P. & Gupta, J.N. 1998. Pb, Sr and Nd isotopic systematics of the carbonatites of Sung Valley, Meghalaya, northeast India: Implications for contemporary plume-related mantle source characteristics. *Journal of Petrology*, **39**, 1875–1884, https://doi.org/10.1093/petroj/39.11-12.1875

Veksler, I.V. & Keppler, H. 2000. Partitioning of Mg, Ca, and Na between carbonatite melt and hydrous fluid at 0.1–0.2 GPa. *Contributions to Mineralogy and Petrology*, **138**, 27–34, https://doi.org/10.1007/PL00007659

Veksler, I.V., Nielsen, T.F.D. & Sokolov, S.V. 1998. Mineralogy of crystallized melt inclusions from Gardiner and Kovdor ultramafic alkaline complexes: implications for carbonatite genesis. *Journal of Petrology*, **39**, 2015–2031.

Viladkar, S.G. 1981. The carboantites of Amba Dongar, Gujarat, India. *Bulletin of the Geological Society of Finland*, **53**, 17–28.

Viladkar, S.G. 1984. Alkaline rocks associated with the carbonatites of Amba Dongar, Chhota Udaipur, Gujarat, India. *The Indian Mineralogist*, Sukheswala Volume, 130–135.

Viladkar, S.G. 1996. *Geology of the Carbonatite–Alkalic Diatreme of Amba Dongar, Gujarat.* Monograph published by GMDC Science and Research Centre, Ahmedabad, India.

Viladkar, S.G. 2012. *Evolution of Calciocarbonatite Magma: Evidence from the Sövite and Alvikite Association in the Amba Dongar Complex, India.* InTech Open Access, Rijeka, Croatia.

Viladkar, S.G. 2015. Mineralogy and geochemistry of fenitized nephelinites of the Amba Dongar complex, Gujarat. *Journal of the Geological Society of India*, **85**, 87–97, https://doi.org/10.1007/s12594-015-0196-5

Viladkar, S.G. & Dulski, P. 1986. Rare earth element abundances in carbonatites, alkaline rocks and fenites of the Amba Dongar complex, Gujrat, India. *Neues Jahrbuch Fur Mineralogie-Monatshefte*, **1**, 37–48.

Viladkar, S.G. & Schidlowski, M. 2000. Carbon and oxygen isotope geochemistry of the Amba Dongar carbonatite complex, Gujarat, India. *Gondwana Research*, **3**, 415–424, https://doi.org/10.1016/S1342-937X(05)70299-9

Viladkar, S.G. & Wimmenauer, W. 1992. Geochemical and petrological studies on Amba Dongar carbonatites (Gujarat, India). *Chemie der Erde*, **52**, 277–291.

Vladykin, N.V. 2009. Potassium alkaline lamproite–carbonatite complexes: petrology, genesis, and ore reserves. *Russian Geology and Geophysics*, **50**, 1119–1128.

Von Eckermann, H. 1961. The petrogenesis of the Alnö alkaline rocks. *Bulletin of the Geological Institution of the University of Uppsala*, **40**, 25–36.

Wallace, M.E. & Green, D.H. 1988. An experimental determination of primary carbonatite magma composition. *Nature*, **335**, 343–346, https://doi.org/10.1038/335343a0

White, W.M. & Duncan, R.A. 1996. Geochemistry and geochronology of the Society Islands: New evidence for deep mantle recycling. *In*: Basu, A. & Hart, S. (eds) *Earth Processes: Reading the Isotopic Code.* American Geophysical Union, Geophysical Monograph Series, **95**, 183–206, https://doi.org/10.1029/GM095p0183

Woolley, A.R. 1982. A discussion of carbonatite evolution and nomenclature, and the generation of sodic and potassic fenites. *Mineralogical Magazine*, **46**, 13–17.

Woolley, A.R. & Kempe, D.R.C. 1989. Carbonatites: nomenclature, average chemical compositions, and element distribution. *In*: Bell, K. (ed.) *Carbonatites: Genesis and Evolution*. Unwin Hyman, London, 1–14.

Woolley, A.R. & Kjarsgaard, B.A. 2008. Paragenetic types of carbonatite as indicated by the diversity and relative abundances of associated silicate rocks: evidence from a global database. *The Canadian Mineralogist*, **46**, 741–752, https://doi.org/10.3749/canmin.46.4.741

Wyllie, P.J. & Huang, W.L. 1975. Influence of mantle CO_2 in the generation of carbonatites and kimberlites. *Nature*, **257**, 297–299, https://doi.org/10.1038/257297a0

Yaxley, G.M. & Brey, G.P. 2004. Phase relations of carbonate-bearing eclogite assemblages from 2.5 to 5.5 GPa: implications for petrogenesis of carbonatites. *Contributions to Mineralogy and Petrology*, **146**, 606–619.

Mineralogy, geochemistry and geochronology of mafic magmatic enclaves and their significance in evolution of Nongpoh granitoids, Meghalaya, NE India

MOHD. SADIQ[1], RAVI K. UMRAO[1]*, B. B. SHARMA[1], S. CHAKRABORTI[1], S. BHATTACHARYYA[1] & A. KUNDU[2]

[1]*Geological Survey of India, North Eastern Region, Shillong 793003, India*

[2]*Geological Survey of India, State Unit: Haryana, NH 5P, NIT, Faridabad 121001, India*

**Correspondence: georavigsi@gmail.com*

Abstract: The Cambrian Nongpoh granitoids, intrusive into the Precambrian gneissic complex and metasediments of the Shillong Group, represent a major phase of granitic magmatism in the Shillong Plateau. The Nongpoh granitoids comprise diorite, granodiorite, porphyritic and grey granites. Porphyritic granite is the dominant lithology exposed in the Nongpoh granitoids, and contains three types of enclaves, viz. xenoliths of gneissic rocks, dark grey porphyry and biotite-rich microgranular enclaves. The mafic magmatic enclaves (MMEs) of various dimensions and shapes, including rounded, ellipsoidal, rectangular, angular to subangular, and stretched bodies, were produced by evolving nature and contrasting kinematics of interacting felsic and mafic magmas. The biotite-rich enclaves are due to the injection of syn-magmatic mafic dykes in felsic magma. The various textural assemblages and sub-linear variations between silica and major oxides, chemical mixing and element diffusion suggest that multistage magma hybridization was the key process during the evolution of the Nongpoh granitoids. The 501 Ma age obtained by chemical (U–Th–Pb) dating of monazite in MME ascertained the age of the magma hybridization event in the Nongpoh granitoids, which is equivalent to the igneous activity at 500 Ma during the amalgamation of Eastern Gondwana.

Mafic magmatic enclaves (MMEs) were originally defined by Didier & Roques (1959) as ellipsoidal-shaped rocks, a few centimetres to some decimetres in size, and showing sharp contacts with host rocks (Cid *et al.* 2002). MMEs are reported as field evidence for magma mixing and/or mingling, commonly in alkali granite, monzogranite, granodiorite or granite rocks (Vernon *et al.* 1983; Didier 1984; Willey 1984; Barbarin & Didier 1992). Moreover, textures like rapakivi, anti-rapakivi, quartz ocelli, acicular and mixed apatite morphologies, inclusion zones in feldspars, titanite ocelli, K-feldspar megacrysts, sieved plagioclase, and poikilitic textures in MMEs are interpreted as textural evidence of magma mixing and mingling (Hibbard 1991; Fernandez & Barbarin 1991; Baxter & Feely 2002). MMEs are studied to understand the genesis and evolution of granitoid rocks, and have been identified in igneous provinces around the world, having been studied for several decades (Didier 1973; Vernon 1984; Rock 1987; Castro 1990; Dorais *et al.* 1990; Didier & Barbarin 1991; Elburg 1996*a*, *b*, *c*; Wiebe *et al.* 1997; Kumar *et al.* 2004, 2005; Yang *et al.* 2004, 2005, 2006, 2007*a*, *b*; Kumar & Rino 2006; Kumar 2010; Kumar & Pieru 2010; Cheng *et al.* 2012). Here, in the present paper, the term MMEs is used for mafic magmatic enclaves (Didier & Barbarin 1991) irrespective of the grain size of the various enclaves present in the Nongpoh granitoids.

Here, we report, for first time, on the evidence of magma hybridization in Nongpoh granitoids based on field, textural and geochemical studies. This study presents petrogenesis of Nongpoh granitoids using detailed field, petrological, geochemical studies and U–Th–Pb monazite dating of MMEs in Nongpoh granitoids in the Ri-Bhoi District, Meghalaya, India.

Geological setting

The Shillong (Meghalaya) Plateau is an uplifted horst-like feature covering an area of about 40 000 km^2 (Fig. 1a) (Saha *et al.* 2010). It mainly comprises basic granulite, amphibolite, a gneissic complex, the Shillong Group of metasediments, various intrusives (viz. granite plutons, ultramafic–alkaline–carbonatite complexes and the Sylhet Trap), Cretaceous–Palaeogene sediments and recent alluvium (Fig. 1a). The Sylhet Trap and alkaline provinces of the Shillong Plateau are thought to be related with the Kerguelen mantle plume (Storey *et al.* 1992; Sadiq *et al.* 2014*a*). Several syn- to late-tectonic granitoids (Kyrdem, Mylliem, Nongpoh, South Khasi, Rongjeng and Sindhuli granitoids) occur as

From: SENSARMA, S. & STOREY, B. C. (eds) 2018. *Large Igneous Provinces from Gondwana and Adjacent Regions*. Geological Society, London, Special Publications, **463**, 171–198.
First published online July 6, 2017, https://doi.org/10.1144/SP463.2

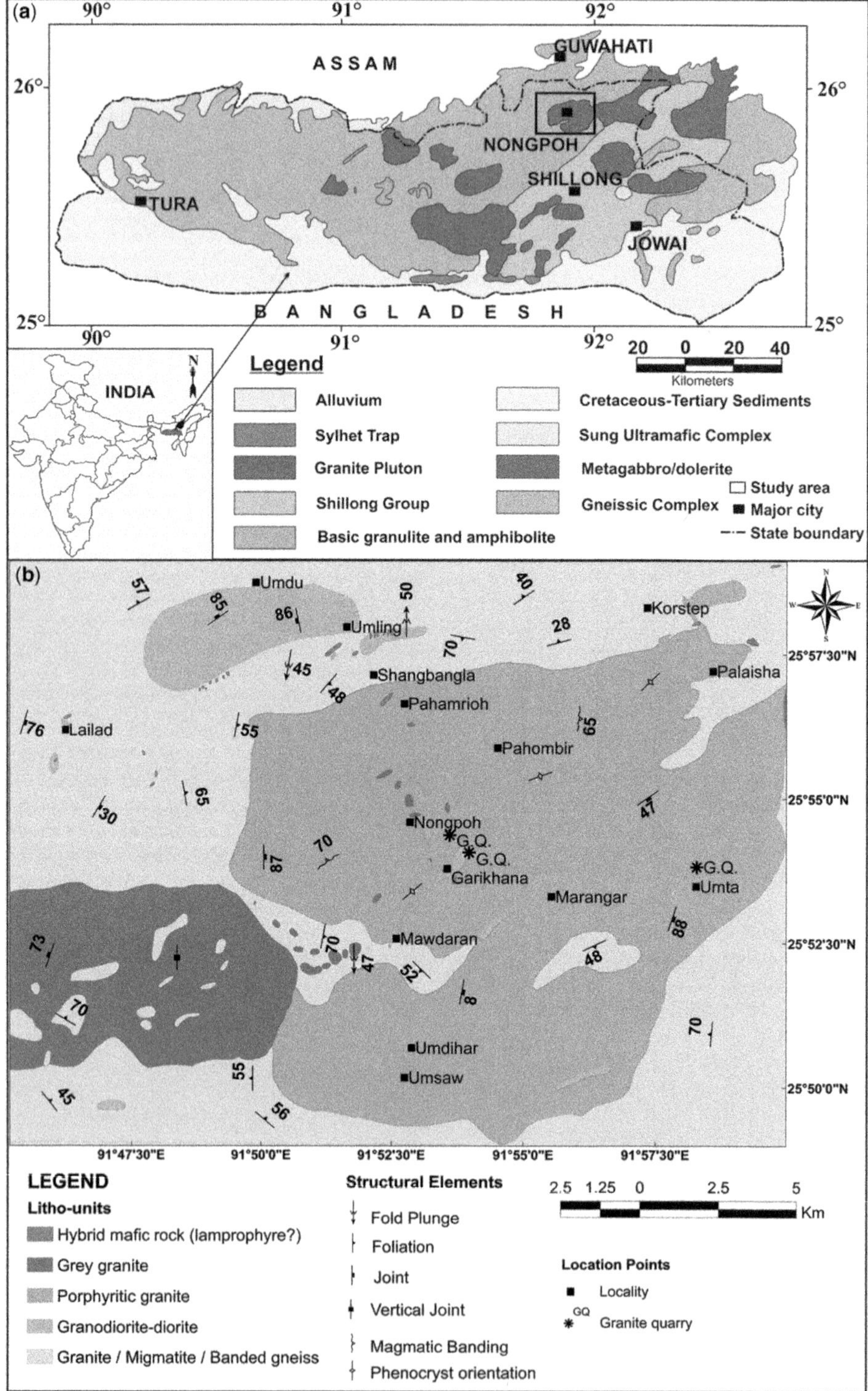

Fig. 1. (**a**) Simplified geological map of the Meghalaya (Shillong) Plateau (modified after Ghosh *et al.* 1991, 2005). Inset shows the geographical location of Meghalaya in NE part of India. (**b**) Geological map of the Nongpoh granitoids, Ri-Bhoi District, Meghalaya, India (modified after Sarkar *et al.* 2007).

discordant plutons and batholiths intruding the Shillong Group of metasedimentary rocks and gneissic complex of the Shillong Plateau (Fig. 1a) (Ghosh *et al.* 2005; Sadiq *et al.* 2014*b*), and their emplacement is controlled by pre-existing fractures and lineaments (Evans 1964; Nandy 1980). Neoproterozoic–Cambrian granitoid emplacement was a major phase of granitic magmatism in the Shillong Plateau and correlated to a Pan-African event during the amalgamation of the Eastern Gondwana supercontinent (Ghosh *et al.* 2005; Yin *et al.* 2010).

The Nongpoh granitoids are intrusive into the Precambrian gneissic complex and metasediments of the Shillong Group (Fig. 1a). The Nongpoh granitoids comprise diorite, granodiorite and granite variants (viz. porphyritic and grey granites) (Fig. 1b). Porphyritic granite is pink and grey coloured, and coarse grained, and comprises K-feldspar, plagioclase, quartz, biotite, amphibole and pyrite. Apophyses of the granite within gneisses and caught-up patches/rafts and xenoliths of older gneisses of variable dimensions, and MMEs of variable shape and size within the Nongpoh granitoids, are observed (Fig. 1b). The K-feldspar phenocrysts in porphyritic granite show a general east–west alignment near the contact of Precambrian gneisses at Shangbangla and Pahamrioh, indicative of the alignment of feldspar phenocrysts during flow in a semi-solid condition, but randomly orientated phenocrysts are also common (Sadiq *et al.* 2014*b*). However, evidence of a tectonic foliation is also observed and is parallel to the regional strike (NE–SW) of the foliation in gneissic rocks. The longer axis of flesh and white coloured K-feldspar phenocrysts sometimes range up to 10 cm (Sadiq *et al.* 2014*b*). The medium-grained, light to dark grey coloured grey granite, exposed around Nongpoh, Garikahana and west of Mawdaran villages of the Ri-Bhoi District, Meghalaya, comprises plagioclase, quartz, biotite, microcline, orthoclase and hornblende.

Three granitic events have been identified on the basis of field relationships around the Nongpoh, Garikhana and Umta villages, viz.: (a) older grey granite – occurring as enclaves in porphyritic granite and showing enrichment of quartzo-feldspathic material along the boundary of the enclave; (b) porphyritic granite – occurring as enclaves in younger grey granites; and (c) younger grey granite – occurring as dykes in porphyritic granite and containing enclaves of porphyritic granite, and also occurring as tongues in porphyritic granite.

Enclave diversity

The MMEs in Nongpoh porphyritic granite occur as circular, rounded, ellipsoidal, rectangular, angular to subangular and stretched bodies orientated in a preferred direction (Fig. 2a–i). Although the size of MMEs ranges from few centimeters up to 10 m and commonly from 5 to 30 cm, a few were noted ranging from 1 to 2 m along the long axis in and around the Pahamrioh, Umta and Mawkibari villages of the Ri-Bhoi District (Fig. 2a–i). Stretched, lensoidal mafic enclaves are also observed (Fig. 2b, d & i) that are elongated along the direction of primary magmatic foliation and alignment of feldspar phenocrysts and hornblende in the MMEs (Fig. 2a & d). A lack of a chilled margin between the MMEs and the host granite suggests a lesser temperature difference between the felsic host and the mafic enclave (Troll *et al.* 2004; Ahmad 2011). The contacts between the mafic enclave and the Nongpoh granitoids are mainly sharp, crenulated and irregular, occasionally rimmed by a quartzofeldspathic margin, and rarely diffused and partially gradational. Angular and tabular MMEs (Fig. 2f) indicate that they were not completely molten when engulfed in the host granitoid magma, and might be formed either due to magma mingling and not mixing (Liankun & Kuriong 1991) or heterogeneous mixing of granitic and mafic magma (Barbarin & Didier 1992). Tabular, square and lensoidal K-feldspar phenocrysts/megacrysts were frequently observed within MMEs and near the mafic enclave and granitoid contacts. Sword-shaped MMEs with a chilled zone 1–3 cm in thickness and corroded margins resulted from the quenching of hot mafic magma against relatively cold felsic magma (Fig. 2h) (Chen *et al.* 1991).

The MMEs hosted in porphyritic granite are fine to medium grained and porphyritic in nature. In porphyritic granite, the double enclave or clots of mafic-rich microgranular enclaves occurring in relatively felsic and/or less mafic porphyritic enclaves is indicative of a multistage process of hybridization (Fig. 2e). The post-emplacement discrete shears are recorded in porphyritic granite, which has imprints on mafic enclaves. Disjointed MMEs disrupted with angular margins may result due to late shear affecting the ductile/semi-solid granitoid host and a relatively solid MME (Fig. 3b) (Barbarin & Didier 1991). Raft-like MMEs with a sharp contact on one side and a diffused and/or gradational contact on the other side indicate diffusional mixing and attempt to reach equilibrium with their host (Fig. 3c) (Barbarin & Didier 1991). Numerous dendritic back-veins of felsic melt into mafic components are noticed within the mixing zone (Figs 2b, e & 3a).

A number of isolated mafic bodies, melanocratic to hyper-melanaocratic in nature, are observed within basement gneiss and in association with porphyritic granite, granodiorite-diorite near the Umling, Lailad and Umling-Lambran villages. In places, the mafic rock is found in isolation around Lailad (Nongkhyllem Wildlife Sanctuary). A diffused contact is observed between porphyritic granite and the mafic body to the east of Pahamrioh village. The

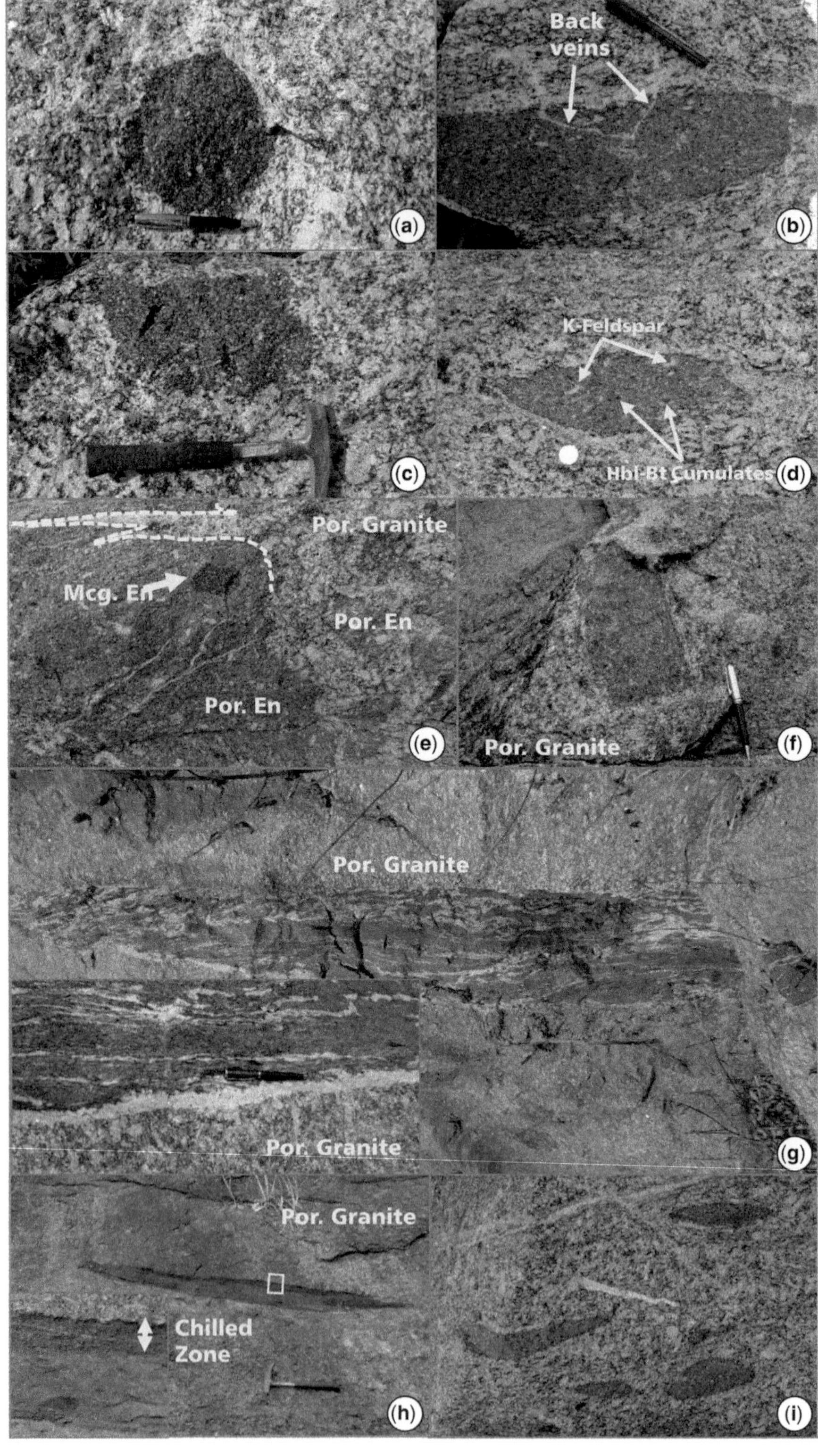
(a)
Back veins
(b)
(c)
K-Feldspar
Hbl-Bt Cumulates
(d)
Por. Granite
Mcg. En
Por. En
Por. En
(e)
Por. Granite
(f)
Por. Granite
Por. Granite
(g)
Por. Granite
Chilled Zone
(h)
(i)

phenocrysts of K-feldspar were also observed inside the hybrid rock near the contact between the two. These mafic bodies occur reclusely or as swarms in a linear pattern sympathetic/parallel to the directions of major lineaments (north–south, NE–SW and NW–SE) in the area. The dimensions of individual mafic bodies range from few metres up to 200 m wide and from a few metres to few hundred metres in length.

The mafic bodies are dark grey, massive, coarse to very-coarse grained, comprising biotite/phlogopite, plagioclase, hornblende and quartz. In places, medium-grained mafic rock is also observed. These mafic rocks usually are composed of 30–40% biotite, 20–30% plagioclase, 20–25% hornblende with small amount of K-feldspar, and a range of accessory minerals, such as clinopyroxene, quartz, apatite, titanite, epidote, monazite and opaques. Microscopically, it is coarse grained, inequigranular and exhibits a hypidiomorphic texture. Sarkar *et al.* (2007) described these mafic rock as early magmatic differentiates of calc-alkaline Nongpoh granitoids, while Sadiq & Umrao (2013) described them as plutonic kersantite similar to those reported only from European Hercynides (Rock 1991). However, a similar rock type has been observed in a granite quarry as a magmatic enclave forming a part of a polygenic swarm in porphyritic granite. Detailed petrographical studies have also indicated the presence of a mixing texture (viz. acicular and tubular apatite, resorbed plagioclase and poikilitic hornblende, biotite, and microcline) in these rocks and these are described here as hybrid rocks (HD), produced as a result of the mixing and mingling of felsic granitic and mafic magma.

Analytical techniques

A total of 34 samples – 10 MMEs, seven hybrid rocks (HD), seven granodiorites (GD) and 10 host porphyritic granites (PG) – were collected and analysed for major, trace and REE. Major oxide and trace elements (Ba, Ga, Sc, V, Th, Pb, Ni, Co, Rb, Sr, Y, Zr, Nb, Cr, Cu and Zn) were analysed by a fully automated X-ray fluorescence (XRF) machine, PANalytical model 148 MagiX-2424 using a end window X-ray tube at the Chemical Division, Geological Survey of India, North Eastern Region, Shillong. The REE and other trace elements (Sn, Hf, Ta, Mo, W, Ge, Be and U) were analysed by inductively coupled plasma emission mass spectrometer (ICP-MS) in the Central Chemical Laboratory, Geological Survey of India, Kolkata, India. The details of techniques, procedure, precision and accuracy of these analyses are described in the standard operating procedures (GSI 2011).

The mineral composition study of thin polished sections was determined using a CAMECA SX-100 electron microprobe equipped with an energy dispersive system (EDS) installed at the Geological Survey of India, Northern Region, Faridabad. Elemental compositional X-ray maps of monazite for Th, U and Y were generated to identify internal compositional variations of monazite grains for further interpretations. Probe points for age dating were decided based on the identified compositional zones. Chemical analysis was carried out at an acceleration potential of 15 kV, 10 nA beam current with a 1 µm beam size for silicate phases, and 200 nA beam current with 2 µm beam size for monazite dating. The age was calculated from the analysis using the inbuilt software of the CAMECA SX 100 electron microprobe, solving the equation of Montel *et al.* (1996). The mean weighted age of the monazite grains was calculated from individual analyses using ISOPLOT (Ludwig 1991).

Petrography and mineral chemistry

General petrographical characteristics of the Nongpoh granitoids

Mineral assemblage of porphyritic granite hosting mafic enclaves are nearly the same except

Fig. 2. Evidence of magma hybridization. (**a**) Magma mingling indicated by a biotite-rich circular mafic enclave. There is the enrichment of quartzofeldspathic material along the boundary of the enclave. (**b**) Ellipsoidal mafic enclave and 'back-veining'. K-feldspar phenocrysts derived from the granitic magma. Lower part of the photograph indicates the thermally rejuvenated granitic magma due to the interaction of hot mafic globules. (**c**) Stretched lensoidal mafic enclave with K-feldspar phenocrysts derived from the granitic magma. Phenocrysts are orientated along the boundary of the enclave. (**d**) Subangular mafic microgranular enclave in porphyritic granite with a crenulated boundary. The mafic enclave contains a few K-feldspar phenocrysts. (**e**) Composite MME (biotite-rich microgranular enclave within a porphyritic enclave) and felsic back-veins in mafic enclaves. The presence of a tongue of porphyritic granite along with a lobate and cuspate margin of the MME indicates its early emplacement in the latter. (**f**) Block prismatic MME with a fine-grained sharp margin, embedded within porphyritic granite. (**g**) Dyke-shaped mafic enclave in the porphyritic granite. There is enrichment of quartzofeldspathic minerals along the boundary and stretched felsic veins within the enclave. Phenocrysts of granite are nearly at right-angles to the dyke-like mafic enclave. (**h**) Late injection of a sword-like composite MME with a chilled zone and corroded margins cutting across the primary magmatic foliation defined by K-feldspar phenocrysts of porphyritic granite, (**i**) Elongated monogenic swarm of MMEs of various shapes (viz., boomerang, eye and ellipsoidal shaped). Hbl, hornblende; Bt, biotite; Por, porphyritic; Mcg, microgranular; En, enclave.

Fig. 3. (**a**) Disjointed MME disrupted and fluxed by later pegmatite veins in the NE part of the photograph. Note the NW–SE-trending thin veins developing from the northern side of the east–west-trending sheared porphyritic granitic vein. (**b**) Raft-like MME orientated parallel to boudinaged synplutonic mafic dykes. (**c**) Porphyritic MME with abundant biotite and corroded plagioclase phenocryst. Note that the curved and lobate margins diffused in places indicate emplacement of mafic magma before solidification of the granitoid magma. (**d**) Rapakivi texture represented by a plagioclase rim over a K-feldspar megacryst. (**e**) Quartz ocelli observed in a mafic enclave. Quartz is surrounded by mainly biotite, amphibole and feldspar.

accessories but differ in modal percentage. Microscopically, the Nongpoh porphyritic granite consists of plagioclase, microcline, quartz and orthoclase as major minerals, and biotite and hornblende occur as minor minerals. Apatite, titanite, allanite, zircon and magnetite are the accessory phases. The rock mainly shows a porphyritic texture, while hypidiomorphic granular, myrmekitic and perthitic textures are also present. The modal estimation of a thin-section sample shows 20% quartz, 40% plagioclase, 26% K-feldspar, 12% biotite, and 1–2% opaque and accessory minerals.

The grey granite is a medium to dark grey, medium-grained, homophanous granite exposed at the northern part of Nongpoh. It comprises quartz, plagioclase, orthoclase and biotite as major minerals, with sulphides, apatite, titanite and opaques as accessory minerals. The younger grey granite is comparatively rich in titanite and allanite.

Granodiorite is composed of plagioclase, microcline, biotite, quartz and hornblende in decreasing abundance, with accessory titanite, zircon, apatite, allanite and opaques. Micro-perthite, flame perthite, flame antiperthite and myrmekite intergrowth is common (Fig. 4a, b). Plagioclase, biotite and microcline grains are euhedral to subhedral. Biotite and hornblende show sharp and corroded boundaries. Inclusions of apatite and hornblende are also noticed in biotite.

The MMEs are broadly dioritic to monzodiorite, with a hypidiomorphic granular texture; they consist of plagioclase, biotite and hornblende as major minerals, and microcline, orthoclase, quartz and clinopyroxene as minor minerals. Titanite, magnetite, apatite and zircon are accessory minerals, and occur mainly as poikilitic inclusions in feldspars, hornblende and biotite. Monazite, hercynite, crichtonite, margarite and thorite also occur as accessory minerals in MMEs.

Petrographical evidence of magma hybridization

Rapakivi texture and quartz ocelli were observed in the field and during microscopic studies (Fig. 3d, e). A rapakivi texture and a quartz–hornblende ocellar texture form due to the mixing of mafic and felsic systems, and the undercooling of mafic magma. A rapakivi texture is produced when undercooling of mafic magma results to form a rim of plagioclase on the growth surfaces of K-feldspar derived from the granitic system (Hibbard 1981). Abundant acicular and tubular apatite crystals occur as oikocrysts in microcline, perthitic microcline, plagioclase, hornblende, biotite and rarely in titanite (Fig. 4c–f). It indicates rapid crystallization or quenching of mafic magma upon its incorporation into the felsic host. Stubby apatite crystals are few, and occur as isolated and enclosed in microcline, plagioclase biotite and hornblende (Fig. 4c, d). Mostly microcline and, in places, plagioclase and hornblende are poikilitic, and enclose oikocryst plagioclase, acicular apatite, zircon, titanite and biotite inclusions (Fig. 4e, f). Resorbed xenocryst plagioclase crystals with a corroded boundary indicate magma mixing due to the in-equilibrium condition of the magma (Fig. 4f).

A growth rim of albitic composition over calcic plagioclase is ubiquitous. The occurrence of acicular apatite and opaques are found to be restricted in the albitic rim, but absent in the core of plagioclase (Fig. 5a). This may be due to the replacement of early formed anorthitic plagioclase and amphibole by Na-rich late liquid. Calcium coming out from early formed minerals has contributed in the formation of titanite, occurring as a rim around opaque, amphiboles and/or plagioclase (Fig. 5b). However, zoned spindle-shaped titanite is also present in porphyritic granite. A late sodic rim encloses the quenched apatite needles and other opaque minerals formed due to the release of Ca from plagioclase and the incorporation of P, F and necessary elements from the late liquid at the marginal part in a closed system. A large anorthitic rim developed over different plagioclase grains (Fig. 5c), and a rim of quartz resorbing and enclosing altered and fresh plagioclase grains (Fig. 5d) indicates the influence of late calcic fluids.

Euhedral biotite crystals are common, blade-shaped biotite is occasional, but resorbed anhedral to subhedral biotite crystals sharing a crenulated and curved contact with K-feldspars are noticed. Biotite in MMEs is replaced by hornblende and encloses early magmatic biotite. Biotite grains in MMEs with oikocrysts such as apatite, zircon, epidote, microcline and plagioclase are common. Hornblende is subhedral and anhedral, and in places is poikilitic, with biotite, microcline, quartz, stubby and acicular apatite, and zircon inclusions. Curved and lobate contacts between hornblende and poikilitic microcline with biotite, and acicular apatite, may indicate crystallization of microcline from late magma and the replacement of early magmatic hornblende (Fig. 5f). Moreover, late magmatic hornblende is also noticed where it is replacing and enclosing resorbed microcline and plagioclase (Fig. 6a). A few millimetre-sized hornblende clots with titanite and plagioclase, crudely rimmed by biotite in MMEs, may be due to segregation of early formed dense phases of mafic magma (Reid & Hamilton 1987) or early-recrystallized large amphibole phenocrysts of mafic magma (Fig. 6b). The intergrain triple point contact relationship of hornblende crystals within the mafic clots and plagioclase crystals within felsic matrix indicate deformation and recrystallization of hornblende and feldspar phenocrysts giving rise to a recovery texture. Nearly ocellar/sub-ophitic titanite feldspars (both plagioclase and microcline) associated with quartz and devoid of biotite association are recorded. Microcline occurs as euhedral to subhedral, resorbed

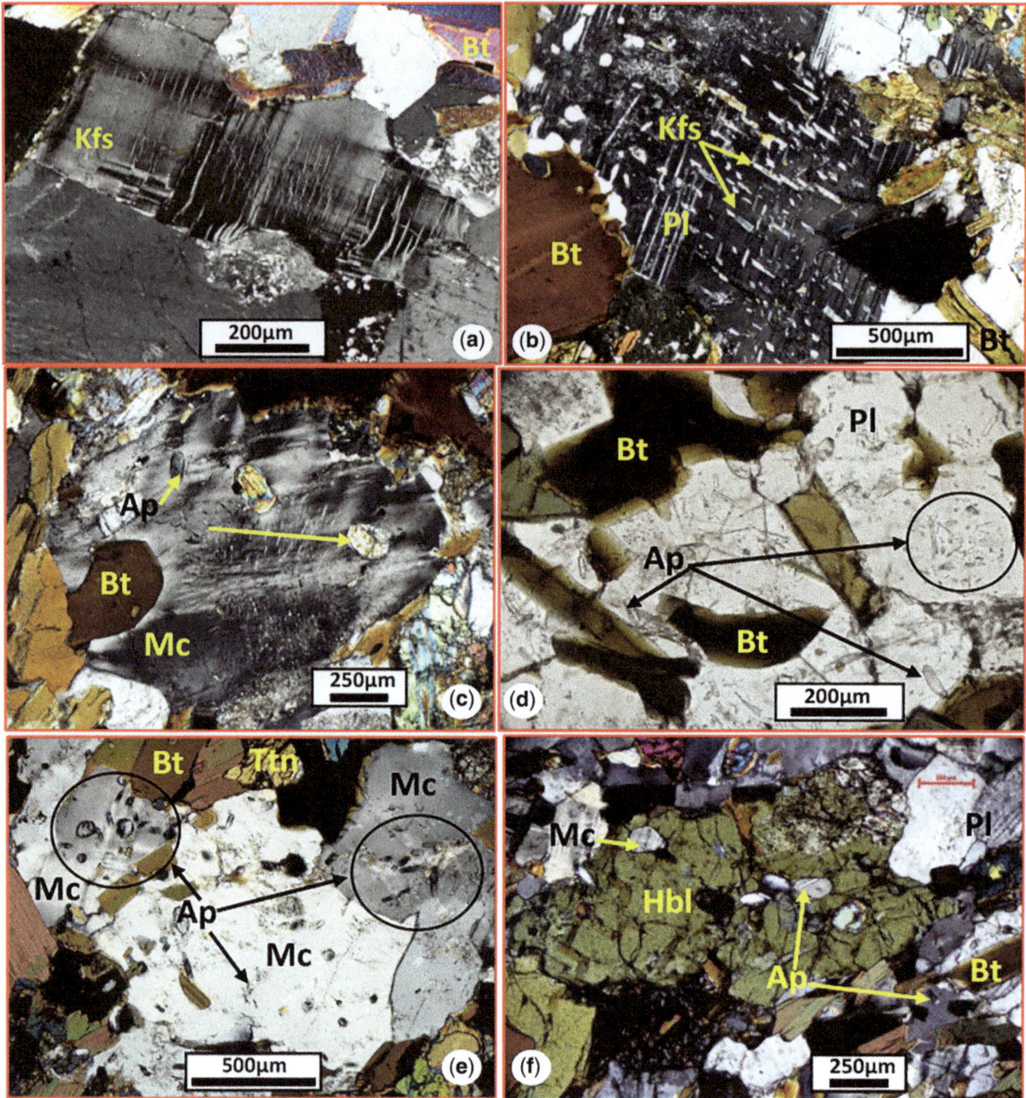

Fig. 4. (**a**) Flames of plagioclase are being developed perpendicular to the boundary of K-feldspar. (**b**) Anti-perthite represented by plagioclase with exsolved specks of K-feldspar later altered by biotite. (**c**) Oikocrysts biotite, hornblende and apatite are poikilitically enclosed within perthitic potassic feldspar. (**d**) Biotite, tubular and acicular apatite poikilitically enclosed in plagioclase. (**e**) Chadacryst microcline with oikocrysts of biotite and stubby, tubular and acicular apatite. (**f**) Poikilitic hornblende containing microcline with and without apatite needle, biotite and apatite. Kfs, K-feldspar; Hbl, hornblende; Pl, plagioclase; Mc, microcline; Bt, biotite; apt, apatite; Ttn, titanite.

anhedral and as oikocrysts in large plagioclase, hornblende and biotite. Perthitic microcline encloses numerous plagioclase grains, with an albitic rim-in optical continuity present (Fig. 6c). Moreover, biotite with quenched apatite also occurs in microcline (Fig. 6d). Quartz occurs as interstitials and as ocellus. Resorbed quartz surrounded by hornblende and biotite is noticed (Fig. 6e). Ocellus quartz with a partially corroded boundary is also surrounded by biotite, hornblende, plagioclase and K-feldspars (Fig. 6f). This forms as a result of the dissolution of felsic quartz in hybrid magma and the consequent growth of mafic minerals during undercooling of the MME magma (e.g. Reid *et al.* 1983; Vernon *et al.* 1983; Vernon 1984, 1990; Kumar & Pieru 2010). Ocellar quartz and

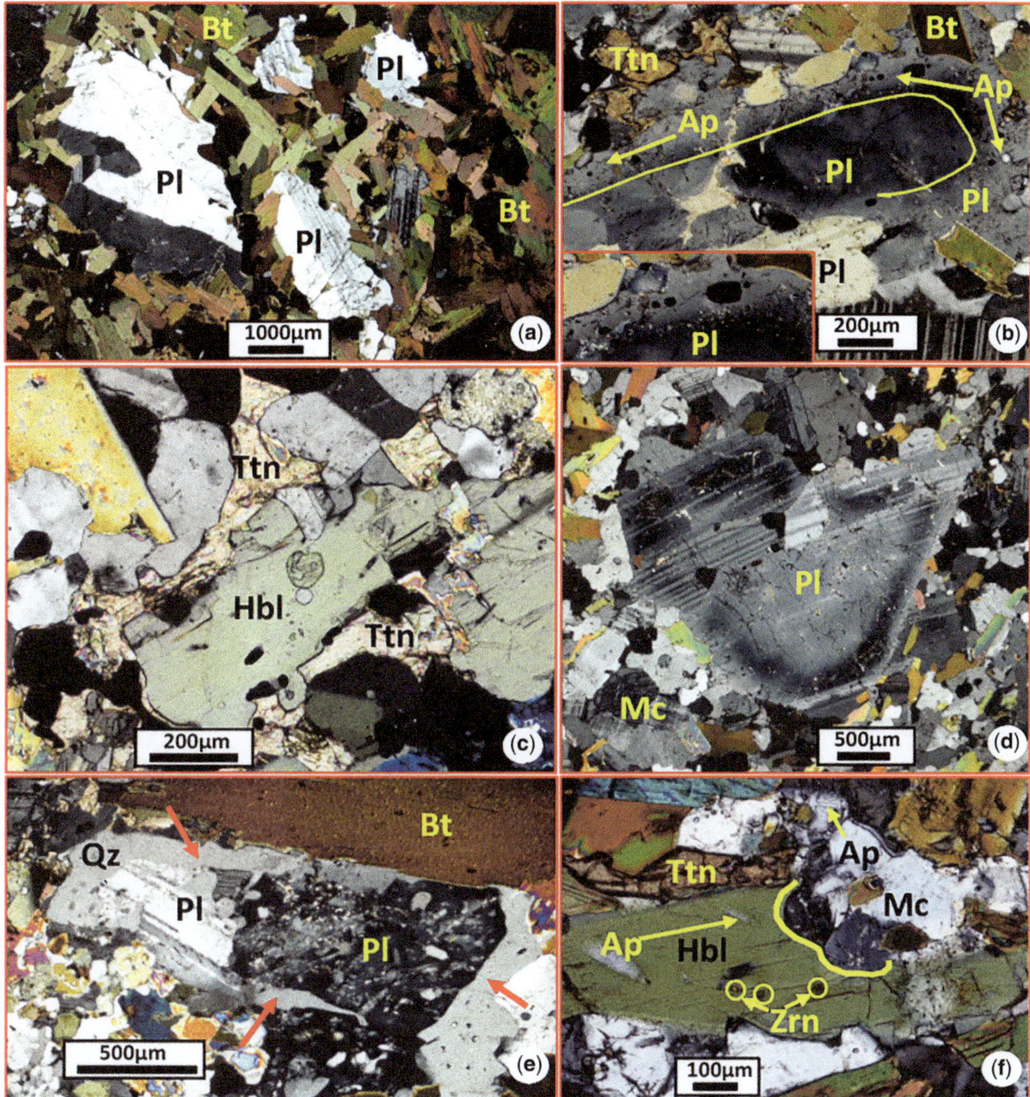

Fig. 5. (**a**) Corroded plagioclase xenocryst (?) showing irregular boundary; being swallowed by the surrounding biotite crystals. (**b**) Sodic rim developed over early formed calcic plagioclase with restricted presence of acicular apatite and abundance of opaque minerals in the marginal part. (**c**) Titanite forming at the contact of plagioclase opaque and plagioclase. (**d**) Dark (anorthitic) rim over plagioclase grains in optical continuity indicates later magmatic influx. Note the lack of apatite crystals, as in (c). (**e**) Altered and spongy plagioclase grains enclosed in a late magmatic quartz rim marked by red arrows. (**f**) Poikilitic hornblende with apatite and zircon inclusions, having lobate contact with poikilitic microcline comprise biotite and randomly orientated tubular and acicular apatite crystals. Bt, biotite; Pl, plagioclase; Ap, apatite; Ttn, titanite; Hbl, hornblende; Mc, microcline; Qz, quartz; Zrn, zircon.

xenocrysts of plagioclase in MMEs favour the mineral disequilibration and mechanical transfer of quartz and plagioclase from the granitic host to the MMEs (Hibbard 1981, 1991; Barbarin 1990; Kumar & Pieru 2010). Late magmatic poikilitic quartz is also noticed with biotite, K-feldspar and apatite oikocrysts (Fig. 4c).

Geochemistry

Geochemical characteristics of granitoids and enclaves

The SiO_2 content in porphyritic granite (PG), granodiorite (GD), hybrid dioritic rocks (HD) and MMEs

Fig. 6. (**a**) Hornblende–biotite–titanite clots/cumulates in granodioritic MME. The triple-point junction between hornblende crystals and the biotite rim is present. (**b**) Early (Hbl-1) and late (Hbl-2) magmatic hornblendes were distinguished; also Hbl-2 has replaced and enclosed plagioclase in the central part of photomicrograph, whereas it replaced microcline in the NE part. The late magmatic hornblende is further enclosed by the titanite. (**c**) Perthitic microcline showing a poikilitic texture with numerous plagioclase inclusions (with Ab rim) in optical continuity. (**d**) Microcline enclosed by biotite crystals and exolved plagioclase, also the quenched apatite inclusions in biotite. (**e**) Quartz–hornblende, biotite ocelli in MME. (**f**) Ocellus quartz surrounded by biotite, hornblende, titanite and feldspars. Bt, biotite; Pl, plagioclase; Hbl, hornblende; Ttn, titanite; Mc, microcline; Ap, apatite; Qz, quartz.

has a wide range, and displays values of 61.83–68.83 wt%, 56.80–62.11 wt%, 52.91–56.64 wt% and 46.80–55.51 wt%, respectively (Table 1). $Fe_2O_3^T$, MgO, TiO_2, and P_2O_5 against SiO_2 shows sub-linear and negative correlation trends for host PG, GD and HD, and for MMEs to certain extent (Fig. 7). Negative correlation of TiO_2, $Fe_2O_3^T$ and MgO with increasing SiO_2 (Fig. 7) may partially be controlled by biotite fractionation during internal differentiation of MMEs (Kumar & Rino 2006). The K_2O content in PG, GD, HD and MME shows a scattered pattern, and indicates a poor but negative correlation with increasing SiO_2, except for HD (Fig. 7), and may correspond to the assimilation of K-feldspar in the aforesaid litho-units. Moreover, the Na_2O content in all of the samples exhibits a positive correlation with SiO_2 (Fig. 7). The scattered pattern of K_2O, Al_2O_3 and, to a certain extent, CaO in PG, GD, HD and MME may be due to a combined effect of diffusion and mineral sorting during magma mixing and mingling (Kumar & Pieru 2010). The MnO against SiO_2 diagram shows a distinctive cluster of different rock types (Fig. 7). The nearly linear variations may be due to the hybridization or mixing of mafic and felsic magma end members in various proportions (Kumar & Rino 2006; Ahmad 2011). It is interesting to note that the chemical variation diagrams show that the HD component is intermediate between the host granitoids and the MMEs.

PG, GD, HD and MMEs follow the boundary (separating tholeiitic and calc-alkaline) trend in the calc-alkaline field in AFM plots (Fig. 8a). The A/CNK (molar $Al_2O_3/CaO + Na_2O + K_2O$) v. A/NK (molar $Al_2O_3/(Na_2O + K_2O)$) diagram (Maniar & Piccoli 1989) shows the metaluminous nature of the MMEs and Nongpoh granitoids (Fig. 8b).

The host PG and GD contain a higher SiO_2 content and lower content of mafic oxides (viz. $Fe_2O_3^T$, MnO, MgO and TiO_2) compared to that of MMEs and HD, and vice versa. The P_2O_5, TiO_2 and MnO contents are moderately higher in MMEs than those of PG and GD, which may correspond to high modal apatite and titanite. SiO_2 v. alkalis plots (Peacock 1931) suggest a progressive enrichment in alkalis from PG to GD and HD to MMEs (Fig. 9a).

In a Peccerillo & Taylor (1976) diagram, MMEs and granitoids plot in the shoshonite field, except for four samples of PG and one sample of GD (Fig. 9b). It is observed that shoshonitic magma is supposed to characterize the transitional stage between calc-alkaline (I-type) and alkaline (A-type) magmas formed during late to post-collisional environments (Liégeois *et al.* 1998), and may be an evolved source from high-K calc-alkaline magma (Kumar & Pieru 2010). Moreover, shoshonite magma can be generated by melded processes of fractional crystallization and magma mixing of juxtaposed magma types (Kumar & Pieru 2010).

CIPW normative minerals were calculated using the 'GCD tool kit' software and aew presented in Table 2. Orthoclase has been documented as highest amongst all the normative minerals (MME, 31.95 wt%; HD, 24.72 wt%; GD, 26.83 wt%; and PG, 26.95 wt%). Anorthite content is similar in MMEs and granitoids, whereas the Ab wt% progressively increases from MMEs to host granitoids. The average normative quartz of the MMEs is 2.80 wt%; of HD is 7.14 wt%; of GD is 17.86 wt%; and of PG is 22.61 wt%, which suggests a relatively more mafic nature of the MMEs and HD, followed by the host GD and PG. Normative olivine appeared in only the MME-4 and MME-5 samples due to an enrichment of mafic minerals. Average normative titanite (MME, 2.24 wt%; HD, 2.33 wt%; GD, 1.63 wt%; and PG, 0.74 wt%) and apatite (MME, 1.75 wt%; HD, 1.48 wt%; GD, 1.55 wt%; and PG, 0.99 wt%) is similar to modal titanite in these litho-units. Normative diopside, hypersthene and hematite are least in PG and most in MMEs. Rutile and ilmenite do occur in minor quantities.

Trace elements and REE variations

Binary plots of trace elements v. silica are presented, and the concentrations of MME, HD, GD and PG are marked as outlined areas in Figure 10. Progressive enrichments of U and Th with increasing SiO_2 content can be noted for all samples, while high fields strength elements (HFSEs) Hf and Zr show considerable depletion in two isolated clusters, viz.: PG, GD and HD; and MME. Progressive depletion with increasing SiO_2 in V and Co from enclaves to host content suggests the mafic nature of the MMEs. Rb, Sr v. SiO_2 show a regular variation, but the deviation noted in data for Rb and Sr in a few samples of GD, HD and MME samples as their Rb and Sr content is more than the host PG. Such a remarkable increase in Rb content suggests a mechanical/chemical transfer of K from the host felsic magma (rich in K-feldspar) to the enclave-forming magma. The influx of K-rich fluids into MMEs leads to the formation of biotite, which contributes to the high Rb (Chen *et al.* 2009). Y shows the least variation amongst the samples, but a poorly developed negative correlation between Y and SiO_2 observed for the host and enclave samples may correspond to the early crystallization of titanite and apatite, and is consistent with the depletion of P_2O_5 with increasing SiO_2. Ni is highest in enclaves, and depleted in GD and PG, and decreases with increasing SiO_2, which is compatible with the mafic nature of hybrid rocks and MMEs. The Ba content is progressively depleted in samples of MMEs to HD, and shows a negative correlation with SiO_2. The concentrations of Nb show least variation amongst the four litho-units, but show isolated clusters of enclaves and host granitoids, and are depleted with increasing silica. The concentration of Pb is scattered when plotted against SiO_2.

Chondrite-normalized REE patterns of MME and host porphyritic granite indicate REE equilibration between them, and show highly fractionated LREE and nearly flat HREE patterns with a negative Eu

Table 1. *Major and trace element data of porphyritic granite (PG), granodiorite (GD), hybrid rock (HD) and mafic magmatic enclave (MME)*

Sample	MME-1	MME-2	MME-3	MME-4	MME-5	MME-6	MME-7	MME-8	MME-9	MME-10	HD-1	HD-2	HD-3	HD-4	HD-5	HD-6	HD-7
SiO_2	53.85	52.29	51.34	47.25	49.37	55.51	52.55	51.55	52.72	46.80	56.64	54.50	53.24	53.38	52.91	53.72	53.45
TiO_2	0.88	1.01	1.39	2.17	1.65	0.59	0.96	0.73	0.90	2.22	1.01	1.08	1.35	1.43	0.80	0.85	0.93
Al_2O_3	11.89	10.65	10.79	10.40	11.33	12.26	11.49	11.40	11.55	12.95	10.30	8.28	10.60	9.75	15.50	11.43	11.03
$Fe_2O_3^T$	6.71	7.51	6.47	12.20	7.47	4.87	7.69	7.26	6.83	9.78	6.35	7.21	7.81	7.28	7.31	7.11	7.43
MnO	0.11	0.12	0.10	0.35	0.13	0.10	0.12	0.12	0.12	0.10	0.11	0.12	0.13	0.11	0.12	0.11	0.11
MgO	8.98	10.15	11.54	10.73	10.92	2.59	9.01	8.89	9.28	10.36	9.58	16.11	10.38	11.86	5.75	9.45	10.64
CaO	5.42	6.96	4.87	2.55	5.37	2.29	8.03	8.32	8.62	6.54	5.89	3.03	5.44	5.62	6.33	8.29	6.30
Na_2O	0.96	0.86	0.74	0.32	1.04	2.21	1.01	1.62	1.30	0.82	1.16	0.85	1.36	0.92	2.00	1.33	1.29
K_2O	7.19	5.58	5.75	8.25	5.90	6.21	4.09	3.98	3.94	4.96	5.04	4.20	4.11	4.60	3.43	3.74	4.52
P_2O_5	0.75	0.72	0.81	1.73	1.08	0.11	0.52	0.54	0.52	1.58	0.63	0.36	0.72	0.74	0.60	0.60	0.65
LOI	0.79	1.16	3.58	1.88	2.77	3.47	2.89	3.18	2.65	1.05	0.82	3.03	2.82	2.66	3.24	2.05	1.98
Ba	1960	1848	1624	1666	1533	860	1648	2023	1625	2696	1504	1023	1515	1998	1923	2265	2065
Rb	419	213	192	857	251	369	146	68	110	145	180	119	184	154	112	79	115
Sr	660	524	627	110	787	291	769	1392	841	580	628	406	491	402	1239	1367	1186
Y	23	31	29	210	39	42	29	28	30	47	28	17	39	37	38	29	26
Zr	657	475	691	877	1014	281	452	335	452	394	524	403	514	459	347	380	434
Nb	24	24	30	96	34	15	11	8	11	27	22	17	26	15	5	4	14
Th	11	16	7	13	8	22	15	30	14	21	18	16	15	24	35	21	23
Pb	39	39	25	20	25	115	27	55	40	9	29	30	24	35	43	57	56
Ga	17	21	15	22	19	39	15	16	16	23	15	16	19	20	22	19	19
Zn	113	114	98	387	144	127	109	103	92	125	98	105	124	102	122	100	102
Cu	10	38	58	57	67	13	41	54	68	44	55	59	28	33	28	55	68
Ni	283	256	149	94	140	13	162	169	159	192	206	476	218	146	69	127	181
V	238	250	271	362	335	143	261	257	263	430	229	186	250	283	222	309	267
Cr	489	670	411	112	362	37	249	376	331	170	517	1150	428	492	42	399	235
Hf	15.36	13.45	22.32	14.17	18.12	7.70	14.88	7.40	13.17	1.51	13.92	11.52	15.32	14.88	2.21	8.94	11.66
Sc	26	26	19	71	33	26	39	29	43	37	26	17	23	24	31	32	30
Ta	0.33	0.40	0.74	1.99	0.40	bd	0.50	bd	0.60	0.49	1.38	9.15	1.33	0.68	0.69	0.42	0.84
Co	32	40	31	36	29	8	32	40	35	60	33	44	32	39	40	31	37
Be	1.72	2.06	2.16	2.64	1.71	4.00	4.22	3.23	3.95	2.10	1.94	4.53	4.55	5.69	2.51	2.20	3.31
U	2.73	1.12	0.58	6.37	1.16	6.67	0.80	2.02	1.55	1.51	1.69	2.29	2.26	2.58	1.71	1.83	2.06
W	bd	bd	bd	bd	bd	225.59	bd	0.57	0.71	0.84	bd	1.05	0.91	0.91	0.89	0.89	0.87
Sn	1.95	4.70	6.23	10.81	5.07	6.79	3.87	1.48	2.69	3.15	2.81	2.92	3.51	3.73	3.35	1.38	1.27
Mo	3.70	0.73	bd	bd	2.07	7.69	1.13	1.99	1.65	1.43	bd	1.01	1.68	2.07	1.45	0.84	1.57
La	67.60	77.60	74.33	281.04	110.38	94.92	85.92	142.22	122.56	121.49	74.81	45.53	97.85	67.96	112.44	158.37	134
Ce	122.99	139.61	135.13	494.75	197.13	160.39	157.41	256.93	200.42	222.38	137.10	81.50	173.21	116.70	175.01	287.07	238.89
Pr	15.16	17.98	17.48	71.05	24.42	19.97	18.78	30.62	26.00	30.09	16.72	9.82	22.16	15.73	28.32	36.25	30.38
Nd	55.87	68.16	64.44	269.15	90.74	69.67	62.62	101.36	83.73	91.12	59.67	18.34	70.31	49.78	84.33	213.39	177.21
Sm	10.35	12.47	12.19	52.79	16.33	13.11	12.3	19.69	17.06	19.18	11.47	10.17	14.65	10.89	18.23	20.36	16.93
Eu	2.09	2.47	2.46	11.66	3.02	2.79	2.62	3.79	3.22	4.27	2.26	1.33	3.08	2.49	4.52	4.19	3.55
Gd	7.74	9.39	9.62	47.83	11.65	11.35	10.34	13.53	11.98	16.85	8.84	6.22	12.55	10.43	17.54	13.84	12.14
Tb	0.88	1.15	1.19	7.04	1.43	1.38	1.27	1.48	1.39	2.14	1.08	0.81	1.61	1.47	2.32	1.48	1.36
Dy	4.23	5.59	6.01	39.47	6.93	7.64	6.46	6.56	6.74	10.42	5.20	4.36	8.40	7.86	11.25	6.40	6.04
Ho	0.70	1.01	1.04	7.49	1.25	1.42	1.18	1.11	1.19	1.84	0.94	0.79	1.51	1.41	2.03	1.09	1.03
Er	1.85	2.77	2.91	22.29	3.60	4.29	3.42	3.23	3.39	4.90	2.55	2.45	4.28	3.86	5.12	2.92	2.84
Tm	bd	bd	bd	3.66	0.61	0.70	0.54	bd	0.50	0.73	bd	bd	0.65	0.58	0.85	bd	bd
Yb	1.75	2.65	2.77	18.78	3.69	4.57	3.24	2.90	3.02	4.24	2.42	2.42	3.91	3.33	4.56	2.80	2.66
Lu	bd	bd	bd	2.90	0.61	0.73	0.52	bd	bd	0.67	bd	bd	0.61	0.53	0.86	bd	bd
Nd/Th	5.08	4.26	9.21	20.70	11.34	3.17	4.17	3.38	5.98	4.34	3.32	1.15	4.69	2.07	2.41	10.16	7.70

Sample	GD-1	GD-2	GD-3	GD-4	GD-5	GD-6	GD-7	PG-1	PG-2	PG-3	PG-4	PG-5	PG-6	PG-7	PG-8	PG-9	PG-10
SiO_2	59.72	56.80	56.89	59.74	62.74	62.11	61.87	61.83	62.57	66.82	63.91	63.70	64.33	65.94	64.78	65.07	68.83
TiO_2	0.95	1.00	1.14	0.99	0.88	0.83	0.79	0.68	0.73	0.50	0.63	0.59	0.63	0.56	0.56	0.61	0.50
Al_2O_3	12.60	13.35	11.09	11.00	12.05	13.09	13.08	14.36	14.21	13.38	14.26	14.80	13.11	13.98	14.10	13.52	12.17
$Fe_2O_3^T$	7.72	6.54	7.00	6.62	5.30	5.14	5.35	5.10	5.19	3.72	4.33	4.14	4.42	3.55	3.88	4.16	3.59
MnO	0.09	0.11	0.12	0.13	0.11	0.08	0.12	0.09	0.09	0.07	0.08	0.07	0.09	0.06	0.08	0.09	0.07
MgO	4.81	5.53	8.07	6.00	4.14	4.55	4.78	3.13	3.45	2.68	2.25	2.34	2.68	2.29	2.22	2.68	2.16
CaO	3.19	4.52	4.88	4.97	3.69	3.18	4.70	4.15	3.28	2.98	3.93	3.69	3.46	2.91	3.14	3.13	2.67
Na_2O	2.07	1.63	1.19	1.69	1.78	2.26	2.11	2.58	1.97	2.66	2.59	2.78	2.55	2.86	2.65	2.78	2.75
K_2O	4.57	5.69	5.22	3.97	4.49	4.44	4.64	4.90	5.66	4.65	4.61	4.88	3.97	4.22	4.67	3.85	4.19
P_2O_5	0.49	0.93	0.87	0.76	0.56	0.39	0.62	0.58	0.51	0.32	0.49	0.48	0.38	0.35	0.41	0.39	0.29
LOI	1.09	1.57	1.40	1.69	2.50	1.24	1.81	0.96	1.27	1.03	1.74	1.29	2.65	1.26	1.20	1.58	1.38
Ba	1088	1675	1792	924	1652	1041	1929	1930	2063	1477	1962	1853	1182	1281	1424	1125	1008
Rb	234	218	267	234	187	207	240	205	215	195	161	191	208	235	223	227	249
Sr	345	1750	1030	590	760	407	681	992	935	708	1002	955	598	553	580	542	439
Y	50	54	42	35	48	52	40	65	66	47	53	52	65	42	102	48	37
Zr	768	802	490	693	505	618	487	591	547	396	509	457	432	325	393	365	317
Nb	37	20	22	31	22	30	19	26	24	22	19	18	22	16	31	18	17
Th	70	31	4	9	14	64	21	53	63	43	38	36	36	43	54	39	34
Pb	34	69	57	40	63	34	37	47	66	49	51	50	46	49	47	35	45
Ga	21	24	22	24	28	25	25	23	27	21	35	39	33	38	26	32	31
Zn	148	116	129	122	96	127	112	119	145	93	94	90	84	82	91	89	94
Cu	23	31	32	16	34	21	14	56	36	23	26	27	27	19	26	14	12
Ni	19	106	228	80	50	16	84	18	26	6	13	19	31	21	19	28	18
V	125	234	274	242	180	120	172	145	145	101	112	116	98	79	89	96	73
Cr	32	85	208	278	99	31	147	27	37	25	bd	17	48	41	24	48	32
Hf	17.47	1.52	10.45	15.27	10.54	15.97	12.67	13.18	15.57	10.88	11.21	8.38	9.56	10.61	9.50	11.57	8.97
Sc	12	28	33	26	25	14	21	17	22	14	13	11	15	11	20	14	12
Ta	1.74	bd	bd	bd	bd	1.96	bd	0.94	3.38	2.60	bd	bd	bd	bd	bd	bd	bd
Co	13	26	25	24	14	5	18	10	12	6	19	10	15	8	16	24	18
Be	2.12	3.84	2.87	4.60	3.21	2.13	3.19	1.95	3.64	2.03	2.63	2.64	3.79	4.37	3.95	4.68	3.90
U	2.89	1.59	1.14	3.35	1.86	2.27	3.62	4.49	3.96	4.50	2.22	2.10	4.61	5.06	5.41	5.22	5.68
W	bd	1.30	bd	bd	0.63	bd	51.88	bd	1.67	bd	126.23	101.82	108.4	109.52	106.41	141.05	154.44
Sn	2.17	2.96	1.45	3.90	2.47	2.57	1.82	7.28	4.91	5.55	1.81	1.32	2.46	3.38	6.03	4.06	3.16
Mo	1.85	1.89	1.17	2.50	2.08	bd	3.35	bd	1.48	bd	2.99	5.94	1.63	8.77	6.45	12.07	8.72
La	211.08	210.99	117.14	66.35	151.31	181.82	157.74	209.64	259.6	152.88	173.17	142.92	137.54	160.86	157.26	158.66	107.30
Ce	379.90	254.01	199.19	143.96	245.79	325.05	227.16	378.26	462.04	274.25	291.96	232.37	213.99	261.56	255.25	248.39	171.13
Pr	42.70	45.18	26.95	20.52	32.41	36.18	28.73	43.52	53.15	31.41	38.11	29.13	26.71	31.58	31.96	30.42	20.80
Nd	159.28	157.26	97.09	76.07	113.81	134.62	99.90	166.13	204.08	117.6	133.05	98.35	91.86	103.70	108.25	101.16	70.46
Sm	28.60	23.28	15.31	13.12	17.84	24.88	16.30	29.20	34.47	21.95	21.97	16.07	15.72	15.93	19.12	15.62	11.76
Eu	4.78	4.96	3.15	2.87	3.47	4.07	3.30	5.11	6.12	3.79	3.86	2.85	2.80	2.98	3.12	2.81	2.12
Gd	18.77	19.00	12.41	10.59	14.92	16.50	13.51	19.74	23.28	14.57	17.85	13.37	13.67	13.17	17.52	13.05	10.24
Tb	2.25	2.00	1.35	1.30	1.74	1.97	1.54	2.40	2.83	1.77	2.00	1.51	1.68	1.50	2.32	1.53	1.25
Dy	10.64	8.68	5.99	6.26	8.26	9.40	7.60	11.95	13.57	8.86	9.53	7.75	9.04	7.66	13.22	8.11	6.70
Ho	1.88	1.51	1.06	1.13	1.48	1.73	1.37	2.13	2.41	1.61	1.67	1.43	1.74	1.44	2.56	1.53	1.25
Er	5.15	4.21	2.93	3.22	4.14	4.97	4.05	6.38	6.55	4.76	4.75	4.13	5.09	4.14	7.86	4.52	3.75
Tm	0.83	0.67	bd	0.53	0.67	0.79	0.64	1.12	1.07	0.83	0.70	0.65	0.85	0.67	1.26	0.74	0.59
Yb	4.76	4.19	2.91	3.35	4.14	4.43	4.04	6.68	6.70	5.14	4.20	3.91	5.25	4.26	7.49	4.72	4.01
Lu	0.75	0.66	bd	0.52	0.63	0.71	0.64	1.11	1.07	0.86	0.64	0.61	0.80	0.65	1.15	0.74	0.63
Nd/Th	2.28	5.07	24.27	8.45	8.13	2.10	4.76	3.13	3.24	2.73	3.50	2.73	2.55	2.41	2.00	2.59	2.07

‘bd’ indicates that a component is below detection. LOI, loss on ignition.

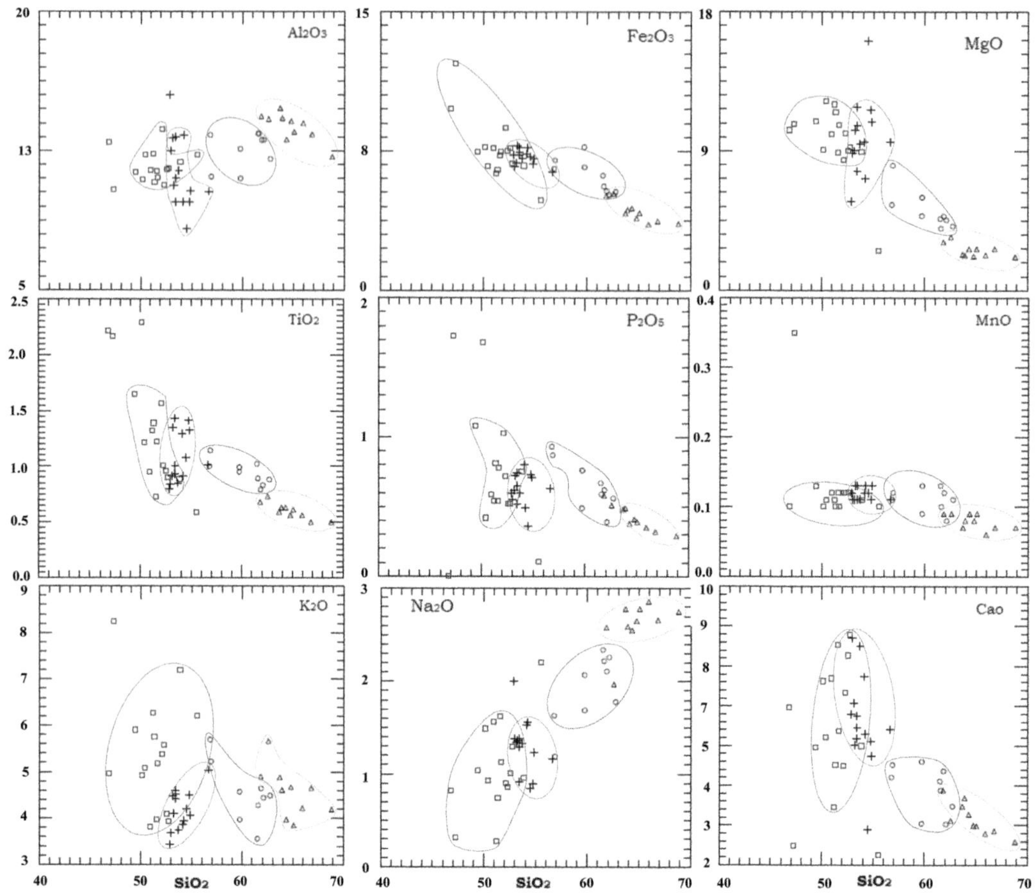

Fig. 7. Harkers' variation diagrams – SiO_2 v. major oxides plotted for MME (□), HD (+), GD (○) and PG (Δ).

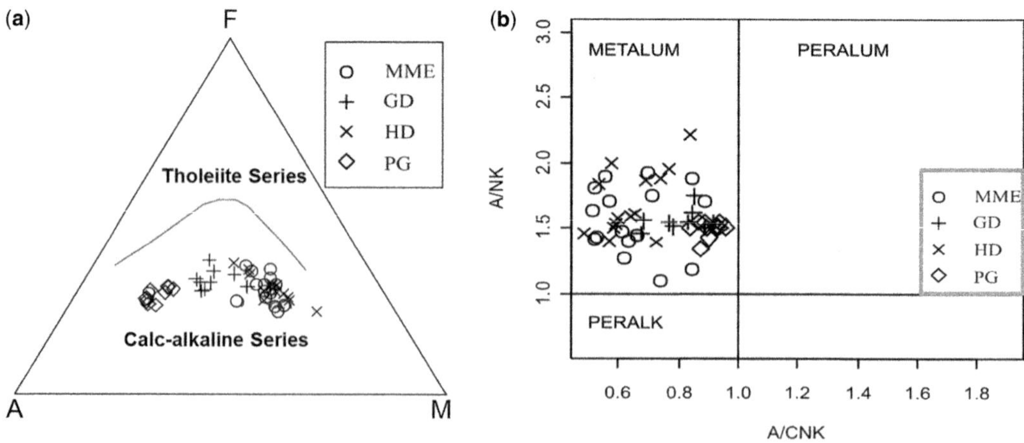

Fig. 8. (**a**) A/CNK v. A/NK diagram (Maniar & Piccoli 1989), showing the metaluminous nature of enclaves and granitoids. (**b**) AFM triangular diagram (Irvine & Baragar 1971) showing enclaves and granitoids following the boundary line separating tholeiitic and calk-alkaline series.

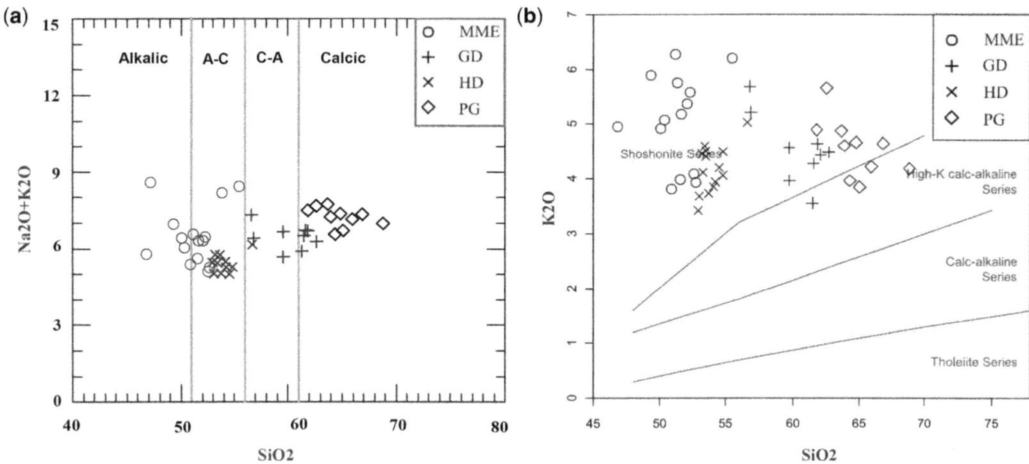

Fig. 9. (**a**) SiO_2 v. alkalis plot after Peacock (1931) showing progressive enrichment in alkalis from granite to enclave. (**b**) SiO_2–K_2O plot after Peccerillo & Taylor (1976).

anomaly (Fig. 11). MME and host porphyritic granite show a low and moderate degree of negative Eu anomaly, viz. Eu* = 0.63–0.69 and Eu* = 0.52–0.63, respectively. A negative europium anomaly is an indication of the early removal of plagioclase feldspar from the melt by crystal fractionation. Moreover, the degree of Eu* is slightly higher in the host porphyritic granite than in the MMEs, which indicates oxidation of an existing Eu^{2+} in the plagioclase of the MME to Eu^{3+}, which is in a higher oxidation state due to an interaction with more evolved fO_2-rich felsic granitic melt. The re-equilibration process in Eu is mainly contributed by the oxygen buffering in an open system. This is further supported by field evidence, as MME contains mechanically transferred xenocrysts of plagioclase and/or chemical interaction of felsic host granitoids and the MMEs. HREE enrichment in the MMEs can be explained by the concentration of hornblende, apatite and monazite or, more likely, by preferential partitioning between felsic and mafic liquids (Watson 1976; Ryerson & Hess 1978).

Monazite geochronology by microprobe dating

Monazite, a LREE-bearing phosphate mineral (Ce, La, Nd, Th)PO_4, occurs as an accessory phase in biotite-rich MMEs in the Nongpoh granitoids. In addition to LREE, monazite also contains Th as major element, Y as minor, with U and Pb concentrations as trace components. Three grains of monazite, with sizes ranging from 25 to 40 µm, were selected for chemical (U–Th–Pb) dating by microprobe (Fig. 12a–c). The multi-element X-ray mapping of the three grains (*c.* 40, 40 and 25 µm) for Th, U and Y have been studied for compositional variations in monazite grains (Fig. 12d–l). X-ray maps indicate distinguishable patchy concentrations of Th and Y (Fig. 12d, f & j–l), and minor but patchy variations in the distribution of the U concentration throughout monazite grains (Fig. 12g–i). Apart from the variation in concentration of elements within individual monazite grains, these compositional maps indicate the absence of zoning. However, geochronological results are not strongly affected by variations in monazite composition (Williams *et al.* 2006). The domains of homogeneous composition in monazite grains were demarcated from X-ray maps and targeted for chemical dating (Williams *et al.* 2006; Jeere & Kundu 2010). The chemical composition of the monazite grains present in the MMEs and computed ages with errors are shown in Table 3. In monazite, the Th concentration varies from 6.51 to 16.53 wt% (average 9.98 wt%), the Y concentration varies from 0.61 to 3.46 wt% (average 2.41 wt%) and the U concentration varies from 0.20 to 0.65 wt% (average 0.35 wt%). The rare earth element (REEs), namely La_2O_3, concentration varies from 9.31 to 13.25 wt% (11.82 wt%), the Ce_2O_3 concentration varies from 21.46 to 28.96 wt% (25.72 wt%), the Pr_2O_3 concentration varies from 2.96 to 3.76 wt% (3.43 wt%), the Nd_2O_3 concentration varies from 10.29 to 13.05 wt% (11.79 wt%), the Gd_2O_3 concentration varies from 2.84 to 3.64 wt% (3.38 wt%) and the Dy_2O_3 concentration varies from 0.18 to 0.63 wt% (0.88 wt%). Monazite is highly rich in LREE concentrations. The total LREE in three grains of

Table 2. *Normative calculations corresponding to Table 1 samples*

Mineral	PG-1	PG-2	PG-3	PG-4	PG-5	PG-6	PG-7	PG-8	PG-9	PG-10	Average (PG)
Q	16.94	18.72	24.47	21.38	19.17	24.42	24.49	22.64	24.56	29.29	2.13
C	0.00	0.10	0.00	0.00	0.00	0.00	0.25	0.00	0.02	0.00	0.01
Or	28.96	33.45	27.48	27.24	28.84	23.46	24.94	27.60	22.75	24.76	30.06
Ab	21.83	16.67	22.51	21.92	23.52	21.58	24.20	22.42	23.52	23.27	8.69
An	13.13	12.94	10.83	13.67	13.49	12.60	12.15	12.79	12.98	8.49	12.85
Di	1.29	0.00	0.31	0.58	0.00	0.19	0.00	0.00	0.00	1.09	7.66
Hy	7.20	8.59	6.53	5.34	5.83	6.59	5.70	5.53	6.68	4.88	19.39
Ol	0.00	0.00	0.00	0.00	0.00	0.00	0.00	0.00	0.00	0.00	1.74
Il	0.19	0.19	0.15	0.17	0.15	0.19	0.13	0.17	0.19	0.15	0.24
Hm	5.10	5.19	3.72	4.33	4.14	4.42	3.55	3.88	4.16	3.59	7.49
Tn	1.42	0.00	1.03	1.33	1.18	1.30	0.00	0.08	0.00	1.03	2.79
Ru	0.00	0.63	0.00	0.00	0.03	0.00	0.49	0.44	0.51	0.00	0.22
Ap	1.37	1.21	0.76	1.16	1.14	0.90	0.83	0.97	0.92	0.69	1.55
Sum	97.43	97.69	97.80	97.11	97.50	95.64	96.74	96.51	96.30	97.24	94.84

Mineral	GD-1	GD-2	GD-3	GD-4	GD-5	GD-6	GD-7	GD-8	GD-9	Average (GD)
Q	17.79	11.11	11.91	19.21	23.36	19.64	17.85	20.23	19.67	2.07
C	0.00	0.00	0.00	0.00	0.00	0.00	0.00	0.00	0.00	0.01
Or	27.01	33.63	30.85	23.46	26.54	26.24	27.42	21.04	25.29	29.42
Ab	17.52	13.79	10.07	14.30	15.06	19.12	17.85	19.80	18.79	9.10
An	11.59	12.30	9.50	10.70	11.63	12.46	12.51	15.63	14.04	12.78
Di	0.00	0.77	4.30	4.71	0.30	0.00	3.48	0.00	0.00	8.77
Hy	11.98	13.42	18.11	12.76	10.17	11.33	10.29	11.51	9.99	18.93
Ol	0.00	0.00	0.00	0.00	0.00	0.00	0.00	0.00	0.00	1.62
Il	0.19	0.24	0.26	0.28	0.24	0.17	0.26	0.28	0.21	0.24
Hm	7.72	6.54	7.00	6.62	5.30	5.14	5.35	6.18	5.60	7.48
Tn	0.73	2.15	2.47	2.07	1.86	0.54	1.61	1.32	1.90	2.70
Ru	0.55	0.00	0.00	0.00	0.00	0.52	0.00	0.34	0.00	0.19
Ap	1.16	2.20	2.06	1.80	1.33	0.92	1.47	1.59	1.40	1.53
Sum	96.24	96.15	96.52	95.91	95.77	96.09	98.09	97.91	96.89	94.84

(Continued)

Table 2. (*Continued*)

Mineral	HD-1	HD-2	HD-3	HD-4	HD-5	HD-6	HD-7	HD-8	HD-9	HD-10	HD-11	HD-12	HD-13	HD-14	Average (HD)
Q	9.17	5.35	7.01	5.85	8.63	6.49	4.59	9.96	7.83	7.51	9.08	8.02	5.55	4.91	3.48
C	0.00	0.00	0.00	0.00	0.00	0.00	0.00	0.00	0.00	0.00	0.00	0.00	0.00	0.00	0.01
Or	29.79	24.82	24.29	27.19	20.27	22.10	26.71	23.28	26.06	22.81	23.99	26.59	21.81	26.42	29.24
Ab	9.82	7.19	11.51	7.79	16.92	11.25	10.92	13.20	11.68	12.95	10.41	7.62	11.68	11.34	10.00
An	8.01	6.37	10.68	8.89	23.19	14.17	10.96	17.79	16.91	8.28	10.70	9.16	17.01	16.75	12.71
Di	10.90	2.35	5.77	7.48	1.54	15.96	10.31	3.87	4.82	14.97	4.69	7.14	14.92	7.44	8.44
Hy	18.81	39.04	23.18	26.07	13.61	16.14	21.72	16.19	16.87	16.97	24.95	25.76	15.25	19.04	17.79
Ol	0.00	0.00	0.00	0.00	0.00	0.00	0.00	0.00	0.00	0.00	0.00	0.00	0.00	0.00	1.12
Il	0.24	0.26	0.28	0.24	0.26	0.24	0.24	0.28	0.28	0.26	0.28	0.24	0.24	0.24	0.25
Hm	6.35	7.21	7.81	7.28	7.31	7.11	7.43	7.68	7.68	7.37	7.11	6.82	6.64	6.84	7.41
Tn	2.18	2.32	2.96	3.21	1.63	1.78	1.98	1.88	2.10	2.84	2.88	3.16	1.76	1.93	2.27
Ru	0.00	0.00	0.00	0.00	0.00	0.00	0.00	0.00	0.00	0.00	0.00	0.00	0.00	0.00	0.24
Ap	1.49	0.85	1.71	1.75	1.42	1.42	1.54	1.16	1.23	1.89	1.68	1.73	1.30	1.70	1.47
Sum	96.75	95.76	95.18	95.73	94.78	96.66	96.39	95.29	95.45	95.84	95.77	96.23	96.15	96.36	94.41

Mineral	MME-1	MME-2	MME-3	MME-4	MME-5	MME-6	MME-7	MME-8	MME-9	MME-10	MME-11	MME-12	MME-13	MME-14	MME-15	MME-16	Average (MME)
Q	1.12	2.48	1.78	0.00	0.00	2.04	11.61	6.17	6.54	2.28	2.88	4.71	0.00	1.71	0.00	1.49	2.93
C	0.00	0.00	0.00	0.45	0.00	0.00	0.00	0.02	0.00	0.00	0.00	0.00	0.00	0.00	0.00	0.00	0.03
Or	42.49	32.98	33.98	48.76	34.87	30.61	36.70	31.79	24.17	22.58	23.52	23.28	29.31	29.08	30.02	37.05	31.18
Ab	8.12	7.28	6.26	2.71	8.80	9.56	18.70	7.62	8.55	13.20	13.71	11.00	6.94	12.61	7.87	2.37	9.26
An	6.90	8.72	9.14	1.35	8.82	9.73	5.19	17.33	14.74	12.96	12.08	14.04	17.01	8.69	14.33	13.84	11.34
Di	9.70	14.06	4.11	0.00	4.30	8.05	2.95	0.00	14.66	13.17	18.37	17.64	6.31	6.94	5.66	0.00	7.33
Hy	17.87	18.77	26.84	17.49	19.61	22.89	5.08	21.07	15.65	19.08	13.63	14.94	12.60	19.5	23.55	29.94	18.70
Ol	0.00	0.00	0.00	6.47	3.92	0.00	0.00	0.00	0.00	0.00	0.00	0.00	7.21	0.00	2.99	0.00	1.46
Il	0.24	0.26	0.21	0.75	0.28	0.21	0.21	0.26	0.26	0.26	0.26	0.26	0.21	0.21	0.24	0.24	0.27
Hm	6.71	7.51	6.47	12.20	7.47	7.47	4.87	8.74	7.69	7.69	7.26	6.83	9.78	7.72	6.70	6.29	7.65
Tn	1.86	2.19	3.14	0.00	3.69	2.72	1.17	0.00	2.03	2.00	1.46	1.88	5.17	5.35	2.67	0.56	2.27
Ru	0.00	0.00	0.00	1.78	0.00	0.00	0.00	1.44	0.00	0.00	0.00	0.00	0.00	0.00	0.00	0.97	0.30
Ap	1.78	1.71	1.92	4.09	2.56	1.85	0.26	2.44	1.23	1.39	1.28	1.23	0.00	3.98	0.99	1.28	1.75
Sum	96.78	95.89	93.84	96.05	94.32	95.13	86.75	96.88	95.50	94.60	94.44	95.81	94.53	95.73	95.01	94.02	94.49

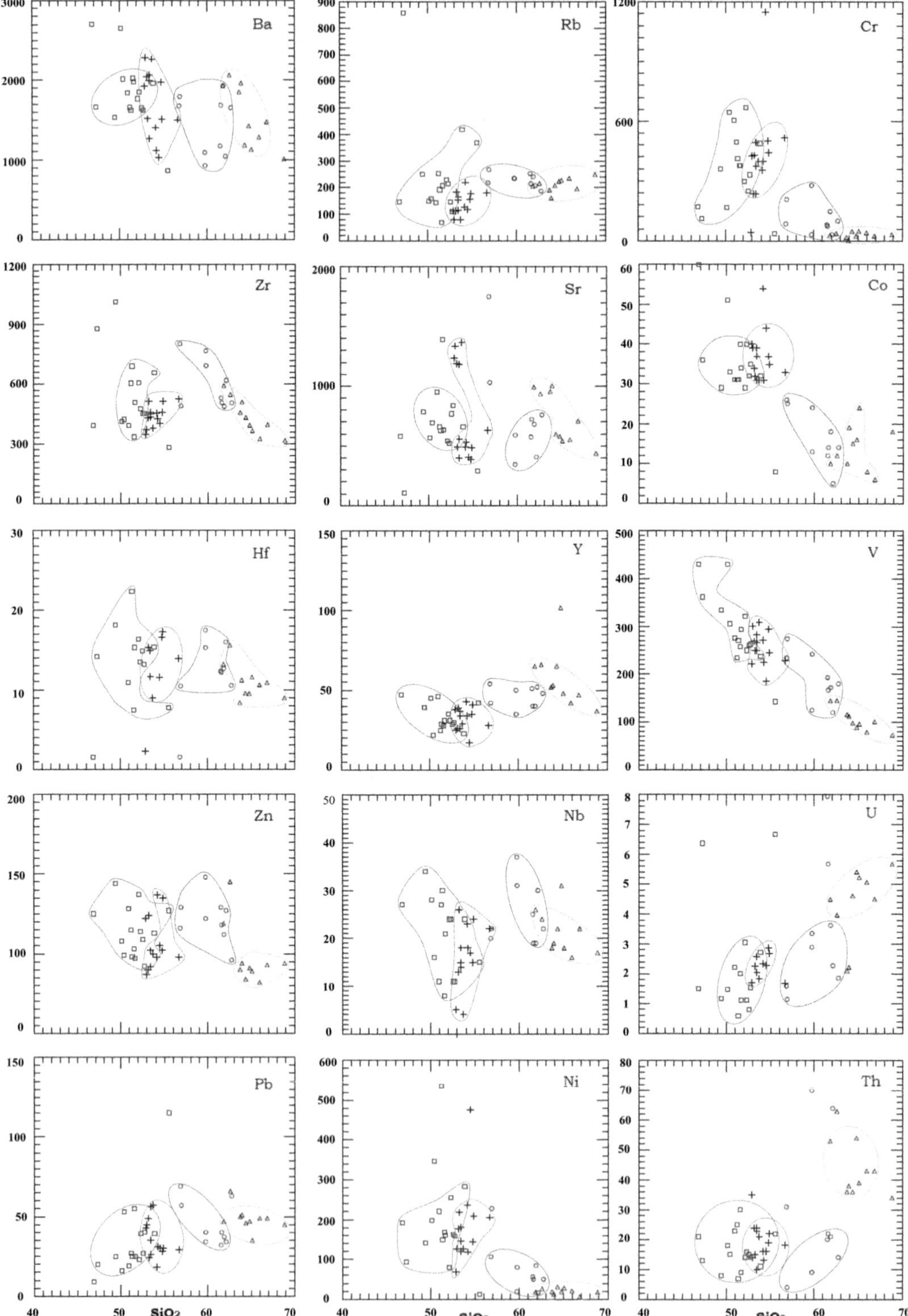

Fig. 10. Binary plots of trace elements v. SiO_2 v. trace elements plotted for MME (□), HD (+), GD (○) and PG (Δ).

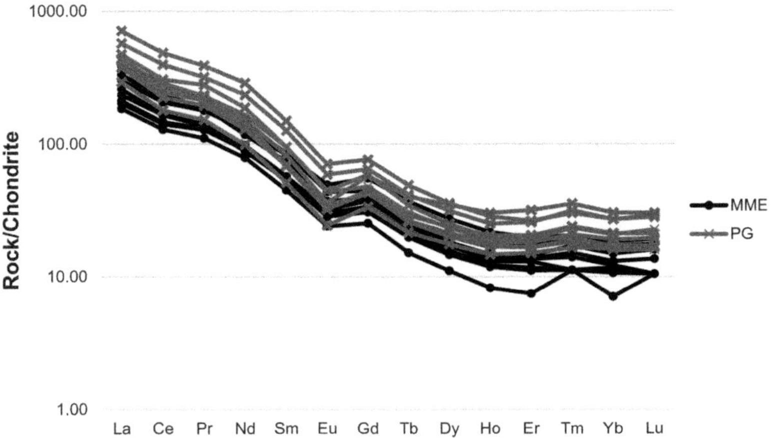

Fig. 11. Chondrite-normalized (McDonough & Sun 1995) REE diagram for porphyritic granite and MMEs showing inclined LREE and flat HREE patterns.

monazite (La_2O_3, Ce_2O_3, Pr_2O_3, Nd_2O_3 and Gd_2O_3) varies from 46.86 to 62.38 wt% (average 56.13 wt%).

Weighted average and probability density plots for analysed ages in three monazite grains yielded an age of 501 ± 5 Ma with a 95% probability of fit (Fig. 13a, b). The bars in Figure 13b represent 2σ, short-term random errors determined by propagating uncertainties associated with each trace element (Th, U, Pb) through the age equation (Williams *et al.* 2006). The shape of monazite grains and its association with primary crystallized minerals of the MMEs suggest that the microprobe chemical date is the crystallization date of dioritic MMEs, as the chances of monazite re-equilibration with granitic magma is minimal.

Discussion

The occurrence of polygenic swarms of MMEs, composite MMEs, disrupted syn-magmatic dykes, petrographical studies and geochemical evidence strongly suggest magma hybridization during the evolution of Nongpoh granitoids. The circular and rounded shapes of mafic enclaves with sharp or crenulated contacts were probably produced by the mingling or heterogeneous mixing of hot mafic magma within cooler felsic magma. This indicates that semi-liquid mafic fractions were emplaced before the solidification of liquefied granitic magma, which is also evidenced by the swerving of phenocrysts along enclave margins and tongues of granite in MMEs (Figs 2a & 3a). The presence of a rounded MME with K-feldspar phenocrysts swerving around it also suggests the mingling and emplacement of MMEs before solidification of the granitic magma. Sword-like MMEs with chilling zones, and raft- and dyke-like MMEs cutting across the primary magmatic foliation, indicate emplacement of the mafic material after the solidification of the granitic magma. The double MME (i.e. presence of a microgranular enclave in a porphyritic enclave) is direct evidence of undercooling of a hotter mafic magma in a relatively cooler hybrid magma followed by undercooling of composite mafic globules into a relatively cooler felsic host (Fig. 2e). This strongly suggests that a hybrid magma zone existed at a relatively deeper level in the magma chamber and was subsequently in-fluxed by a fresh mafic magma, prior to the commencement of the mingling process (Kumar 2010). The presence of a granitic tongue and cuspate margin (convex towards composite porphyritic MME) suggest multistage magma hybridization: that is, the emplacement of a hotter mafic magma globule (microgranular enclave) into a hybrid magma (porphyritic enclave) and later injection of a fresh felsic magma as tongues in composite MMEs. Syn-magmatic mafic dykes are emplaced along the weak zones produced by cooling and crystallization of granitoids melt (Kumar 2010) which is fragmented and disrupted due to the mechanical interaction with the felsic host, and with increasing viscosity contrast the shapes of enclaves become rounded to angular (Pitcher 1991; Barbarin & Didier 1992). Elongated MMEs and the alignment of mafic minerals parallel to those of the host porphyritic granite suggest their crystallization from mafic magma that was coeval with the enclosing granitic magma. The numerous dendritic back-veining of felsic melt into mafic magma is due to the injection of low-viscosity mafic magma into crystallizing more viscous felsic magmas. The latent heat of injecting

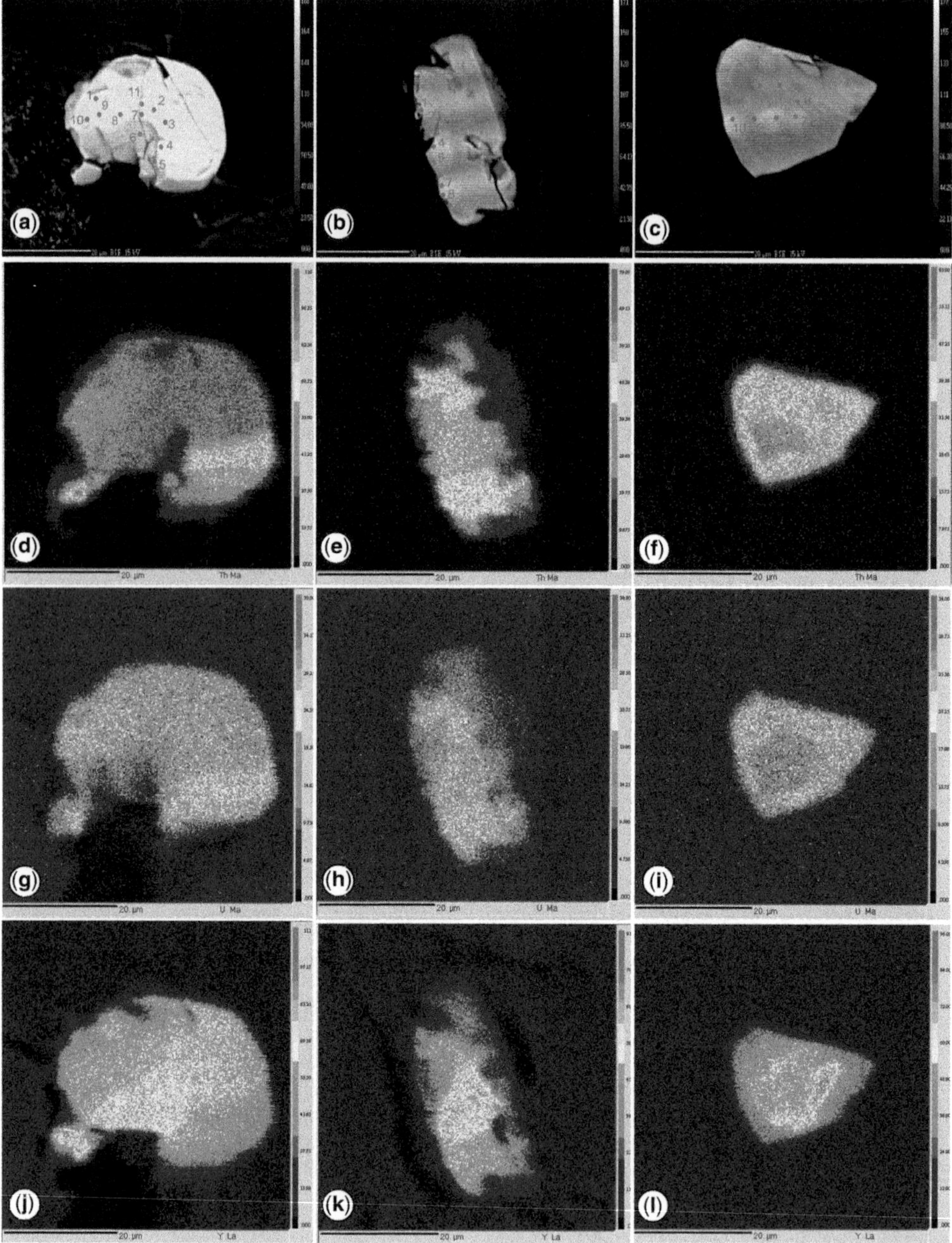

Fig. 12. (**a**)–(**c**) BSE images of monazite grains with analysis points in MMEs. (a) Points selected randomly. (b) Points selected in a linear pattern. (c) Points selected in a grid pattern. (**d**)–(**l**) Compositional maps of monazite grains show variations in (d–f) Th concentrations, (g–i) U concentrations and (j–l) Y concentrations within one, two and three monazite grains, respectively.

syn-plutonic mafic dykes probably induces local remelting of the host magma, and the resultant melts can penetrate the crystallizing mafic magma as back-veining, as observed (Jayananda *et al.* 2009).

The presence of coexisting mixed apatite morphologies (viz., stubby, tubular and acicular apatite) suggests magma mixing. Acicular apatite crystals occur as a result of quenching of mafic magma and

Table 3. *Representative electron microprobe compositions of monazite grains present in MMEs*

Dataset/ point	P_2O_5	ThO_2	K_2O	CaO	Y_2O_3	SiO_2	La_2O_3	Ce_2O_3	Pr_2O_3	Nd_2O_3	Gd_2O_3	Dy_2O_3	PbO	UO_2	Total	Age (Ma)	Age error
1/1	26.53	10.19	0.00	1.43	2.92	2.12	11.04	24.55	3.28	11.58	3.33	0.50	0.26	0.44	98.17	515	29
1/2	29.18	8.91	0.00	1.33	3.20	1.44	11.35	25.79	3.45	12.12	3.49	0.56	0.22	0.43	101.47	499	31
1/3	28.66	8.89	0.00	1.29	3.24	1.54	11.50	25.82	3.56	11.96	3.43	0.55	0.21	0.41	101.06	489	30
1/4	29.18	8.87	0.00	1.28	3.30	1.46	11.31	25.58	3.49	12.23	3.47	0.57	0.22	0.43	101.37	504	31
1/5	24.30	16.53	0.00	1.66	2.59	3.69	9.31	21.46	2.96	10.29	2.84	0.33	0.42	0.65	97.02	527	22
1/6	27.48	8.63	0.00	1.31	3.46	1.85	11.02	24.04	3.17	11.69	3.37	0.54	0.21	0.44	97.20	492	30
1/7	29.13	8.98	0.01	1.33	3.20	1.47	11.53	25.77	3.53	12.12	3.51	0.61	0.22	0.43	101.83	498	30
1/8	28.60	9.43	0.01	1.26	2.80	1.58	11.67	26.00	3.50	12.05	3.41	0.51	0.22	0.40	101.42	489	30
1/9	26.94	11.20	0.00	1.44	2.57	2.11	11.12	24.94	3.31	11.60	3.29	0.48	0.28	0.46	99.74	523	28
1/10	28.08	12.11	0.01	1.52	2.67	2.25	11.09	24.35	3.34	11.46	3.24	0.50	0.29	0.47	101.36	503	26
1/11	28.99	8.87	0.00	1.31	3.17	1.54	11.57	25.64	3.52	12.15	3.44	0.56	0.22	0.43	101.40	495	31
2/1	27.67	8.02	0.00	0.75	0.61	1.98	13.13	28.96	3.76	13.05	3.48	0.18	0.16	0.20	101.95	439	34
2/2	27.54	9.53	0.00	1.03	1.24	2.06	12.32	27.76	3.71	12.50	3.47	0.36	0.19	0.19	101.89	447	30
2/3	27.61	9.73	0.00	0.85	1.50	2.15	12.18	26.64	3.59	12.07	3.48	0.39	0.22	0.23	100.65	490	30
2/4	28.12	9.62	0.00	0.88	1.96	2.00	12.27	27.61	3.66	12.67	3.64	0.42	0.21	0.25	103.32	478	30
2/5	28.54	10.37	0.00	1.48	2.96	1.80	11.05	24.89	3.44	11.70	3.50	0.63	0.26	0.47	101.07	503	28
2/6	27.63	10.57	0.01	1.47	3.29	1.85	10.99	24.90	3.51	11.84	3.43	0.56	0.25	0.47	100.78	493	27
2/7	27.23	10.59	0.00	1.53	3.34	1.88	11.16	25.13	3.40	11.79	3.55	0.59	0.25	0.48	100.91	490	27
2/8	26.39	14.02	0.00	1.54	2.14	2.72	10.94	24.15	3.18	11.10	3.17	0.39	0.34	0.49	100.58	508	24
2/9	23.97	12.29	0.01	1.36	1.89	3.71	11.46	24.35	3.27	11.24	3.15	0.30	0.28	0.40	97.68	491	25
3/1	29.50	9.85	0.00	0.95	2.10	2.04	13.13	26.91	3.51	11.80	3.31	0.28	0.21	0.22	103.80	463	29
3/2	30.17	6.51	0.00	1.89	2.28	0.58	13.21	27.31	3.49	12.10	3.49	0.45	0.15	0.06	101.67	532	40
3/3	29.76	7.02	0.00	1.96	2.20	0.64	13.25	28.00	3.45	12.01	3.52	0.41	0.16	0.06	102.44	507	38
3/4	30.09	7.73	0.00	2.09	2.29	0.73	12.73	26.37	3.37	11.73	3.46	0.47	0.17	0.08	101.31	503	35
3/5	24.91	12.40	0.00	1.13	1.33	3.28	12.39	25.81	3.34	10.92	3.08	0.25	0.29	0.40	99.52	503	26
3/6	29.27	9.95	0.01	1.95	2.49	1.03	11.57	25.06	3.40	11.95	3.41	0.59	0.22	0.14	101.03	501	29
3/7	29.25	10.64	0.00	2.14	2.40	1.36	11.47	24.91	3.35	11.93	3.41	0.51	0.24	0.11	101.73	510	28
3/8	28.25	9.46	0.00	1.26	1.98	1.95	12.66	26.84	3.54	11.60	3.35	0.44	0.23	0.40	101.97	498	30
3/9	27.99	9.52	0.00	1.17	1.73	2.06	12.96	26.70	3.48	11.75	3.35	0.36	0.23	0.35	101.64	511	30
3/10	27.48	9.69	0.00	1.32	1.94	2.65	12.76	26.25	3.45	11.58	3.38	0.45	0.24	0.44	101.61	497	29
3/11	25.89	9.27	0.00	1.68	1.87	6.39	12.16	24.81	3.22	11.04	3.27	0.36	0.23	0.42	100.62	511	30

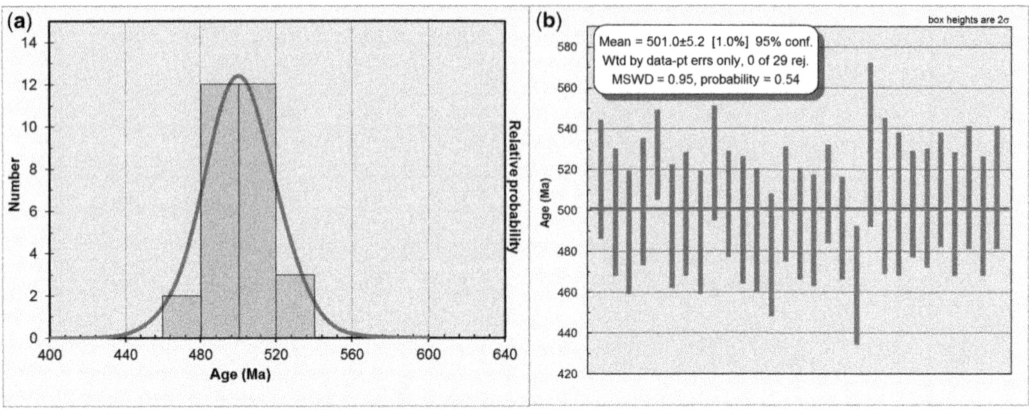

Fig. 13. (**a**) Probability density plot. (**b**) Weighted average plot and histogram of analysed ages of monazite grain.

is typical of magma mixing (Hibbard 1991), and stubby apatite enclosed in K-feldspar in MMEs suggests its mechanical transfer from the felsic host. Inclusion of small plagioclase crystals in large plagioclase, the presence of early formed quenched generated plagioclase, hornblende, biotite, apatite poikilitically enclosed in the large quartz and K-feldspar indicate a hybrid system (Hibbard 1991). Partial resorption or dissolution of early formed plagioclase from a felsic system in MMEs takes place due to superheating by the interaction with a more mafic system and an albitic rim zone on a plagioclase generated during an equilibration stage of a prior mixing stage (Hibbard 1991). Dissolution of more calcic plagioclase by a more sodic melt forms a growth albitic rim, and produced an inclusion zone of acicular apatite and opaque minerals that are absent in the core of the plagioclase. A continuous anorthite rim over an aggregate of plagioclase crystals may have formed when a sodic plagioclase crystal from the felsic system was introduced into the more mafic melt, which is rich in Ca. The presence of mafic clots with a granoblastic texture are reported from various calc-alkaline plutons (Castro & Stephens 1992; Baxter & Feely 2002) and results in recrystallization of early crystallized large amphibole phenocrysts from a mafic magma. There are two distinct compositional variants of hornblende in the MMEs. One is compositionally the same as the host granite (MgO, 9.5–10 wt%; and SiO_2, 42.7–43.4 wt%), and the other has more MgO (15.6–19.8 wt%) and SiO_2 (49.7–57.4 wt%) (Table 4). This could be due to the influx of hornblende into the mafic system from the granitic system during mixing and/or the influx of SiO_2 from the felsic system which modified the silica content in the original hornblende present in the mafic system.

Ocellar quartz is also formed as a result of the mixing of felsic and mafic magma, where quartz crystals from the felsic system were introduced into the mafic system where they become unstable. The marginal solution of the quartz extracts latent heat of crystallization from the adjacent melt and promotes undercooled generated mafic minerals (Baxter & Feely 2002).

Chemical mixing and element diffusion has considerably modified the chemistry of the MMEs. Linear to sub-linear trends in Harker diagrams indicate that MMEs are genetically related to granitoids. Such behaviour on variation diagrams could be interpreted as: (1) fractional-crystallization enclaves with such origin are interpreted as autoliths and are cognate with their host (e.g. Donaire *et al.* 2005; Pascual *et al.* 2008); (2) the same variation trends in many other granitoid suites are thought to have been originated by varying degrees of restite–melt separation (e.g. Chappell *et al.* 1987; Chappell & White 1991; Kumar 2010); and (3) as a result of the mixing of felsic and mafic magma in various proportions (e.g. Kumar & Rino 2006; Ashari *et al.* 2011; Burda *et al.* 2011). Since field and textural characteristics support magma hybridization as the dominant process in the evolution of Nongpoh granitoids, the observed linear trends in variation diagrams for MME, HD, GD and PG indicate that mixing was the predominant factor governing the evolution of the magma (Kumar & Rino 2006; Kumar & Pieru 2010; Burda *et al.* 2011). The high K_2O content at lower SiO_2 levels in MMEs with respect to NG and GD suggest K_2O migration from granitic to mafic magma during mixing and mingling. This is further supported by the presence of K-feldspar phenocrysts in MMEs mechanically transferred from the host porphyritic granite during hybridization as the composition of the porphyritic mafic enclave does not favour the crystallization of megacrysts of K-feldspar, especially in the early stages of crystallization (Vernon 1991). An influx of K can

Table 4. *Mineral composition of amphibole in MMEs*

Dataset/ point	Na_2O	F	Al_2O_3	MgO	SiO_2	K_2O	P_2O_5	Cl	MnO	FeO	TiO_2	CaO	Cr_2O_3	BaO	Total
1/1	1.48	0.77	10.43	9.55	41.97	1.60	0.00	0.37	0.51	18.95	1.04	11.53	0.02	0.00	98.23
2/1	1.51	0.79	10.69	9.48	41.81	1.60	0.02	0.37	0.58	18.69	0.96	11.75	0.09	0.04	98.36
16/1	1.36	0.67	10.24	9.66	43.08	1.52	0.04	0.17	0.38	18.42	0.80	11.79	0.00	0.50	98.64
17/1	1.41	0.84	10.14	9.40	42.67	1.57	0.02	0.22	0.34	19.70	0.84	11.85	0.05	0.26	99.31
23/1	1.54	0.73	10.28	9.91	43.27	1.48	0.05	0.19	0.69	18.80	1.40	12.01	0.00	0.04	100.40
24/1	1.48	0.46	10.36	9.46	42.40	1.62	0.09	0.20	0.38	17.94	1.03	11.94	0.18	0.24	97.78
33/1	1.51	0.62	10.33	9.90	43.08	1.47	0.00	0.18	0.60	18.91	0.77	11.27	0.06	0.00	98.70
34/1	1.46	0.72	10.32	9.60	42.56	1.55	0.00	0.23	0.70	19.11	0.72	11.80	0.02	0.00	98.76
24/1	1.26	0.59	6.75	15.66	49.71	0.73	0.04	0.11	0.25	11.50	0.72	12.24	0.23	0.00	99.79
19/1	1.29	0.59	10.19	10.05	42.84	1.39	0.02	0.21	0.30	18.46	0.88	11.89	0.07	0.00	98.19
20/1	1.29	0.60	10.40	9.60	42.71	1.35	0.00	0.26	0.54	19.17	0.66	11.62	0.04	0.27	98.50
27/1.	1.44	0.63	10.40	9.93	43.35	1.40	0.02	0.22	0.40	18.57	0.68	11.81	0.00	0.00	98.84
28/1	1.55	0.72	10.33	10.11	43.47	1.54	0.02	0.23	0.46	18.54	0.94	11.62	0.00	0.19	99.72
11/1	0.07	0.28	0.73	19.26	56.43	0.05	0.00	0.03	0.22	7.57	0.05	13.11	0.00	0.03	97.84
12/1	0.23	0.27	1.04	19.82	57.46	0.08	0.00	0.02	0.29	6.72	0.00	12.80	0.07	0.00	98.80
24/1	1.05	0.45	5.72	16.86	51.45	0.60	0.04	0.08	0.23	10.11	0.46	12.36	0.07	0.07	99.54
25/1	1.01	0.21	5.17	16.46	51.00	0.65	0.13	0.07	0.15	10.22	0.58	12.35	0.11	0.00	98.09

lead to the formation of biotite which acts as a reservoir for Rb and Ba, and therefore supports the additional dispersal of these elements from the host felsic magma to the enclave-forming magma, resulting in their substantial enrichment in the MMEs (Orsini *et al.* 1991; Blundy & Sparks 1992; Kumar *et al.* 2004; Ahmad 2011). Enrichment of P_2O_5 in the MMEs may be due to P immigration in the mafic system, leading to the crystallization of apatite. The concentration of Nd and Th represents contrasting magmatic sources: Nd enrichment is typical of mantle-derived magmas, while Th is enriched in magmas derived from the upper crust (Burda *et al.* 2011). PG and most of the MMEs show Nd/Th = 2–4, suggesting an interplay between crustal rocks and mantle-derived magma (Bea *et al.* 1999; Burda *et al.* 2011). Besides major and trace elements, enrichment of REEs in MMEs in a few samples could have been achieved by diffusional processes during the solidification of the magma sources (Sahin 2008).

Magma mixing and mingling was the major process in the evolution of the Nongpoh granitoids, which may result in isotopic disequilibrium among the host granitoids and MMEs. The available whole-rock Rb–Sr geochronological date of emplacement of the porphyritic granite of the Nongpoh granitoids was 550 ± 15 Ma with an initial $^{87}Sr/^{86}Sr$ of 0.71 ± 47 (Ghosh *et al.* 1994). Also, K–Ar dating of biotite from granite yielded an age of 464 Ma (Middle Ordovician) (Ghosh *et al.* 1991). This age of biotite may be concordant with the emplacement age of granites, which suggests the absence of a post-emplacement tectonothermal event. However, there are different generations of biotite in granitoids that might not be concordant with the emplacement age of the Nongpoh granitoids. Therefore, these initial geochronological dates by Ghosh *et al.* (1994) may have certain discrepancies largely due to lack of precision in the Rb–Sr method and its inability to detect multiple thermal events (Yin *et al.* 2010), and have scope for re-evaluation, particularly the age of the mixing and/or mingling processes during the evolution of the Nongpoh granitoids.

Chemical (U–Th–Pb) dating of monazite using the microprobe method was chosen to date the age of the magma hybridization event in the Nongpoh granitoids. Monazite geochronology is an accurate and precise primary method for dating monazite from igneous and metamorphic terrains to obtain meaningful ages from small and/or irregular domains of discrete age that may be unattainable by other dating techniques. Thus, in spite of being less precise than isotope geochronology, monazite geochronology by electron microprobe dating essentially generates reliable and comparable results to more precise isotope geochronology techniques (Williams *et al.* 2006). The three grains of monazite contained LREE and Th as major elements, Y as a minor element, and U and Pb concentrations as trace components. The weighted average and probability density plots for the three monazite grains yielded an age of 501 ± 5 Ma with a 95% probability of fit. The association of monazite grains with primary crystallized minerals of MME and its shape suggest that the microprobe chemical date is the crystallization date of the dioritic MME, as the chances of monazite re-equilibration with granitic magma are minimal. This age for the monazite grains is correlatable with the age of igneous activity at 500 Ma during the amalgamation of the Eastern Gondwana supercontinent (Yin *et al.* 2010). Gondwanaland shows a more than 10 000 km-long chain of 530–500 Ma granitic magmatism over a Laurentian, European and Asian assembly of continents along the northern margin, and with a trench along the western and southern margins due to intense deformation and epeirogenic uplift of cratons (Veevers 2004).

Conclusions

The detailed field observations and the petrographical, mineralogical, geochemical and chemical dating studies of the Nongpoh granitoids has provided information about the genesis and evolution of the mafic magmatic enclaves (MMEs). The Nongpoh granitoids represent a complex amalgamation of hybridization processes during the evolution of the granitoids and mafic rocks, including the MMEs:

- This study, for the first time, has documented mafic injections during different stages of crystallization of the host magma of the Nongpoh granitoids. The early injections resulted in MMEs, whereas later injections led to the syn-plutonic dykes.
- The injected syn-plutonic mafic dykes during the crystallization of the host magma supplied the supplementary heat required for local re-melting of the host magma and crustal reworking processes.
- Hybridization or mixing of mafic and felsic magma in various proportions is characterized by negative correlation of mafic oxides (TiO_2, $Fe_2O_3^T$ and MgO) with increasing SiO_2 and the nearly linear variations, chemical mixing and K_2O diffusion. However, a linear distribution of major and trace element data with SiO_2 as a differentiation index also suggests derivation from a single parental magma, which is hybridized. The increase in HFSEs and the overlapping of normalized REE also point towards formation from a single parentage. This indicates that multistage magma mixing and/or mingling was a major process during the evolution of Nongpoh granitoids.

- These hybridized mafic injections formed the terminal event during the Cambrian and its age of crystallization has been dated to 501 ± 5 Ma by chemical (U–Th–Pb) dating using the microprobe method, and is similar to the plutonic and tectonothermal events in other parts of India, SW Australia and Antarctica, documenting the Neoproterozoic–Cambrian Pan-African events that climaxed in the formation of the Gondwana supercontinent.

The authors express their gratitude towards Director General, Geological Survey of India, Kolkata, and Deputy Director General and Head of Department, Geological Survey of India, North Eastern Region, Shillong for continuous support and permission to publish the work. Sincere thanks to Md Naushad and Shri Amit Srivastava for providing technical support with the manuscript. Mrs Sonalika Joshi and Mrs Chinchu VS are also acknowledged for carrying out EPMA analysis at Geological Survey of India, Faridabad. We are grateful to two anonymous reviewers for their constructive comments, and to Dr Bryan C. Storey and Dr S. Sensarma, the volume editors, for their careful editorial handling that lead to significant improvement of the paper.

References

AHMAD, M. 2011. Enclaves in granitoids of north of Jonnagiri schist belt, Kurnool district, Andhra Pradesh: Evidence of magma mixing and mingling. *Journal of the Geological Society of India*, **77**, 557–573.

ASHARI, E.A., HASSANZADEH, J. & VALIZADEH, M.V. 2011. Geochemistry of microgranular enclaves in Aligoodarz Jurassic arc pluton, western Iran: implications for enclave generation by rapid crystallization of cogenetic granitoid magma. *Mineralogy and Petrology*, **101**, 195–216.

BARBARIN, B. 1990. Plagioclase xenocrysts and mafic magmatic enclaves in some granitoids of the Sierra Nevada batholith, California. *Journal of Geophysical Research*, **95**, 17 747–17 756.

BARBARIN, B. & DIDIER, J. 1991. Macroscopic features of mafic microgranular enclaves. *In*: DIDIER, J. & BARBARIN, B. (eds) *Enclaves and Granite Petrology*. Developments in Petrology, **13**. Elsevier, Amsterdam, 253–261.

BARBARIN, B. & DIDIER, J. 1992. Genesis and evolution of mafic microgranular enclaves through various types of interaction between co-existing felsic and mafic magmas. *Transactions of the Royal Society of Edinburgh: Earth Sciences*, **83**, 145–153.

BAXTER, S. & FEELY, M. 2002. Magma mixing and mingling textures in granitoids: examples from Galway granite, Connemara, Ireland. *Mineralogy and Petrology*, **66**, 63–74.

BEA, F., MONTERO, P. & MOLINA, F. 1999. Mafic precusrsors, peraluminous granitoids and late lamprophyres in the Avila batholith: a model for the generation of Variscan batholiths Iberia. *Journal of Geology*, **107**, 399–419.

BLUNDY, J.D. & SPARKS, R.S.J. 1992. Petrogenesis of mafic inclusions in granitoids of the Adamelo Massif, Italy. *Journal of Petrology*, **33**, 1039–1104.

BURDA, J., GAWĘDA, A. & KLÖTZLI, U. 2011. Magma hybridization in the Western Tatra Mountains granitoid intrusion (S-Poland, Western Carpathians). *Mineralogy and Petrology*, **103**, 19–36.

CASTRO, A. 1990. Microgranular enclaves of the Quintana granodiorite (Los Pedroches batholith) petrogenetic significance. *Revista de la Sociedad Geológica de España*, **3**, 7–21.

CASTRO, A. & STEPHENS, W.E. 1992. Amphibole-rich polycrystalline clots in calc-alkaline granitic rocks and their enclaves. *The Canadian Mineralogist*, **30**, 1093–1112.

CHAPPELL, B.W. & WHITE, A.J.R. 1991. Restite enclaves and restite model. *In*: DIDIER, J. & BARBARIN, B. (eds) *Enclaves and Granite Petrology*. Developments in Petrology, **13**. Elsevier, Amsterdam, 375–391.

CHAPPELL, B.W., WHITE, A.J.R. & WYBORN, D. 1987. The importance of residual source material (restite) in granite petrogenesis. *Journal of Petrology*, **28**, 1111–1138.

CHEN, B., HE, J.B. & MA, X.H. 2009. Petrogenesis of mafic enclaves from the north Taihang Yanshanian intermediate to felsic plutons: evidence from petrological, geochemical, and zircon Hf–O isotopic data. *Science in China Series D: Earth Sciences*, **52**, 1331–1344, https://doi.org/10.1007/s11430-009-0130-z

CHEN, Y., CHAPPELL, B.W. & WHITE, A.J.R. 1991. Mafic enclaves of some I-type granites of the Palaeozoic Lachlan fold belt, south eastern Australia. *In*: DIDIER, J. & BARBARIN, B. (eds) *Enclaves and Granite Petrology*. Developments in Petrology, **13**. Elsevier, Amsterdam, 113–123.

CHENG, Y., SPANDLER, C., MAO, J. & RUSK, B.G. 2012. Granite, gabbro and mafic microgranular enclaves in the Gejiu area, Yunnan Province, China: a case of two-stage mixing of crust- and mantle-derived magmas. *Contributions to Mineralogy and Petrology*, **164**, 659–676.

CID, J.P., NARDI, L.V.S. & GISBERT, P.E. 2002. Textural relations of lamprophyric mafic microgranular enclaves and petrological implications for the genesis of potassic syenitic magmas: the example of Piquiri syenite, southern Brazil. *Pesquisas em Geociências*, **29**, 21–30.

DIDIER, J. 1973. *Granites and Their Enclaves: The Bearing of Enclaves on the Origin of Granites*. Developments in Petrology, **3**. Elsevier, Amsterdam.

DIDIER, J. 1984. The problem of enclaves in granitic rocks: a reviews of recent ideas on their origin. *In*: XU, K.Q. & TU, G.C. (eds) *Proceedings of International Symposium on Geology of Granites and their Metallogenetic Relations, Nanjing University, Nanjing China, October 26–30, 1982*. Science Press, Beijing, 137–144.

DIDIER, J. & BARBARIN, B. 1991. Enclaves and granite petrology. *In*: DIDIER, J. & BARBARIN, B. (eds) *Enclaves and Granite Petrology*. Developments in Petrology, **13**. Elsevier, Amsterdam, 603–625.

DIDIER, J. & ROQUES, M. 1959. Sur les enclaves des granites du Massif Central Français. *Comptes rendus de l'Académie des sciences, Paris* [On the enclaves of the granites of the French Massif Central. Proceedings of the Académie des sciences, Paris], **228**, 1839–1841.

DONAIRE, T., PASCUAL, E., PIN, C. & DUTHOU, J.L. 2005. Microgranular enclaves as evidence of rapid cooling in granitoid rocks: the case of the Los Pedroches granodiorite, Iberian Massif, Spain. *Contributions to Mineralogy and Petrology*, **149**, 247–265.

DORAIS, M.J., WHITNEY, J.A. & RODEN, M.F. 1990. Origin of mafic enclaves in the Dinkey Creek Pluton, Central Sierra Neveda Batholith, California. *Journal of Petrology*, **31**, 853–881.

ELBURG, M.A. 1996*a*. U–Pb ages and morphologies of zircon in microgranitoid enclaves and peraluminous host granites: evidence for magma mingling. *Contributions to Mineralogy and Petrology*, **123**, 177–189.

ELBURG, M.A. 1996*b*. Evidence of isotopic equilibration between microgranitoid enclaves and host granodiorite, Warburton Granodiorite, Lachlan Fold Belt, Australia. *Lithos*, **38**, 1–22.

ELBURG, M.A. 1996*c*. Genetic significance of multiple enclave types in a peraluminous ignimbrite suite, Lachlan Fold Belt, Australia. *Journal of Petrology*, **37**, 1385–1408.

EVANS, P. 1964. The tectonic framework of Assam. *Journal of the Geological Society of India*, **5**, 80–96.

FERNANDEZ, A.N. & BARBARIN, B. 1991. Relative rheology of coeval mafic and felsic magmas: nature of resulting interaction processes – shape and mineral fabric of mafic microgranular enclave. *In*: DIDIER, J. & BARBARIN, B. (eds) *Enclaves and Granite Petrology*. Developments in Petrology, **13**. Elsevier, Amsterdam, 263–275.

GHOSH, S., BHALLA, J.K., PAUL, D.K., SARKAR, A., BISHUI, P. K. & GUPTA, S.N. 1991. Geochronology and geochemistry of granite plutons from East Khasi Hills, Meghalaya. *Journal of the Geological Society of India*, **37**, 331–342.

GHOSH, S., CHAKRABORTY, S., PAUL, D.K., BHALLA, J.K., BISHUI, P.K. & GUPTA, S.N. 1994. New Rb–Sr isotopic ages and geochemistry of granitoids from Meghalaya and their significance in middle to late Proterozoic crustal evolution. *Indian Minerals*, **48**, 33–44.

GHOSH, S., FALLICK, A.E., PAUL, D.K. & POTTS, P.J. 2005. Geochemistry and origin of Neoproterozoic granitoids of Meghalaya, northeast India: implications for linkage with amalgamation of Gondwana Supercontinent. *Gondwana Research*, **8**, 421–432.

GSI 2011. *Geological Survey of India Operating Procedure – XRF*, Issue No. 01. Geological Society of India, Bangalore, India, http://www.portal.gsi.gov.in/gsiDoc/pub/sop_chemical_lab_all_version2.zip [last accessed 8 November 2016].

HIBBARD, M.J. 1981. The magma mixing origin of mantled feldspar. *Contributions to Mineralogy and Petrology*, **76**, 158–170.

HIBBARD, M.J. 1991. Textural anatomy of twelve magma-mixed granitoid system. *In*: DIDIER, J. & BARBARIN, B. (eds) *Enclaves and Granite Petrology*. Developments in Petrology, **13**. Elsevier, Amsterdam, 431–444.

IRVINE, T.N. & BARAGAR, W.R.A. 1971. A guide to the chemical classification of the common volcanic rocks. *Canadian Journal of Earth Sciences*, **8**, 523–548.

JEERE, D.S. & KUNDU, A. 2010. Microprobe geochronology of monazite from chromitite of Precambrian mafic–ultramafic rocks of Vagde area, Shindhudurg district, Maharashtra, India. *Indian Journal of Geosciences*, **67**, 131–140.

JAYANANDA, M., MIYAZAKI, T., GIREESH, R.V., MAHESHA, N. & KANO, T. 2009. Synplutonic mafic dykes from late Archaean granitoids in the eastern Dharwar craton, Southern India. *Journal of the Geological Society of India*, **73**, 117–130.

KUMAR, S. 2010. Mafic to hybrid microgranular enclaves in the Ladakh batholith, northwest Himalaya: Implications on calc-alkaline magma chamber processes. *Journal of the Geological Society of India*, **76**, 5–25.

KUMAR, S. & PIERU, T. 2010. Petrography and major elements geochemistry of microgranular enclaves and Neoproterozoic granitoids of south Khasi, Meghalaya: evidence of magma mixing and alkali diffusion. *Journal of the Geological Society of India*, **76**, 345–360.

KUMAR, S. & RINO, V. 2006. Mineralogy and geochemistry of microgranular enclaves in Palaeoproterozoic Malanjkhand granitoids, Central India: evidences of magma mixing, mingling, and chemical equilibration. *Contributions to Mineralogy and Petrology*, **152**, 591–609.

KUMAR, S., RINO, V. & PAL, A.B. 2004. Field evidence of magma mixing from microgranular enclaves hosted in Palaeoproterozoic Malanjkhand granitoids, central India. *Gondwana Research*, **7**, 539–548.

KUMAR, S., PIERU, T., RINO, V. & LYNGDOH, B.C. 2005. Miocrogranular enclaves in Neoproterzoic granitoids of south Khasi Hills, Meghalaya plateau, Northeast India: field evidences of interacting coeval mafic and felsic magmas. *Journal of the Geological Society of India*, **65**, 629–633.

LIANKUN, S. & KURIONG, Y. 1991. A two stage crust–mantle interaction model for mafic microgranular enclaves in the Daning granodiorite pluton, Guangxi, China. *In*: DIDIER, J. & BARBARIN, B. (eds) *Enclaves and Granite Petrology*. Developments in Petrology, **13**. Elsevier, Amsterdam, 95–110.

LIÉGEOIS, J.P., NAVEZ, J., HERTOGEN, J. & BLACK, R. 1998. Contrasting origin of post-collisional high-K calc-alkaline and shoshonitic v. alkaline and peralkaline granitoids: the use of sliding normalization. *Lithos*, **45**, 1–28.

LUDWIG, K.R. 1991. *ISOPLOT; A Plotting and Regression Program for Radiogenic-Isotope Data; Version 2.53*. United States Geological Survey, Open-File Report, **91-445**.

MCDONOUGH, W.F. & SUN, S.-S. 1995. The composition of the Earth. *Chemical Geology*, **120**, 223–253.

MANIAR, P.D. & PICCOLI, P.M. 1989. Tectonic discrimination of granitoids. *Geological Society of America Bulletin*, **101**, 635–643.

MONTEL, J.M., FORET, S., VESCHAMBRE, M., NICOLLET, C. & PROVOST, A. 1996. Electron microprobe dating of monazite. *Chemical Geology*, **131**, 37–53.

NANDY, D.R. 1980. Tectonic pattern in northeastern India. *Indian Journal of Earth Sciences*, **7**, 103–107.

ORSINI, J.B., COCIRTA, C. & ZORPI, M.J. 1991. Genesis of mafic microgranular enclaves through differentiation of basic magmas, mingling and chemical exchanges with their host granitoid magmas. *In*: DIDIER, J. & BARBARIN, B. (eds) *Enclaves and Granite Petrology*. Developments in Petrology, **13**. Elsevier, Amsterdam, 445–464.

Pascual, E., Donaire, T. & Christian, P. 2008. The significance of microgranular enclaves in assessing the magmatic evolution of a high-level composite batholith: a case on the Los Pedroches batholith, Iberian massif, Spain. *Geochemical Journal*, **42**, 177–198.

Peacock, M.A. 1931. Classification of igneous rock series. *Journal of Petrology*, **39**, 54–67.

Peccerillo, A. & Taylor, S.R. 1976. Geochemistry of Eocene calc-alkaline volcanic rocks from the Kastamonu area, northern Turkey. *Contributions to Mineralogy and Petrology*, **58**, 68–81.

Pitcher, W.S. 1991. Synplutonic dykes and mafic enclaves. *In*: Didier, J. & Barbarin, B. (eds) *Enclaves and Granite Petrology*. Developments in Petrology, **13**. Elsevier, Amsterdam, 383–392.

Reid, J.B., Jr. & Hamilton, M.A. 1987. Origin of Sierra Nevadan granite: evidence from small scale composite dikes. *Contributions to Mineralogy and Petrology*, **96**, 441–454.

Reid, J.B., Jr., Evans, O.C. & Fates, D.G. 1983. Magma-mixing in granitic rocks of the Central Sierra Neveda, California. *Earth and Planetary Science Letters*, **66**, 243–261.

Rock, N.M.S. 1987. The nature and origin of lamprophyres: an overview. *In*: Fitton, J.G. & Upton, B.G.J. (eds) *Alkaline Igneous Rocks*. Geological Society, London, Special Publications, **30**, 191–226, https://doi.org/10.1144/GSL.SP.1987.030.01.09

Rock, N.M.S. 1991. *Lamprophyres*. Blackie. Glasgow.

Ryerson, F.J. & Hess, P.C. 1978. Implications of liquid–liquid distribution coefficients to mineral–liquid portioning. *Geochimica et Cosmochimica Acta*, **42**, 921–932.

Sadiq, M. & Umrao, R.K. 2013. Occurrence of lamprophyre bodies around Umling and Lailad villages, Ri-Bhoi districts, Meghalaya. *In*: *National Conference on Sedimentation and Tectonics with Special Reference to Energy Resources of North-East India & XXX Convention Indian Association of Sedimentologists*, Manipur University, Canchipur, Imphal, Manipur, India, 105–106.

Sadiq, M., Ranjith, A. & Umrao, R.K. 2014*a*. REE mineralization in the carbonatites of the Sung valley ultramafic–alkaline–carbonatite complex, Meghalaya, India. *Open Geosciences*, **6**, 457–475.

Sadiq, M., Umrao, R.K. & Dutta, J.C. 2014*b*. Occurrence of rare earth elements in parts of Nongpoh granite, Ri-Bhoi district, Meghalaya. *Current Science*, **106**, 162–165.

Saha, A., Ganguly, S., Ray, J. & Chaterjee, N. 2010. Evaluation of phase chemistry and petrochemical aspects of Samchampi–Samteran differentiated alkaline complex of Mikir Hills, northeastern India. *Journal of Earth System Science*, **119**, 675–699.

Sahin, S.Y. 2008. Geochemistry of mafic microgranular enclaves in the Tamdere quartz monzonite, south of Dereli/Giresun, Eastern Pontides, Turkey. *Chemie der Erde – Geochemistry*, **68**, 81–92.

Sarkar, S., Khonglah, M.A., Khan, M.A., Kumar, T.B.R. & Ray, J.N. 2007. Polymodal occurrence of early mafic differentiate associated with Mid-Proterozoic calc-alkaline plutons of Meghalaya. *Journal of the Geological Society of India*, **70**, 53–58.

Storey, M., Kent, R.W. *et al.* 1992. Lower Cretaceous volcanic rocks on continental margins and their relationship to the Kerguelen plateau. *In*: Wise, S.W., Jr., Schlich, R. *et al.* (eds) *Proceedings of the Ocean Drilling Program, Scientific Results, Volume 120*. Ocean Drilling Program, College Station, TX, 33–53.

Troll, V.R., Donaldson, C.H. & Emeleus, C.H. 2004. Pre-eruptive magma mixing in ash-flow deposits of the Tertiary Rum igneous center, Scotland. *Contributions to Mineralogy and Petrology*, **147**, 722–739.

Veevers, J.J. 2004. Gondwanaland from 650–500 Ma assembly through 320 Ma merger in Pangea to 185–100 Ma breakup: supercontinental tectonics via stratigraphy and radiometric dating. *Earth-Science Reviews*, **68**, 1–132.

Vernon, R.H. 1984. Microgranitoid enclaves: globules of hybrid magma quenched in a plutonic environment. *Nature*, **304**, 438–439.

Vernon, R.H. 1990. Crystallization and hybridism in microgranitoids enclave magmas: microstructural evidence. *Journal of Geophysical Research*, **95**, 17 849–17 859.

Vernon, R.H. 1991. Interpretation of microstructure of microgranitoids enclaves. *In*: Didier, J. & Barbarin, B. (eds) *Enclaves and Granite Petrology*. Developments in Petrology, **13**. Elsevier, Amsterdam, 277–292.

Vernon, R.H., Etheridge, M.A. & Wall, V.J. 1983. Magma mixing in the development of metaluminous granitoid suites of the Moruya batholith around Tuross Head, N. S. W. *In*: *Lithosphere Dynamics and Evolution of Continental Crust*. Geological Society of Australia, Abstracts, **9**, 185.

Watson, E.B. 1976. Two liquid partition coefficients; Experimental data and geochemical implications. *Contributions to Mineralogy and Petrology*, **56**, 119–134.

Wiebe, R.A., Smith, D., Sturn, M. & King, E.M. 1997. Enclaves in the Cadillac mountain granite (Coastal Maine): samples of hybrid magma from the base of the chamber. *Journal of Petrology*, **38**, 393–426.

Willey, R.A. 1984. Sources of granitoid magmas at convergent plate boundaries. *Physics of the Earth and Planetary Interiors*, **35**, 12–18.

Williams, M.L., Jercinovic, M.J., Goncalves, P. & Mahan, K. 2006. Format and philosophy for collecting, compiling, and reporting microprobe monazite ages. *Chemical Geology*, **225**, 1–15.

Yang, J.H., Wu, F.Y., Chung, S.L., Chu, M.F. & Wilde, S. A. 2004. Multiple sources for the origin of granites: geochemical and Nd/Sr isotopic evidence from the Gudaoling granite and its mafic enclaves, northeast China. *Geochimica et Cosmochimica Acta*, **68**, 4469–4483.

Yang, J.H., Wu, F.Y., Chung, S.L., Wilde, S.A., Chu, M. F., Lo, C.H. & Song, B. 2005. Petrogenesis of early Cretaceous intrusions in the Sulu ultrahigh-pressure orogenic belt, east China and their relationship to lithospheric thinning. *Chemical Geology*, **222**, 200–231.

Yang, J.H., Wu, F.Y., Chung, S.L., Wilde, S.A. & Chu, M. F. 2006. A hybrid origin for the Qianshan A-type granite, northeast China: geochemical and Sr–Nd–Hf isotopic evidence. *Lithos*, **89**, 89–106.

Yang, J.H., Wu, F.Y. & Wilde, S.A. 2007*a*. Petrogenesis of Late Triassic granitoids and their enclaves with implications for post-collisional lithospheric thinning of the Liaodong Peninsula, North China Craton. *Chemical Geology*, **242**, 155–175.

Yang, J.H., Wu, F.Y., Wilde, S.A., Xie, L.W., Yang, Y.H. & Liu, X.M. 2007*b*. Tracing magma mixing in granite genesis: in situ U–Pb dating and Hf-isotope analysis of zircons. *Contributions to Mineralogy and Petrology*, **135**, 177–190.

Yin, A., Dubey, C.S., Webb, A.A.G., Kelty, T.K., Grove, M., Gehrels, G.E. & Burgess, W.P. 2010. Geologic correlation of the Himalayan orogen and Indian craton: part 1. Structural geology, U–Pb zircon geochronology, and tectonic evolution of the Shillong Plateau and its neighboring regions in NE India. *Geological Society of America Bulletin*, **122**, 336–359.

Regional volcanism of northern Zealandia: post-Gondwana break-up magmatism on an extended, submerged continent

N. MORTIMER[1]*, P. B. GANS[2], S. MEFFRE[3], C. E. MARTIN[4], M. SETON[5], S. WILLIAMS[5], R. E. TURNBULL[1], P. G. QUILTY[3], S. MICKLETHWAITE[6], C. TIMM[7], R. SUTHERLAND[8], F. BACHE[9], J. COLLOT[10], P. MAURIZOT[10], P. ROUILLARD[10] & N. ROLLET[11]

[1]*GNS Science, Dunedin, New Zealand*

[2]*Department of Earth Science, University of California, Santa Barbara, USA*

[3]*Department of Earth Sciences, University of Tasmania, Hobart, Australia*

[4]*Department of Geology, University of Otago, Dunedin, New Zealand*

[5]*School of Geosciences, University of Sydney, Australia*

[6]*School of Earth, Atmosphere and Environment, Monash University, Melbourne, Australia*

[7]*GNS Science, Lower Hutt, New Zealand*

[8]*School of Geography, Environment and Earth Sciences, Victoria University of Wellington, New Zealand*

[9]*Santos Ltd., Adelaide, Australia*

[10]*Service Geologique de Nouvelle Calédonie, Nouméa, New Caledonia*

[11]*Geoscience Australia, Canberra, Australia*

Correspondence: n.mortimer@gns.cri.nz

Abstract: Volcanism of Late Cretaceous–Miocene age is more widespread across the Zealandia continent than previously recognized. New age and geochemical information from widely spaced northern Zealandia seafloor samples can be related to three volcanotectonic regimes: (1) age-progressive, hotspot-style, low-K, alkali-basalt-dominated volcanism in the Lord Howe Seamount Chain. The northern end of the chain (*c.* 28 Ma) is spatially and temporally linked to the 40–28 Ma South Rennell Trough spreading centre. (2) Subalkaline, intermediate to silicic, medium-K to shoshonitic lavas of >78–42 Ma age within and near to the New Caledonia Basin. These lavas indicate that the basin and the adjacent Fairway Ridge are underlain by continental rather than oceanic crust, and are a record of Late Cretaceous–Eocene intracontinental rifting or, in some cases, speculatively subduction. (3) Spatially scattered, non-hotspot, alkali basalts of 30–18 Ma age from Loyalty Ridge, Lord Howe Rise, Aotea Basin and Reinga Basin. These lavas are part of a more extensive suite of Zealandia-wide, 97–0 Ma intraplate volcanics. Ages of northern Zealandia alkali basalts confirm that a late Cenozoic pulse of intraplate volcanism erupted across both northern and southern Zealandia. Collectively, the three groups of volcanic rocks emphasize the important role of magmatism in the geology of northern Zealandia, both during and after Gondwana break-up. There is no compelling evidence in our dataset for Late Cretaceous–Paleocene subduction beneath northern Zealandia.

Supplementary material: Trace element compositions of zircons and whole-rock chemical compositions obtained by previous studies are available at: https://doi.org/10.6084/m9.figshare.c.3850975

Zealandia is a 4.9×10^6 km^2 continent that was formerly a part of Gondwana and now lies 94% submerged in the SW Pacific Ocean (Luyendyk 1995; Mortimer *et al.* 2017) (Fig. 1). It consists of a basement of Cambrian–Early Cretaceous terranes and batholiths (Austral Superprovince) and a cover of Late Cretaceous–Holocene sedimentary basins (Zealandia Megasequence: Mortimer *et al.* 2014*b*). A major tectonic feature of Zealandia is the modern-day Pacific–Australia plate boundary (Alpine Fault–Hikurangi Trench in Fig. 1). This divides Zealandia into northern and southern parts, and is responsible

From: Sensarma, S. & Storey, B. C. (eds) 2018. *Large Igneous Provinces from Gondwana and Adjacent Regions*. Geological Society, London, Special Publications, **463**, 199–226.
First published online August 16, 2017, https://doi.org/10.1144/SP463.9

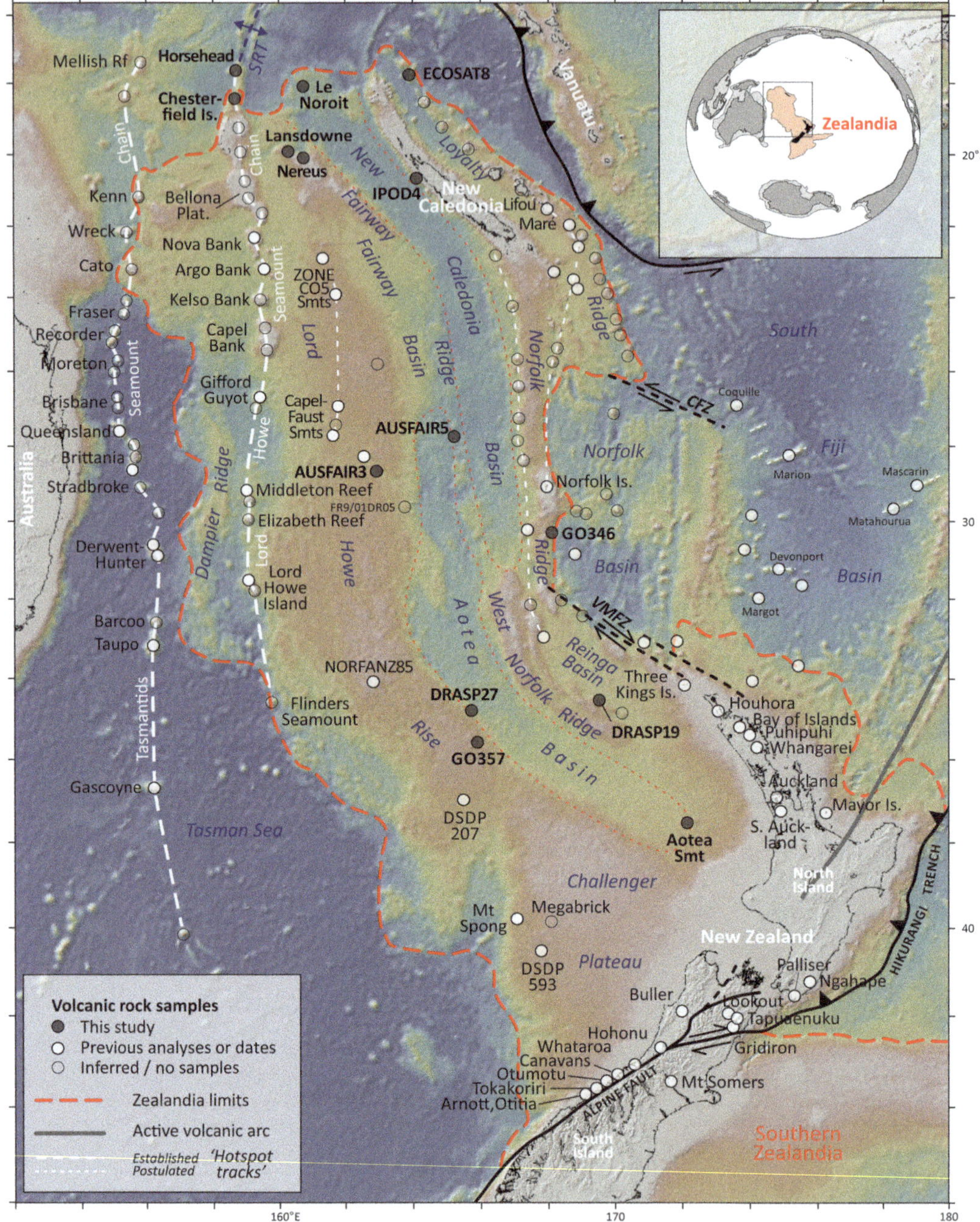

Fig. 1. Volcanic rock samples of northern Zealandia and the surrounding oceanic crust. Existing analysed samples from Cole (1986), Nelson *et al.* (1986), McDougall & Duncan (1988), Weaver & Smith (1989), Quilty (1993), Exon *et al.* (2004), Hoernle *et al.* (2006), Panter *et al.* (2006), Collot *et al.* (2009), Heap *et al.* (2009), Tulloch *et al.* (2009), Timm *et al.* (2010), Mortimer *et al.* (1998, 2007, 2008*a*, *b*, 2014*a*) and Scott *et al.* (2015). VMFZ, Vening Meinesz Fracture Zone; CFZ, Cook Fracture Zone.

for Miocene–Holocene subduction-related volcanic rocks on the North Island of New Zealand. In addition to Neogene subduction-related volcanism, Zealandia has a record of different styles and compositions of protracted, scattered, low-volume volcanism that is not related to subduction (Cole 1986;

Weaver & Smith 1989; Panter *et al.* 2006; Tulloch *et al.* 2009; Timm *et al.* 2010). This intraplate volcanism erupted in New Zealand sedimentary basins from the Late Cretaceous to the Holocene (Mortimer *et al.* 2014*b*). Zealandia intraplate volcanic rocks share a broadly similar age range and composition with those in formerly adjacent West Antarctica and Australia (Johnson *et al.* 1989). These now-dispersed lavas have previously been described as a Diffuse Alkaline Magmatic Province (DAMP: Finn *et al.* 2005).

The datasets used to characterize and explain Zealandia intraplate magmatism have, to date, mostly come either from onland New Zealand or from islands and dredges in southern Zealandia, which lies on the present-day Pacific Plate (e.g. Cole 1986; Gamble *et al.* 1986; Herzer *et al.* 1989; Weaver & Smith 1989; Baker *et al.* 1994; Tappenden 2003; Nicholson & Black 2004; Cook *et al.* 2004; Hoernle *et al.* 2006; Panter *et al.* 2006; Sprung *et al.* 2007; Coombs *et al.* 2008; Tulloch *et al.* 2009; Timm *et al.* 2010; McCoy-West *et al.* 2010; van der Meer *et al.* 2013; Scott *et al.* 2015). In contrast, examples of intraplate magmatism from offshore northern Zealandia, on the present-day Australian Plate are far fewer (e.g. Green 1973; Baubron *et al.* 1976; McDougall *et al.* 1981; Mortimer *et al.* 1998; Timm *et al.* 2010; Dadd *et al.* 2011; Higgins *et al.* 2011; Nicholson *et al.* 2011). In large part, this is because New Zealand and its subantarctic islands have the most accessible igneous rock occurrences. Most published accounts of intraplate volcanism in both northern and southern Zealandia preceded the use of the name Zealandia as a continent, and were equivocal as to the geological setting of the Lord Howe Rise and Norfolk Ridge. As such, the older literature on the intraplate magmatism of the Zealandia continent lacks context and is not comprehensive.

In this paper we present 18 Ar/Ar ages, two U–Pb ages, six micropalaeontological ages, 22 whole-rock geochemical analyses and 19 Nd isotope analyses from Late Cretaceous to Miocene volcanic rock samples on the Loyalty, Fairway and Norfolk ridges, Lord Howe Rise, and in the New Caledonia, Aotea and Reinga basins (Fig. 1). These samples were collected in rock dredges on the GEORSTOM, AUSFAIR, IPOD, DRASP and ECOSAT cruises (Table 1) (see Monzier & Vallot 1983; Colwell *et al.* 2006; Collot *et al.* 2013; Bache *et al.* 2014*b*; Seton *et al.* 2016*b*, respectively).

Our new dataset adds considerably to the existing meagre collection of igneous rocks from northern Zealandia. In turn, this enables a more complete picture of the magmatism across all of Zealandia to be presented and explained. This paper builds on and extends syntheses of intraplate volcanism by Finn *et al.* (2005), Hoernle *et al.* (2006), Timm *et al.* (2010) and Bryan *et al.* (2012), and the Zealandia stratigraphic–magmatic framework of Mortimer *et al.* (2014*b*). It is a companion paper to the dating of Lord Howe Rise lavas by Higgins *et al.* (2011), Reinga Basin sedimentary rock interpretations of Browne *et al.* (2016) and investigations of the South Rennell Trough spreading centre by Seton *et al.* (2016*a*).

Northern Zealandia framework

The three main emergent areas of northern Zealandia are New Zealand's North Island, the west coast of New Zealand's South Island and New Caledonia (Fig. 1). Continental basement of Palaeozoic–Mesozoic accreted terranes is exposed in all three places. Several small islands, island groups, reefs and atolls pockmark northern Zealandia. The largest of these are the Three Kings, Norfolk, Lord Howe, Chesterfield and Loyalty islands. The submerged part of northern Zealandia is dominated by the 2000 × 300 km Lord Howe Rise. East and west of this lie the Dampier and Norfolk ridges, and to the SE are the Challenger Plateau and onland New Zealand. The New Caledonia Trough, a composite Cretaceous and Eocene rift basin (Sutherland *et al.* 2010), is the deepest part of the Zealandia continent and comprises two separate sedimentary basins: the New Caledonia and Aotea basins (these, rather than the single trough, are shown in Fig. 1). There has been debate as to whether the basins are floored by continental or oceanic crust (Klingelhoefer *et al.* 2007) and some of the new data reported in this paper bear on this.

The exposed bedrock geology of the Three Kings, Norfolk, Lord Howe, Chesterfield and Loyalty islands consists solely of intraplate volcanic rocks. Some islands and seamounts form linear chains and have been proposed as hotspot tracks. The 2600 km-long, north–south-trending Tasmantids are entirely constructed on oceanic crust, except for Cato seamount which impinges on the Zealandia continent (Exon *et al.* 2006). In contrast the 1900 km-long, north–south-trending Lord Howe Seamount Chain is entirely constructed on Zealandia continental crust, except possibly for Horsehead Seamount at its northernmost end (Fig. 1). The Tasmantids in the northern Tasman Sea have been shown by McDougall & Duncan (1988) and Quilty (1993) to get younger to the south, and these authors presumed a similar younging for the Lord Howe chain. Three other postulated north-south-trending hotspot tracks are the 500 km-long Capel–Faust ZONECO5 seamounts (van de Beuque *et al.* 1998; Exon *et al.* 2004; Dadd *et al.* 2011), a 900 km linear north-south alignment of seamounts on the eastern Norfolk Ridge (Rigolot 1988) and an 1100 km alignment of seamounts along the western Norfolk Ridge (Fig. 1).

Table 1. *Location data for samples described in this paper*

Cruise and dredge	Latitude (°S)	Longitude (°E)	Depth (m)	Site description	Rocks recovered
Lord Howe Seamount Chain					
ECOSAT DR15	17.6918	158.4713	1850–2600	Horse Head Seamount	Basalts, limestone
ECOSAT DR16	18.7083	158.3688	2500–2800	Chesterfield Plateau, NW side	Basalts, limestone
New Caledonia basin margins					
ECOSAT DR17	20.1399	160.7699	650–1200	Lansdowne Bank, Fairway Ridge	Altered andesites
ECOSAT DR18	19.9178	160.1923	800–1500	Nereus Reef, Fairway Ridge	Altered andesites
ECOSAT DR11	17.9856	160.7284	2250–2500	Le Noroit Seamounts	Trachyte cores in Mn nodules
AUSFAIR DR03	28.4219	162.7899	1450–1700	Fault scarp, Lord Howe Rise	Rhyolites
AUSFAIR DR05	27.7127	165.2894	*c.* 2900	Southern Fairway Ridge, east flank	Andesite breccia
IPOD DR4	20.6418	164.1222	*c.* 710	Offshore from Koumac, New Caledonia	Autobrecciated dacite, limestone
North Zealandia scattered seamounts					
GEORSTOM 357 D1	35.6585	165.9749	770–1250	Seamount, SE Lord Howe Rise	Gabbro xenolith in alkali basalt
GEORSTOM 346 D1	30.4769	168.0898	1840–2300	Seamount, eastern Norfolk Ridge	Vesicular ol-plag basalt
DRASP d02	37.5588	171.9598	1389–1570	Aotea Seamount	Basanite
DRASP d19	34.5218	169.6479	1715–1777	Seamount, Reinga Basin	Amygdaloidal basalt
DRASP d27B	34.7359	165.6869	1981–2221	Seamount, western side Aotea Basin	Basalts, mudstones
ECOSAT DR08	17.583	164.0078	1300–1500	Seamount, Loyalty Ridge	Amygdaloidal basalt breccia

Marine expedition acronyms, ships and cruise numbers are as follows: DRASP (Dredging Reinga and Aotea basins to constrain seismic Stratigraphy and Petroleum systems) R/V *Tangaroa* TAN1312, November 2013; AUSFAIR (AUStralia-FAIRway basin bathymetry and sampling survey) N/O *Marion Dufresne* MD153, February 2006; ECOSAT (Eastern COral SeA Tectonics) R/V *Southern Surveyor* SS2012v06, November 2012; IPOD (Investigation of Post-Obduction Deposits) R/V *l'Alis*, August 2012; GEORSTOM III SUD N/O *Le Noroit*, November 1975.

Methods

Methods and analytical laboratories used for whole-rock geochemistry varied depending on the sample size. Large (*c.* >20 g) samples were crushed in a tungsten carbide ring mill and analysed by X-ray fluorescence (XRF) methods at Spectrachem Analytical, Lower Hutt, New Zealand, and by fused glass bead inductively coupled plasma mass spectrometry (ICP-MS) methods at Washington State University, Pullman, USA (see Mortimer *et al.* 2010 for methods). Small (*c.* 5–10 g) samples were crushed in agate mortars, and analysed by XRF and fused bead ICP-MS methods at Washington State University. Two very small (*c.* <5 g) samples were analysed by laser ICP-MS scanning of polished mounts at the University of Tasmania. Except for the latter two samples, we report Ba, Cr, Cu, Ni, V, Sc and Zn as analysed by XRF instead of ICP-MS.

Argon geochronology at the University of California Santa Barbara followed methods described in Mortimer *et al.* (2014*a*). U–Pb laser ICP-MS geochronology at the University of Tasmania followed methods described in Sack *et al.* (2011). Laser ICP-MS pyroxene geochemistry methods at the University of Otago, and Nd isotope methods at the University of Otago followed methods described in Mortimer *et al.* (2014*a*) and Weis *et al.* (2006), respectively.

Samples prefixed 'P' are archived in GNS Science's National Petrology Reference Collection, as are thin sections, rock powders and mineral separates. Location information, sample descriptions and images, and analytical data are stored in the online Petlab database (http://pet.gns.cri.nz: Strong *et al.* 2016).

Data and results

In this section we describe sample petrography, age and composition under three main geological–geographical headings. Sample locations are given in Table 1, a dating summary in Table 2, and full geochemical and Nd isotope data in Tables 3 and 4. Full geochronological, mineral composition and micropalaeontological data are given in the Supplementary material. A recurring theme in our dataset is the difficulty in extracting reliable primary geochemical and geochronological signals. The samples still yield

Table 2. *Reported ages of rocks dated in this study*

Cruise and dredge No.	GNS No.	Lab. No.	Dated material	Quality	Age ± 2σ (Ma)
Lord Howe Seamount Chain					
ECOSAT DR15A Horsehead	na	na	Forams	na	Late Early Miocene?
ECOSAT DR15B Horsehead	P82218Bi	SB66-25	Ar/Ar plagioclase	Low	48 ± 3?
ECOSAT DR15B Horsehead	P82218C	SB66-26	Ar/Ar groundmass	Medium	27.2 ± 0.5
ECOSAT DR15C Horsehead	na	Q1135	Forams	na	Pleistocene
ECOSAT DR16Ai Chesterfield	P82221	SB66-28	Ar/Ar groundmass	Medium	28.1 ± 1.0
ECOSAT DR16Aiv Chesterfield	P82224	SB66-32	Ar/Ar groundmass	Low	>23
New Caledonia basin margins					
ECOSAT DR11Ei Le Noroit	na	na	U–Pb zircon	na	63.8 ± 3.1
ECOSAT DR11Ev Le Noroit	na	na	U–Pb zircon	na	65.5 ± 4.2
ECOSAT DR17A Lansdowne	na	na	Forams	na	Early Miocene
ECOSAT DR17Aii Lansdowne	P82230	SB66-3	Ar/Ar groundmass	Low	>78
ECOSAT DR17Aiii Lansdowne	P82231	SB66-4	Ar/Ar groundmass	Low	>81
ECOSAT DR17Gi Lansdowne	na	na	Forams	na	Late Oligocene–Early Miocene
ECOSAT DR18Bii Nereus	P82240	SB66-5	Ar/Ar groundmass	Low	>58
ECOSAT DR18Ci Nereus	na	na	Forams	na	Early Miocene
AUSFAIR DR05-B1 Fairway Ridge	P81400	SB65-7	Ar/Ar hornblende	High	74.1 ± 0.3
IPOD DR4-VRAC2 New Caledonia	P84022	SB67-65,66,67	Ar/Ar groundmass, plagioclase	Low	42 ± 5?
North Zealandia scattered seamounts					
ECOSAT DR08Ai Loyalty Ridge	P82194	SB66-44	Ar/Ar plagioclase	High	24.6 ± 0.3
GEORSTOM 357 D1 Lord Howe Rise	P57145	SB63-5	Ar/Ar plagioclase	High	27.0 ± 0.3
GEORSTOM 346 D1 Norfolk Ridge	P78644	SB61-108	Ar/Ar groundmass	High	18.1 ± 0.2
DRASP d19C Reinga Basin	P83198	SB67-62	Ar/Ar groundmass	Low	25.5 ± 2.5
DRASP d27C Aotea Basin	P83225	SB67-29,60	Ar/Ar groundmass, plagioclase	Medium	27.7 ± 1.2
DRASP d27C Aotea Basin	P83226	SB67-31	Ar/Ar groundmass	Medium	29.5 ± 1.5
DRASP d02C Aotea Seamount	P83160	SB67-24	Ar/Ar groundmass	Medium	22.5 ± 0.2

na, not applicable. All quoted numerical age uncertainties in this table and in the text are ±2σ. High, medium and low quality of Ar/Ar samples is explained in Mortimer *et al.* (2014*a*).

Table 3. *Whole-rock geochemical data for this study*

Dredge	GNS #	Approx wt.	Rock description
Lord Howe Seamount Chain			
ECOSAT DR15Bi	P82218A	10 g	Cpx plag basalt clast in limestone
ECOSAT DR15Bii	P82218B	10 g	Cpx plag basalt clast in limestone, dark reddish brown
ECOSAT DR16Ai	P82221	<5 g	Aphyric subtrachytic textured lava. Clay alteration
ECOSAT DR16Aiv	P82224	<5 g	Sparsely plag pptic lava. Dk brown clay alteration
Lord Howe Island LH2	GD6060	250 g	Doleritic plag-cpx basalt, zeol amygdaloidal
New Caledonia Basin margins			
ECOSAT DR17Aiv	P82232	<5 g	V strongly altered reddish volc sandstone, basaltic devitrified?
ECOSAT DR17Av	P82233	<5 g	Reddish plag-porphyritic andesite, strongly altered
ECOSAT DR18Bi	P82239	800 g	Strong hydroth altered (zeol) red aphyric devitrified andesite
ECOSAT DR18Bii	P82240	50 g	Strong hydroth altered red plag-pptic devitrified andesite
ECOSAT DR18Di	P82246	20 g	Strong hydroth altered orange red plag-ol basalt
ECOSAT DR11Ei	na	<5 g	Grey aphyric trachyte corestone to Mn nodule
ECOSAT DR11Ev	na	<5 g	Grey aphyric trachyte corestone to Mn nodule
AUSFAIR DR03-D1	P81396	300 g	Plag-hbl-bi lava, mafics altered
AUSFAIR DR03-G1	P81397	250 g	Plag-altd mafic lava w qtz or zeol amygdules
AUSFAIR DR05-B1	P81400	100 g	Plag cpx hbl oliv andesite clast in volcanic breccia
Isolated North Zealandia seamounts			
GEORSTOM 357 D1	P57144	50 g	Alkali basalt host to gabbro xenolith
GEORSTOM 346 D1	P78644	200 g	Vesicular ol-plag basalt
DRASP d02C	P83160	500 g	Fresh ol+cpx lava. Minor clay alteration.
DRASP d19C	P83198	500 g	Dark brownish grey olivine micro-phyric basalt. Cc amygdules
DRASP d27C	P83226	2 kg	Sparsely cpx porphyritic, highly vesicular basalt. Clay altered
IPOD DR4-VRAC2	P84022	>100g	Sparsely plag-cpx porphyritic amygdaloidal dacite fresh glass
ECOSAT DR08Ai	P82194	>100g	Ol-plag pptic basalt clast in breccia. Cc & zeol amygdules

useful provisional data, despite small sample size, secondary alteration of mafic minerals and glass to clays, and the presence of zeolites, phosphates and carbonate in amygdules, veins and matrix patches. Concentrations of incompatible large-ion lithophile elements (LILEs), such as Rb, Sr, Th and K, are very vulnerable to such secondary alteration and we have avoided using these elements in our interpretations. We have been conservative in our evaluations of the primary composition of the rocks, placing more emphasis on relatively immobile high field strength elements (HFSEs), such as Nb, Zr and Ti. We have also increased many statistically geochronological age uncertainties where mineralogy and degassing behaviour indicates poor sample quality.

A number of samples display petrographical evidence for low-temperature alteration or weathering but we are confident, except where otherwise noted, that most of the ages reported here are primary eruption and crystallization ages, not alteration ages. This is justified by petrographical characteristics of most samples (e.g. pristine igneous zoning, and twinning in both phenocrystic and groundmass plagioclase), relatively high radiogenic yields of most samples, and apparent K/Ca ratios from both groundmass and plagioclase separates that are typical of igneous compositions, not hydrothermal phases. Thus, we attribute most of the complexities in the argon spectra to the fact that we are dating very low K/Ca materials, and to issues with argon loss and reactor-induced recoil from very-fine-grained groundmass samples.

Lord Howe Seamount Chain

The Lord Howe Seamount Chain consists of at least 19 definable volcanic centres (Fig. 1) (Missegue & Collot 1987; McDougall & Duncan 1988; Quilty 1993). Prior to this study, lavas had been sampled at Lord Howe Island, Middleton Reef and Gifford Guyot, all in the southern part of the chain (McDougall *et al.* 1981; Mortimer *et al.* 2010; Dadd *et al.* 2011). We present analytical results of samples from the two northernmost seamounts in the chain: the informally named Horsehead Seamount (ECOSAT DR15); and from a dredge site approximately 50 km north of the Chesterfield Islands (ECOSAT DR16). Both sites yielded only small, altered samples, and we acknowledge the data are poor quality and interpretations provisional.

Horsehead. From ECOSAT dredge 15, approximately 5 kg of hard cream-coloured shallow-water, Mn-crusted limestone was recovered. Several small (0.5–3 cm diameter) pebbles of lava were present between the Mn crust and limestone, with some fully enclosed by the limestone. P82218A and Bi are approximately 1.5 × 1 × 1 cm red-brown plagioclase–pyroxene porphyritic lava pebbles, and P82218C is a grey lava pebble. No pebbles were fresh, all showed extensive clay alteration.

Laboratory	SiO_2	TiO_2	Al_2O_3	$Fe_2O_3^T$	MnO	MgO	CaO	Na_2O	K_2O	P_2O_5	LOI	Total
WSU												
WSU												
WSU												
WSU												
SCA, WSU	48.39	2.78	15.23	11.25	0.14	5.67	8.55	3.63	1.47	0.51	2.16	99.78
WSU												
WSU												
WSU	50.35	0.56	17.9	6.83	0.14	5.24	4.01	5.46	1.76	0.46	7.11	99.81
WSU												
WSU												
UTAS	58.78	0.77	16.38	6.66	0.03	0.59	4.48	4.70	5.97	1.64		100.00
UTAS	58.66	0.96	17.57	6.83	0.03	0.46	3.17	4.58	6.92	0.82		100.00
WSU	72.17	0.28	14.13	2.20	0.07	0.58	0.68	2.90	5.29	0.10	1.56	94.76
WSU	67.29	0.78	14.55	5.15	0.07	1.17	0.61	3.29	4.31	0.25	2.46	99.93
WSU	50.44	1.20	19.38	7.02	0.05	2.43	6.75	3.81	1.19	1.19	6.10	92.26
SCA, WSU	34.78	1.99	10.04	10.55	0.14	10.47	16.97	2.61	1.41	0.93	9.30	99.19
OU	47.02	2.44	19.84	10.14	0.16	2.87	9.63	3.53	1.38	0.90	2.15	100.06
SCA	39.87	3.77	11.69	15.39	0.18	8.65	12.20	4.26	1.44	1.27	1.04	99.75
SCA	45.06	2.07	13.49	13.80	0.16	8.76	10.32	2.79	0.95	0.63	1.33	99.36
SCA	42.95	2.11	15.95	12.61	0.26	3.12	11.19	3.75	1.28	2.70	2.96	98.88
SCA, OU	65.07	0.49	13.41	5.87	0.16	1.51	3.80	2.98	1.65	0.12	4.33	99.39
WSU	53.40	3.27	16.86	7.27	0.08	2.16	9.72	2.35	0.38	1.26	2.92	99.68

The argon spectrum of groundmass from P82218C was fairly flat (slightly hump-shaped) for the first 75% of gas released, with good K/Ca ratios (0.03–0.06) and high radiogenic yields (90%–95%) for most of the spectrum (Fig. 2a). Ages and K/Ca ratios both dropped at the highest temperatures due to recoil. We used the broad flat, central step at 27.2 ± 0.5 Ma as a preferred (igneous) age for the sample, expanding the statistical uncertainty to reflect the broad hump shape and the likely minor influence of both argon loss and recoil. The argon spectrum of plagioclase from Horsehead P82218Bi started with ages of *c.* 45 Ma and then climbed to *c.* 75 Ma. K/Ca ratios were 0.013–0.050, reasonable for igneous plagioclases. Radiogenic yields for the low-temperature steps were high. However, there were no good isochrons and we suspect there is excess argon in the higher temperature steps of this plagioclase. At face value, a preferred age might be *c.* 48 ± 3 Ma (arbitrary error assigned to bound the ages of the four low-temperature steps) but our confidence in this is low, with an older interpretation possible. We do not use P82218Bi to date the Lord Howe Seamount chain volcanism.

The hard limestone from Horsehead (DR15A) enclosing some volcanic clasts gave a tentative late Early Miocene age, in agreement with the dated clasts from Horsehead being older. Soft white limestone from Horsehead (DR15C) gave a Pleistocene age.

Because of the small sample size, only ICP-MS trace elements could be obtained from two Horsehead clasts P82218A and 82218C (Table 3). The lack of major element data means they cannot be plotted in Figure 3a, b, but petrography and Sc content indicate a likely basaltic composition (Winchester & Floyd 1977). Zr content and convex-up multi-element-normalized patterns with peaks at Ta and Nb match those of typical intraplate basalts (Figs 3c, d & 4a). Horsehead is arguably the only Lord Howe Seamount Chain volcano to have erupted on oceanic, rather than continental, crust. Yet, it has the second lowest initial ε_{Nd} of all four Lord Howe Seamount Chain volcanoes for which we have data (Fig. 5). The initial ε_{Nd} value falls within the range of the enriched mid-ocean ridge basalts (E-MORBs) of the South Rennell Trough immediately to the north and is not dissimilar to a Lord Howe Island lava (Table 4; Fig. 5).

Chesterfield Islands. From ECOSAT dredge 16, approximately 0.2 kg of separate, pebble-sized pieces of lava and limestone were obtained. P82221 is a 4 × 2 × 1 cm angular piece of subtrachytic, sparsely olivine-phyric basalt. There is fresh plagioclase in the groundmass, but, otherwise, the groundmass and olivine are completely replaced by clays. P82224 consists of three approximately 1 cm pieces of olivine–plagioclase porphyritic altered basalt with a subtrachytic groundmass; clay fills amygdules.

Table 3. *Continued.*

Dredge	As	Ba	Ce	Cr	Cs	Cu	Dy	Er	Eu	Ga	Gd	Hf	Ho	La	Lu	Nb
Lord Howe Seamount Chain																
ECOSAT DR15Bi		122	38.3		0.43		5.00	2.63	1.86		5.12	4.88	1.01	17.0	0.33	22.8
ECOSAT DR15Bii		149	39.1		0.5		5.58	2.76	2.02		6.01	5.84	1.07	23.9	0.33	27.5
ECOSAT DR16Ai		52	29.9		0.89		6.03	3.13	1.94		6.06	3.41	1.21	17.9	0.39	14.8
ECOSAT DR16Aiv		60	22.9		0.53		8.13	3.89	2.86		9.37	3.11	1.61	26.9	0.43	13.6
Lord Howe Island LH2	<1	343	73.4	160	0.08	51	5.66	2.56	2.74	25	6.95	6.29	1.03	36.7	0.29	41.1
New Caledonia Basin margins																
ECOSAT DR17Aiv		157	11.1		17.5		2.78	1.59	0.88		2.64	1.23	0.60	5.51	0.21	1.36
ECOSAT DR17Av		183	30.6		5.78		5.50	2.94	1.72		5.65	3.12	1.12	13.2	0.40	3.98
ECOSAT DR18Bi	13	15	11.5	61	0.23	77	2.95	1.91	0.9	19	2.85	0.97	0.66	7.15	0.27	0.81
ECOSAT DR18Bii		1770	52.5		24.6		4.47	2.34	1.9		5.69	2.18	0.87	31.1	0.28	1.36
ECOSAT DR18Di		136	19.7		5.00		4.95	3.18	1.22		5.24	1.73	1.14	20.4	0.41	1.04
ECOSAT DR11Ei	36	308	63.2	8	6.80	248	11.2	6.73	1.50	25	11.1	10.4	2.32	34.8	0.88	7.77
ECOSAT DR11Ev	8	455	63.1	3	5.70	433	10.2	5.71	1.61	27	11.2	15.4	2.04	24.6	0.64	10.0
AUSFAIR DR03-D1		846	49.9	6	2.39	18	2.49	1.45	0.75	14	2.65	5.23	0.49	23.1	0.26	7.94
AUSFAIR DR03-G1		913	48.7	5	2.53	5	5.63	3.29	1.34	16	5.57	6.29	1.18	23.3	0.53	8.97
AUSFAIR DR05-B1		244	35.7	29	0.21	70	3.82	2.12	1.46	18	4.16	3.71	0.78	18.8	0.29	5.92
Isolated North Zealandia seamounts																
GEORSTOM 357 D1		386	106	526	0.42	38	8.07	2.64	4.18	18	11.6	8.19	1.28	55.2		73.2
GEORSTOM 346 D1			57.1		0.39		8.92	5.87	3.30		8.47	4.41		56.2	0.57	39.9
DRASP d02C	5	432	207	205		71				24				120		128
DRASP d19C	2	281	104	345		74				19				57		62
DRASP d27C	51	346	43	272		172				19				33		41
IPOD DR4-VRAC2	3	60	11.1	<1	0.58	4	4.12	2.81	0.75	13	3.31	2.41	0.93	4.25	0.46	0.62
ECOSAT DR08Ai	51	142	14.7	395	0.11	7	2.06	0.96	1.10	16	2.24	6.94	0.39	5.39	0.12	54.6

Major elements as wt% oxides, trace elements are ppm. WSU, Washington State University; SCA, Spectrachem Analytical; UTAS, pptic, porphyritic; qtz, quartz; zeol, zeolite.

The Ar/Ar degassing spectrum of Chesterfield sample P82221 groundmass was similar to Horsehead P82218C. However, P82221 gave not quite as flat a spectrum, and also slightly lower K/Ca ratios and radiogenic yields. We used the flat step in the middle of spectrum to give an interpreted preferred age of 28.1 ± 1.0 Ma (uncertainty boosted to greater than just statistical uncertainty because of the sharper hump). The argon release spectrum of the Chesterfield P82224 groundmass was the most difficult to interpret of the three Lord Howe seamount chain groundmass samples. The spectrum was more hump-shaped, with no real flat top; thus, there was more evidence for both argon loss and recoil. We regard the age of the top of the hump, *c.* 23 Ma, as a minimum age (Table 2).

As with Horsehead, because of the small clast size, only ICP-MS trace element data were obtained. Although the multi-element-normalized patterns are also convex-up, both analysed Chesterfield samples (P82221 and 82224) have about half the concentration of LILEs and HFSEs than the Horsehead lavas (Fig. 4a), and their Nb/Yb ratio is lower (Fig. 3d). Although the lack of major elements again hinders assignment of a rock name, they have the features of transitional rather than alkaline basalts. Chesterfield lava P82221 has a higher initial ε_{Nd} value when compared to that of the analysed Horsehead lava (Fig. 5). The reasons for these chemical and isotopic differences are explored in the Discussion later in this paper.

New Caledonia Basin margins

A group of lavas of distinct chemistry and age occur around the margins of the New Caledonia and Fairway basins in the northern part of northern Zealandia (Figs 1, 3, 4b & 5). These samples have not been dredged from well-preserved, upstanding volcanic seamounts. Instead, they are samples of seismic acoustic basement exposed as relatively subdued, current-swept areas (e.g. Le Noroit, IPOD sites), or on fault scarps that expose rocks older than seismic sedimentary cover (e.g. Nereus Reef, Lansdowne Bank, AUSFAIR sites).

Le Noroit seamounts. Although described as seamounts, a seismic line (Daniel *et al.* 1977, fig. 4) shows that they consist of a broad (*c.* 100 × 100 km), partly faulted basement high that has not been completely covered by the sediments of the New Caledonia and d'Entrecasteaux basins. ECOSAT dredge 11 was taken from the highest seamount. It yielded no substantial solid basement rock but only manganese crusts (one up to 11 cm thick) and a few dozen approximately 2–5 cm-diameter

Nd	Ni	Pb	Pr	Rb	Sc	Sm	Sr	Ta	Tb	Th	Tm	U	V	Y	Yb	Zn	Zr	GNS No.
18.4		7.04	4.13	10.7	33.3	4.91	431	2.21	0.85	2.23	0.36	1.13		29.9	2.15		195	P82218A
23.5		21.2	5.42	12.0	32.4	5.96	446	2.58	0.96	2.43	0.37	1.62		24.2	2.25		269	P82218B
21.1		1.34	4.70	11.1	33.6	5.41	334	1.02	0.98	1.28	0.43	0.64		36.0	2.47		131	P82221
34.0		1.48	7.43	21.6	34.6	8.49	291	1.59	1.40	1.38	0.51	0.59		49.9	2.79		111	P82224
35.1	72	7.00	8.83	27.3	19.0	7.66	690	2.30	1.00	3.81	0.35	0.90	275	27.4	2.06	102	270	GD6060
8.46		5.31	1.76	78.2	23.2	2.35	402	0.99	0.44	1.02	0.22	0.81		16.7	1.38		46	P82232
21.8		12.4	4.71	74.8	37.8	5.76	258	2.36	0.91	2.82	0.43	1.02		30.9	2.56		122	P82233
9.60	65	4.50	2.01	3.60	14.4	2.51	115	0.06	0.45	1.00	0.27	0.39	174	23.9	1.59	92	39	P82239
36.7		6.87	8.91	207	22.8	7.10	512	0.08	0.80	4.80	0.31	1.07		26.4	1.76		80	P82240
19.2		8.67	4.16	69.1	30.2	4.32	114	0.07	0.77	1.79	0.44	1.59		46.8	2.55		72	P82246
43.9	22	21.9	10.2	113	10.8	10.7	153	0.63	1.75	13.1	0.94	3.43	136	71.1	6.16	232	374	na
40.5	5	36.2	8.91	115	17.6	10.4	134	0.69	1.66	12.7	0.75	2.42	96	54.3	4.68	555	606	na
18.4	7	11.1	5.11	122	3.40	3.39	181	0.64	0.41	11.4	0.23	1.91	38	13.4	1.55	55	195	P81396
26.5	5	12.9	6.49	108	12.7	6.02	293	0.65	0.92	9.51	0.52	2.45	45	29.7	3.31	105	239	P81397
24.0	34	5.24	5.68	10.9	13.7	4.85	1400	0.39	0.64	2.01	0.31	0.43	167	26.1	1.81	173	122	P81400
54.7	61	3.71	12.6	31.5	17.2	13.3	917	4.56	1.68	6.67	0.30	1.77	136	33.1	1.56	135	347	P57144
41.4			9.86			8.29		2.70		4.80		2.41		63.4	4.55		187	P78644
	184	9		40	16		800			13		4	253	42		155	458	P83160
	236	6		24	20		579			6		1	174	27		121	208	P83198
	114	3		19	28		629			4		5	225	32		153	149	P83226
8.57	9	3.69	1.63	14.7	20	2.57	117	0.05	0.61	0.61	0.42	0.36	24	24.5	3.04	81	83.9	P84022
8.45	25	0.85	1.80	3.9	29.8	2.14	787	4.22	0.36	2.40	0.13	0.23	209	9.10	0.80	45	350	P82194

University of Tasmania; OU, University of Otago. bi, biotite; cc, calcite; cpx, clinopyroxene; hbl, hornblende; ol, olivine; plag, plagioclase;

Table 4. *Nd isotope data for samples selected from this study (first 10 rows), and from Mortimer* et al. *(2010, 2014*a) *and Seton* et al. *(2016*a) *(last nine rows)*

Location	Dredge	GNS No.	$^{147}Sm/^{144}Nd$	$^{147}Sm/^{144}Nd$ (measured)	±2 SE	Age (Ma)	$^{147}Sm/^{144}Nd$ (initial)
Lord Howe Seamount Chain							
Horsehead	ECOSAT DR15Bi	P82218A	0.1609	0.512739	7	28	0.512711
Chesterfield	ECOSAT DR16Ai	P82221	0.1545	0.512826	7	27	0.512799
Lord Howe Island	na	GD6060	0.1303	0.512735	6	7	0.512729
Middleton Reef	NORFANZ DR49	P69722	0.1271	0.512883	8	12	0.512873
New Caledonia basin margins							
Le Noroit	ECOSAT DR11Fi	P82196	0.1283	0.512917	6	64	0.512863
Lansdowne Bank	ECOSAT DR17Aiv	P82232	0.1672	0.512896	7	100	0.512787
Nereus Reef	ECOSAT DR18Bi	P82239	0.1574	0.512847	6	100	0.512744
Nereus Reef	ECOSAT DR18Bi	P82239 dup	0.1574	0.512837	8	100	0.512734
Nereus Reef	ECOSAT DR18Bii	P82240	0.1164	0.512951	9	100	0.512875
Nereus Reef	ECOSAT DR18Bii	P82240 dup	0.1164	0.512950	8	100	0.512874
Eocene–Oligocene back-arc basin basalts							
Rennell Ridge SW side	GEORSTOM3 DR301A	P78604	0.2537	0.513012	8	38	0.512949
South Rennell Trough	GEORSTOM3 DR308A	P78613	0.2018	0.512996	8	28	0.512959
D'Entrecasteaux Ridge	GEORSTOM3 DR316-23	P78622	0.1669	0.513101	9	40	0.513057
Rennell Ridge NE side	ECOSAT DR03Ai	P82182	0.1503	0.512798	8	40	0.512759
West Torres Plateau	ECOSAT DR04Aii	P82188 rind	0.2006	0.512881	7	35	0.512835
West Torres Plateau	ECOSAT DR04Aii	P82188 inside	0.1973	0.512804	6	35	0.512759
East Laperouse Ridge	ECOSAT DR06A	P82193	0.1606	0.512727	8	35	0.512690
South Rennell Trough	ECOSAT DR14Ei	P82204	0.1916	0.51264	7	28	0.512605
South Rennell Trough	ECOSAT DR14Eiii	P82206	0.1539	0.512659	8	28	0.512631

The Chondritic Uniform Reservoir (CHUR) values used are: $^{143}Nd/^{144}Nd = 0.512638$ (present day), $^{147}Sm/^{144}Nd = 0.1967$, $\lambda^{147}Sm = 6.54 \times 10^{-12}\ a^{-1}$. na, not applicable.

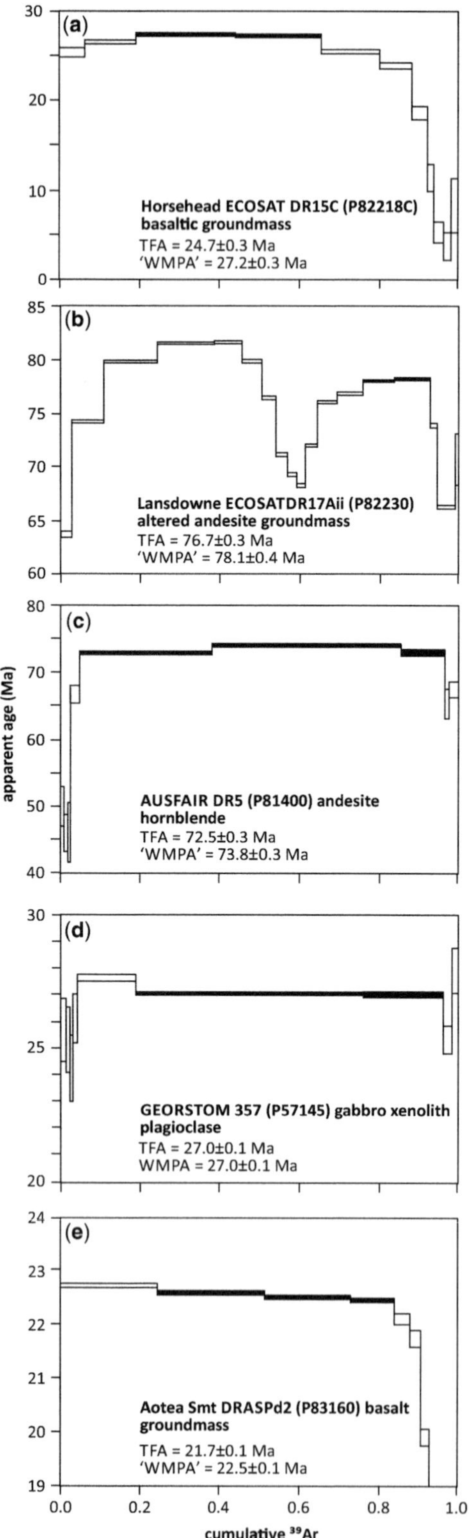

Fig. 2. Ar/Ar degassing spectra of selected samples. The height of rectangles in this figure are $\pm 1\sigma$ but reported ages are $\pm 2\sigma$. Black rectangles are those used to interpret the age. For degassing spectra, isochron plots, K/Ca plots and raw data for all samples, see the Supplementary material.

manganese nodules. The nodules contained <1 cm angular kernels of pale grey, green and red lava in their cores, as well as various calcareous and phosphatic rocks.

Automated scanning electron microscopy (SEM) of 18 polished manganese nodules identified small (7–30 μm), rare ($n = 15$) zircons in two of the lava kernels which were then dated by *in situ* laser ablation ICP-MS (LA-ICP-MS) methods. For DR11Ei, 10 of the 11 zircons were Paleocene in age and one was Middle Jurassic (Fig. 6). The Paleocene zircons gave an intercept age of 63.8 ± 3.1 Ma that we interpret as the crystallization age of the lava. The single 168 Ma zircon is probably a xenocryst incorporated into the lava from underlying (continental) crust. The four zircons found in DR11Ev collectively give an intercept age of 65.5 ± 4.2 Ma, within the error of DR11Ei.

The chemistry of the two Le Noroit lavas reveals them to be trachytes, slightly more siliceous than trachyandesite and showing an iron-enrichment trend (Fig. 3a, b). Their overall high incompatible element content (Figs 3c & 4b) indicates they are part of an alkaline suite. We explain their relatively low Eu, Nb, Ta and Ti content in part due to fractionation of plagioclase, amphibole, biotite, ilmenite and titanite. The initial ε_{Nd} of Le Noroit lava DR11Fi (Table 4) is slightly more radiogenic than the Lord Howe Seamount Chain lavas (Fig. 5).

Lansdowne Bank and Nereus Reef. Launay *et al.* (1977, fig. 2) and Collot *et al.* (2008, figs 11 & 13) presented seismic profiles across the northern Fairway Ridge. Two dredges were taken from the northern Fairway Ridge on the ECOSAT cruise, one on the steep, NE side of the Lansdowne Bank (DR17) and another on the steep, NE side of Nereus Reef (DR18) (Fig. 1). Although 65 km apart, similar lithologies occur in the two dredges and we describe them together.

Lansdowne Bank yielded 18 small pieces of hard, dark green-grey, fine- to medium-grained aphyric lava and volcaniclastic sandstone. About half of these were >10 mm in size (the largest being 15 × 10 × 5 mm), and the other half <5 mm. All pieces were angular in shape. The fact that some samples have clean broken faces and very thin (<0.5 mm) crusts on non-broken faces suggested they probably were broken off *in situ* rock outcrops. In thin section, the lavas contain many secondary minerals including

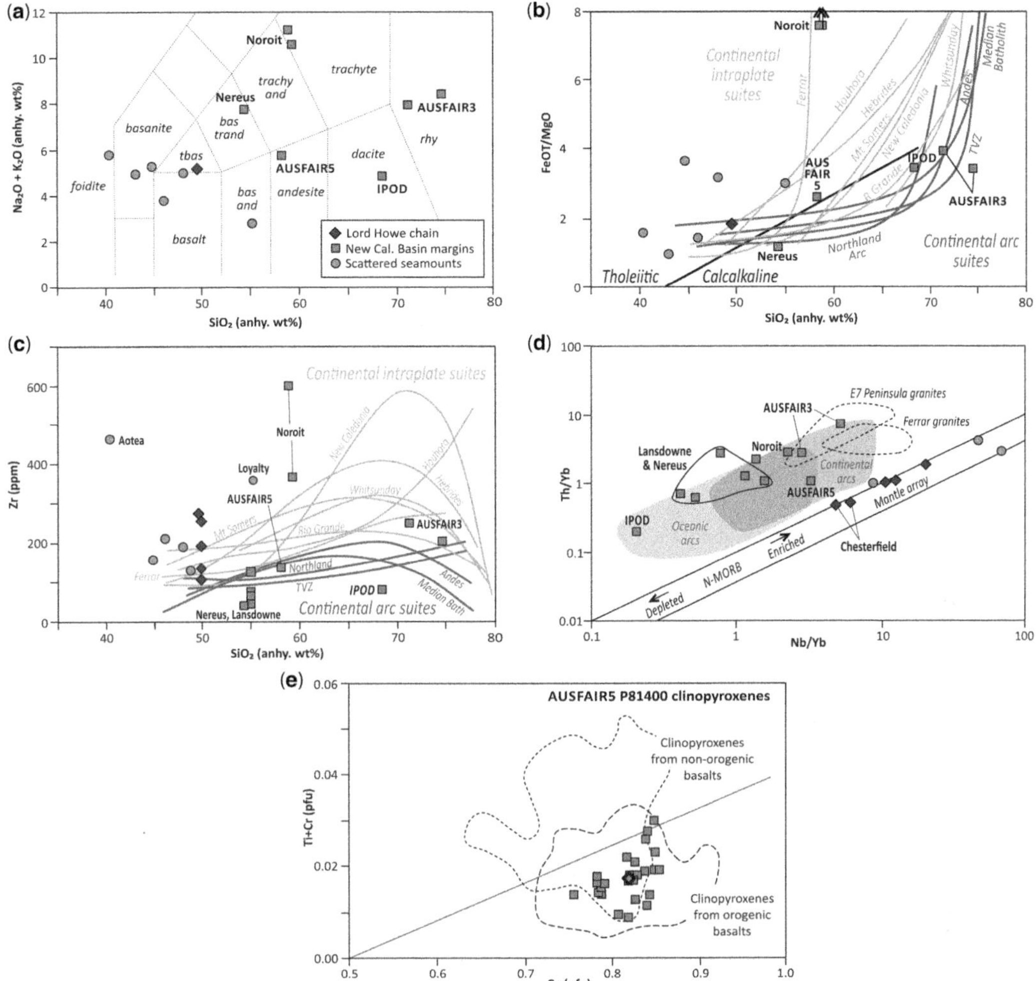

Fig. 3. Binary geochemistry plots of new analyses. **(a)** Whole-rock anhydrous SiO_2 v. anhydrous $Na_2O + K_2O$ from Le Maitre (1989). **(b)** Whole-rock anhydrous SiO_2 v. FeO^T/MgO, tholeiitic and calc-alkaline dividing line from Miyashiro (1974). Lines are second-order polynomial curve fits to various subduction-related and intraplate datasets from McMillan *et al.* (2000), Nicholson *et al.* (2011), Steiner & Streck (2014), and the Petlab (http://pet.gns.cri.nz and Georoc (http://georoc.mpch-mainz.gwdg.de/georoc/ databases. **(c)** Whole-rock anhydrous SiO_2 v. Zr. Note that, on the basis of petrography and Sc content, four Lord Howe Seamount Chain basalts for which there are no SiO_2 analyses are plotted at $SiO_2 = 50$ wt% and four northern Fairway Ridge basaltic andesites are plotted at $SiO_2 = 55$ wt%. Lines are second-order polynomial curve fits to various subduction-related and intraplate datasets from same sources as (b). **(d)** Whole-rock Nb/Yb v. Th/Yb after Pearce & Peate (1995). Reference Edward VII Peninsula and Ferrar Antarctic intraplate granites from Storey *et al.* (1988) and Weaver *et al.* (1992). **(e)** Clinopyroxenes are from andesite P81400; reference fields are from Leterrier *et al.* (1982).

chlorite and zeolite. From Nereus Reef, a dozen angular pieces of red coloured, hydrothermally altered, veined and brecciated plagioclase–augite porphyritic lava were dredged, the largest was DR18Bi (10 × 8 × 6 cm) and the smallest 3–4 cm in size. Despite the obvious alteration of the Lansdowne and Nereus samples, three samples were chosen for Ar/Ar geochronology of groundmass separates and four samples for whole-rock geochemistry. Because of the small sample sizes, different samples had to be selected for the different kinds of analysis.

All three Ar/Ar dated samples from the northern Fairway Ridge had complex gas-release spectra in the form of two humps (e.g. Fig. 2b; see the Supplementary material for all three spectra). The low-temperature humps represent degassing of

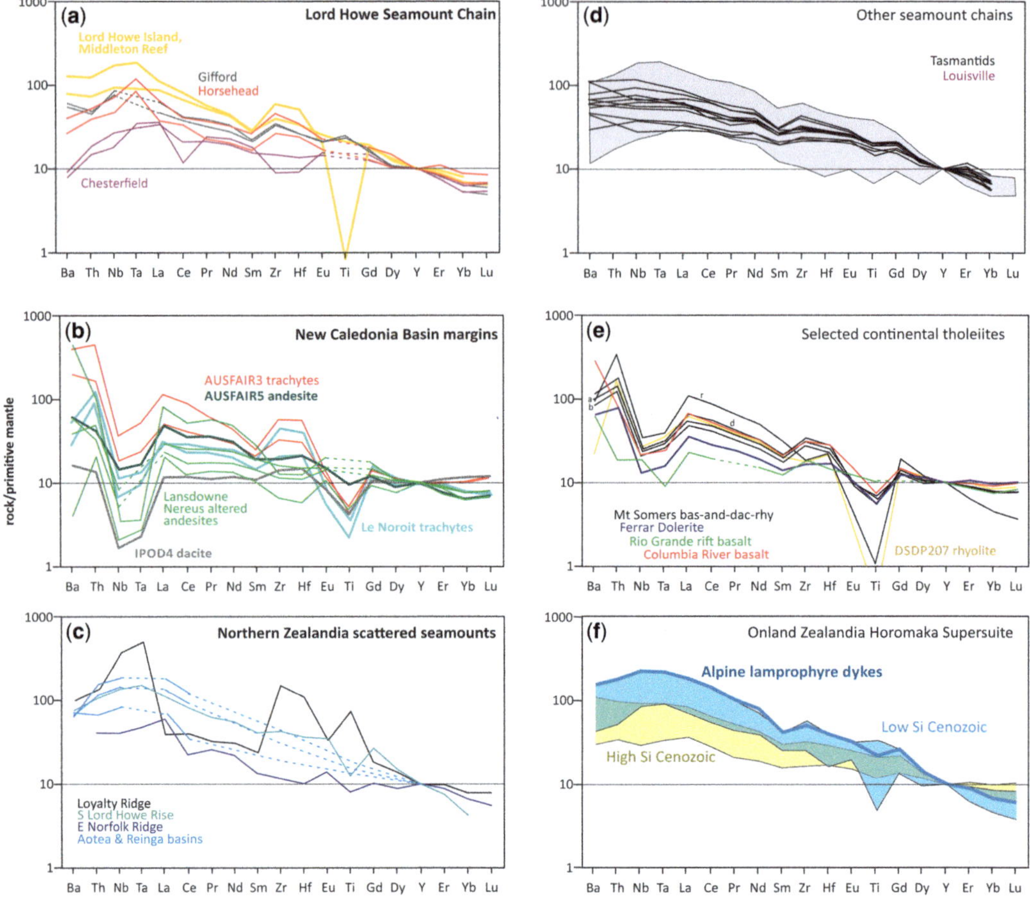

Fig. 4. Multi-element diagrams normalized to primitive mantle of Sun & McDonough (1989). All analyses have been double normalized to $Y_n = 10$ for better comparison between variably differentiated samples. Cs, Rb, U, K, Pb, Sr and P have been omitted because of substantial secondary alteration effects to their concentrations. **(a)** Lord Howe Seamount Chain lavas (Mortimer *et al.* 2010; Dadd *et al.* 2011; this study). **(b)** Lavas from periphery of New Caledonia Basin (this study). **(c)** Northern Zealandia scattered seamounts (this study). **(d)** Seamount chain reference data from Eggins *et al.* (1991) and Beier *et al.* (2011). **(e)** Continental tholeiite reference data from McMillan *et al.* (2000), Tappenden (2003) and Steiner & Streck (2014). **(f)** Cenozoic southern Zealandia data from Timm *et al.* (2010).

high K/Ca material, probably clays, and show the combined effects of argon loss and reactor-induced recoil. The higher-temperature humps had higher K/Ca (0.4–0.8) probably corresponding to degassing of groundmass adularia of hydrothermal origin. The ages in the high-temperature domains climbed to *c.* 78 Ma in Lansdowne P82230, *c.* 81 Ma in Lansdowne P82231 and *c.* 58 Ma in Nereus P82240. Based on the abundance of secondary minerals in thin sections, even the oldest ages in the high-temperature humps are likely to be partly or wholly alteration ages. As such, we interpret these as minimum ages for the stratigraphic ages of the lavas (Table 2), which could actually be as old as Permian or Early Cretaceous.

Supporting age information was obtained from Lansdowne Bank limestone DR17Gi. This is a separate sample of hard foraminiferal limestone (not enclosing, or in contact with, any of the lavas); *Miogypsinoides* suggests an earliest Miocene, or possibly latest Oligocene, age and a shallow-water, tropical palaeoenvironment. From Nereus Reef, foraminiferal limestone DR18Ci contains approximately 30% red clasts similar to the altered volcanics described above. The limestone contains a juvenile *Lepidocyclina*, which indicates an Early–Middle Miocene age and a palaeowater depth shallower than about 100 m.

The small sample size and extreme secondary alteration of the northern Fairway Ridge lavas

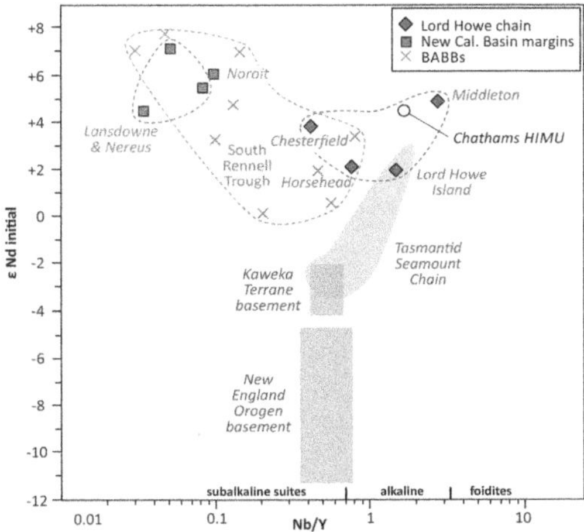

Fig. 5. Whole-rock Nb/Y v. initial ε_{Nd} of northern Zealandia lavas. Lord Howe Seamount Chain and New Caledonia Basin margins described in this paper. Back-arc basin basalts (BABBs) described in Mortimer *et al.* (2014*a*) and Seton *et al.* (2016*a*). The degree of alkalinity on the *x*-axis is after Winchester & Floyd (1977), the Tasmantid Seamount field is from Eggins *et al.* (1991), and the basement fields (that span range of initial ε_{Nd} from 100 to 0 Ma) from Mortimer *et al.* (2008*b*) and Price *et al.* (2015).

present considerable difficulty in the interpretation of their primary geochemistry. Based on petrography and Sc content, all five samples appear to be altered basaltic andesites and andesites. Zr varies from 39 to 122 ppm, and Ba, Th, Nb, La and Ce show even more inter-sample variation (Table 3; Figs 3c & 4b), but to what extent this variation is primary or secondary is hard to assess. The five northern Fairway ridge lavas overlap in trace element composition with the two Le Noroit lavas, but show more extreme negative depletion in Nb and Ta and, as expected, lower Nb/Yb ratios (Fig. 3d). The initial ε_{Nd} of the Nereus and Lansdowne lavas (Table 4; Fig. 5) overlap those of the South Rennell Trough spreading centre, the Lord Howe Seamount Chain and Le Noroit seamounts.

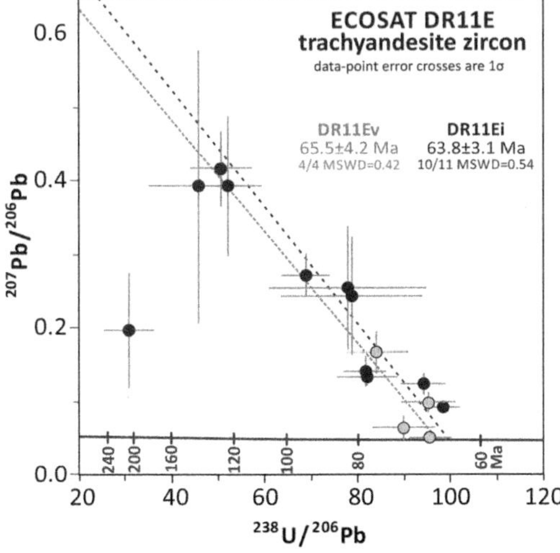

Fig. 6. U–Pb Tera-Wasserburg plot of the Le Noroit zircons.

AUSFAIR5, southern Fairway Ridge. Despite its subdued bathymetric expression south of 26°S, Collot *et al.* (2009) showed that the Fairway Ridge continues as a well-defined magnetic and structural feature that is co-linear with the West Norfolk Ridge and divides the New Caledonia Basin from the Aotea Basin (Fig. 1). Dredge site 5 on the AUSFAIR cruise (Colwell *et al.* 2006) was made on the steep, eastern side of this part of the Fairway Ridge, and is labelled as 'Northern West Norfolk Ridge' on the seismic profile of figure 5 in Exon *et al.* (2007). The dredged rocks included palagonitic volcanic breccias, some carbonate cemented. The largest and freshest volcanic clast was chosen for study, an approximately 5 × 4 × 2 cm angular plagioclase–augite–hornblende–olivine porphyritic andesite (AUSFAIR-DR5B, P81400). Hornblende is primary and there is no biotite.

Hornblende from AUSFAIR5 sample P81400 was Ar/Ar dated. The gas-release spectrum yielded a 'pseudo-plateau' for approximately 90% of gas released (Fig. 2c). We used the step at the top of the central hump, with the highest precision and radiogenic yield, to give an age of 74.1 ± 0.3 Ma, which we interpret as the age of crystallization of the lava.

Compositionally, the P81400 is a medium-K andesite, and has moderate Fe and Zr for its SiO_2 content (Fig. 3a–c). On a multi-element-normalized diagram (Fig. 4c), the pattern has a prominent negative Nb–Ta anomaly and is similar in shape to the two central Lord Howe Rise trachytes (see below) but is somewhat less fractionated. Analyses of fresh clinopyroxenes from P81400 overlap those of reference basalts from orogenic and non-orogenic settings (Fig. 3e).

AUSFAIR3, central Lord Howe Rise. Although not spatially on the margins of the New Caledonia Basin, the AUSFAIR3 dredge site is on a fault scarp near the edge of the Fairway Basin (Fig. 1) (Colwell *et al.* 2006). Higgins *et al.* (2011) reported U–Pb zircon ages from two lava samples from AUSFAIR-DR3: DR3D1 was described as a trachyte and gave an age of 96.9 ± 0.7 Ma, and DR3G1 was described as a latite (potassic trachyandesite) and gave an age of 74.1 ± 0.7 Ma. In this paper we present the first whole-rock geochemical data of these lavas (Table 3). At face value, Figure 3a indicates they are rhyolites. However, as previously noted by Higgins *et al.* (2011), the lavas contain secondary quartz but no quartz phenocrysts. Because hydrothermal alteration may have increased the silica content of the lavas, we conservatively and loosely refer to both lavas as trachytes. Their high to extreme K_2O content (4.3 and 5.3 wt%) classifies them as shoshonitic and they have the highest LILE (Ba, Th) concentrations of our dataset (Fig. 4b). The two AUSFAIR3 lavas are grouped with the New Caledonia Basin margin lavas because they are of Late Cretaceous age, show pronounced negative Nb and Ta anomalies on normalized multi-element diagrams (Fig. 4c), and are not basalts. The AUSFAIR3 lavas show strong compositional similarities with the 97 Ma high-K to shoshonitic rhyolite described from DSDP 207 on the Lord Howe Rise (Figs 1 & 4e) (Tulloch *et al.* 2009).

IPOD4, offshore New Caledonia. Southwest of New Caledonia, the seafloor descends to the floor of the New Caledonia Basin and is one of the steepest large submarine slopes within Zealandia (with 30° slopes in some places). Just outside the fringing reef near Koumac, near the top of the slope, dredge 4 of the IPOD cruise (IPOD4 in Fig. 1) recovered several decimetre-sized pieces of glassy, autobrecciated and agglomerated vesicular lava. No volcanic edifice was visible in multibeam bathymetry. Some fractures in the lavas were lined with thin Mn crusts and then further infilled with limestone. Some samples in the dredge consisted of angular lava clasts in a micritic limestone. A thin section of IPOD DR4-VRAC2 (P84022) revealed variably devitrified glass with sparse plagioclase phenocrysts and a thin Mn rind.

Two glass and one plagioclase separates from IPOD P84022 were dated by Ar/Ar methods. Degassing spectra from both glass separates were similar, both climbing gradually (one from *c.* 40 up to 45 Ma, the other from *c.* 45 up to 48 Ma) and then plummeting with dropping K/Ca. These spectra showed classic combined recoil and low-temperature Ar-loss features. In the degassing of plagioclase from the same sample, the spectrum was satisfactorily flattish, with most steps within error. The weighted mean plateau age from the first three steps (73% of gas) was 40.0 ± 1.1 Ma. The glass ages had higher apparent precision, but glass in submarine settings has a tendency to trap a non-atmospheric (excess argon) component and thus give spuriously old ages. The plagioclase had very low K/Ca, and tiny signals, but seemed reliable. All things considered, we very provisionally conclude that the ages of all the separates are in rough agreement and that an age of 42 ± 5 Ma should be reported as the age of crystallization of the lava. Morgans (2014) reported a Late Pliocene age for limestone in cracks in the IPOD4 lava, not inconsistent with an Eocene age of eruption.

P84022 is the only glass in our geochemical dataset and, as such, gives reasonably reliable primary compositions (even so, we note it has 4.3 wt% LOI (loss on ignition)). It is a medium-K dacite that has low Zr, Nb and Ta, and the flattest normalized trace element composition (Fig. 4c) of our dataset but only has a small negative Eu anomaly. In this regard, it is curiously basalt-like and very different from the other siliceous igneous rocks shown in Figure 4c, d, which, as expected, have higher trace element concentrations. Based on its Nb/Yb ratio (Fig. 3d), P84022 is derived from very depleted mantle.

Scattered northern Zealandia seamounts

The third group of lavas is not part of either the Lord Howe Seamount Chain or the acoustic basement. Instead, they were sampled from isolated, sometimes partly eroded, volcanic edifices, recognizable as such in multibeam bathymetry or crossing seismic lines. Our new samples are from the Loyalty Ridge, southern Lord Howe Rise, Aotea Basin and Norfolk Ridge area. Similar isolated volcanic centres and cones in northern Zealandia have been described and sampled by van de Beuque *et al.* (1998), Exon *et al.* (2004), Mortimer *et al.* (2010) and Dadd *et al.* (2011), and in southern Zealandia by Timm *et al.* (2010).

ECOSAT8, Loyalty Ridge. ECOSAT dredge 8 was made on an unnamed seamount towards the NW end of the Loyalty Ridge (Fig. 1). Sample DR08Ai (P82194) was an altered plagioclase porphyritic basalt with approximately 30% zeolite and calcite amygdules. The degassing of plagioclase from P82194 gave an excellent flat argon spectrum with a well-defined plateau and isochron. K/Ca ratios were reasonably high (0.013), and radiogenic yields were good for most of the spectrum (60–95%). We interpret the 24.6 ± 0.3 Ma weighted mean plateau age of the sample (700–1150°C steps, 98% of gas) as the age of crystallization of the basalt.

The whole-rock chemistry of this low-K, high-Ti alkali basalt shows the effects of secondary alteration and sediment infiltration. This is particularly noticeable in terms of the high CaO, P_2O_5, LOI and As, and low K_2O. We also note the strong and unusual decoupling/depletion of the rare earth elements (REEs) relative to HFSEs Nb, Ta, Zr and Hf (Fig. 4c), and speculate that this is due to a low REE content in amygdaloidal minerals. Despite this, it is clear from the multi-element-normalized pattern in Figure 4c that P82194 is of ocean island basalt (OIB)-type affinity and is derived from very enriched mantle (rightmost point in Fig. 3d).

GO357, southern Lord Howe Rise. A prominent volcanic edifice on the southern Lord Howe Rise was dredged on the GEORSTOM III SUD cruise (GO357 in Fig. 1). A seismic line across the seamount is shown in Bentz (1974, fig. 10) and Launay *et al.* (1977, fig. 2). Geochemical analyses of the dredged calcite amygdaloidal alkali basalt from GO357 (P57144) and of (petrologically unrelated) gabbro basement xenoliths in the basalt (P57145) were presented by Mortimer (2004).

For the current study, we performed Ar/Ar dating of a plagioclase separate from the xenolith (P57145). This gave a reasonably flat spectrum, although most of the gas came out over a fairly narrow temperature range (Fig. 2d). We interpret the weighted mean plateau age of 27.0 ± 0.3 Ma as the age of rapid cooling of the heated xenolith after eruption and therefore to date, within error, the eruptive age of the enclosing lava.

Like the Loyalty Ridge lava, P57144 has a high LOI and CaO content (Table 3). Except for a small negative Ti anomaly, P57145 has the smooth, convex-up normalized-multi-element pattern typical of intraplate ocean island low-K alkali basalts (Fig. 4c). It has the second highest Nb/Yb ratio of our dataset (Fig. 3d), so is derived from very enriched mantle.

GO346, eastern Norfolk Ridge. A small seamount on the eastern flank of the southern Norfolk Ridge lies approximately 150 km south of Norfolk Island (Fig. 1). It was dredged on the GEORSTOM III SUD cruise (site GO346 in Fig. 1). Subsequently, the Sonne-7 cruise shot a reflection seismic line across the seamount (Hinz 1979). Sample GO346D1 (P78644) is a plagioclase–olivine microporphyritic basalt.

Step heating of a groundmass separate from P78644 gave a simple argon release spectrum with an excellent flat plateau for first 58% of gas. The ages decreased at higher temperatures, probably due to recoil effects. Our preferred age for the crystallization of the basalt is the weighted mean plateau age of 18.1 ± 0.2 Ma. P78644 has the lowest Ti and Nb/Y content of the basalts in this seamount group and is a transitional (tholeiitic to mildly alkaline) low-K basalt. This is reflected in its flattish multi-element-normalized pattern (Fig. 4c).

DRASP19, Reinga Basin. The Reinga Basin lies approximately 70 km east of, and parallel to, the West Norfolk Ridge (Fig. 1). DRASP dredge d19C (DRASP19 in Fig. 1) was made on the NE edge of a subdued high on the west side of the basin. The area around the dredge site was not fully surveyed by multibeam bathymetry but, seemingly, the feature is part of a low, shield-like, volcano (Bache *et al.* 2014*b*, figure on p. 21). Sample d19C (P83198) is a plagioclase–augite–olivine basalt showing extensive clay alteration of olivine and groundmass.

Ar/Ar dating of Reinga Basin P83198 groundmass gave results that were difficult to interpret (see the Supplementary material). Most gas was released at very low temperatures, probably from clays. K/Ca ratios are adequate (0.13–0.21), but the signals were small. Ages monotonically decreased from 26 Ma, indicating a major recoil issue. There was no good plateau or isochron. For a loosely constrained age, we used the first two steps but increased the uncertainty: 25.5 ± 2.5 Ma. Foraminifera in two separate limestone samples from the same dredge were dated by Browne *et al.* (2016) as Early Oligocene and Early–Middle Miocene: that is, close to the lava age. Thin sections show that neither limestone contains volcaniclastic detritus.

No ICP-MS trace element data are available for P83198 but the high TiO_2 and Nb/Y ratio, and the partial but convex-up multi-element-normalized diagram (Fig. 4c), show that the lava is a classic sodic alkali basalt.

DRASP27, west side of the Aotea Basin. At the foot of slope of the SE Lord Howe Rise, a meandering submarine canyon impinges on an ovoid-shaped 10 × 5 km low plateau before debouching on the floor of the western Aotea Basin (Bache *et al.* 2014*b*, figure on p. 23). The plateau has an irregular top but, based on the volcanic rocks recovered, a volcanic–volcaniclastic unit may have been sampled. Sample d27B (P83225; DRASP27 on Fig. 1) is a holocrystalline, coarse-grained basalt comprising interlocking grains of titanaugite, plagioclase and olivine (the latter altered to clay minerals). In contrast, sample d27C (P83226) is a highly vesicular olivine porphyritic basalt.

The overall Ar/Ar gas-release spectrum of P83225 groundmass was fairly flat at *c.* 27 Ma for the first 80% of gas released, then dropped to 23 Ma. K/Ca ratios were good (0.3–0.6). As a reasonable interpretation, we take the weighted mean pseudoplateau age for the first 80% of gas and increase the uncertainty, giving an age of 27.0 ± 0.5 Ma. For plagioclase from the same sample, the spectrum was fairly flat, but descended slightly from 28.5 to 25.0 Ma. K/Ca ratios are approximately 0.025, and radiogenic yield is good at around 55%. The statistical weighted mean plateau age was 27.7 ± 1.2 Ma. It is reassuring to see agreement in age between the two different materials from the same sample. As a preferred crystallization age for P83225, we select the plagioclase age of 27.7 ± 1.2 Ma.

Groundmass from P83226 degassed in a typical hump-shaped fashion, indicating both low-temperature argon loss and recoil. Ages climbed from 26 to 30 Ma, flattened and then descended to 13 Ma at high temperatures. The central pseudo-plateau part of the spectrum gave an age of 29.4 ± 0.5 Ma. Given that it is not strictly a plateau and we are unsure as to which part of the spectrum is most reliable, we report the age as 29.5 ± 1.5 Ma: that is, overlapping or possibly a little older than P83225. Palynomorphs from mudstone d27H from the same dredge gave an early Teurian (66–60 Ma) age (Browne *et al.* 2016), suggesting a stratigraphic relationship of Paleocene mudstone overlain by Oligocene lava.

No ICP-MS trace element data are available for Aotea Basin sample P83226 but the Nb/Y ratio and the partial multi-element-normalized diagram (Fig. 4c) show that the lava is a mildly alkaline basalt, slightly more enriched than the subalkaline P78644.

DRASP2, Aotea Seamount. This prominent seamount at the southern end of the Aotea Basin has long been speculated to be of volcanic origin (Brodie 1965). It is approximately 50 × 15 × 0.8 km in size, elongated in an ENE direction. DRASP cruise dredge d02 from the western end of the seamount obtained plagioclase–titanaugite–olivine basalts (Bache *et al.* 2014*b*, figure on p. 14), thus confirming the volcanic origin. In thin section, olivine is completely altered to clay but the groundmass appears fresh and unaltered.

The argon dating of groundmass from sample d02C (P83160) gave reliable results. As expected, given the very fine-grained matrix, most gas came out at fairly low temperatures. Ages decreased from 22.7 to 22.4 Ma in the first 85% of gas released, then plummeted to ages as low as 7 Ma in the highest temperature steps (associated with degassing of Ca-rich phases). Such a spectrum is typical of recoil. For most of the spectrum, K/Ca is high (1–3), as is the radiogenic signal (85%–92%). Given the evidence for recoil, and monotonically descending ages, we interpret the crystallization age of the lava as 22.5 ± 0.2 Ma.

No ICP-MS trace element data are available for Aotea Seamount sample P83160 but the Nb/Y ratio >2, TiO_2 >3 wt% and the partial multi-element-normalized diagram (Fig. 4c) show that the lava is probably a strongly alkaline to nephelenitic basalt.

Discussion

Lord Howe Seamount Chain

Volcanic rocks have now been sampled and dated from five of the 19 Lord Howe Seamount Chain centres. Limestones dated using foraminifera have been sampled from another two (Figs 1 & 7). Our samples from Horsehead and Chesterfield seamounts, although small and poor in quality, are important as they provide data from the northernmost end of the chain. A linear regression through all Lord Howe Seamount chain lava ages give an average southwards rate of younging of approximately 60 mm a^{-1}, similar to the subparallel Tasmantids seamount chain (McDougall & Duncan 1988; Quilty 1993) and to alignments of 35–2 Ma volcanic centres in mainland Australia (west of Fig. 1) (Knesel *et al.* 2008; Sutherland *et al.* 2012; Davies *et al.* 2015). All chains show a deflection from linearity at *c.* 26–23 Ma (Figs 1 & 7) (Kalnins *et al.* 2015). This indicates, to a first approximation, that the sources of the volcanism of both the Tasmantid and Lord Howe chains were: (a) approximately fixed relative to each other; and (b) to some degree coeval. The sparse dating, sometimes only of post-volcanic guyot limestones, does not allow precise bracketing of the age range of volcanism at any one seamount.

The volcanism along the Tasmantid and Lord Howe chains marks the northwards passage of the Australian Plate over sources of magmatism that are approximately fixed in the mantle. Recent absolute and relative plate motion models (Steinberger *et al.* 2004; Wessel & Kroenke 2008) indicate good agreement between predictions from Indo-Atlantic and Pacific hotspots (Fig. 7b). These updated predictions are affirmatively tested with the measured age progression along the Lord Howe chain (Fig. 7b).

All Lord Howe Seamount Chain samples lie along the mantle array on a Nb/Yb v. Th/Yb diagram (Fig. 3c). Pearce & Norry (1979) used the Zr/Y ratio of basalts to explore the petrogenesis of alkali and tholeiitic basalts. The Zr/Y ratio of most lavas along the Lord Howe Seamount Chain is between 8 and 14 (Fig. 7a), typical of alkali basalts and their differentiates and of the Tasmantids. The notable exception is Chesterfield, which has Zr/Y = 3. The Chesterfield Islands and Bellona platforms represent the largest volume of Lord Howe chain eruptions: they have very shallow-water depths (45–80 m), an area of 16 000 km^2 and resulted from the coalescence of five volcanic centres (Missegue & Collot 1987). The high emplacement rate for the Early Miocene Chesterfield–Bellona eruptive pulse fits with the low Zr/Y ratio, which can indicate a high degree of mantle melting. Published geochemical data from the intraoceanic Tasmantid Seamount Chain (Figs 4d & 7a) indicate a range of alkaline–subalkaline (including picritic) compositions and therefore also a range of melting regimes with time (Eggins *et al.* 1991).

Despite all the Lord Howe Seamount Chain volcanoes erupting through continental crust, all lavas have high initial ε_{Nd} values and point to a minimal degree of crustal contamination (to low ε_{Nd}) by older basement (Fig. 5). The overall high initial ε_{Nd} and multi-element-normalized patterns for the Lord Howe Seamount Chain (with pronounced humps

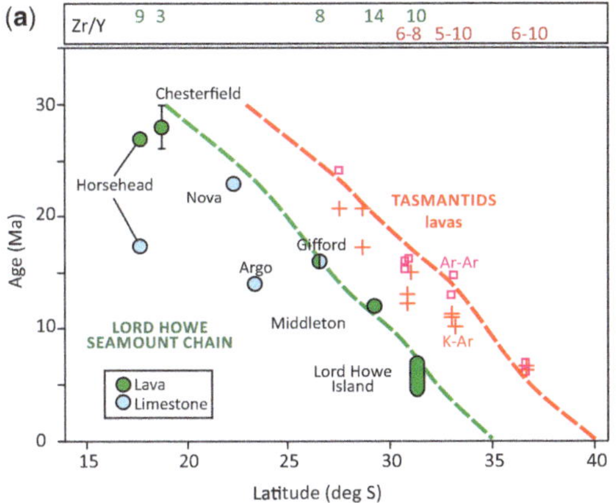

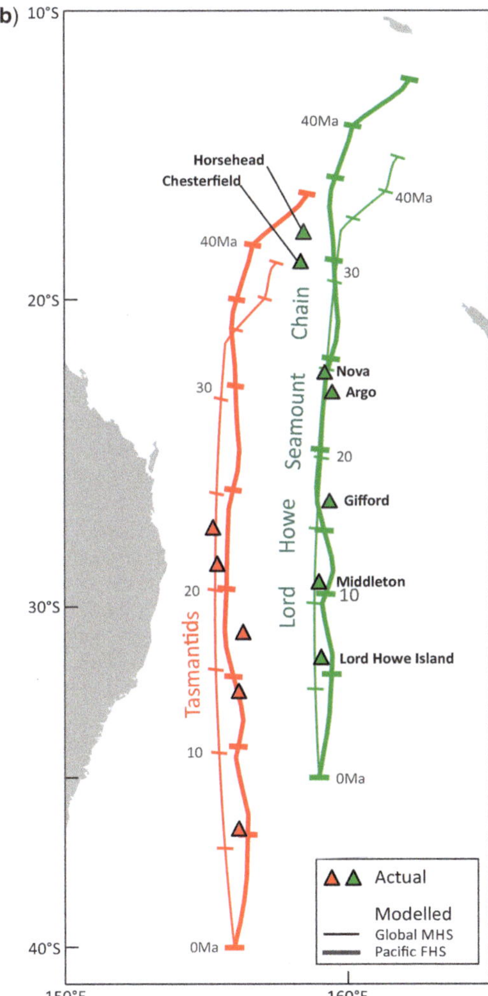

Fig. 7. **(a)** Co-variation of latitude, age and mean Zr/Y of dated Lord Howe Seamount Chain and Tasmantid lavas (McDougall *et al.* 1981; Eggins *et al.* 1991; Quilty 1993; Mortimer *et al.* 2010; Dadd *et al.* 2011; this study). **(b)** Predicted Tasmantid and Lord Howe seamount trails based on the absolute motion of the Australian Plate, anchored at the oldest Lord Howe Island age. Modelled tracks were computed using two alternative absolute reference frames, one based on global moving hotspot predictions (Global MHS: Steinberger *et al.* 2004) and the other computed using a fixed hotspot assumption for the Pacific Plate only (Pacific FHS: Wessel & Kroenke 2008). The plotted Lord Howe Seamount Chain trails neglect any relative motion across the South Rennell Trough. Because of the way they are derived, the computed lines do not resolve changes in absolute motion on timescales of less than 10 myr. Triangles show the locations of actual dated samples.

at Nb and Ta) resemble typical low-silica basalts described from the Chatham Islands by Panter *et al.* (2006) and Timm *et al.* (2010). The low silica Chatham Island, and other southern Zealandia locations, lavas also generally have HIMU-type (high $\mu = U/Pb$) isotopic compositions as opposed the EM-type (enriched mantle) high-silica basalts of Timm *et al.* (2010), which have negative slopes descending from Ba (Figs 3d & 4f). Tasmantids isotope data have been interpreted as relating to an EM-I-type mantle plume (Eggins *et al.* 1991), seemingly quite different from the Lord Howe Seamount Chain (Fig. 5).

The space–time relationship of the Lord Howe Seamount Chain to the South Rennell Trough is interesting in the spatial coincidence of hotspot volcanism and back-arc spreading. The approximate pole of rotation of the *c.* 45–28 Ma South Rennell Trough spreading is located at the south end of the South Rennell Trough: that is, at Horsehead Seamount (Seton *et al.* 2016*a*). This age range is bracketed by the 48 ± 3 and 27 Ma lava ages from Horsehead, and the E-MORBs of the South Rennell spreading centre overlap in geochemical and Nd isotopic composition with the Horsehead and Chesterfield lavas (Fig. 5). This match in space, time and composition suggests a genetic relationship between back-arc basin spreading and a mantle plume. A plume control on ridge location has been suggested for the western Galapagos by Sinton *et al.* (2003); and Jellinek *et al.* (2003) showed that theoretically, lithospheric separation can capture some or all of a nearby ascending deep-mantle plume. The South Rennell Trough and Lord Howe Seamount Chain may be another example of plume-related spreading in a back-arc setting (the Eocene–Miocene arc and trench being located to the north and east, possibly under Vanuatu).

No other age-progressive seamount chains

In addition to the Tasmantid and Lord Howe seamount chains, three other putative north–south-trending age-progressive seamount chains have been identified on the eastern Australian Plate (thin, white, dashed lines in Fig. 1). If they showed similar age progressions to the Lord Howe and Tasmantids chains, then they would all be expected to become younger to the south at a rate of approximately 1.8 Ma per degree of latitude. The existence of the easternmost chain, proposed by Rigolot (1988) linking 3 Ma Norfolk Island with 10 Ma seamounts on the Loyalty Ridge, is not supported by the intervening 23–25 Ma potassic volcanics (Fig. 8c). The 900 km linear chain along the west side of the Norfolk Ridge may yet show a north–south age progression and has been sampled to test this hypothesis (Mortimer *et al.* 2015). The 500 km-long Capel–Faust seamount chain may be a relatively short track on the Australian Plate (Dadd *et al.* 2011) but precise age data are still lacking to establish this.

With available age data, age-progressive hotspot-style volcanism can only be demonstrated for the Lord Howe and Tasmantid seamount chains (Fig. 8c), restricted to the western part of northern Zealandia. We regard all other widely scattered, Late Cretaceous–Holocene intraplate volcanism in Zealandia (Fig. 8) as derived from asthenospheric and/or lithospheric sources unrelated to postulated deep-mantle plumes (Hoernle *et al.* 2006; Timm *et al.* 2010).

Continental crust of the Fairway Ridge and New Caledonia Basin

The New Caledonia Basin is the longest and most submerged sedimentary basin in Zealandia (Fig. 1). Whereas New Caledonia is underlain by continental crust, no drillhole has yet penetrated basement in the basin or the Fairway Ridge that bounds the basin to the west. Based on geophysical data, a variety of hypotheses of oceanic crust or rifted continental crust have been proposed for the New Caledonia Basin and the Fairway Ridge (see summaries by Lafoy *et al.* 2005; Klingelhoefer *et al.* 2007).

Our new dredge data provide the first direct samples of the New Caledonia Basin and Fairway Ridge acoustic basement. We regard it as significant that the lavas dredged from the Le Noroit seamounts, Lansdowne Bank, Nereus Reef and the AUSFAIR5 site are andesites, basaltic trachyandesites and trachytes. As such, these features cannot be basaltic oceanic crust. The high ε_{Nd} values for the northern Fairway Ridge and Le Noroit lavas (Fig. 4) do not necessarily argue against continental crust basement. Eastern Zealandia Mesozoic greywacke terranes can have high initial ε_{Nd} (Price *et al.* 2015), and lavas that pass through continental crust do not inevitably assimilate it (e.g. Timm *et al.* 2010). The Jurassic zircon in one of Le Noroit lavas does, however, indicate that part of the New Caledonia Basin probably is underlain by continental crust. Middle–Late Jurassic detrital zircons have been found in the Boghen and Central basement terranes of New Caledonia (Adams *et al.* 2009). Our data support a thinned continental crust origin for the New Caledonia Basin and Fairway Ridge (Lafoy *et al.* 2005). In a wider context (Fig. 8a), we regard the *c.* 88 Ma siliceous volcanics of the Nouméa Basin in New Caledonia (Nicholson *et al.* 2011) as a related synrift continental igneous suite, along with the 101 Ma Houhora Complex of New Zealand's Three Kings Islands, West Coast South Island granitoids and the Mount Somers Volcanic Group in southern Zealandia (Fig. 1) (Tulloch *et al.* 2009).

Late Cretaceous–Eocene subduction?

It is important for global plate circuits to establish whether a Late Cretaceous–Palaeogene subduction zone existed off or along the eastern edge of northern Zealandia. So far, this remains controversial (see the review by Matthews *et al.* 2015). The lack of preservation of clear bathymetric volcanic chains and basins means that tectonic models have to be based on geochemical interpretations of dredged lavas. This is fraught with difficulty because mantle melting processes rarely have a direct connection to surficial tectonic regimes, and even compositionally distinctive lavas such as shoshonites, boninites and adakites can form in a variety of melting regimes. Subalkaline basalts and andesites commonly erupt in orogenic (subduction-related) settings. However, non-orogenic basalts and andesites, with broadly similar whole-rock compositions to orogenic lavas, are well known, especially from continental rift settings (e.g. Hebrides, Columbia River and Sonora: Morrison 1978; Morris *et al.* 2000; Till *et al.* 2009).

Amid the plethora of geochemical discrimination diagrams, negative Nb and Ta anomalies on multi-element-normalized plots are considered indicators of the subduction-related process (Baier *et al.* 2008 and references therein). The known partitioning of Ti, Nb and Ta into titanite, FeTi oxides and amphibole could, in principle, lead to depletion of these elements in andesites, dacites and rhyolites through crystal fractionation as magmas move and change between mantle and the surface. However, an example of 97 Ma lavas from the Mount Somers Volcanic Group in the South Island (Fig. 1) shows that double-normalizing to yttrium partially corrects for fractional crystallization in the basalt, andesite, dacite and rhyolite spectrum. As such, the double-normalized element concentrations of the Mt Somers rhyolite in Figure 4e are graphically 'lowered' to be close to those of the basalt, and it is seen that and Nb and Ta anomalies do not substantially deepen with increasing SiO_2. There are further complications, however, in that: (a) long-lived earlier subduction (e.g. Cambrian–Early Cretaceous subduction under Gondwana: Mortimer *et al.* 2014*b*)

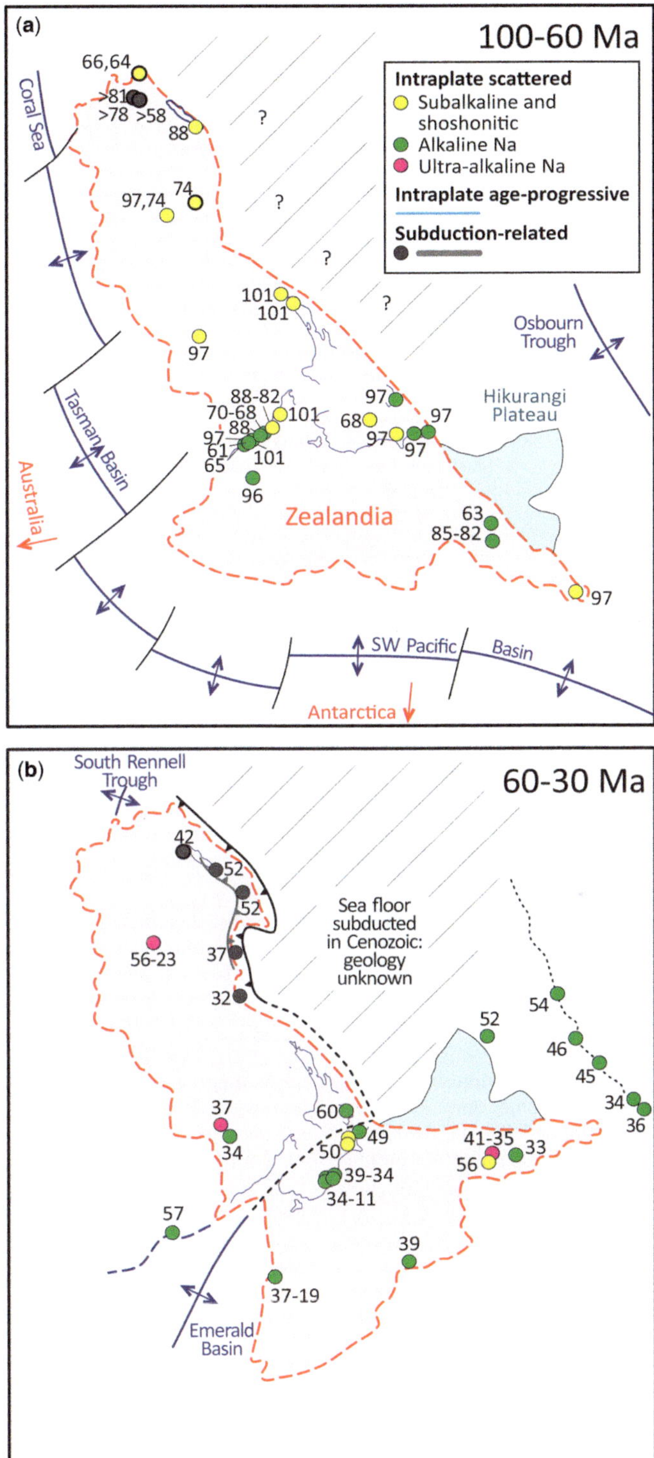

Fig. 8. Age and composition of volcanic rocks of Zealandia summarized on schematic palaeogeographical reconstructions at (a) 60 Ma, (b) 30 Ma.

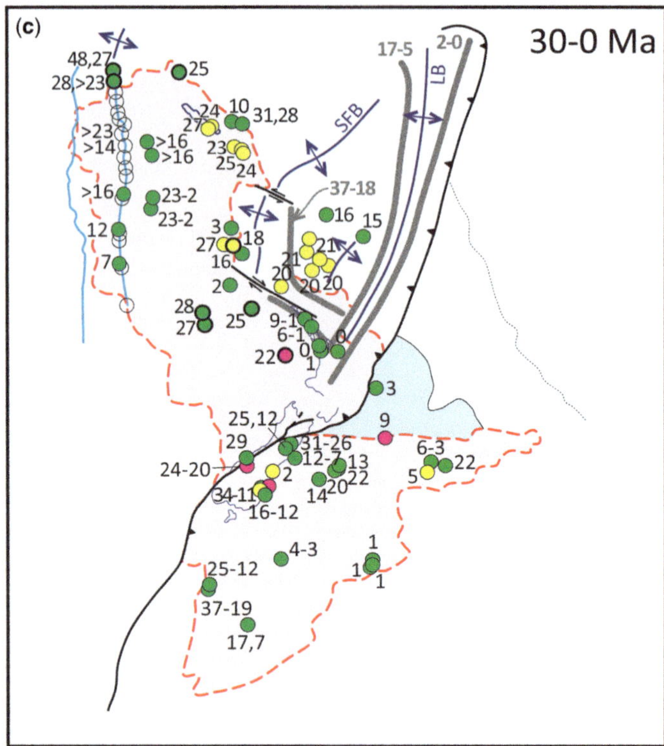

Fig. 8. (*Continued*) **(c)** 0 Ma. Data are from the sources listed in the Figure 1 caption plus Beier *et al.* (2011), Nicholson *et al.* (2011) and Mortimer *et al.* (2012). Ages are rounded to the nearest 1 Ma. Some closely spaced onland occurrences have been combined and/or simplified for plotting at this generalized map scale. Northern Zealandia is fixed. SFB, South Fiji Basin; LB, Lau Basin.

may pre-condition the lithospheric mantle such that, when it later melts in an intracontinental setting, the resulting basalts may have negative Nb and Ta anomalies; (b) melts of earlier subduction-related basement terranes, plutons or lavas in an intraplate setting may yield siliceous lavas with negative Nb and Ta anomalies; and (c) assimilation of very large amounts of continental crust may, in some cases, impose negative Nb and Ta anomalies on lavas. Thus, relying on negative Nb and Ta anomalies from a few samples to infer palaeosubduction may give erroneous and misleading results. This longstanding and important issue cannot be resolved in this paper.

So, with all the above caveats, what can be made of the small samples of very altered Late Cretaceous–Eocene intermediate to siliceous volcanic rocks dredged from around the New Caledonia Basin? Taken at face value, the negative Nb and Ta anomalies of the lavas from around the edge of the New Caledonia Basin could be interpreted as having been acquired as a result of Late Cretaceous–Eocene subduction under northern Zealandia (e.g. Nicholson & Black 2004; Schellart *et al.* 2006; Nicholson *et al.* 2011). This is supported by the high initial ε_{Nd} values of the Lansdowne, Nereus and Le Noroit lavas, indicating melting of depleted mantle (arguably mantle wedge), and by the relatively high Th/Yb at a given Nb/Yb, Pb and LILE concentrations (Figs 3d & 5). Comparing the iron enrichment and SiO_2 v. Zr trends of subduction and intraplate suites from the SW Pacific and selected parts of the world (Fig. 3b, c), we find that, on balance, all continental intraplate igneous suites (even subalkaline ones) tend to have higher FeO^T/MgO and Zr than subduction-related suites at the same SiO_2 content, even though the best-fit lines conceal a huge amount of intra-suite and intra-region variation. The Lansdowne, Nereus, AUSFAIR3, AUSFAIR5 and IPOD lavas mainly plot along low FeO^T/MgO and Zr trends (Fig. 3b). They also plot well off the mantle array, and in the field of continental and oceanic arcs on a Nb/Yb v. Th/Yb diagram (Fig. 3d). However, a subduction interpretation for each individual seafloor site is made less certain by various factors: samples from every site are few (from one to three per dredge) and thus show a limited compositional range; and the lack of primitive basalts adds complications to the geochemical interpretation, the two Le Noroit samples are extremely small, the Lansdowne and Nereus lava samples are extremely small and altered. There is also a general absence of independent supporting evidence such as identifiable volcanic chains, accretionary wedges or palaeotrenches. The samples lie 250–500 km from the eastern edge of Zealandia and so, if they do represent a continental arc, the arc–trench gap is quite wide.

An alternative view (and our preferred view) is that the lavas are not subduction-related but are of continental rift (i.e. intraplate) origin. In this scenario, they would have acquired their Nb and Ta anomalies via melting of, or interaction with, existing continental crust or from a relict Gondwana slab. The high Zr of the AUSFAIR3 and Le Noroit lavas, and high FeO^T/MgO of the AUSFAIR5 lava (Fig. 3b, c), suggest they are likely to be intracontinental, non-orogenic lavas. The AUSFAIR3 lavas also overlap the field of intraplate A-type granites from Antarctica (Fig. 3d). Taken at face value, the chemistry of the Nereus and Lansdowne lavas would seem to be the most consistent with a subduction-related setting. But these rocks are hydrothermally altered and yield only minimum Ar/Ar ages. They could be samples of pre-Late Cretaceous (e.g. Darran Suite: Mortimer *et al.* 2014*b*) magmatism along the Gondwana margin.

The *c.* 42 Ma dacite dredged by the IPOD cruise from offshore Koumac is especially challenging to interpret. The glass age is speculative and may be affected by either excess argon or argon loss. The trace element (including REE) concentrations are puzzlingly basalt-like, not dacite-like. The lack of a prominent Eu anomaly suggests little fractionation, so possibly the dacite is an anatectic melt of mafic crust (if so, the low Sr/Y indicates a garnet-free source). Taken at face value, the low Nb/Yb, Nb, Y + Nb and Zr contents (Figs 3c, d & 4c) do indicate a subduction-related origin, although some continental rift granites can have the relatively low Y + Nb content of P84022 (Förster *et al.* 1997). The fresh glassy dacite does not have any known onshore correlatives (basalts in the Late Eocene Pandope flysch of the Poya Terrane nappe are metamorphosed). Cluzel *et al.* (2005) reported dates of 27 and 24 Ma from two rare granitoid stocks in onland New Caledonia that also showed multi-element-normalized patterns with negative Nb and Ta anomalies.

Distribution and causes of Zealandia intraplate magmatism

Diffuse Alkaline Magmatic Province. Finn *et al.* (2005) outlined the extent of a Cenozoic Diffuse Alkaline Magmatic Province (DAMP) in the eastern Australian Plate and in Antarctica, now dispersed over an area of approximately 40×10^6 km^2. Our new data in the context of a Zealandia continent (Mortimer *et al.* 2017) support the overall concept but, at the same time, require some modification to the DAMP as defined and explained by Finn *et al.* (2005).

First, it should be emphasized that widespread intraplate magmatism commenced across Zealandia in the Late Cretaceous (at *c.* 101 Ma: Tulloch *et al.* 2009) and is not just a Cenozoic phenomenon (Fig. 9). Second, the better definition of continent–ocean boundaries in the SW Pacific region (Mortimer *et al.* 2017) gives a whole new perspective and clarity to the magmatism. DAMP activity is mainly restricted to areas of continental crust, not the intervening oceanic crust. As such, the DAMP is better described as low-volume magmatism mainly scattered across an approximately 12×10^6 km^2 area of formerly contiguous Australia, Zealandia and Antarctica continental crust.

Both age-progressive and non-age-progressive magmatism were lumped into the SW Pacific DAMP by Finn *et al.* (2005). From our analysis, we agree that the age, geochemical and isotopic ranges of the Lord Howe and Tasmantid seamount chains fall within those of other Zealandia intraplate magmatism, yet only the seamount chains show an age progression. Finally, although alkaline suites do indeed dominate sampled rocks (the 'A' in DAMP), rocks of subalkaline and ultra-alkaline composition are also widespread across Zealandia (Fig. 8) (Timm *et al.* 2010).

The challenge for petrogenetic models is to explain the close juxtaposition in space and time between lavas of very different major element, trace element and isotopic composition. Tulloch *et al.* (2009) observed a general pattern of early Late Cretaceous Zealandia rhyolites and granites being of I-type character, and late Late Cretaceous rhyolites and granites being of more A-type character. They attributed this change to progressively thinning crust allowing mantle-derived magmas to reach the surface. Although it is a generalization, we see a similar change in Figure 8 with subalkaline lavas tending to be more abundant in the Cretaceous, and alkaline and ultra-alkaline lavas in the Cenozoic.

Finn *et al.* (2005) suggested that long-lived Palaeozoic–Mesozoic subduction would likely lead to enrichment of Nb and Ta in the subcontinental lithospheric mantle, whereas conventional wisdom indicates that the opposite happens, with these elements being retained in rutile or aluminous clinopyroxene, subducted to greater depths and lost from the subcontinental lithospheric mantle (e.g. Baier *et al.* 2008). Models of Zealandia intraplate magmatism petrogenesis (e.g. Weaver & Smith 1989; Finn

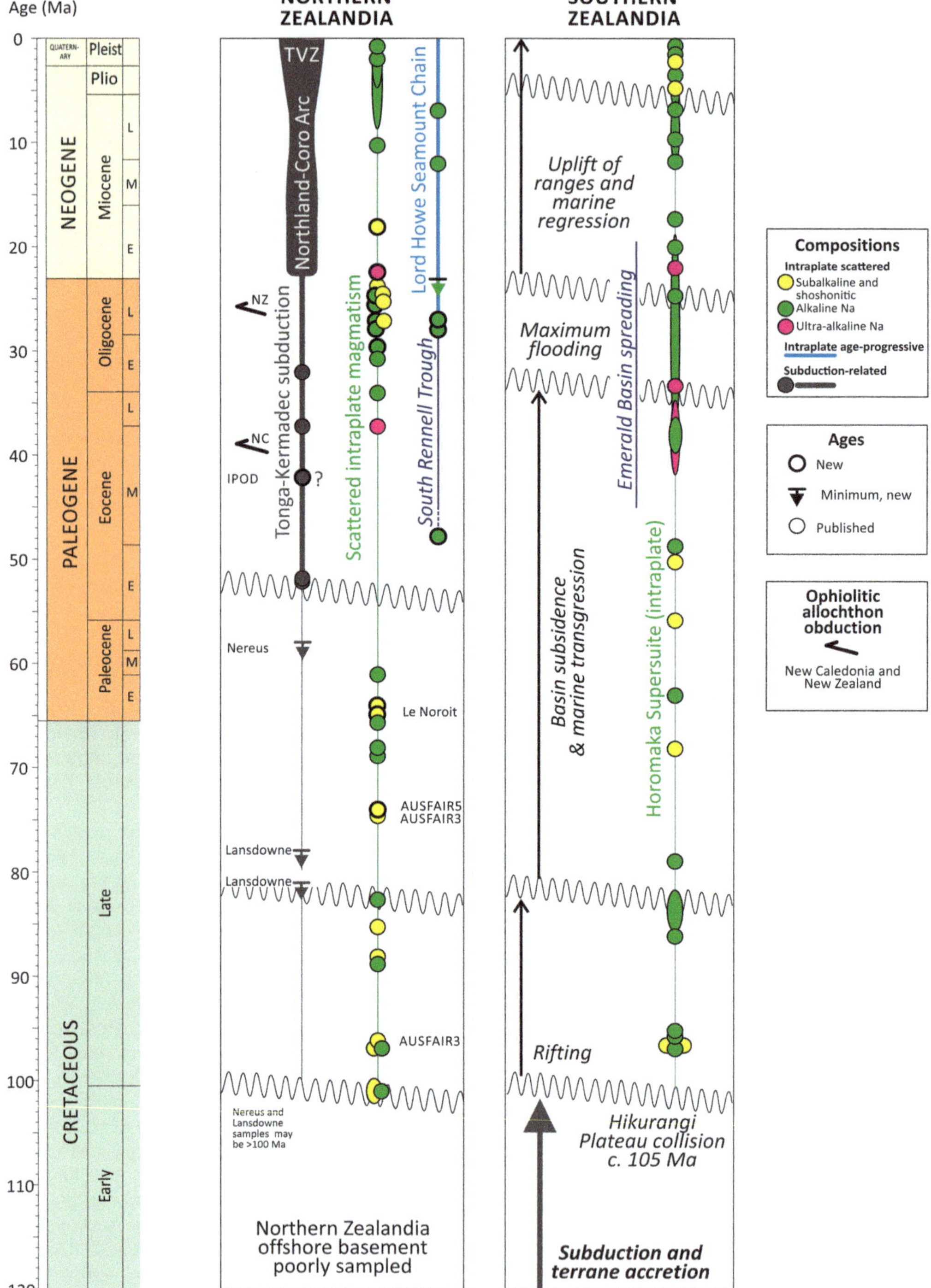

Fig. 9. Summary of the Late Cretaceous–Holocene magmatic chronology of northern and southern Zealandia. Northern Zealandia has three compositional groups of magmatic rocks, southern Zealandia just one. Rock compositions are generalized to represent the main types. Data are from the sources listed in the Figure 1 caption. TVZ,Taupo Volcanic Zone.

et al. 2005; Panter *et al.* 2006; Hoernle *et al.* 2006; Sprung *et al.* 2007; Timm *et al.* 2010) are all variations of the edge-driven convection model of King & Anderson (1998). An important observation we can add to this debate is that the features (spatial distribution, age and compositional range) of the non-age-progressive igneous rocks are broadly similar in northern and southern Zealandia (Fig. 9). This is despite the two halves of the continent having had very different histories. Since the Late Cretaceous, southern Zealandia has been locked in a passive margin relationship with the Hikurangi Plateau and Pacific Plate. In contrast, the greater bathymetric relief of northern Zealandia suggests more variation in lithospheric thickness. As the crustal and lithospheric structure of Zealandia becomes better known (Mortimer *et al.* 2017), and more volcanoes are sampled and dated, it may be possible more accurately to identify steps in lithospheric thickness and find some spatial control to the intraplate magmatism (cf. Davies *et al.* 2015).

Siliceous Large Igneous Province. As well as the DAMP, another concept that has been introduced to Australasian Mesozoic geology is that of a SLIP (Silicic Large Igneous Province), arising largely from studies of the Cretaceous Whitsunday volcanics in Australia (Bryan 2007; Bryan *et al.* 2012). Bryan (2007, fig. 1) included the Lord Howe Rise in the Whitsunday SLIP, mainly on the basis of rhyolite in DSDP 207 but also on the inferred extent of Cretaceous volcaniclastic rift basins on the Lord Howe Rise. Subsequent seismic and petrological work (Collot *et al.* 2009; Bache *et al.* 2014*a*; this study) has identified more basins and more silicic igneous rocks in northern Zealandia. While these might seem to support the expansion of a SLIP from mainland Australia onto the Lord Howe Rise, there are a few things to consider: (1) many of the northern Zealandia silicic volcanics ages are younger than the 95 Ma Whitsunday cut-off proposed by Bryan (2007); (2) some of the rocks may possibly be subduction-related, not intraplate (e.g. Nereus and Lansdowne, see above); (3) the rocks may not occur in sufficient volume to define a Large Igneous Province; and (4) Late Cretaceous silicic igneous rocks are not restricted to northern Zealandia but occur in southern Zealandia as well; so is all of Zealandia part of a Late Cretaceous SLIP?

In terms of exploring a causal relationship between SLIP magmatism and continental break-up, it should also be noted that, at the latitude of the Whitsunday Islands (currently 20°–26°S in coastal Australia), seafloor spreading (Zealandia–Gondwana separation) did not actually start until *c.* 62 Ma, rather than the oft-quoted *c.* 85 Ma for Zealandia–Gondwana separation to the south (Gaina *et al.* 1999). Thus, the hiatus between the end of SLIP magmatism and spreading is approximately 33 myr rather than approximately 10 myr.

Space–time–composition patterns. Numerous studies, such as those of Cole (1986), Weaver & Smith (1989), Finn *et al.* (2005) and Timm *et al.* (2010), have searched without success for space–time–composition patterns in subsets of Zealandia intraplate lavas – the Horomaka Supersuite of Mortimer *et al.* (2014*b*). We have compiled all known occurrences of lavas in Figure 8 and show them on schematic palinspastic reconstructions appropriate for their age. Once again, no obvious age trends emerge but some subtle patterns are present: for example, the aforementioned slightly greater abundance of subalkaline lavas in the early Late Cretaceous and ultra-alkaline lavas in the Cenozoic.

Although we have not attempted to calculate volumes or fluxes, it is apparent from Figures 8 and 9 that lavas of Late Cretaceous and Oligocene–Neogene age are more common than those of Eocene age. The Late Cretaceous pulse can be understood as magmatism associated with synrift deformation prior to, during and immediately after Zealandia break-up from Gondwana. The Oligocene–Neogene pulse of intraplate magmatism is more difficult to explain as it does not coincide with any major plate motion change. To the south of Zealandia, the Emerald Basin opened from *c.* 45 Ma and this is also when a subduction phase initiated off northern Zealandia (Sutherland *et al.* 2010), propagating south to form a volcanic arc in New Zealand by *c.* 23 Ma (Cole 1986) (Fig. 8b, c). It is simply possible that the younger rocks are more easily sampled, by virtue of their higher stratigraphic positions. By the same token, the Late Cretaceous synrift magmatism in Zealandia may be far more common, and is therefore under-represented in Figures 8a and 9. Magnetic anomalies on the Campbell Plateau and either side of the Aotea Basin (Sutherland 1999, fig. 3) may represent such Late Cretaceous magmatism.

Conclusions

New sampling, dating, and geochemical and isotopic analysis of small and altered volcanic rock samples from the submerged northern Zealandia continent, combined with earlier work, reveal three volcanotectonic regimes (Figs 8 & 9):

- Age-progressive, Oligocene–Pliocene, alkaline volcanism of the Lord Howe Seamount Chain – Inception of the Lord Howe Seamount Chain as a linear feature began at *c.* 28 Ma. Prior to that, the position of the Lord Howe plume was coincident with the South Rennell Trough, a back-arc spreading centre which was active from *c.* 45 to

c. 28 Ma. The Lord Howe Seamount Chain is partly coeval with the nearby Tasmantids Seamount Chain. Although basalts from each chain are derived from the melting of different sorts of geochemical and isotopic mantle, positions of both plumes are approximately fixed in the mantle and are well predicted from plate kinematic modelling. Age-progressive, hotspot-type volcanism is not known from anywhere else in Zealandia.
- Late Cretaceous–Palaeogene, generally subalkaline intermediate to silicic volcanism on and near ridges and rises around the New Caledonia Basin – Our new samples are of >78 Ma to *c.* 42 Ma age but similar silicic suites elsewhere in northern Zealandia are as old as 101 Ma. The samples support a continental crust basement for the New Caledonia Basin and Fairway Ridge. The tectonic setting of eruption of most of these lavas, based on geochemistry, is intraplate continental. The very highly altered Nereus and Lansdowne lavas may have a subduction-related chemistry, but their stratigraphic age is older than their Late Cretaceous hydrothermal overprint. There is no unambiguous volcanic evidence in our dataset for Late Cretaceous–Paleocene subduction beneath northern Zealandia. The *c.* 42 Ma IPOD4 dacite seems to have the clearest subduction-related geochemistry but the correct interpretation of negative Nb and Ta anomalies in continental lavas is a longstanding problem that we cannot resolve.
- Alkaline, ultra-alkaline and subalkaline volcanism, scattered across both northern and southern Zealandia with no clear spatial or compositional pattern – Our new northern Zealandia samples are of 30–18 Ma age, and reinforce a late Cenozoic pulse in what is otherwise low-volume and long-lived (*c.* 97–0 Ma) intraplate volcanism. Unlike the Lord Howe Seamount Chain lavas, this volcanism is unrelated to any mantle plume. The dispersed magmatism mostly coincides with areas of continental crust within Zealandia, not the surrounding oceanic crust.
- To date, most of our knowledge of intraplate volcanism on Zealandia has come from the southern part of the continent. The sampling and analysis of offshore northern Zealandia volcanic rocks described in this paper provides a more complete picture of syn- and post-Gondwana break-up magmatism on the entire Zealandia continent.

We thank the captains and crews of N/O *Le Noroit*, N/O *Marion Dufresne*, R/V *Southern Surveyor*, N/O *l'Alis* and R/V *Tangaroa*, without whom the samples described in this paper would not have been obtained. Neville Orr, Ben Durrant, John Simes, Belinda Smith Lyttle, Craig Fraser, Charles Knaack and Diane Johnson provided technical assistance. Information, discussions and/or encouragement from Rick Herzer, Greg Browne, Mark Lawrence, Robert Crookbain, Andy Tulloch and Volkmar Damm are acknowledged. Comments by Ron Hackney, Takehiko Hashimoto and three anonymous reviewers helped improve earlier versions of the manuscript. The analytical work was supported by core funding to GNS Science from the New Zealand Ministry for Business, Employment and Innovation.

References

ADAMS, C.J., CLUZEL, D. & GRIFFIN, W.L. 2009. Detrital-zircon ages and geochemistry of sedimentary rocks in basement Mesozoic terranes and their cover rocks in New Caledonia, and provenances at the Eastern Gondwanaland margin. *Australian Journal of Earth Sciences*, **56**, 1023–1047.

BACHE, F., MORTIMER, N., SUTHERLAND, R., COLLOT, J., ROUILLARD, P., STAGPOOLE, V. & NICOL, A. 2014*a*. Seismic stratigraphic record of transition from Mesozoic subduction to continental breakup in the Zealandia sector of eastern Gondwana. *Gondwana Research*, **26**, 1060–1078.

BACHE, F., SUTHERLAND, R. *ET AL.* 2014*b*. *Tangaroa TAN1312 voyage report: Dredging Reinga and Aotea basins to constrain seismic Stratigraphy and Petroleum systems (DRASP), NW New Zealand.* GNS Science Report **2014/05**. GNS Science, Lower Hutt, New Zealand.

BAKER, J., GAMBLE, J.A. & GRAHAM, I.J. 1994. The age, geology, and geochemistry of the Tapuaenuku Igneous Complex, Marlborough, New Zealand. *New Zealand Journal of Geology and Geophysics*, **37**, 249–268.

BAIER, J., AUDETAT, A. & KEPPLER, H. 2008. The origin of the negative niobium tantalum anomaly in subduction zone magmas. *Earth and Planetary Science Letters*, **267**, 290–300.

BAUBRON, J.C., GUILLON, J.H. & RÉCY, J. 1976. Géochronologie par la méthode K/Ar du substrat volcanique de l'île de Maré, archipel des Loyauté (Sud-Ouest Pacifique). *Bulletin BRGM, Comptes Rendus de l'Académie des Sciences*, 3 Series, **2**, 165–176.

BEIER, C., VANDERKLUYSEN, L., REGELOUS, M., MAHONEY, J.J. & GARBE-SCHÖNBERG, D. 2011. Lithospheric control on geochemical composition along the Louisville Seamount Chain. *Geochemistry, Geophysics, Geosystems*, **12**, Q0AM01, https://doi.org/10.1029/2011GC003690

BENTZ, F.P. 1974. Marine geology of the southern Lord Howe Rise, southwest Pacific. *In*: BURK, C.A. & DRAKE, C.L. (eds) *The Geology of Continental Margins.* Springer, New York, 537–547.

BRODIE, J.W. 1965. Aotea Seamount, Eastern Tasman Sea. *New Zealand Journal of Geology and Geophysics*, **8**, 510–517.

BROWNE, G.H., LAWRENCE, M.J.F. *ET AL.* 2016. Stratigraphy of Reinga and Aotea basins, NW New Zealand: constraints on regional correlations and reservoir character from dredge samples. *New Zealand Journal of Geology and Geophysics*, **59**, 396–415.

BRYAN, S.E. 2007. Silicic Large Igneous Provinces. *Episodes*, **30**, 20–31.

BRYAN, S.E., COOK, A., ALLEN, C.M., SIEGEL, C., PURDY, D., GREENTREE, J. & UYSAL, T. 2012. Early-mid Cretaceous tectonic evolution of eastern Gondwana: from silicic LIP magmatism to continental rupture. *Episodes*, **35**, 142–152.

Cluzel, D., Bosch, D., Paquette, J.-L., Lemennicier, Y., Montjoie, P. & Menot, R.-P. 2005. Late Oligocene post-obduction granitoids of New Caledonia: a case for reactivated subduction and slab break-off. *Island Arc*, **14**, 254–271.

Cole, J.W. 1986. Distribution and tectonic setting of late Cenozoic volcanism in New Zealand. *Royal Society of New Zealand Bulletin*, **23**, 7–20.

Collot, J., Geli, L., Lafoy, Y., Vially, R., Cluzel, D., Klingelhoefer, F. & Nouzé, H. 2008. Tectonic history of northern New Caledonia Basin from deep offshore seismic reflection: relation to late Eocene obduction in New Caledonia, southwest Pacific. *Tectonics*, **27**, TC6006, https://doi.org/10.1029/2008TC002263

Collot, J., Herzer, R.H., Lafoy, Y. & Geli, L. 2009. Mesozoic history of the Fairway-Aotea Basin: implications for the early stages of Gondwana fragmentation. *Geochemistry, Geophysics, Geosystems*, **10**, Q12019, https://doi.org/10.1029/2009GC002612

Collot, J., Rouillard, P., Juan, C., Patriat, M., Lenault, Y., Privat, A. & Maurizot, P. 2013. *Rapport de Mission de la Campagne Océanographique IPOD (Investigating Post Obduction Deposits) a bord du N/O Alis du 1er au 18 Aout 2012*. Rapport Service Géologique de Nouvelle Calédonie, 2013, https://doi.org/10.17600/12100080

Colwell, J., Foucher, J.-P., Logan, G. & Balut, Y. 2006. *Partie 2, Programme AUSFAIR (Australia–Fairway basin bathymetry and sampling survey) Cruise Report*. Les rapports de campagnes a la mer, MD 153/AUS-FAIR–ZoNéCo 12 and VT 82/GAB on board R/V Marion Dufresne. OCE/2006/05. Institut Polaire Français Paul Emile Victor, Plouzané, France.

Cook, C., Briggs, R.M., Smith, I.E.M. & Maas, R. 2004. Petrology and geochemistry of intraplate basalts in the South Auckland Volcanic Field, New Zealand: evidence for two coeval magma suites from distinct sources. *Journal of Petrology*, **46**, 473–503.

Coombs, D.S., Adams, C.J., Roser, B.P. & Reay, A. 2008. Geochronology and geochemistry of the Dunedin Volcanic Group, eastern Otago, New Zealand. *New Zealand Journal of Geology and Geophysics*, **51**, 195–218.

Dadd, K.A., Locmelis, M., Higgins, K. & Hashimoto, T. 2011. Cenozoic volcanism of the Capel-Faust Basins, Lord Howe Rise, SW Pacific Ocean. *Deep-Sea Research Part II: Topical Studies in Oceanography*, **58**, 922–932.

Daniel, J., Jouannic, C., Larue, B. & Recy, J. 1977. Interpretation of D'Entrecasteaux Zone (north of New Caledonia). *In*: *International Symposium on Geodynamics in South-West Pacific*, 27 August–2 September 1976, Nouméa. Editions Technip, Paris, 117–124.

Davies, D.R., Rawlinson, N., Iaffaldano, G. & Campbell, I.H. 2015. Lithospheric controls on magma composition along Earth's longest continental hotspot track. *Nature*, **525**, 511–514.

Eggins, S.M., Green, D.H. & Falloon, T.H. 1991. The Tasmantid Seamounts: shallow melting and contamination on an EM1 mantle plume. *Earth and Planetary Science Letters*, **107**, 448–462.

Exon, N.F., Quilty, P.J., Lafoy, Y. & Auzende, J.-M. 2004. Miocene volcanic seamounts on northern Lord Howe Rise: lithology, age, ferromanganese crusts, and origin. *Australian Journal of Earth Sciences*, **51**, 291–300.

Exon, N.F., Hill, P.J., Lafoy, Y., Heine, C. & Bernardel, G. 2006. Kenn Plateau off northeast Australia: a continental fragment in the southwest Pacific jigsaw. *Australian Journal of Earth Sciences*, **53**, 541–564.

Exon, N.F., Lafoy, Y., Hill, P.J., Dickens, G.R. & Pecher, I. 2007. Geology and petroleum potential of the Fairway Basin in the Tasman Sea. *Australian Journal of Earth Sciences*, **54**, 629–645.

Finn, C.A., Muller, R.D. & Panter, K.S. 2005. A Cenozoic diffuse alkaline magmatic province (DAMP) in the southwest Pacific without rift or plume origin. *Geochemistry, Geophysics, Geosystems*, **6**, Q02005, https://doi.org/10.1029/2004GC000723

Förster, H.-J., Tischendorf, G. & Trumbull, R.B. 1997. An evaluation of the Rb v. (Y + Nb) discrimination diagram to infer tectonic setting of siliceous igneous rocks. *Lithos*, **40**, 261–293.

Gaina, C., Muller, R.D., Royer, J.-Y. & Symonds, P. 1999. Evolution of the Louisiade triple junction. *Journal of Geophysical Research*, **104**, 12 927–12 939.

Gamble, J.A., Morris, P.A. & Adams, C.J. 1986. The geology, petrology and geochemistry of Cenozoic volcanic rocks from the Campbell Plateau and Chatham Rise. *Royal Society of New Zealand Bulletin*, **23**, 344–365.

Green, T.H. 1973. Petrology and geochemistry of basalts from Norfolk Island. *Journal of the Geological Society of Australia*, **20**, 259–272.

Heap, A.D., Hughes, M. *et al.* 2009. *Seabed Environments and Subsurface Geology of the Capel and Faust Basins and Gifford Guyot, Eastern Australia – Post Survey Report*. Geoscience Australia Record, **2009/22**.

Herzer, R.H., Challis, G.A., Christie, R.H.K., Scott, G. H. & Watters, W.A. 1989. The Urry Knolls, late Neogene alkaline basalt extrusives, southwestern Chatham Rise. *Journal of the Royal Society of New Zealand*, **19**, 181–193.

Higgins, K., Hashimoto, T., Fraser, G., Rollet, N. & Colwell, J. 2011. Ion microprobe (SHRIMP) U–Pb dating of Upper Cretaceous volcanics from the northern Lord Howe Rise, Tasman Sea. *Australian Journal of Earth Sciences*, **58**, 195–207.

Hinz, K. 1979. *Report of the Sonne Southeast Asia Cruise 1978: Cruise SO-7-16.10.1978–22.12.1978. BGR Report* **6300/79**. Bundesanstalt für Geowissenschaften und Rohstoffe, Hanover, Germany.

Hoernle, K., White, J.D.L. *et al.* 2006. Cenozoic intraplate volcanism on New Zealand: upwelling induced by lithospheric removal. *Earth and Planetary Science Letters*, **248**, 350–367.

Jellinek, A.M., Gonnermann, H.M. & Richards, M.A. 2003. Plume capture by divergent plate motions: implications for the distribution of hotspots, geochemistry of mid-ocean ridge basalts, and estimates of the heat flux at the core–mantle boundary. *Earth and Planetary Science Letters*, **205**, 361–378.

Johnson, R.W., Knutson, J. & Taylor, S.R. 1989. *Intraplate Volcanism in Eastern Australia and New Zealand*. Australian Academy of Sciences, Melbourne, Australia.

Kalnins, L.M., Cohen, B.E., Fitton, J.G., Mark, D.F., Richards, F.D. & Barfod, D.N. 2015. The East

Australian, Tasmantid, and Lord Howe Volcanic Chains: possible mechanisms behind a trio of hotspot trails. *Abstract DI41A-2591 presented at the American Geophysical Union Fall Meeting 2015*, 14–18 December 2015, San Francisco, CA, USA.

King, S.D. & Anderson, D.L. 1998. Edge-driven convection. *Earth and Planetary Science Letters*, **160**, 289–296.

Klingelhoefer, F., Lafoy, Y., Collot, J., Cosquer, E., Géli, L., Nouzé, H. & Vially, R. 2007. Crustal structure of the basin and ridge system west of New Caledonia (southwest Pacific) from wide-angle and reflection seismic data. *Journal of Geophysical Research*, **112**, B11102, https://doi.org/10.1029/2007JB005093

Knesel, K.M., Cohen, B.E., Vasconcelos, P.M. & Thiede, D.S. 2008. Rapid change in drift of the Australian plate records collision with Ontong Java plateau. *Nature*, **454**, 754–757.

Lafoy, Y., Brodien, I., Vially, R. & Exon, N.F. 2005. Structure of the basin and ridge system west of New Caledonia (Southwest Pacific): a synthesis. *Marine Geophysical Researches*, **26**, 37–50.

Launay,J., Dupont, J., Lapouille, A., Ravenne, C. & de Broin, C.E. 1977. Seismic traverses across the northern Lord Howe Rise and comparison with the southern part (south-west Pacific). *In*: *International Symposium on Geodynamics in South-West Pacific*, 27 August–2 September 1976, Nouméa. Editions Technip, Paris, 155–164.

Le Maitre, R.W. 1989. *A Classification of Igneous Rocks and Glossary of Terms*. Blackwell Scientific, Oxford.

Leterrier, J., Maury, R.C., Thonon, P., Girard, D. & Marchal, M. 1982. Clinopyroxene composition as a method of identification of the magmatic affinities of paleo-volcanic series. *Earth and Planetary Science Letters*, **59**, 139–2154.

Luyendyk, B.P. 1995. Hypothesis for Cretaceous rifting of east Gondwana caused by subducted slab capture. *Geology*, **23**, 373–376.

Matthews, K.J., Williams, S.E., Whittaker, J.M., Müller, R.D., Seton, M. & Clarke, G.L. 2015. Geologic and kinematic constraints on Late Cretaceous to mid Eocene plate boundaries in the southwest Pacific. *Earth-Science Reviews*, **140**, 72–107.

McCoy-West, A.J., Baker, J.A., Faure, K. & Wysoczanski, R. 2010. Petrogenesis and origins of mid-Cretaceous continental intraplate volcanism in Marlborough, New Zealand: implications for the long-lived HIMU magmatic mega-province of the SW Pacific. *Journal of Petrology*, **51**, 2003–22045.

McDougall, I. & Duncan, R.A. 1988. Age progressive volcanism in the Tasmantid Seamounts. *Earth and Planetary Science Letters*, **89**, 207–220.

McDougall, I., Embleton, B.J.J. & Stone, D.B. 1981. Origin and evolution of Lord Howe Island. *Journal of the Geological Society of Australia*, **28**, 155–176.

McMillan, N.J., Dickin, A.P. & Haag, D. 2000. Evolution of magma source regions in the Rio Grande rift, southern New Mexico. *Geological Society of America Bulletin*, **112**, 1582–1593.

Missegue, F. & Collot, J.-Y. 1987. Etude geophysique du Plateau des Chesterfields (Pacifique sud-ouest). Resultats preliminaires de la campagne Zoneco de N/O Coriolis. *Comptes Rendus de l'Académie des Sciences*, Series II, **304**, 279–283.

Miyashiro, A. 1974. Volcanic rock series in island arcs an active continental margins. *American Journal of Science*, **274**, 321–355.

Monzier, M. & Vallot, J. 1983. *Rapport preliminaire concernent les dragages realises lors de la campagne GEORSTOM III SUD (1975). Office de la Recherche Scientifique et Technique Outre-Mer (ORSTOM)*. Centre de Noumea Geologie- Geophysique Rapport **2-83**.

Morgans, H.E.G. 2014. *Foraminiferal biostratigraphy from New Caledonia IPOD dredge samples*. GNS Science Consultancy Report **2014/223**. GNS Science, Lower Hutt, New Zealand.

Morris, G.A., Lartson, P.B. & Hooper, P.R. 2000. 'Subduction style' tmagmatism in a non-subduction setting: the Colville Igneous Complex, NE Washington state, USA. *Journal of Petrology*, **41**, 43–67.

Morrison, M.A. 1978. The use of 'immobile' trace elements to distinguish the palaeotectonic affinities of metabasalts: applications to the Paleocene basalts of Null, Skye, northwest Scotland. *Earth and Planetary Science Letters*, **39**, 407–416.

Mortimer, N. 2004. Basement gabbro from the Lord Howe Rise. *New Zealand Journal of Geology and Geophysics*, **47**, 501–507.

Mortimer, N., Herzer, R.H., Gans, P.B., Parkinson, D.L. & Seward, D. 1998. Basement geology from Three Kings Ridge to West Norfolk Ridge, southwest Pacific Ocean: evidence from petrology, geochemistry and isotopic dating of dredge samples. *Marine Geology*, **148**, 135–162.

Mortimer, N., Herzer, R.H., Gans, P.B., Laporte-Magoni, C., Calvert, A. & Bosch, D. 2007. Oligocene–Miocene tectonic evolution of the South Fiji Basin and Northland Plateau, SW Pacific Ocean: Evidence from petrology and dating of dredged rocks. *Marine Geology*, **237**, 1–24.

Mortimer, N., Dunlap, W.J., Palin, J.M., Herzer, R.H., Hauff, F. & Clark, M. 2008*a*. Ultra-fast early Miocene exhumation of Cavalli Seamount, Northland Plateau, Southwest Pacific Ocean. *New Zealand Journal of Geology and Geophysics*, **51**, 29–42.

Mortimer, N., Hauff, F. & Calvert, A.C. 2008*b*. Continuation of the New England Orogen, Australia, beneath the Queensland Plateau and Lord Howe Rise. *Australian Journal of Earth Sciences*, **55**, 195–209.

Mortimer, N., Gans, P.B., Palin, J.M., Meffre, S., Herzer, R.H. & Skinner, D.N.B. 2010. Location and migration of Miocene–Quaternary volcanic arcs in the SW Pacific region. *Journal of Volcanology and Geothermal Research*, **190**, 1–10.

Mortimer, N., Gans, P.B., Hauff, F. & Barker, D.H.N. 2012. Paleocene MORB and OIB from the Resolution Ridge, Tasman Sea. *Australian Journal of Earth Sciences*, **59**, 953–964.

Mortimer, N., Gans, P.B., Palin, J.M., Herzer, R.H., Pelletier, B. & Monzier, M. 2014*a*. Eocene and Oligocene basins and ridges of the Coral Sea–New Caledonia region: Tectonic link between Melanesia, Fiji, and Zealandia. *Tectonics*, **33**, 1386–1407, https://doi.org/10.1002/2014TC003598

Mortimer, N., Rattenbury, M.S. *et al.* 2014*b*. High-level stratigraphic scheme for New Zealand rocks. *New Zealand Journal of Geology and Geophysics*, **57**, 402–419.

MORTIMER, N., PATRIAT, M. *ET AL.* 2015. The VESPA research cruise (Volcanic Evolution of South Pacific Arcs): a voyage of discovery to the Norfolk, Loyalty and Three Kings Ridges, northeast Zealandia. *Geoscience Society of New Zealand Miscellaneous Publication*, **143A**, 98–99.

MORTIMER, N., CAMPBELL, H.J. *ET AL.* 2017. Zealandia: Earth's hidden continent. *GSA Today*, **27**, https://doi.org/10.1130/GSATG321A.1

NELSON, C.S., BRIGGS, R.M. & KAMP, P.J.J. 1986. Nature and significance of volcanogenic deposits at the Eocene/Oligocene boundary, Hole 593, Challenger Plateau, Tasman Sea. *In*: KENNETT, J.P., VON DER BORCH, C.C. *ET AL.* (eds) *Initial Reports of the Deep Sea Drilling Project, Volume 90*. Ocean Drilling Program, College Staion, TX, 1175–1187.

NICHOLSON, K.N. & BLACK, P.M. 2004. Cretaceous to early Tertiary basaltic volcanism in the Far North of New Zealand: geochemical associations and their tectonic significance. *New Zealand Journal of Geology and Geophysics*, **47**, 437–446.

NICHOLSON, K.N., MAURIZOT, P., BLACK, P.M., PICARD, C., SIMONETTI, A., STEWART, A. & ALEXANDER, A. 2011. Geochemistry and age of the Nouméa Basin lavas, New Caledonia: evidence for Cretaceous subduction beneath the eastern Gondwana margin. *Lithos*, **125**, 659–674.

PANTER, K.S., BLUSZTAJN, J., HART, S.R., KYLE, P.R., ESSER, R. & MCINTOSH, W.C. 2006. The origin of HIMU in the SW Pacific: evidence from intraplate volcanism in southern New Zealand and Subantarctic Islands. *Journal of Petrology*, **47**, 1673–1704.

PEARCE, J.A. & NORRY, M.J. 1979. Petrogenetic implications of Ti, Zr, Y, and Nb variations in volcanic rocks. *Contributions to Mineralogy and Petrology*, **69**, 33–47.

PEARCE, J.A. & PEATE, D.W. 1995. Tectonic implications of the composition of volcanic arc magmas. *Annual Review of Earth and Planetary Sciences*, **23**, 251–285.

PRICE, R.C., MORTIMER, N., SMITH, I.E.M. & MAAS, R. 2015. Whole-rock geochemical reference data for Torlesse and Waipapa terranes, North Island, New Zealand. *New Zealand Journal of Geology and Geophysics*, **58**, 213–228.

QUILTY, P.G. 1993. Tasmantid and Lord Howe seamounts: biostratigraphy and palaeoceanographic significance. *Alcheringa*, **17**, 27–53.

RIGOLOT, P. 1988. Prolongement meridional des grandes structures geologiques de Nouvelle-Caledonie et decouverte de monts sous-marins interpretes comme un jalon dans un nouvel alignement de hot-spot. *Comptes Rendu Academie Sciences Paris*, Serie II, **307**, 965–972.

SACK, P.J., BERRY, R.F., MEFFRE, S., FALLOON, T.J., GEMMELL, J.B. & FRIEDMAN, R.M. 2011. In situ location and U–Pb dating of small zircon grains in igneous rocks using laser ablation-inductively coupled plasma-quadrupole mass spectrometry. *Geochemistry, Geophysics, Geosystems*, **12**, https://doi.org/10.1029/2010GC003405

SCHELLART, W.P., LISTER, G.S. & TOY, V.G. 2006. A Late Cretaceous and Cenozoic reconstruction of the Southwest Pacific region: tectonics controlled by subduction and slab rollback processes. *Earth-Science Reviews*, **76**, 191–233.

SCOTT, J.M., TURNBULL, I.M., SAGAR, M.W., TULLOCH, A.J., WAIGHT, T.E. & PALIN, J.M. 2015. Geology and geochronology of the Sub-Antarctic Snares Islands/Tini Heke, New Zealand. *New Zealand Journal of Geology and Geophysics*, **58**, 202–212.

SETON, M., MORTIMER, N. *ET AL.* 2016*a*. Melanesian back-arc basin and arc development: constraints from the eastern Coral Sea. *Gondwana Research*, **39**, 77–95.

SETON, M., WILLIAMS, S., MORTIMER, N., MEFFRE, S. & MICKLETHWAITE, S. 2016*b*. *Voyage Report for SS2012V06 Eastern Coral Sea Tectonics (ECOSAT), R/V Southern Surveyor, October-November 2012*. GNS Science Report **2016-49**. GNS Science, Lower Hutt, New Zealand.

SINTON, J., DETRICK, R., CANALES, J.P., ITO, G. & BEHN, M. 2003. Morphology and segmentation of the western Galápagos Spreading Center, 90.5°–98°W: plume-ridge interaction at an intermediate spreading ridge. *Geochemistry, Geophysics, Geosystems*, **4**, 8515, https://doi.org/10.1029/2003GC000609,12.

SPRUNG, P., SCHUTH, S., MÜNKER, C. & HOKE, L. 2007. Intraplate volcanism in New Zealand: the role of fossil plume material and variable lithospheric properties. *Contributions to Mineralogy and Petrology*, **153,** 669–687

STEINBERGER, B., SUTHERLAND, R. & O'CONNELL, R.J. 2004. Prediction of Emperor–Hawaii seamount locations from a revised model of global plate motion and mantle flow. *Nature*, **430**, 167–173.

STEINER, A. & STRECK, M.J. 2014. The Strawberry Volcanics: generation of 'orogenic' andesites from tholeiite within an intra-continental volcanic suite centered on the Columbia River flood basalt province, USA. *In*: GÓMEZ-TUENA, A., STRAUB, S.M. & ZELLMER, G.F. (eds) *Orogenic Andesites and Crustal Growth*. Geological Society, London, Special Publications, **385**, 281–302, https://doi.org/10.1144/SP385.12

STOREY, B.C., HOLE, M.J., PANKHURST, R.J., MILLAR, I.L. & VENNUM, W. 1988. Middle Jurassic within-plate granites in West Antarctica and their bearing on the break-up of Gondwanaland. *Journal of the Geological Society, London*, **145**, 999–1007, https://doi.org/10.1144/gsjgs.145.6.0999

STRONG, D.T., TURNBULL, R.E., HAUBROCK, S.E. & MORTIMER, N. 2016. Petlab: New Zealand's national rock catalogue and geoanalytical database. *New Zealand Journal of Geology and Geophysics*, **59**, 475–481.

SUN, S.-S. & MCDONOUGH, W.F. 1989. Chemical and isotopic systematics of oceanic basalts: implications for mantle composition and processes. *In*: SAUNDERS, A. D. & NORRY, M.J. (eds) *Magmatism in the Ocean Basins*. Geological Society, London, Special Publications, **42**, 313–345, https://doi.org/10.1144/GSL.SP.1989.042.01.19

SUTHERLAND, F.L., GRAHAM, I.T., MEFFRE, S., ZWINGMANN, H. & POGSON, R.E. 2012. Passive-margin prolonged volcanism, east Australian plate: outbursts, progressions, plate controls and suggested causes. *Australian Journal of Earth Sciences*, **59**, 983–1005.

SUTHERLAND, R. 1999. Basement geology and tectonic development of the greater New Zealand region: an interpretation from regional magnetic data. *Tectonophysics*, **308**, 341–362.

SUTHERLAND, R., COLLOT, J. *ET AL.* 2010. Lithosphere delamination with foundering of lower crust and mantle caused permanent subsidence of New Caledonia Trough and transient uplift of Lord Howe Rise during Eocene and Oligocene initiation of Tonga–Kermadec subduction, western Pacific. *Tectonics*, **29**, TC2004, https://doi.org/10.1029/2009TC002476

Tappenden, V. 2003. *Magmatic response to the evolving New Zealand Margin of Gondwana during the Mid–Late Cretaceous*. PhD thesis, University of Canterbury, Canterbury, New Zealand.

Till, C.B., Gans, P.B., Spera, F.J., MacMillan, I. & Blair, K.D. 2009. Perils of petrotectonic modeling: a view from southern Sonora, Mexico. *Journal of Volcanology and Geothermal Research*, **186**, 160–168.

Timm, C., Hoernle, K. *et al.* 2010. Temporal and geochemical evolution of the Cenozoic intraplate volcanism of Zealandia. *Earth-Science Reviews*, **98**, 38–64.

Tulloch, A.J., Ramezani, J., Mortimer, N., Mortensen, J., van den Bogaard, P. & Maas, R. 2009. Cretaceous felsic volcanism in New Zealand and Lord Howe Rise (Zealandia) as a precursor to final Gondwana break-up. *In*: Ring, U. & Wernicke, B. (eds) *Extending a Continent: Architecture, Rheology and Heat Budget*. Geological Society, London, Special Publications, **321**, 89–118, https://doi.org/10.1144/SP321.5

van de Beuque, S., Auzende, J.-M., Lafoy, Y. & Missegue, F. 1998. Tectonique et volcanisme tertiaire sur la ride de Lord Howe (Sud-Ouest Pacifique). *Comptes Rendus de l'Académie des Sciences*, Series IIA, **326**, 663–669.

van der Meer, Q.H.A., Scott, J.M., Waight, T.E., Sudo, M., Schersten, A., Cooper, A.F. & Spell, T.L. 2013. Magmatism during Gondwana break-up: new geochronological data from Westland, New Zealand. *New Zealand Journal of Geology and Geophysics*, **56**, 229–242.

Weaver, S.D. & Smith, I.E.M. 1989. New Zealand intraplate volcanism. *In*: Johnson, R.W., Knutson, J. & Taylor, S.R. (eds) *Intraplate Volcanism in Eastern Australia and New Zealand*. Cambridge University Press, Cambridge, 57–188.

Weaver, S.D., Adams, C.J., Pankhurst, R.J. & Gibson, I.L. 1992. Granites of Edward VII Peninsula, Marie Byrd Land: anorogenic magmatism related to Antarctic–New Zealand rifting. *Transactions of the Royal Society of Edinburgh: Earth Sciences*, **83**, 281–290.

Weis, D., Kieffer, B. *et al.* 2006. High-precision isotopic characterization of USGS reference materials by TIMS and MC-ICP-MS. *Geochemistry, Geophysics, Geosystems*, **7**, Q08006.

Wessel, P. & Kroenke, L.W. 2008. Pacific absolute plate motion since 145 Ma: An assessment of the fixed hot spot hypothesis. *Journal of Geophysical Research: Solid Earth*, **113**, B06101, https://doi.org/10.1029/2007JB005499

Winchester, J.A. & Floyd, P.A. 1977. Geochemical discrimination of different magma series and their differentiation products using immobile elements. *Chemical Geology*, **20**, 325–343.

Modelling basalt weathering at elevated CO_2 concentrations: implications for terminal to post-magmatic rifting in the Deccan Traps, Kachchh, India

KAUSHIK MITRA[1], SOUVIK MITRA[1], SAIBAL GUPTA[1]*, SATADRU BHATTACHARYA[2], PRAKASH CHAUHAN[2] & NIRMALA JAIN[2]

[1]*Department of Geology & Geophysics, Indian Institute of Technology, Kharagpur 721 302, India*

[2]*Space Applications Centre, Indian Space Research Organization, Ahmedabad 380 015, India*

**Correspondence: saibl2008@gmail.com*

Abstract: Deccan volcanism was synchronous with rifting along the west coast of India. Pre- and synmagmatic rifting has been widely reported in the Deccan Volcanic Province, but extension post-dating magmatism, and predating India–Eurasia collision, is less well known. A recent study in the Kachchh area of western India documented weathering of basalts to kaolinite at the base of Cenozoic rift basins, with rift flanks relatively less altered to smectites, and this was attributed to post-magmatic rifting. This study models basalt weathering under open- and closed-system conditions to simulate rainwater interacting with basalts either on topographical slopes (within rifts) or on flat-topped hills (flow tops). Both systems were modelled under p_{CO_2} conditions ranging from low, present-day values to higher values more appropriate for the end-Cretaceous–early Paleocene time, after basalt emplacement. The results show that if p_{CO_2} exceeded values of $10^{-2.5}$, basalts would be altered to kaolinite in both open and closed systems. Existing p_{CO_2} estimates in the aftermath of Deccan volcanism fall below this value, implying that the differential basalt weathering was more likely to have been caused by terminal to post-magmatic rifting. This indicates that extensional tectonics along the Indian west coast in the Kachchh region continued even after cessation of Deccan volcanism.

Chemical weathering is the ‘adjustment process’ that rocks formed at temperature and pressure conditions higher than those existing on the Earth’s surface undergo at the surface to form stable clay minerals, iron and aluminium oxides, and hydroxides (Anderson *et al.* 2007; Depetris *et al.* 2014). Lithology, climate, vegetation, relief and run-off are some of the most important factors dictating the rate and intensity of the chemical weathering of rocks (Benedetti *et al.* 1994; Drever 1994; White & Blum 1995; Gislason *et al.* 1996; Brady *et al.* 1999; Moulton *et al.* 2000; Dessert *et al.* 2001, 2003; Hinsinger *et al.* 2001; Das *et al.* 2005). Lithology is a major control on weathering (e.g. Dupré *et al.* 2003), which is dealt with in greater detail below. Of the other factors that determine the rate and extent of weathering, climate is the most important parameter that controls weathering, controlled mainly by temperature and rainfall (Keller 1956; White & Blum 1995), which in turn also influences vegetation. The effect of relief is seen in pedogenetic processes that are controlled primarily by water availability at low elevation sites, and by temperature at high elevations (Nagelschmidt *et al.* 1940; Rasmussen *et al.* 2010). Chemical weathering is generally initiated by slightly acidic water and gases, which attack thermodynamically unstable minerals on the surface of the Earth (Drever 2005). The extent of chemical alteration is strongly dependent on temperature, the amount of water available to leach rocks and the acidity of the available water (Carroll 2012; Millot 2013).

Meybeck (1986), Amiotte Suchet *et al.* (1995) and Dupré *et al.* (2003) underlined the effect of lithology in controlling weathering, demonstrating that granites and gneisses are 5–10 times less easily erodible than basalts. As a consequence, basalts, which occupy only about 5% of the land surface (about 6.8×10^6 km^2: Dessert *et al.* 2003; Self *et al.* 2006; Rasmussen *et al.* 2010), are responsible for over 30% of the atmospheric CO_2 drawdown (e.g. Amiotte Suchet *et al.* 2003; Dessert *et al.* 2003; Kent & Muttoni 2013). Thus, an elevated concentration of CO_2 in the atmosphere enhances the rate of basalt weathering (e.g. Kent & Muttoni 2013; Li & Long 2014). It is in this context that the eruption of continental flood basalts (CFBs) becomes particularly important, as a significant proportion of silicate material that is a potential CO_2 sink is added to the Earth’s surface. At the same time, the eruption of CFBs is also responsible for the addition of a significant proportion of

From: Sensarma, S. & Storey, B. C. (eds) 2018. *Large Igneous Provinces from Gondwana and Adjacent Regions*. Geological Society, London, Special Publications, **463**, 227–241.
First published online July 17, 2017, https://doi.org/10.1144/SP463.8

CO_2 (1792 ppmv) into the atmosphere (Dessert *et al.* 2001).

The Deccan Traps in India are one of the best-known CFB provinces in the world, and erosion of these basalts has been suggested to have had a significant climatic impact in lowering atmospheric CO_2 content through the Cenozoic (Dessert *et al.* 2001). One of the primary debates associated with the Deccan Traps has been its temporal relationship with the break-up of India from the Seychelles microcontinent at the Cretaceous–Tertiary transition (e.g. Collier *et al.* 2008; Bhattacharya & Yatheesh 2015). Recently, Mitra *et al.* (2016) suggested that subaerially erupted Deccan Trap basalts in the Kachchh area of western India underwent post-eruption rifting and subsidence in the early Tertiary time, which they tentatively correlated with the India–Seychelles break-off. Basalts on the flank and floor of the Cenozoic rift basin were altered to kaolinite, unlike the smectitic alteration generally observed in the adjacent flat-topped basaltic flows outside the rift, and, indeed, in most other parts of the Deccan Traps (e.g. Salil *et al.* 1997; Deshmukh *et al.* 2012; Greenberger *et al.* 2012). Mitra *et al.* (2016) attributed this differential weathering to rift-generated topography, and supported their argument with geochemical models using present-day atmospheric CO_2 concentrations. Since basalt weathering is a strong function of rainwater acidity, which in turn is dependent on atmospheric carbon dioxide content, it is important to see if the observed weathering products are also consistent with calculations using increased atmospheric CO_2 concentrations, as were likely in the aftermath of Deccan volcanism. If this can be demonstrated through geochemical models using elevated p_{CO_2} levels in the atmosphere, the proposition of a phase of rifting that post-dated Deccan volcanism in western India by Mitra *et al.* (2016) can be supported. This is what we investigate using geochemical modelling in this study.

Geological background

The Deccan Traps cover much of central and western India (Fig. 1). It is an archetypal Large Igneous Province (LIP) that erupted at *c.* 65.5 Ma (Fig. 1) (Courtillot *et al.* 1986; Baksi 1987, 2014; Pande 2002). The eruption of the Deccan Traps has commonly been attributed to an underlying, deep-seated mantle plume that is, at present, located below the Réunion Islands in the Indian Ocean (e.g. Morgan 1981; Richards *et al.* 1989; Campbell & Griffiths 1990). Some workers, however, have disputed the plume theory, and attribute the eruption of the Deccan Trap lavas to continental rifting (e.g. Mahoney 1988; Sheth 1999, 2005). The controversy centres around the evidence for syn- or pre-volcanic rifting, which has been used to argue if eruption of the Deccan Traps initiated continental break-off or was its consequence. Interestingly, there has been little discussion on the evidence for a phase of continental rifting that operated during the terminal phase of Deccan volcanism at *c.* 61–60 Ma (e.g. Sheth *et al.* 2001; Shukla *et al.* 2001) and continued into the post-volcanic period, prior to the India–Eurasia collision at *c.* 55 Ma (Patriat & Achache 1984; Klootwijk *et al.* 1992).

In a recent study, Mitra *et al.* (2016) argued that deposition of the Cenozoic Matanomadh Formation in Kachchh, in the state of Gujarat, western India (see Fig. 1) occurred within a rift which was generated on Deccan basalts that were already intensely weathered. The implication was that a new Cenozoic basin was generated by extensional tectonics that post-dated subaerial eruption of the Deccan Traps, and that this phase of extension post-dated, and was therefore unrelated to, Deccan Trap volcanism. Mitra *et al.* (2016) tentatively correlated this with the India–Seychelles break-up. At Matanomadh, unaltered Deccan Trap basalts grade upwards into a zone where the basalts form spheroidally weathered boulders and host clay minerals; this is the basal litho-unit of the Matanomadh Formation (Saxena 1977; Biswas 1992). Parts of the Matanomadh Formation contain hydrous sulphates that have been considered similar to hydrous sulphate localities on the Martian surface (Mitra *et al.* 2014; Bhattacharya *et al.* 2016). Eocene–Recent sedimentary rocks overlie both the basalts and the Matanomadh Formation (Biswas 1992). Basalts on the flanks of the Cenozoic basins are visibly less altered to smectitic assemblages, while those forming the clay boulders at the base of the Cenozoic succession are more intensely altered to kaolinite. Mitra *et al.* (2016) attributed this differential weathering to the effect of acidic rainwater flowing down the rift flanks and slopes, while the limited amounts of water retained on the horizontal flow tops were only sufficient to transform the basaltic mineralogy to smectites. This weathering pattern was interpreted to reflect a temporal hiatus between the eruption of the Deccan volcanics and the onset of Cenozoic extension in Kachchh, and, therefore, Mitra *et al.* (2016) tentatively suggested that the rifting was a far-field effect of the India–Seychelles break-up that must have occurred off the Indian west coast to the south of Kachchh in the early Paleocene. As mentioned earlier, this study did not take into account the potential increase in the extent of chemical weathering of basalts if CO_2 concentrations in the atmosphere were significantly higher, as in the end-Cretaceous, following the eruption of the Deccan Trap volcanics, and the fact that the terminal phase of Deccan magmatism may have overlapped with the onset of this late rifting event.

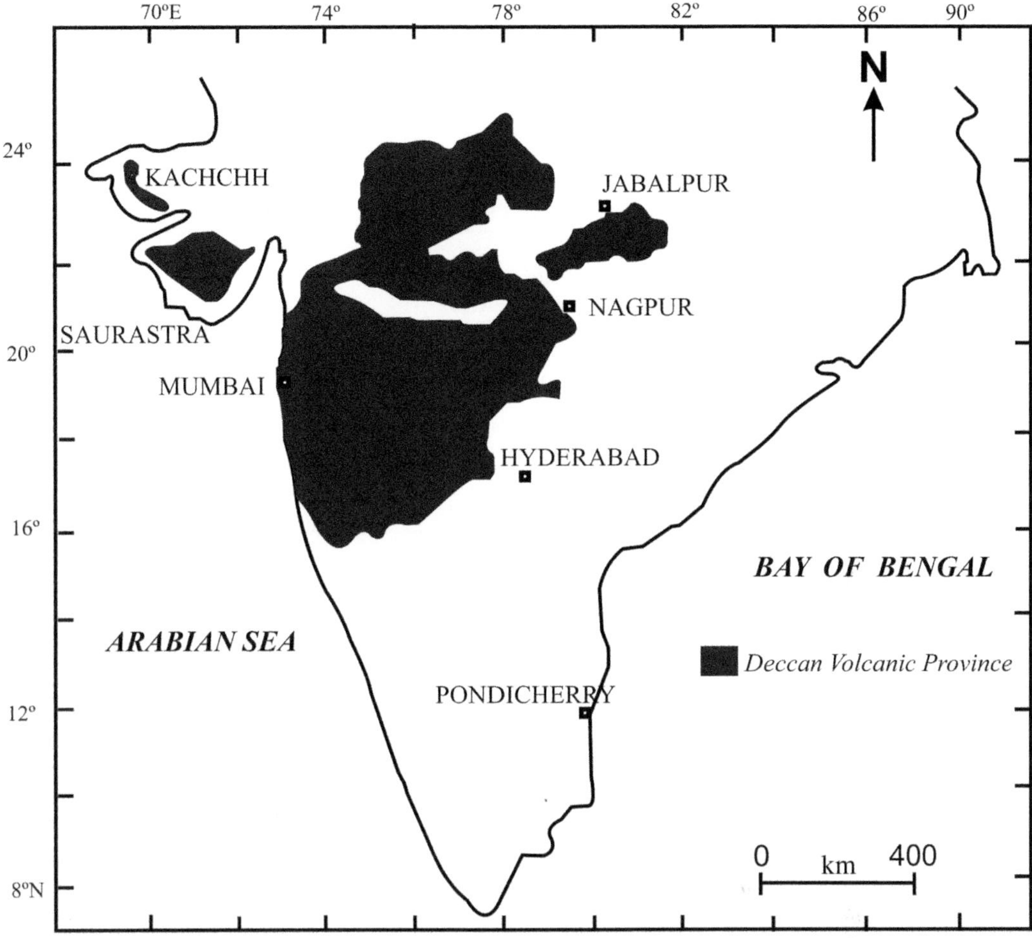

Fig. 1. Map of Peninsular India showing the extent of the Deccan Traps over central and western India (modified after Siddaiah & Kumar 2009). Note the position of Kachchh in western India.

Geochemical modelling

The problem being investigated in this study is whether the differential weathering of basalt to either smectite group minerals or kaolinite can be attributed to topographical differences alone, as interpreted by Mitra *et al.* (2016), or if this can be influenced more strongly by the amount of CO_2 in the atmosphere which controls rainwater acidity. The composition of rainwater plays a vital role in basalt weathering and needs to be considered when determining the climate of any region (Carroll 2012). The pH of rainwater is about 5.8 at 25°C and normal carbon dioxide fugacity, f_{CO_2}, is $10^{-3.5}$, under present-day atmospheric conditions (see Garrells & Mackenzie 1971; Bethke 2007; Carroll 2012). The slight acidity of rainwater is attributed to the chemical equilibrium it establishes with the atmospheric CO_2 (Carroll 2012). However, the acidity of rainwater may reach lower pH levels in areas of active volcanism. Murata (1966) and Harding & Miller (1982) reported pH values in the range of 3–4 within 1 km of the active summit of Kilauea in Hawaii. Rainwater with similar acidity has also been reported in the vicinity of active volcanoes in Nicaragua (e.g. Johnson & Parnell 1986) and Japan (e.g. Unoki & Itoi 2005). Thus, natural rainwater may show a substantial change in acidity during volcanic activity. At the time of eruption of the Deccan Traps, major changes in atmospheric conditions are postulated to have occurred worldwide (e.g. Font *et al.* 2014; Self *et al.* 2014). Volatile emanations associated with Deccan volcanism have been suggested to have created conditions of 'mock aridity' in segments of the Deccan Volcanic Province (Khadkikar *et al.* 1999), and have been correlated with major foraminiferal mass extinctions across the Cretaceous–Tertiary (K–T) boundary in NE India (Gertsch *et al.* 2011). Thus, a possible increase in rainwater acidity at the time of emplacement of the Deccan Trap lava flows needs to be taken into account in order to make more robust predictions about the products of basalt weathering.

Methodology

REACT, a module included in the software package Geochemists' Workbench (GWB) 10.0.4. (Bethke 2007), was used to model thermodynamic equilibrium reaction paths for the weathering of basalt by rainwater. The methodology adopted was essentially similar to that used by Mitra *et al.* (2016), although the input parameters were varied to simulate a variety of environmental conditions. An expanded variant of the Lawrence Livermore National Laboratory database, V8 R6+, has been used to perform the thermodynamic calculations. The oxide composition of the uppermost Deccan flow basalt (a tholeiite) from Shukla *et al.* (2001) has been used as an input parameter to calculate the mineral wt% of the basalt (see (a) in Table 1). Since a minor proportion of the basalts exposed in Kachchh are also of alkaline affinity, the calculations were also repeated using a model alkali basalt composition (from Paul *et al.* 2008) (see (a) in Table 2). Standard CIPW Norm calculations were employed on the oxide composition, and the mineral assemblage obtained thereby has been used as the representative of the Deccan basalt composition for all purposes in the present study (see (b) in Tables 1 & 2). The data in Table 1 were used by Mitra *et al.* (2016) in their calculations; weathering with the alkali basalt composition in Table 2 was also modelled to encompass differences arising from primary compositional variations of the starting basalt. Apatite has been omitted from the calculations owing to its low weight fraction in basalt (*c.* 0.37%) and its relatively low solubility, owing to which it usually remains unaltered during basalt weathering (Catalano 2013). Kachchh, being a region in the tropical zone, is expected to have experienced heavy rainfall in the past; evidence of a humid, tropical climate in Cenozoic times is preserved in the form of lignite units within the Matanomadh Formation, and also higher up in the Eocene part of the succession (Biswas 1992; Mathews *et al.* 2013; Saraswati *et al.* 2014). Therefore, a high water : rock ratio (10^5 : 1) is used to simulate high rainfall conditions in Kachchh. The average rainwater composition from Garrells & Mackenzie (1971) was taken to titrate basalt at ambient temperature (25°C) and pressure (1 atm). Calculations at higher temperatures more applicable to conditions of tropical weathering (i.e. >25°C) could not be carried out owing to limitations of the software package (GWB) used in our study. Moreover, an increase in the weathering temperature is not expected to affect the weathering product composition substantially, apart from enhancing reaction kinetics. In order to allow convergence of the geochemical model, species such as Al^{3+}, Fe^{2+} and $Ti(OH)_4$, that were practically absent, had to be allotted negligibly small concentrations (1 ppb) in rainwater.

The initial pH of the system has been assigned a value of 5.7, which is the pH of rainwater equilibrating with atmospheric CO_2 (Carroll 2012). The concentration of bicarbonate (HCO_3^-) was calculated by assigning equilibrium with atmospheric CO_2,

Table 1. *(a) Composition of Flow F9 of Deccan basalt from Kachchh, calculated from XRF and ICP-MS analysis (Shukla* et al. *2001). (b) Mineral assemblages represented as wt%, calculated from the CIPW Norm. The data used are similar to those used by Mitra* et al. *(2016)*

(a)		(b)	
Constituents	Wt%	Minerals	Wt%
SiO_2	45.78	Albite	18.53
TiO_2	1.56	Anorthite	36.13
Al_2O_3	17.03	Orthoclase	1
Fe_2O_3	14.4	Diopside	23.69
MnO	0.21	Olivine	12.48
MgO	5.99	Magnetite	3.13
CaO	12.05	Ilmenite	2.96
Na_2O	2.19	Apatite	0.37
K_2O	0.17	Quartz	0
P_2O_5	0.16		
Total	99.54	Total	98.32

Table 2. *(a) Composition of alkali basalt from the Dinodhar plug (Sample BH14.4) within the Deccan basalts of Kachchh (after Paul* et al. *2008) used in the modelling calculations. (b) Mineral assemblages represented as wt%, calculated from the CIPW Norm*

(a)		(b)	
Constituents	Wt%	Minerals	Wt%
SiO_2	42.34	Plagioclase	16.3
TiO_2	3.21	Orthoclase	17.61
Al_2O_3	11.06	Nepheline	9.03
Fe_2O_3	13.68	Diopside	25.01
MnO	0.16	Olivine	9.69
MgO	10.21	Ilmenite	0.34
CaO	10.89	Hematite	13.68
Na_2O	2.92	Apatite	1.09
K_2O	2.98	Perovskite	5.16
P_2O_5	0.47		
Total	97.92	Total	97.91

since all dissolved bicarbonates formed during basalt weathering originate from dissolved CO_2 (Dessert *et al.* 2001). The simulations were carried out in CO_2-buffered and -unbuffered systems to replicate open and closed systems, respectively. In the open system, the atmospheric gases are in continuous equilibrium with rainwater, which is assumed to represent a flowing/moving/transient water body, simulating a geological situation of water flowing off a slope. The closed system acts as a proxy to the geological situation where the system and the atmosphere are not in equilibrium, with the gases in the atmosphere not being buffered to the system. A geological setting analogous to the unbuffered/closed system is likely to be a large volume of standing water on the rock body. After initial equilibrium between the atmosphere and the system, the atmospheric gases no longer interact with the system. The water mass is not transient and thus does not facilitate mixing with atmospheric gases, making the whole system unbuffered or closed with respect to the atmosphere. Since acidic oxides are not present to buffer the system, the pH of the fluid rises due to the presence of the basic oxides in the rock.

Kinetically unfavourable minerals (andradite, boehmite, diaspore, epidote, gismondine, greenalite, grossular, mesolite, minnesotaite, muscovite, paragonite, phlogopite, prehnite, scolecite, stilbite, tremolite and natrolite) were suppressed for calculations using both basalt compositions in Tables 1 & 2. This was necessary as including kinetics in the calculations was not possible due to the unavailability of effective surface-area data for minerals.

Weathering mechanism

The most important chemical weathering mechanism for silicates is hydrolysis (decomposition of minerals by water) (Depetris *et al.* 2014). The silicate minerals present in basalt are considered to be salts of weak acids and strong bases, thereby making their hydrolysed solutions alkaline (Millot 2013). A typical example of acid hydrolysis, the incongruent dissolution of albite to kaolinite by CO_2-saturated water (Depetris *et al.* 2014), which is responsible for kaolinite formation is given by the following reaction:

$$2NaAlSi_3O_8(s) + 9H_2O + 2H_2O + 2CO_2(g)$$
$$= Al_2Si_2O_5(OH)_4(s) + 2Na^+(aq)$$
$$+ 2HCO_3^-(aq) + 4H_4SiO_4(aq)$$

The equilibrium constant (K) of the above reaction at 25°C is 7.9×10^{-13}. According to Le Chatelier's principle, a large amount of CO_2 is thus needed in the atmosphere to convert albite into kaolinite.

Results of geochemical modelling

As stated earlier, the methodology adopted was similar to that used by Mitra *et al.* (2016), but with two important modifications to simulate conditions that existed in the present-day Deccan Traps region closer to the end of the Cretaceous. First, Mitra *et al.* (2016) used present-day values of CO_2 fugacity (f_{CO_2} = 400 ppm) in their calculations. To investigate an alteration sequence expected at f_{CO_2} values more relevant to the time of eruption of the Deccan Traps, the f_{CO_2} value of $10^{-2.84}$ (*c.* 1435 ppmv), based on the estimates of elevated pre-Deccan atmospheric carbon dioxide (Dessert *et al.* 2001), was used in the calculations. Secondly, Mitra *et al.* (2016) calculated the predicted clay mineral assemblage for varying water : rock ratios. The preponderance of lignite deposits at the locations in Kachchh, the region for which these conditions were being simulated, indicates tropical, humid conditions. Thus, in this study, the water : rock ratio was consistently assumed to be 10^5 : 1, appropriate for approximating high rainfall conditions (Mitra *et al.* 2016). The results of these calculations are shown in Figures 2 and 3, and are discussed below separately for both open- and closed-system conditions.

Open-system condition

The CO_2-buffered system is representative of an open system where the atmospheric gases are in constant equilibrium with rainwater. The constant supply of CO_2 helps in maintaining the acidity of the system. As basalt weathering tends to neutralize the acidity of the fluid, maintaining acidity of the reactive fluid (rainwater) makes basalt weathering more effective.

The results of open-system weathering obtained in this study for the tholeiite basalt composition (Table 1) are shown in Figure 2a in which the weight of minerals in grams (y-axis on the left) and pH (y-axis on the right) are plotted against the extent of basalt weathering. The pH of the fluid starts from 5.7 and increases to 6.75. Five different minerals – viz., gibbsite (Gbs), hematite (Hem), rutile (Rt), nontronite (Non) and kaolinite (Kln) – are found to precipitate and dissolve during different phases of the weathering process. On the basis of the appearance and disappearance of the different minerals, the complete weathering process has been divided into five segments, segments I–V.

Segment I. This segment covers 0–35% of the whole weathering process. Gibbsite ($Al(OH)_3$), rutile (TiO_2) and hematite (Fe_2O_3) start to form as the first set of alteration minerals. The Al^{3+} concentration in the fluid decreases as gibbsite starts to precipitate. The SiO_2 (aq) concentration in the fluid increases at a

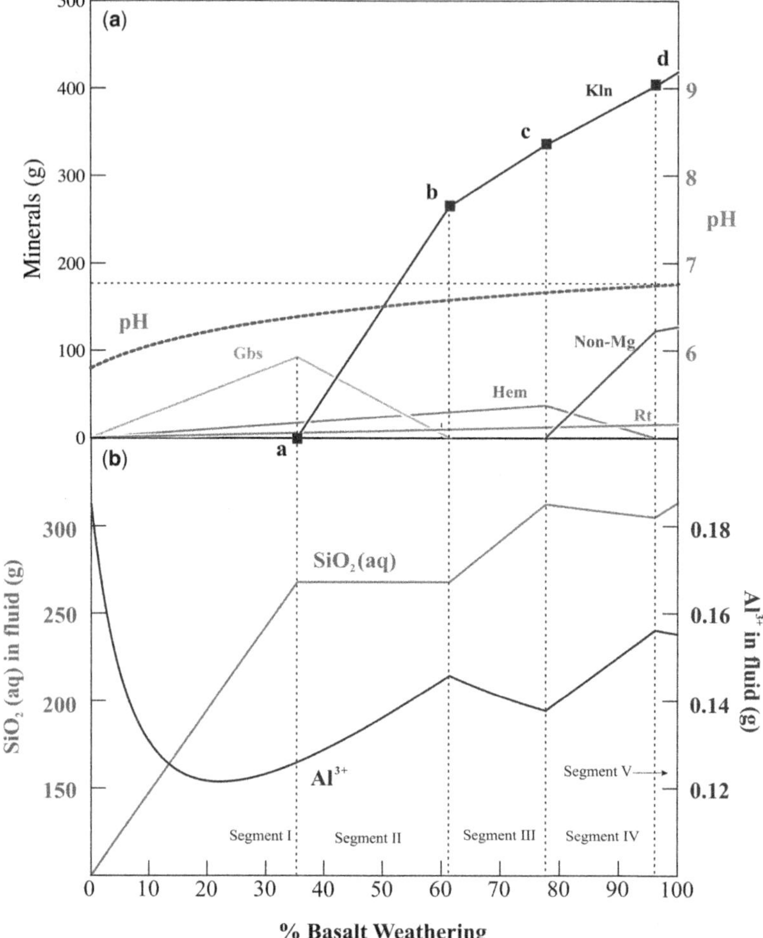

Fig. 2. (**a**) Predicted mineral assemblage and pH; and (**b**) the SiO_2 and Al^{3+} concentration in the fluid formed during near-surface (25°C and 1 atm) rainwater alteration of Deccan basalt at a water : rock ratio of $10^5 : 1$, and $\log(f_{CO_2}) = -2.84$, in an open (i.e. buffered) system. All calculations were performed using REACT.

steady rate with the dissolution/hydrolysis of the silicate minerals in basalt (Fig. 2b). The pH attains a mildly acidic value of approximately 6.4. Gibbsite starts dissolving out at the end of segment I.

Segment II. Segment II is bound between 35 and 61%. It marks the onset of kaolinite formation at point 'a' in Figure 2a until the complete disappearance of gibbsite. Gibbsite reacts with silicic acid (H_4SiO_4) to form kaolinite:

$$2Al(OH)_3(s) + 2H_4SiO_4(aq) = Al_2Si_2O_5(OH)_4(s) + 5H_2O$$

The amount of Al^{3+} (aq) in the fluid increases, while SiO_2 (aq) values remain constant in this segment. Hematite and rutile continue to be produced at rates similar to Segment I.

Segment III. Segment III starts from 61.1% weathering, marking the complete absence of gibbsite (point 'b' in Fig. 2a). There is a sharp change in the slope for kaolinite caused by the reduction of kaolinite production owing to the absence of gibbsite. The amount of Al^{3+} in the fluid also starts to reduce due its consumption in forming kaolinite. The end of Segment III is marked by a decline in production of hematite and the simultaneous production of nontronite-Mg at 78% weathering.

Segment IV. This segment demarcates the decline of hematite, eventually leading to its complete disappearance (point 'c' in Fig. 2a). Nontronite-Mg, a

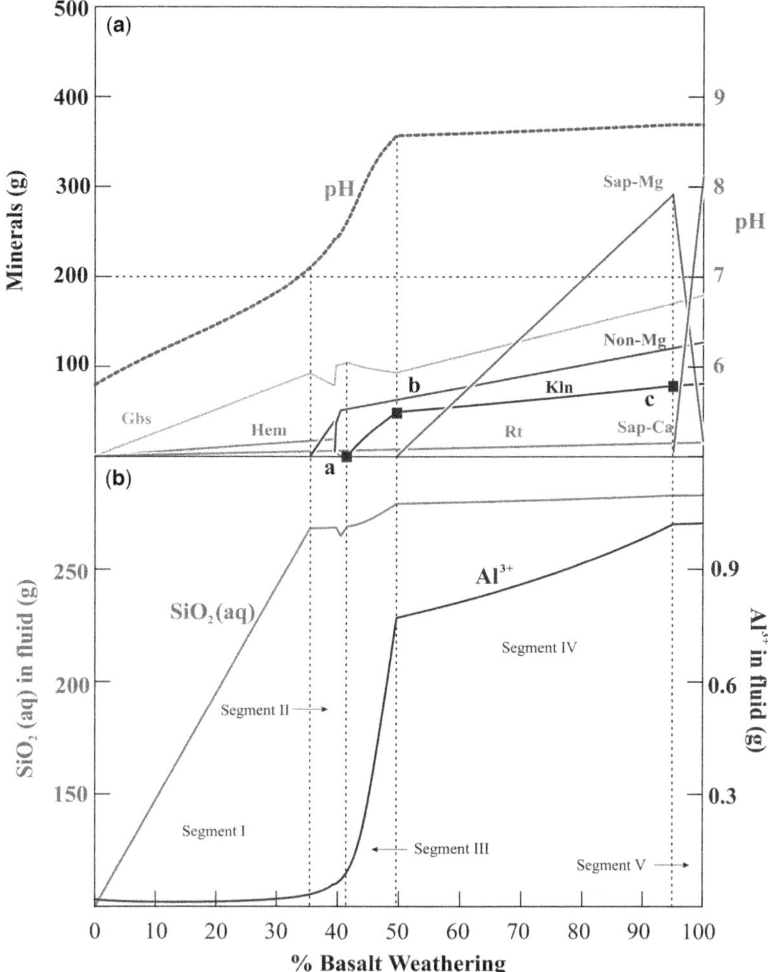

Fig. 3. (**a**) Predicted mineral assemblage and pH; and (**b**) the SiO_2 and Al^{3+} concentration in the fluid formed during near-surface (25°C and 1 atmosphere) rainwater alteration of Deccan basalt at a water : rock ratio of 10^5 : 1, and log (f_{CO_2}) = −2.84, in a closed (i.e. unbuffered) system.

smectite, starts forming while there is a slight decrease in the rate of kaolinite formation, which may be a consequence of a slight increase in Al^{3+}. Segment IV ends at 95% of basalt weathering.

Segment V. This segment represents the final stage (95%) of basalt weathering, marking the complete transformation of basalt into the more stable clay minerals. The final assemblage of minerals includes kaolinite, nontronite-Mg and rutile in decreasing order of abundance. Hematite is completely absent and the production rate of nontronite-Mg is reduced. Al^{3+} concentrations decrease in the fluid. Kaolinite comprises about 75% of the total alteration assemblage. The final pH of the fluid is near neutral (*c.* 6.75). The non-alkalinity of the system helps kaolinite to stabilize as the most abundant mineral in the alteration assemblage.

Closed-system condition

The closed system is represented by CO_2-unbuffered conditions in which the system is not in constant equilibrium with the atmosphere. Figure 3a plots the weights of minerals in grams (*y*-axis on the left) and pH (*y*-axis on the right) against the extent of basalt weathering. The pH of the fluid (Fig. 3b) starts from 5.7 and increases to 8.7, which is more than that calculated for the open system. Gibbsite (Gbs), hematite (Hem), rutile (Rt), nontronite (Non), saponite (Sap) and kaolinite (Kln) are the resulting alteration minerals, which are formed by the unbuffered weathering of basalt. As in the case of the open system, this weathering pattern (Fig. 3a) is also described in terms of five segments, based on the appearance and disappearance of various mineral phases.

Segment I. The region from 0 to 35% of the weathering process is kept within Segment I. Similar to the open system, Segment I of the unbuffered system marks the beginning of the weathering processes. Gibbsite (Gbs), rutile (Rt) and hematite (Hem) are the first minerals to form in the alteration process. The rate and amount of hematite and rutile formed is very limited compared to gibbsite. The pH of the fluid also increases at a much faster rate than that observed during weathering in the buffered condition. The initial pH of the system is equal to that in the previous case, as the rainwater was equilibrated with the same atmosphere. Al^{3+} concentrations are extremely low, while SiO_2 concentrations are found to increase at a constant rate.

Segments II and III. Segment II marks the start of the formation of kaolinite, as well as nontronite, and the end of hematite precipitation (point 'a' in Fig. 3a). Apart from a local maximum, the SiO_2 concentration remains constant over the entire segment. Segment III marks the extent of basalt weathering in which gibbsite reaches its maximum amount followed by its dissolution to make way for kaolinite. A sharp increase in Al^{3+} concentrations is one of the most striking features in this segment, accompanied by an increase in the rate of SiO_2 formation. Importantly, through Segments II and III, the chemistry of the system changes from acidic to alkaline at about 50% of basalt weathering.

Segments IV and V. Point 'b' in Figure 3a demarcates the point from which the rate of formation of kaolinite decreases. A new smectite, saponite, starts forming at the beginning of segment IV. The rate of formation of saponite is extremely rapid and overshoots all other alteration minerals in abundance. The amount of other minerals (nontronite, rutile and kaolinite) increases with the extent of alteration but never exceeds the weight proportion of saponite. Al^{3+} and SiO_2 in the fluid reach an upper limit of concentration that remains almost constant over the remaining period of basalt weathering. Saponite-Ca is formed at the expense of saponite-Mg during the last 5% of the alteration process. The final pH of the system acquires a value of about 8.7, and indicates mildly alkaline conditions.

Discussion

Basalt to smectite v. basalt to kaolinite: p_{CO_2} *or topographical control?*

Kaolinite is typically generated in acidic environments (Ross 1945; Millot 2013). The two most vital parameters for determining the acidity of the system are the nature of the system (open or closed) and the CO_2 concentration in the atmosphere. In the context of the Deccan Trap basalts, the open- and closed-system conditions were interpreted by Mitra *et al.* (2016) to indicate flowing water (compatible with rainfall on a slope) and stagnant water (standing water on flow tops or consumed by reaction with basalt), respectively. Based on their conclusions using present-day atmospheric CO_2 levels (400 ppm), they concluded that formation of kaolinite from basalt was only favoured under open-system conditions, while closed-system conditions would tend to limit basalt weathering to smectite formation. However, the results obtained by Mitra *et al.* (2016) are only valid if it can be demonstrated that differential formation of smectite and kaolinite from basalt weathering is either unaffected by the carbon dioxide content in the atmosphere under closed-system conditions, or if the transformation to kaolinite falls below a threshold critical value even in the post-Deccan atmosphere. Thus, calculations were carried out using variable p_{CO_2} concentrations, in both open and closed systems, to encompass a range of variable CO_2 contents in the atmosphere.

Although there is a general consensus that the end-Cretaceous atmospheric CO_2 concentration values were higher, estimates of the precise values of atmospheric CO_2 concentrations understandably vary (e.g. Dessert *et al.* 2001; Bice *et al.* 2006). Bice *et al.* (2006) estimated the end-Cretaceous atmospheric CO_2 concentrations to vary between 600 and 2400 ppmv, while Dessert *et al.* (2001) suggested that the already high pre-Deccan values of 1435 ppmv may possibly have been enhanced by 1792 ppmv following another eruption of the Deccan Traps. The carbon dioxide fugacities corresponding to 1435, 2400 and 3227 ppmv are $10^{-3.5}$ to values of $10^{-2.84}$, $10^{-2.62}$ and $10^{-2.5}$, respectively. To investigate if these elevated values would affect the resultant alteration mineralogy substantially, various environmental situations were simulated by using normal (present-day) and higher values of CO_2 fugacity (equal to the partial pressure) under both open- and closed-system conditions. All calculations were carried out at high water : rock ratios (10^5 : 1) to simulate high rainfall, tropical conditions. The results of the calculations are shown in Figures 4 and 5.

In an open system, at normal partial pressures of CO_2 ($f_{CO_2} = 10^{-3.5}$) (Fig. 4), kaolinite is the major

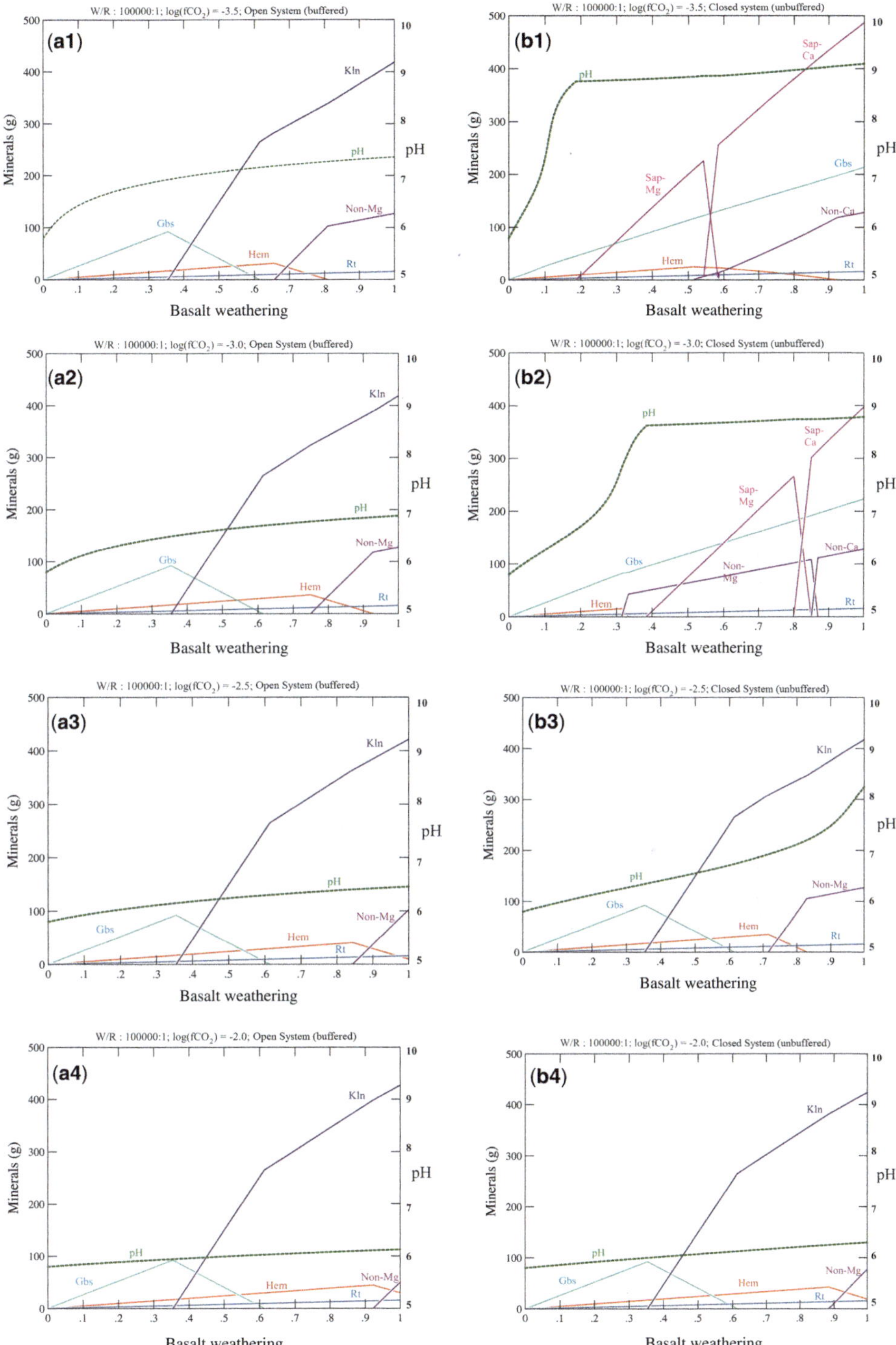

Fig. 4. Predicted alteration mineralogy produced using REACT, by weathering 1 kg of Deccan basalt (with the composition of Flow No. F9 of Shukla *et al.* 2001; see Table 1) at a water : rock ratio of 10^5 : 1 under CO_2-buffered/open (column a) and -unbuffered/closed (column b) conditions. Log (f_{CO_2}) values used are: **(a1)** & **(b1)** −3.5; **(a2)** & **(b2)** −3.0; **(a3)** & **(b3)** −2.5; and **(a4)** & **(b4)** −2.0. (a1) & (b1) are modified after Mitra *et al.* (2016).

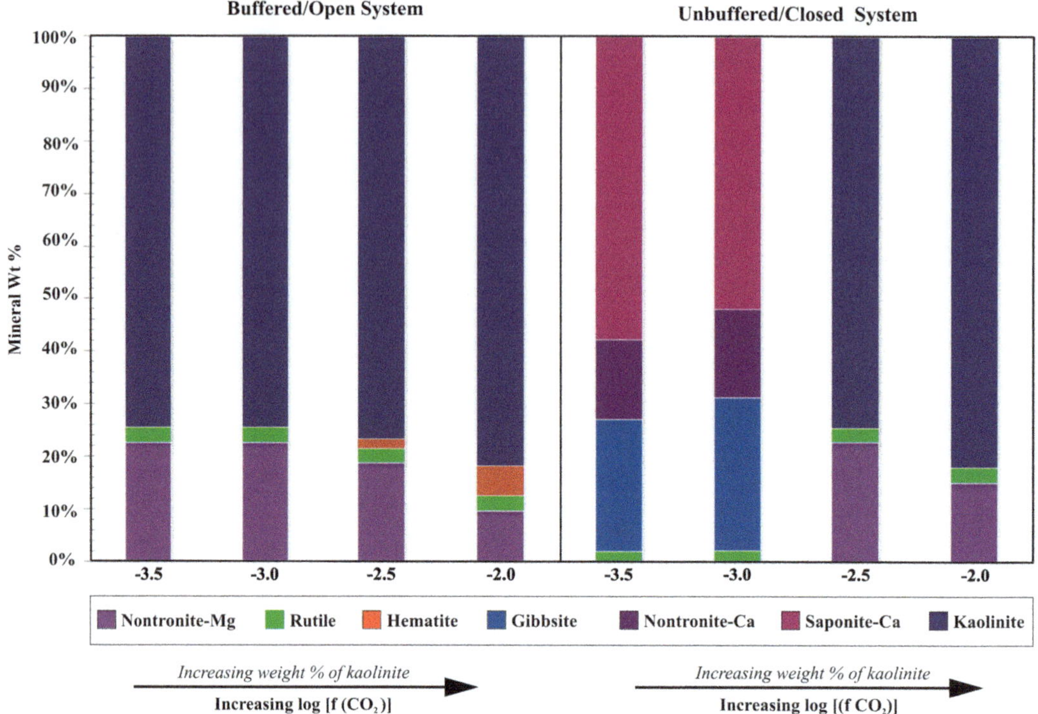

Fig. 5. Calculated weight percentage (wt%) of the alteration mineral assemblage produced when basalt weathers out at a water : rock ratio of 10^5 : 1. The effect of increasing f_{CO_2} in both buffered (i.e. open-system) and unbuffered (i.e. closed-system) conditions is illustrated in the left- and right-hand panes, respectively.

product of basalt weathering, amounting to approximately 74% of the total clay mineral assemblage (including smectite and rutile). The final pH levels rise to slightly alkaline values (*c.* 7.3: Fig. 4a1). Smectites dominate the clay assemblage when basalt weathering is carried out in a closed system at present-day CO_2 concentrations. The pH levels rise much higher in this system, marking the strong alkalinity (pH *c.* 9.1: Fig. 4b1) fostered by basalt weathering without the constant supply of CO_2 from the atmosphere. These results are similar to those obtained by Mitra *et al.* (2016).

To investigate the effect of varying CO_2 content in the atmosphere, a range of calculations were performed with f_{CO_2} values increasing from the present-day value of $10^{-3.5}$ to extreme values of $10^{-2.0}$ (see Fig. 4a2–a4, b2–b4). Increasing concentrations of atmospheric CO_2 help to maintain the acidity of the system, with the lowest values (pH *c.* 6.12) obtained at the highest f_{CO_2} value of $10^{-2.0}$, thereby facilitating the formation and stabilization of kaolinite as the major alteration product. At these values, kaolinite comprises 82% of the altered assemblage formed in an open system (Fig. 4a4), with only minor amounts of associated smectite, hematite and rutile. Under closed-system conditions, however, the results differ. For low f_{CO_2} values varying from $10^{-3.5}$ to $10^{-3.0}$ (Fig. 4b1, b2), the altered mineral assemblage is dominated by smectitic group minerals. Kaolinite starts appearing in the assemblage at an f_{CO_2} value of $10^{-2.98}$. However, as f_{CO_2} values increase from $10^{-2.5}$ to $10^{-2.0}$, kaolinite begins to dominate the weathered products, even under closed-system conditions (Fig. 4b3, b4), comprising approximately 79% of the assemblage at the highest f_{CO_2} value. This demonstrates that kaolinite can be formed as the major alteration product of basalt weathering in high p_{CO_2} conditions, even if it is not buffered to the system continuously (see Fig. 5).

Similar calculations were conducted for the open and closed systems involving the alkali basalt composition from the Dinodhar plug, given in Table 2 (after Paul *et al.* 2008); the results are shown in Figure 6a and b, respectively. As can be seen, kaolinite dominates the clay mineral assemblage under all conditions for the open system, while smectite dominates the assemblage in the closed system at low p_{CO_2} values. In the latter, kaolinite starts appearing in the assemblage at f_{CO_2} values greater than $10^{-2.65}$ (higher than those in the case of the tholeiite basalt, $10^{-2.98}$), and would correspondingly become the dominant alteration phase at even higher

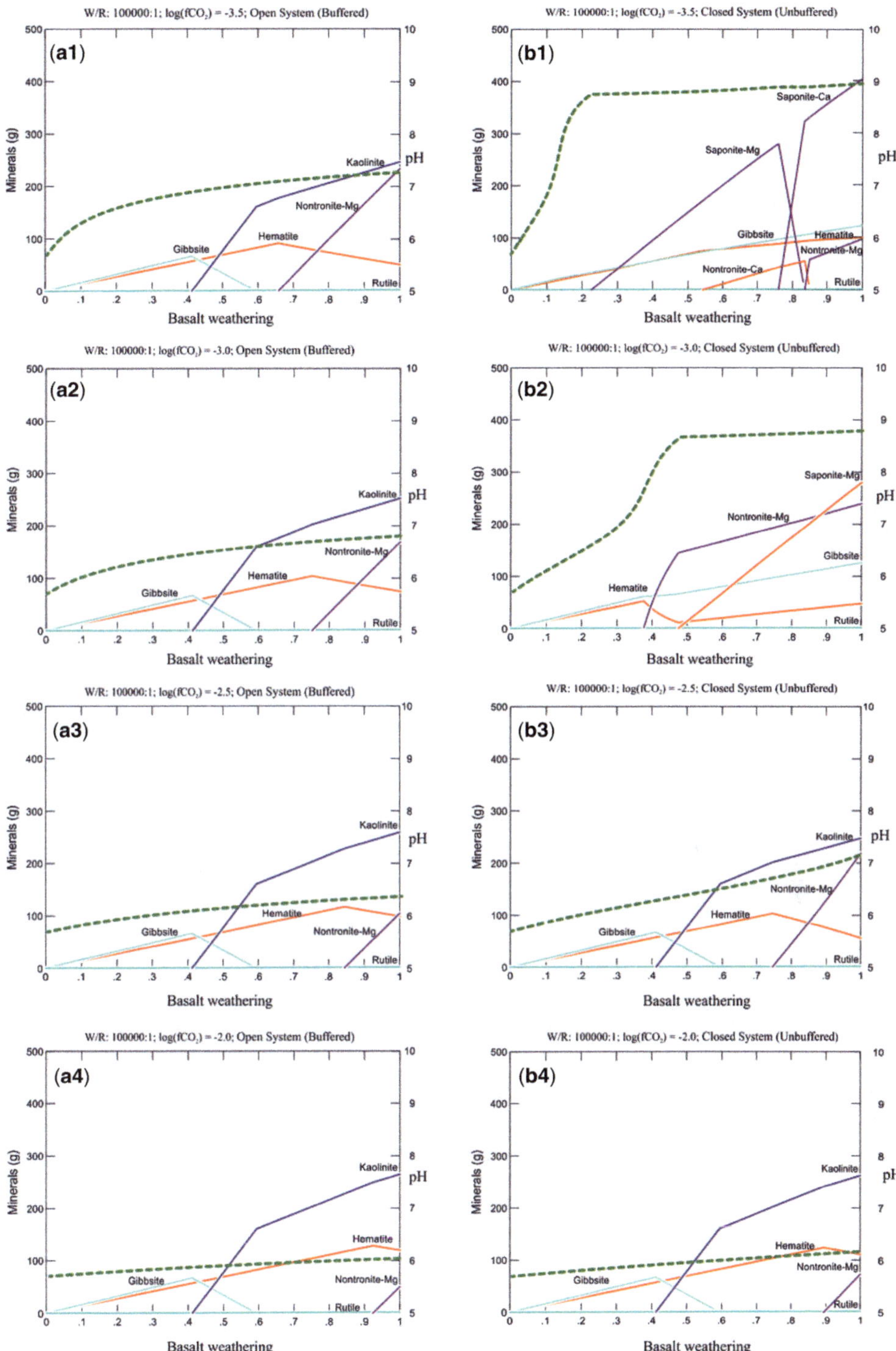

Fig. 6. Predicted alteration mineralogy produced using REACT, by weathering 1 kg of an alkali basalt from the Dinodhar plug (sample BH14.4: Paul *et al.* 2008) (see Table 2) at a water : rock ratio of 10^5 : 1 under CO_2-buffered (column a) and -unbuffered (column b) conditions. Log f_{CO_2} values used are: (**a1**) & (**b1**) −3.5; (**a2**) & (**b2**) −3.0; (**a3**) & (**b3**) −2.5; and (**a4**) & (**b4**) −2.0, similar to those in Figure 4.

f_{CO_2} values. Thus, under closed-system conditions, kaolinite can be the main alteration product if f_{CO_2} values were well above $10^{-2.65}$, irrespective of the composition of the basalt.

The implications of these results are important in the context of the question being addressed in this section. In high rainfall, tropical conditions and at atmospheric p_{CO_2} values of less than $10^{-2.5}$, basalts will be altered to smectites on flow tops (closed-system conditions) and kaolinite on slopes (open-system conditions). However, if p_{CO_2} exceeds $10^{-2.5}$, all basalts, whether on flow tops or slopes, will tend to be altered to kaolinite in preference to smectite. Thus, correlating the alteration of basalts with topography under the high p_{CO_2} conditions that may have existed following Deccan Traps extrusion is critically dependent on the precise atmospheric CO_2 concentration at the time.

Implications for rifting during and post-dating terminal Deccan magmatism in Kachchh

As mentioned earlier, Mitra *et al.* (2016) observed that in the Kachchh area, the Deccan basalts were altered to kaolinite within rift zones that were overlain by Cenozoic sediments, while rift-flank basalt flows were only altered to smectite. They attributed the differential weathering to topographical differences created by post-emplacement rifting of the Traps, implying that extensional tectonics operated along this part of the west coast of India following Deccan Trap eruption, and prior to Indo-Eurasia collision. Since they used present-day p_{CO_2} values in their geochemical models, the calculations were repeated in this study by using a range of higher p_{CO_2} values more appropriate for the end-Cretaceous time. As our results show, at p_{CO_2} values higher than $10^{-2.5}$, the generalization that the basalt to kaolinite transition was facilitated only under open-system conditions existing on rift flanks and slopes cannot be confirmed. Thus, it is critical to know if post-Deccan Trap p_{CO_2} values in the early Paleocene in Kachchh were, indeed, high enough to preclude the conclusion of Mitra *et al.* (2016).

Emplacement of Large Igneous Complexes like the Deccan Traps have the potential to affect the environment and climate substantially, as a consequence of the proportion of volatiles released into the atmosphere (e.g. Self *et al.* 2006, 2014; Gerlach 2011; Font *et al.* 2014). While all these studies unequivocally accept that the atmospheric CO_2 concentration was higher in the end-Cretaceous, and was further increased by the eruption of the Deccan Traps, they are also unanimous in concluding that the volumetric proportion of CO_2 added to the atmosphere was insignificant in the context of the total volume of the atmosphere, with possibly a negligible climatic effect, as suggested earlier by Caldeira & Rampino (1990). Indeed, the upper-bound p_{CO_2} values used in this study, based on the data of Dessert *et al.* (2001), fall just short of the critical $10^{-2.5}$ value. Further, this value was computed assuming that the bulk of the CO_2 was released into the atmosphere within a very short time span (<1 myr: e.g. Schoene *et al.* 2015), without accounting for CO_2 removal by weathering or plants during the period of eruption. Indeed, if the eruption of the Deccan Trap lavas occurred over a time span of 8 myr (e.g. Sheth *et al.* 2001), the proportion of atmospheric CO_2 may never have been close to the calculated critical value of $10^{-2.5}$, which has been determined by the present study. Ironically, basalts erupted within that period would have acted as CO_2 sinks and removed the excess from the atmosphere (Dessert *et al.* 2003; Li & Elderfield 2013).

Based on the available data, therefore, it seems unlikely that, following eruption of the Deccan Traps, the atmospheric p_{CO_2} ever increased to the high levels required for transforming all basalts on flow tops to kaolinite. Indeed, even if this was the case, it would be difficult to explain how adjacent basalts, within and outside the Cenozoic rift basins in Kachchh, would be differentially altered, as argued by Mitra *et al.* (2016). Thus, the transformation of basalt to smectite or kaolinite is, in all probability, not governed by variations in the atmospheric p_{CO_2} value in post-Deccan times. We therefore conclude that topographical slopes, under open-system conditions, were necessary to stabilize kaolinite in preference to smectite during basalt alteration in the Cenozoic rifts of Kachchh. The precise time of initiation of Cenozoic rifting in Kachchh remains equivocal. However, $^{40}Ar/^{39}Ar$ age estimates for the last phase of Deccan-related magmatism from around Bombay are *c.* 60.4 ± 0.6 Ma (Sheth *et al.* 2001), which may be synchronous with or post-date the youngest Deccan lava flow in Kachchh, dated at *c.* 61 ± 1.6 Ma (Shukla *et al.* 2001). Thus, it is possible that the weathering of the Kachchh lava flows may have been initiated in the terminal stages of Deccan volcanism, but continued into the post-Deccan period, and may well have been linked to the India–Seychelles break-up.

Conclusions

The break-up of the India prior to its collision with Eurasia is postulated to have occurred in a sequence of phases extending from the Mesozoic to the early Cenozoic. Based on widespread reports of pre- and synmagmatic rifting, the emplacement of the Deccan Traps has been considered to either post-date or be synchronous with the final stages of these rifting events. On the other hand, extension in the western coastal domain post-dating magmatism but predating India–Eurasia collision is less well known. Mitra *et al.* (2016) documented the weathering of basalts to kaolinite at the base of Cenozoic rift basins in the Kachchh area of western India, although the rift flanks were relatively less altered to smectites. They attributed this to post-magmatic rifting. In this study, basalt weathering is modelled under open- and closed-system conditions to simulate rainwater interacting with basalts either on topographical slopes (within rifts) or on flat-topped hills (flow tops), respectively. Partial pressures of CO_2 (p_{CO_2}) were varied from low, present-day values to higher values that are reportedly more appropriate for end-Cretaceous–early Paleocene times. The results show that if p_{CO_2} exceeded values of $10^{-2.5}$, basalts would be altered to dominantly kaolinitic assemblages in both open and closed systems, and the interpretation of Mitra *et al.* (2016) could then not be definitively ascribed to topographical control. However, existing p_{CO_2} estimates in the aftermath of Deccan volcanism fall below this value, suggesting that the differential basalt weathering observed in Kachchh is most likely to have been caused by post-magmatic rifting. This indicates that extensional tectonics along the Indian west coast in the Kachchh region continued after the cessation of Deccan volcanism, and is in all probability related to far-field stresses that were transmitted inland as the Seychelles microcontinent broke away from the Indian landmass, since this is the last major extensional tectonic event known along the west coast of India prior to its collision with Eurasia.

We thank Sarajit Sensarma for inviting us to contribute to this volume, and for his suggestions towards improving this manuscript. The revised version has greatly benefited from critical but constructive reviews provided by Patricia Craig and an anonymous reviewer. We would also like to express our gratitude to Mr A.S. Kiran Kumar, former Director, Space Applications Centre (SAC) and Chairman, Indian Space Research Organization (ISRO), for his encouragement in this study. Mr Tapan Misra, Director, SAC and Dr P.K. Pal, DD, EPSA, SAC, ISRO are also thanked for their valuable guidance and support. SG and SM were funded through the DOS-ISRO grant 0303615RM401, while SB and KM have been supported by DOS-ISRO grant 0303617RM401, respectively; SAC under the ISRO is gratefully acknowledged for this financial support.

References

Amiotte Suchet, P., Probst, A. & Probst, J.L. 1995. Influence of acid rain on CO_2 consumption by rock weathering: local and global scales. *Water, Air, & Soil Pollution*, **85**, 1563–1568.

Amiotte Suchet, P., Probst, J.L. & Ludwig, W. 2003. Worldwide distribution of continental rock lithology: Implications for the atmospheric/soil CO_2 uptake by continental weathering and alkalinity river transport to the oceans. *Global Biogeochemical Cycles*, **17**, 1038, https://doi.org/10.1029/2002GB001891

Anderson, S.P., Von Blackenburg, F. & White, A.F. 2007. Physical and chemical control on the critical zone. *Elements*, **3**, 315–319.

Baksi, A.K. 1987. Critical evaluation of the age of the Deccan Traps, India: implications for flood-basalt volcanism and faunal extinctions. *Geology*, **15**, 147–150.

Baksi, A.K. 2014. The Deccan Trap – Cretaceous–Paleogene boundary connection; new $^{40}Ar/^{39}Ar$ ages and critical assessment of existing argon data pertinent to this hypothesis. *Journal of Asian Earth Sciences*, **84**, 9–23.

Benedetti, M.F., Menard, O., Noack, Y., Carvalho, A. & Nahon, D. 1994. Water–rock interactions in tropical catchments: field rates of weathering and biomass impact. *Chemical Geology*, **118**, 203–220.

Bethke, C.M. 2007. *Geochemical and Biogeochemical Reaction Modeling*. Cambridge University Press, New York.

Bhattacharya, G.C. & Yatheesh, V. 2015. Plate-tectonic evolution of the deep ocean basins adjoining the Western Continental Margin of India – a proposed model for the early opening scenario. *In*: Mukherjee, S. (ed.) *Petroleum Geosciences: Indian Contexts*. Springer International, Cham, Switzerland, 1–61.

Bhattacharya, S., Mitra, S., Gupta, S., Jain, N., Chauhan, P. & Parthasarathy, G. & AJAI 2016. Jarosite occurrence in the Deccan Volcanic Province of Kachchh, western India: Spectroscopic studies on a Martian analog locality. *Journal of Geophysical Research: Planets*, **121**, 402–431, https://doi.org/10.1002/2015JE004949

Bice, K.L., Birgel, D., Meyers, P.A., Dahl, K.A., Hinrichs, K.U. & Norris, R.D. 2006. A multiple proxy and model study of Cretaceous upper ocean temperatures and atmospheric CO_2 concentrations. *Paleoceanography*, **21**, PA2002, https://doi.org/10.1029/2005PA001203

Biswas, S.K. 1992. Tertiary stratigraphy of Kutch. *Journal of the Palaeontological Society of India*, **37**, 1–29.

Brady, P.V., Dorn, R.I., Brazel, A.J., Clark, J., Moore, R.B. & Glidewell, T. 1999. Direct measurement of the combined effects of lichen, rainfall and temperature on silicate weathering. *Geochimica et Cosmochimica Acta*, **63**, 3293–3300.

Caldeira, K. & Rampino, M.R. 1990. Carbon dioxide emissions from Deccan volcamism and a K/T boundary greenhouse effect. *Geophysical Research Letters*, **17**, 1299–1302.

Campbell, I.H. & Griffiths, R.W. 1990. Implications of mantle plume structure for the evolution of flood basalts. *Earth and Planetary Science Letters*, **99**, 79–93.

Carroll, D. 2012. *Rock Weathering*. Springer Science & Business Media, New York.

Catalano, J.G. 2013. Thermodynamic and mass balance constraints on iron-bearing phyllosilicate formation and alteration pathways on early Mars. *Journal of Geophysical Research: Planets*, **118**, 2124–2136.

Collier, J.S., Sansom, V., Ishizuka, O., Taylor, R.N., Minshull, T.A. & Whitmarsh, R.B. 2008. Age of Seychelles–India break-up. *Earth and Planetary Science Letters*, **272**, 264–277.

Courtillot, V., Besse, J., Vandamme, D., Montigny, R., Jaeger, J.J. & Cappetta, H. 1986. Deccan flood basalts at the Cretaceous/Tertiary boundary? *Earth and Planetary Science Letters*, **80**, 361–374.

Das, A., Krishnaswami, S., Sarin, M.M. & Pande, K. 2005. Chemical weathering in the Krishna Basin and Western Ghats of the Deccan Traps, India: rates of basalt weathering and their controls. *Geochimica et Cosmochimica Acta*, **69**, 2067–2084.

Depetris, P.J., Pasquini, A.I. & Lecomte, K.L. 2014. *Weathering and the Riverine Denudation of Continents*. Springer, Berlin.

Deshmukh, V.V., Ray, S.K., Chandran, P., Bhattacharyya, T. & Pal, D.K. 2012. Speciation of smectites in two shrink-swell soils of Central Peninsular India. *Clay Research*, **31**, 84–93.

Dessert, C., Dupré, B., Francóis, L.M., Schott, J., Gaillardet, J., Chakrapani, G.J. & Bajpai, S. 2001. Erosion of Deccan Traps determined by river geochemistry: impact on the global climate and the $^{87}Sr/^{86}Sr$ ratio of seawater. *Earth and Planetary Science Letters*, **188**, 459–474.

Dessert, C., Dupré, B., Gaillardet, J., François, L.M. & Allegre, C.J. 2003. Basalt weathering laws and the impact of basalt weathering on the global carbon cycle. *Chemical Geology*, **202**, 257–273.

Drever, J.I. 1994. The effect of land plants on weathering rates of silicate minerals. *Geochimica et Cosmochimica Acta*, **58**, 2325–2332.

Drever, J.I. (ed.) 2005. *Surface and Ground Water, Weathering, and Soils*. Treatise on Geochemistry, **5**. Elsevier, Amsterdam.

Dupré, B., Dessert, C. *et al.* 2003. Rivers, chemical weathering and Earth's climate. *Comptes Rendus Geoscience*, **335**, 1141–1160.

Font, E., Fabre, S. *et al.* 2014. Atmospheric halogen and acid rains during the main phase of Deccan eruptions: magnetic and mineral evidence. *In*: Keller, G. & Kerr, A.C. (eds) *Volcanism, Impacts, and Mass Extinctions: Causes and Effects*. Geological Society of America, Special Papers, **505**, 353–368.

Garrells, R.M. & Mackenzie, F.T. 1971. *Evolution of Sedimentary Rocks*. Norton, New York.

Gerlach, T. 2011. Volcanic v. anthropogenic carbon dioxide. *Eos, Transactions of the American Geophysical Union*, **92**, 201–208.

Gertsch, B., Keller, G., Adatte, T., Garg, R., Prasad, V., Berner, Z. & Fleitmann, D. 2011. Environmental effects of Deccan volcanism across the Cretaceous–Tertiary transition in Meghalaya, India. *Earth and Planetary Science Letters*, **310**, 272–285.

Gislason, S.R., Arnorsson, S. & Armannsson, H. 1996. Chemical weathering of basalt as deduced from the composition of precipitation, rivers and rocks in Southwest Iceland: effect of runoff, age of rocks and vegetative/glacial cover. *American Journal of Science*, **296**, 837–907.

Greenberger, R.N., Mustard, J.F., Kumar, P.S., Dyar, M.D., Breves, E.A. & Sklute, E.C. 2012. Low temperature aqueous alteration of basalt: mineral assemblages of Deccan basalts and implications for Mars. *Journal of Geophysical Research: Planets*, **117**, E00J12.

Harding, D. & Miller, J.M. 1982. The influence on rain chemistry of the Hawaiian volcano Kilauea. *Journal of Geophysical Research: Oceans*, **87**, 1225–1230.

Hinsinger, P., Barros, O.N., Benedetti, M.F., Noack, Y. & Callot, G. 2001. Plant-induced weathering of a basaltic rocks: experimental evidence. *Geochimica et Cosmochimica Acta*, **65**, 137–152.

Johnson, N. & Parnell, R.A. 1986. Composition, distribution and neutralization of 'acid rain' derived from Masaya volcano, Nicaragua. *Tellus B*, **38B**, 106–117.

Keller, W.D. 1956. Clay minerals as influenced by environments of their formation. *American Association of Petroleum Geologists Bulletin*, **40**, 2689–2710.

Kent, D.V. & Muttoni, G. 2013. Modulation of Late Cretaceous and Cenozoic climate by variable drawdown of atmospheric pCO_2 from weathering of basaltic provinces on continents drifting through the equatorial humid belt. *Climate of the Past*, **9**, 525–546.

Khadkikar, A.S., Sant, D.A., Gogte, V. & Karanth, R.V. 1999. The influence of Deccan volcanism on climate: insights from lacustrine intertrappean deposits, Anjar, western India. *Palaeogeography, Palaeoclimatology, Palaeoecology*, **147**, 141–149.

Klootwijk, C.T., Gee, J.S., Peirce, J.W., Smith, G.M. & McFadden, P.L. 1992. An early India–Asia contact: paleomagnetic constraints from Ninetyeast Ridge, ODP Leg 121. *Geology*, **20**, 395–398.

Li, G. & Elderfield, H. 2013. Evolution of carbon cycle over the past 100 million years. *Geochimica et Cosmochimica Acta*, **103**, 11–25.

Li, G. & Long, X. 2014. Weathering of Chinese basaltic fields. *Procedia Earth and Planetary Science*, **10**, 69–72.

Mahoney, J.J. 1988. Deccan traps. *In*: Macdougall, J.D. (ed.) *Continental Flood Basalts*. Springer, Dordrecht, The Netherlands, 151–194.

Mathews, R.P., Tripathi, S.M., Banerjee, S. & Dutta, S. 2013. Palynology, palaeoecology and palaeodepositional environment of eocene lignites and associated sediments from Matanomadh Mine, Kutch Basin, western India. *Journal of the Geological Society of India*, **82**, 236–248.

Meybeck, M. 1986. Composition chimique naturelle des ruisseaux non pollués en France. *Sciences géologiques Bulletin*, **39**, 3–77.

Millot, G. 2013. *Geology of Clays: Weathering Sedimentology Geochemistry*. Springer Science & Business Media, New York.

Mitra, S., Gupta, S., Bhattacharya, S., Banerjee, S., Chauhan, S. & Parthasarathy, G. 2014. Jarosite precipitation from acidic saline waters in Kachchh, Gujarat, India: an appropriate Martian analogue? *Abstract P41A-3891 presented at the AGU 2014 Fall Meeting*, 15–19 December 2014, San Francisco, CA, USA.

Mitra, S., Mitra, K., Gupta, S., Bhattacharya, S., Chauhan, P. & Jain, N. 2016. Alteration and submergence of basalts in Kachchh, Gujarat, India: implications for the role of the Deccan Traps in the India–Seychelles break-up. *In*: Mukherjee, S., Misra, A.A., Calvès, G. & Nemčok, M. (eds) *Tectonics of the Deccan Large Igneous Province*. Geological Society, London, Special Publications, **445**, 47–67, https://doi.org/10.1144/SP445.9

Morgan, W.J. 1981. Hotspot tracks and the opening of the Atlantic and Indian Oceans. *The Oceanic Lithosphere*, **7**, 443–487.

Moulton, K.L., West, J. & Berner, R.A. 2000. Solute flux and mineral mass balance approaches to the quantification of plant effects on silicate weathering. *American Journal of Science*, **300**, 539–570.

Murata, K.J. 1966. *An Acid Fumarolic Gas from Kilauea Iki, Hawaii*. United States Geological Survey, Professional Paper, **537-C**.

Nagelschmidt, G., Desai, A.D. & Muir, A. 1940. The minerals in the clay fractions of a black cotton soil and a red earth from Hyderabad, Deccan State, India. *Journal of Agricultural Science*, **30**, 639–653.

Pande, K. 2002. Age and duration of the Deccan Traps, India: a review of radiometric and paleomagnetic constraints. *Proceedings of the Indian Academy of Sciences: Earth and Planetary Sciences*, **111**, 115–124.

Patriat, P. & Achache, J. 1984. India–Eurasia collision chronology has implications for crustal shortening and driving mechanism of plates. *Nature*, **311**, 615–621.

Paul, D.K., Ray, A., Das, B., Patil, S.K. & Biswas, S.K. 2008. Petrology, geochemistry and paleomagnetism of the earliest magmatic rocks of Deccan Volcanic Province, Kutch, Northwest India. *Lithos*, **102**, 237–259.

Rasmussen, C., Dahlgren, R.A. & Southard, R.J. 2010. Basalt weathering and pedogenesis across an environmental gradient in the southern Cascade Range, California, USA. *Geoderma*, **154**, 473–485.

Richards, M.A., Duncan, R.A. & Courtillot, V.E. 1989. Flood basalts and hotspot tracks: plume heads and tails. *Science*, **246**, 103–107.

Ross, C.S. 1945. Minerals and mineral relationships of the clay minerals. *Journal of the American Ceramic Society*, **28**, 173–183.

Salil, M.S., Shrivastava, J.P. & Pattanayak, S.K. 1997. Similarities in the mineralogical and geochemical attributes of detrital clays of Maastrichtian Lameta Beds and weathered Deccan basalt, Central India. *Chemical Geology*, **136**, 25–32.

Saraswati, P.K., Khanolkar, S., Raju, D.S.N., Dutta, S. & Banerjee, S. 2014. Foraminiferal biostratigraphy of lignite mines of Kutch, India: age of lignite and fossil vertebrates. *Journal of Palaeogeography*, **3**, 90–98.

Saxena, R.K. 1977. On the stratigraphic status of Matanomadh Formation, Kutch, India. *The Palaeobotanist*, **24**, 211–214.

Schoene, B., Samperton, K.M. *et al.* 2015. U–Pb geochronology of the Deccan Traps and relation to the end-Cretaceous mass extinction. *Science*, **347**, 182–184.

Self, S., Widdowson, M., Thordarson, T. & Jay, A.E. 2006. Volatile fluxes during flood basalt eruptions and potential effects on the global environment: a Deccan perspective. *Earth and Planetary Science Letters*, **248**, 518–532.

Self, S., Schmidt, A. & Mather, T.A. 2014. Emplacement characteristics, time scales, and volcanic gas release rates of continental flood basalt eruptions on Earth. *In*: Keller, G. & Kerr, A.C. (eds) *Volcanism, Impacts, and Mass Extinctions: Causes and Effects*. Geological Society of America, Special Papers, **505**, 319–337.

Sheth, H.C. 1999. A historical approach to continental flood basalt volcanism: insights into pre-volcanic rifting, sedimentation, and early alkaline magmatism. *Earth and Planetary Science Letters*, **168**, 19–26.

Sheth, H.C. 2005. From Deccan to Réunion: no trace of a mantle plume. *In*: Foulger, G.R., Natland, J.H., Presnall, D.C. & Anderson, D.L. (eds) *Plates, Plumes, and Paradigms*. Geological Society of America, Special Papers, **388**, 477–501.

Sheth, H.C., Pande, K. & Bhutani, R. 2001. ^{40}Ar–^{39}Ar ages of Bombay trachytes: evidence for a Palaeocene phase of Deccan volcanism. *Geophysical Research Letters*, **28**, 3513–3516.

Shukla, A.D., Bhandari, N., Kusumgar, S., Shukla, P.N., Ghevariya, Z.G., Gopalan, K. & Balaram, V. 2001. Geochemistry and magnetostratigraphy of Deccan flows at Anjar, Kachchh. *Journal of Earth System Science*, **110**, 111–132.

Siddaiah, N.S. & Kumar, K. 2009. Discovery of minamiite from the Deccan Volcanic Province, India: implications for Martian surface exploration. *Current Science*, **97**, 1664–1669.

Unoki, R. & Itoi, R. 2005. Effects of volcanic gases on rain and soil chemistries at Kuju, Japan. *In*: *Proceedings World Geothermal Congress 2005*, 24–29 April 2005, Antalya, Turkey, International Geothermal Association, 1–5.

White, A.F. & Blum, A.E. 1995. Effects of climate on chemical_ weathering in watersheds. *Geochimica et Cosmochimica Acta*, **59**, 1729–1747.

Geochemical and Sm–Nd isotopic constraints on the petrogenesis and tectonic setting of the Proterozoic mafic magmatism of the Gwalior Basin, central India: the influence of Large Igneous Provinces on Proterozoic crustal evolution

JWELLYS D. SAMOM[1]*, TALAT AHMAD[1,2] & A. K. CHOUDHARY[3]

[1]*Department of Geology, University of Delhi, Delhi, 110007, India*

[2]*Jamia Millia Islamia, Jamia Nagar, New Delhi, 110025, India*

[3]*Institute Instrumentation Centre, (IIT), Roorkee, 247667, India*

**Correspondence: 1785jwellys@gmail.com*

Abstract: Palaeoproterozoic mafic magma intruded the Gwalior Group of sediments in the form of gabbroic sills. Huge sills extending more than 120 m in depths are exposed in quarries. They are coarse to medium grained, massive and have chilled margins. They predominantly consist of plagioclase feldspar, pyroxene with accessory quartz, iron oxides and apatite. Ophitic to sub-ophitic, porphyritic and intergrowth textures are common. These are tholeiitic, sub-alkaline magma types which have been generated by varying degrees of partial melting and have experienced fractionation of dominantly pyroxene + olivine ± plagioclase. They are enriched in light rare earth elements (LREEs) and large-ion lithophile elements (LILEs), and exhibit negative Nb and P anomalies and positive Pb anomalies. A mineral–whole-rock Sm–Nd isochron corresponds to an age of 2104 ± 23 Ma with an initial $^{143}Nd/^{144}Nd = 0.509938 \pm 0.000023$ and $\varepsilon Nd^{t} = -0.9$ ($t = 2000$). Model ages (2.6–1.7 Ga) indicate that their mantle sources had a protracted evolution. Mantle source melting may have been triggered by mantle plume in an extensional intra-continental rift tectonic setting, facilitating emplacement of a Large Igneous Province (LIP) in the northern Indian Shield. Thus, Gwalior mafic magmatism represented by the studied sills is the manifestation of one such LIP in the northern Indian Shield. These magmatic activities are contemporaneous with the widespread mafic magmatic activities concomitant with the development of supra-crustal basins on Archaean cratons worldwide and the Indian Shield in particular, where upwelling plumes triggered large-scale crustal extension, breaking-up of supercontinents and emplacement of LIPs.

Proterozoic crust records marked diversity in its composition with many varieties of rocks (Pharaoh *et al.* 1987) that show a transition through time due to the cooling of the mantle temperature and changing tectonic scenarios (Bickle 1986; Ahmad & Rajamani 1991; Nisbet *et al.* 1993; Polat *et al.* 2011). Proterozoic mafic magmatism associated with extension or intracontinental rifting and subsequent development of basins are reported from various Archaean cratons worldwide (French *et al.* 2008; Hawkesworth *et al.* 2010). Such widespread magmatic activities are preserved in the form of huge sills, dykes and dyke swarms, which may represent remnants of Large Igneous Provinces (LIPs). LIPs commonly occur in large volumes (*c.* 100 000 km^3) in intraplate settings (Ernst & Buchan 2001); they are short-lived magmatic pulse(s) (<1–5 myr) and are generally thought to be related to plume or superplume activities (Ernst *et al.* 2013). Such plume-related episodes of magmatic activities have occurred repeatedly and caused crustal extension, rifting and basin development, and thus are important parameters for crustal re-assemblages or palaeogeographical supercontinental reconstructions (Buchan *et al.* 2000; Ernst & Buchan 2001; Meert 2002, 2003; Rogers & Santosh 2003; Ernst & Srivastava 2008; French *et al.* 2008; Radhakrishna *et al.* 2013*a*, *b*). Thus, LIPs are the mantle-derived products that may act as windows in understanding the crustal evolutionary and geodynamic processes prevalent at that time.

The Indian Peninsular Shield is one such region, and has undergone multiple episodes of magmatism marked by extensive Proterozoic mafic magmatic bodies traversing the Archaean craton in the form of radiating dyke swarms, intrusives (dykes and sills) and LIPs (Drury 1984; Murthy 1987, 1995; Ramchandra *et al.* 1995; Pradhan *et al.* 2008, 2010, 2012; Radhakrishna *et al.* 2013*a*, *b*). LIPs of *c.* 1.9 Ga are reported from the southern Indian Peninsula, such as the Pulivendla sills of the Cuddapah Basin, the BD2 swarms of southern Bastar, the *c.* 2.1 Ga Mahbubnagar LIP of the eastern Dharwar Craton, and widespread NW- and east–west-trending mafic dykes of Shimoga and the NE Dharwar region

From: Sensarma, S. & Storey, B. C. (eds) 2018. *Large Igneous Provinces from Gondwana and Adjacent Regions*. Geological Society, London, Special Publications, **463**, 243–268.
First published online July 10, 2017, https://doi.org/10.1144/SP463.10

of the Dharwar Craton (French *et al.* 2008; Meert *et al.* 2011; Khanna *et al.* 2013). Similar LIPs of 1.89–1.87 Ga are also preserved in both the Kalahari and Superior cratons (French *et al.* 2008). Although there have been no reports of LIPs in the northern Indian Craton, which may be due to fewer data on their extent, synchronous mafic magmatic events recorded in neigbouring cratons and sedimentary basins plus the reported plume-related magmatic activities responsible for mafic magmatism in the Aravalli Craton and western Himalaya (Ahmad & Rajamani 1991; Ahmad & Tarney 1991, 1994; Mallikharjuna Rao 2004) may possibly represents similar LIPs. Due to this widespread synchronous mafic magmatic activity on a global scale, French *et al.* (2008) discussed the possibility of extension due to lithospheric thinning caused by enriched plume activity.

The Central Indian Tectonic Zone (CITZ) runs nearly east–west (ENE–WSW) in central India, dividing the Indian Peninsular Shield into the North and South Indian cratons (Radhakrishna 1989; Acharyya 2003; Absar *et al.* 2009). Four main Archaean cratons (the Aravalli, the Bundelkhand, the Bastar and Dharwar cratons) of complex nature occur to the north and south of the CITZ (Fig. 1a). The Proterozoic intracratonic sedimentary basins developed during cratonization of these Archaean cratons (Chakraborty 2006; Malone *et al.* 2008; Patranabis-Deb *et al.* 2008) are found along or within the periphery of these cratons (Fig. 1a). Rift-related mafic intrusive rocks preserved in Gwalior, Bijawar, Cuddapah and Bastar (Abujhmar) associated with the development of the basins are synsedimentary sills and flows (Ramakrishnan & Vaidyanadhan 2008). Each of the basins, despite individual histories, has similar lithostratigraphy, correlation, structure and depositional processes (Radhakrishna 1987; Kale & Phansalkar 1991). The Gwalior Basin, in the northern part of central India, is one such Palaeoproterozoic intra-cratonic basin deposited over the NW part of the Bundelkhand Craton (Fig. 1b). The Gwalior Basin has remained poorly studied; the area has been neglected specifically in terms of studies of mafic magmatic rocks over decades. The data generated by an earlier worker are of preliminary nature both in quantity as well as quality (Bajpai 1935). This has prompted us to study these mafic rocks, to characterize them, understand their nature and comprehend the tectonic evolution of this basin. Thereafter, it is of interest to understand and discuss the possible origin of these intrusive rocks of the Gwalior Basin in the context of other globally reported contemporary intraplate magmatism: LIPs during the Proterozoic period.

In this contribution, we report a new and comprehensive set of geochemical data comprising major oxides, trace elements and rare earth elements, coupled with whole-rock Rb–Sr, Sm–Nd isotopes and individual mineral isotopic studies for mafic magmatic rocks of the Gwalior Basin. A Sm–Nd mineral–whole-rock isochron age and Nd model ages are reported for the first time. These new ages for Gwalior mafic magmatism, and considering the existing widespread contemporaneous mafic activity in this part of the northern Indian Shield, may possibly document/indicate the remnants of a LIP in the northern Indian Shield. We discuss the petrogenesis and evolutionary history of these rocks to put constraints on the Proterozoic crustal evolution, and make a possible connection with the generation of LIPs in the northern Indian Shield.

Geological setting

Gwalior Basin is a weakly metamorphosed east–west-trending intracratonic Palaeoproterozoic sedimentary basin that occupies the NW periphery of the Archaean Bundelkhand Craton (3.5/3.2–2.5 Ga, SIMS U–Pb: Joshi *et al.* 2017) (Fig. 1c), with the Gwalior Group of rock exposed over an area of about 2400 km^2 (Absar *et al.* 2009). The strike length of the basin is 80 km along the east–west direction and 25 km in width (Ramakrishnan & Vaidyanadhan 2008). The bedding dips (*c.* 8°) stretching all along the exposure suggest that there was no post-depositional tectonic effect in the Gwalior Basin (Chakraborty & Paul 2014). Gwalior sediments, as well as the mafic intrusives, show little deformation at the outcrop scale, although there are some regional faults (the Son Narmada North faults and the NE–SW-flowing Sind River) and there are records of crustal movement along these. The Bundelkhand massif is composed mainly of granitoid intrusions; however, the basement components, the Bundelkhand Granite Gneisses Complex (BGGC), occur as enclaves in the younger granitoids or as separate outcrops; the granitoid intrusion are insignificantly deformed relative to the basement rocks (Mondal *et al.* 2002; Absar *et al.* 2009). The Mesoproterozoic Kaimur Group (sandstone and shales) of the Vindhyan Supergroup conceal the north, NW and western part of the basin (Fig. 1b). In the east, the NE–SW-flowing Sind River delineates the easternmost boundary of the basin, representing a regional fault (Absar *et al.* 2009). The lithostratigraphic classification of the Gwalior Group as given by Absar *et al.* (2009) is presented in Table 1.

The 1 km-thick Gwalior Group of sediments are deposited over the uneven surface of the Archaean BGGC and comprise two main formations: the lower clastic arenaceous–argillaceous Par Formation and the upper chemogenic Morar Formation (Absar *et al.* 2009). The lower basal Par Formation is just 200 m thick, and comprises quartz pebble

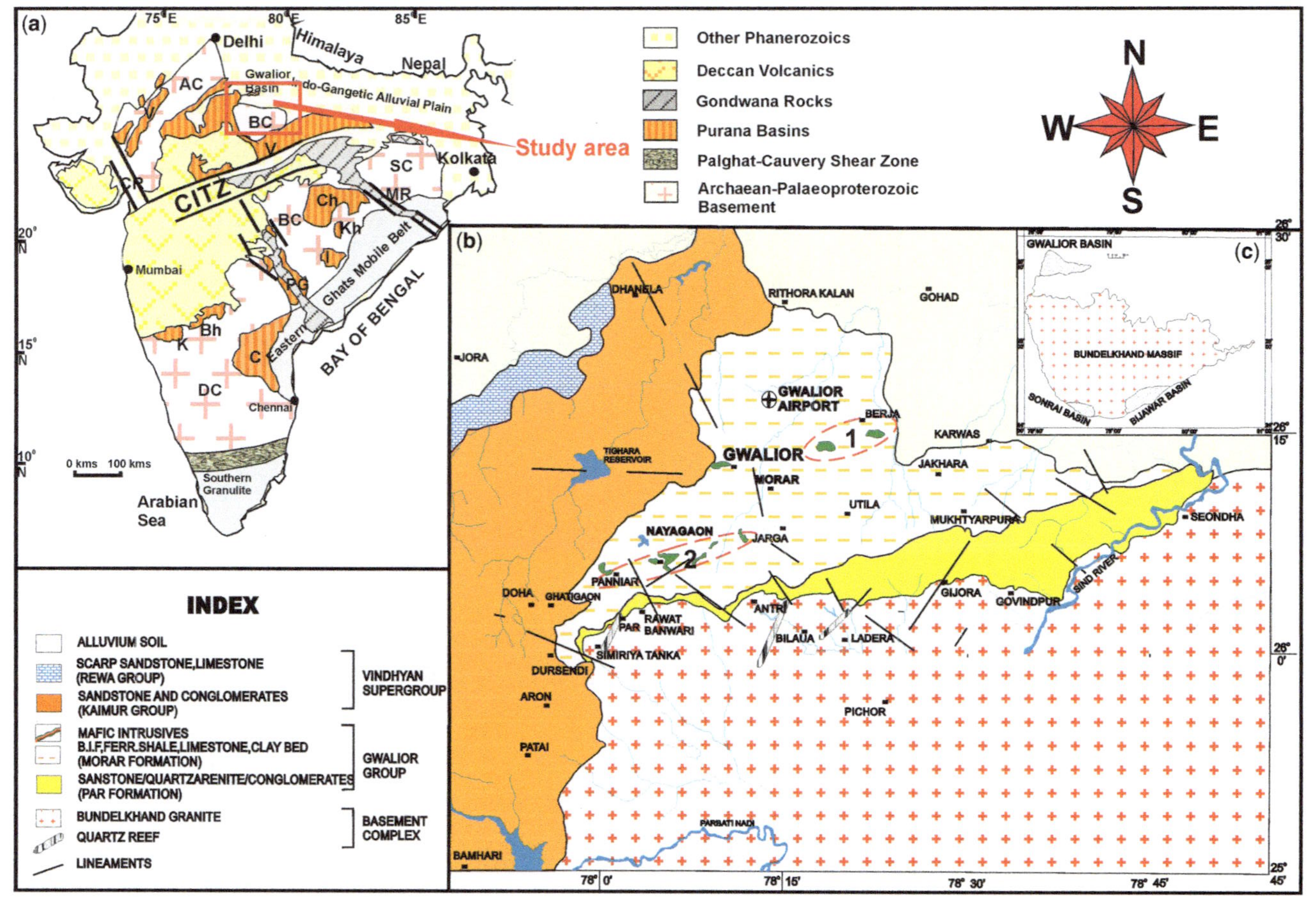

Fig. 1. (**a**) Distribution of the Proterozoic platformal basins in Indian Peninsula Shield. The major tectonic features are also shown (after Naqvi & Rogers 1987; Chakraborty *et al.* 2010). The Proterozoic basins are: V, Vindhyanchal; Ch, Chhattisgarh; Kh, Khariar; PG, Pranhita Godavari; C, Cuddapah; K, Kaladgi; Bh, Bhima. Major tectonic features are: CITZ, Central Indian Tectonic Zone; CR, Cambay Rift; MR, Mahanadi Rift. (**b**) Geological map of the Gwalior Basin (modified after Absar *et al.* 2009); dotted ovals 1 and 2 are the marked areas from where the samples were collected. (**c**) Inset shows the location of the Bundelkhand massif, CITZ and the three sedimentary basins within the Bundelkhand Craton.

Table 1. *Stratigraphic succession of the Gwalior Group, Gwalior District, Madhya Pradesh (modified after Absar* et al. *2009)*

Group	Formation	Rock unit
Kaimur Group of the Vindhyan Supergroup		Matrix-rich sandstone, quartz arenite, shale BIF Pebble breccia conglomerate
	Unconformity	
Gwalior Group	**Morar**	Mafic intrusive rock
		Banded iron formation (BIF) with impersistent limestone horizons
		Clay bed
		Black siliceous shale
	Par	Quartzarenite, black shale
		Arkosic sandstone, lithic sandstone, ferruginous sandstone, intercalated sand–mud
		Quartz pebble conglomerate
		Quartz pebble boulder conglomerate
	Unconformity	
Bundelkhand Granite–Gneiss Complex (BGGC)		Pink granites, grey granites, Quartz reefs and mafic dykes with xenolithic older gneisses and metasedimentary enclaves

conglomerate, granular sandstone and sandstone–shale heterolithic sediments, quartz arenite, and shale; while the 800 m-thick overlying chemogenic Morar Formation consists mainly of banded iron formation (BIF), black siliceous shale, ferrugineous shale with limestone in subordinate volume, and a kaolinitic clay bed at the base (Absar *et al.* 2009; Chakraborty & Paul 2014). The upper Morar Formation occupies the northern part and covers most of the outcrop of the basin, whilst the Par sandstone occupies the southern side (Fig. 1b). Massive intrusive bodies of mafic magmatic rocks are mainly represented by huge gabbro sills of Palaeoproterozoic age found traversing the Morar Formation. The Gwalior Mafic Intrusives (GMI) rocks (gabbroic sills) are massive and display a variation from finer-grained micro-gabbro to coarse-grained gabbro. Based on earlier reports, the depositional age of the Gwalior sediments is bracketed between 2000 ($^{40}Ar/^{39}Ar$, from dykes within the basement) and 1791 Ma (Rb–Sr, from mafic rocks within the basin) (Crawford & Compston 1969; Mallikharjuna Rao 2004; Absar *et al.* 2009). Again, on the basis of the geochemical signature of the clastic Par Formation, Absar *et al.* (2009) suggested the deposition of these sediments in a shallow-marine submarine basin environment in a stable craton interior platformal setting.

Field observation

A total of 66 representative samples of gabbro sills were collected from the Gwalior Basin (quarries and other exposures in the form of hillocks). The locations of the collected samples are Mau Quarry (78° 11′ 56.5″ N, 26° 15′ 36.1″ E), the Sankarpur area near Transport Nagar (78° 8′ 8.9″ N, 26° 13′ 54.9″ E), Ramdas Ghati (78° 9′ 25.5″ N, 26° 12′ 38.6″ E), the Nayagaon area, Sultan mines (78° 4′ 47.3″ N, 26° 6′ 39.4″ E), small hillocks at Maharajpura, east of the airport section (78° 13′ 39.8″ N, 26° 16′ 03.9″ E) and Biwari village (78° 13′ 51.11″ N, 26° 15′ 47.5″ E). In the field, the GMI (gabbro sills) occur as huge sills (*c.* 120 m height) exposed in the quarry (Fig. 2a); the depth of the sills extends up to 500 m (borehole data). Variations of grain size with depth are visible with finer gabbros near the contacts, forming chilled margins due to the quenching of magma and baking of the enclosing sediments, reflecting the intrusive nature of the sills. The absence of any vesicular nature of the mafic magmatic rocks and their coarse-grained nature indicate that they cannot be lavas. The majority of these bodies follow the fabric and bedding, and seldom show cross-cutting relationship with the sediments, indicating they are sills rather than dykes, although some younger finer-grained dykes are seen cutting across the whole stratigraphic sequence. Excellent horizontal layering of the BIF lies over the mafic bodies (Fig. 2b), and in some places palaeosols and shale units are exposed in-between the intrusives and overlying BIF (Fig. 2b). The presence of palaeosols indicates weathering of the sediments (shale–clay bed), followed by deposition of the BIF. Calcite veins and hydrothermal alteration, which is typical of *in situ* weathering, and is also known as spheroidal or onion-peel weathering, are locally developed in gabbroic rocks (Fig. 2c). In another quarry, nice horizontal layering of tuffaceous material and heterolithic sediments lie between the sills (Fig. 2d), indicating shallow-level intrusion. The gabbroic

Fig. 2. Field photographs of Gwalior mafic sills. (**a**) Huge sills up to 120 m exposed in the quarry. (**b**) BIF above the sills showing nice horizontal layering, palaeosols between sills and BIF. (**c**) Spheroidal or onion-peel weathering. (**d**) Nice horizontal layering of tuffaceous material in-between the mafic sills. (**e**) Well-preserved magmatic layering of large plagioclase phenocrysts. (**f**) Massive outcrop of medium-grained porphyritic texture with phenocryst of plagioclase.

sills exposed on the hillocks display beautiful horizontal magmatic layering of slightly pale green phenocrysts of plagioclase, alternating with layers of medium-grained mafic phases (pyroxene) and plagioclases (Fig. 2e). It is possible that this layering reflects either the compositional variation within the magma body or the settling of crystal during cooling. The gabbroic bodies also display medium-grained porphyritic texture with phenocryst of plagioclase (Fig. 2f).

Petrography

The GMI rocks of the Gwalior Basin are medium to coarse grained, massive, and exhibit holocrystalline, hypidiomorphic granular textures. They mainly comprise plagioclase and pyroxene (clinopyroxene ± orthopyroxene); occurring both as phenocrysts and as groundmass. Quartz, alkali feldspar, iron oxides (ilmenite/magnetite) and apatite occur as minor and accessory minerals. Secondary minerals, such as amphibole, chlorite, biotite and suassurite, are also observed locally. Ophitic to sub-ophitic textures between plagioclase and relict clinopyroxene phenocryst (Fig. 3a), and intergrowth textures formed due to exsolution, are common (Fig. 3b). Plagioclases are mostly euhedral and are saussuratized in a few places, whereas pyroxenes are subhedral to anhedral, moderately pleochroic, and show fracture and two-sets of cleavages. Alteration in some of the plagioclase phenocrysts is visible as they give greyish or pigmented fine-grained clouded features (sieve texture) (Fig. 3c). Long apatite needles occur between plagioclase crystals and intercumulus grains (Fig. 3b), which suggests that it crystallized during or after the plagioclase crystallization. The alteration products of pyroxene are amphibole, chlorite and a few biotite grains, indicating hydrothermal fluid activity, which is consistent with the field observations. Pyroxene locally occurs as a cluster of phenocrysts, indicating its cumulus nature. Plagioclase show polysynthetic twinning, simple twinning and, in some places, two sets of twins (complex twinning). The accessory minerals, like magnetite and ilmenite (Fe–Ti oxides), are mostly euhedral shaped and some show typical ilmenite skeleton-type or herringbone-type textures (Fig. 3d).

Analytical methods

Major element analyses were carried out using WD-XRF (AXIOUS, PAN Analytical, Almelo, PW4400/40) at the Department of Geology, University of Delhi, India. The accuracy of the analyses was better than 1% for SiO_2 and 2% for most of the other major oxides, 2–5% for minor elements, and better than 10% for trace elements (Longjam & Ahmad 2012). Trace elements including rare earth elements (REEs) were analysed by inductively-coupled plasma mass spectrometer (ICP-MS ELAN DRC-II) at the National Geophysical Research Institute (NGRI), Hyderabad, India. The detailed procedures for the analyses of trace and REE are described in

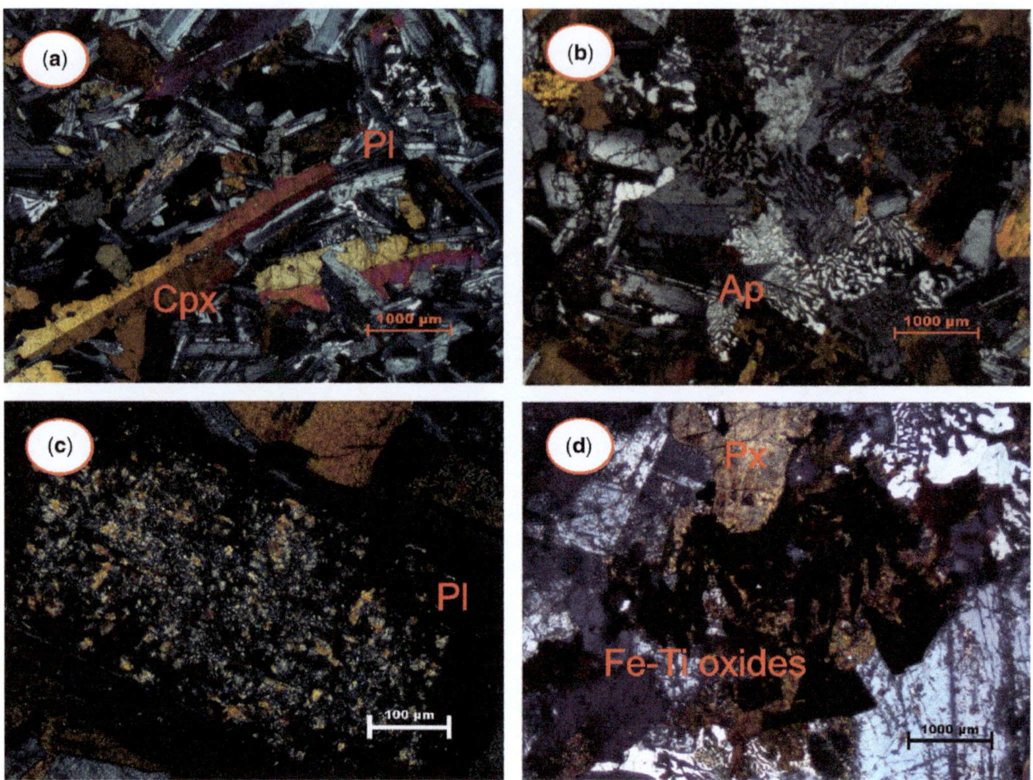

Fig. 3. Photomicrographs of Gwalior mafics. (**a**) Ophitic to sub-ophitic textures between well-developed crystal of plagioclases and pyroxenes. (**b**) Intergrowth textures and long apatite needles hoisted between (**c**) pigmented fine-grained clouded features (sieve texture) in plagioclase due to alteration. (**d**) Opaque mineral showing skeleton- or herringbone-type texture.

Balaram & Gnaneshwar Rao (2003). The precision and accuracy of the ICP-MS was better than 6% for trace and REE (Balaram *et al.* 2006).The international rock standards used for calibration during the analyses were BHVO-2, BCR-2, MB-H and PM-S. Results of major, trace and REE for the studied samples are presented in the Table 2.

Rb–Sr and Sm–Nd isotopic analyses were carried out at the National Facility for Geochronology and Isotope Geology, Indian Institute of Technology, Roorkee, India. The isotopic ratios of $^{87}Sr/^{86}Sr$, $^{147}Sm/^{144}Nd$ and $^{143}Nd/^{144}Nd$ were measured by Thermo Triton T1 mass spectrometer in fixed multicollector mode. Detailed descriptions of analytical procedure are described in detail in Ravikant (2006) and Alam *et al.* (2017). The concentration of the measurement during the analysis was monitored periodically with USGS rock standards. The mean $^{143}Nd/^{144}Nd$ ratio for the standard (GSJ JNdi-1) during the period of analysis was 0.512105 ± 10 (1 SE; quoted value 0.512106: Tanaka *et al.* 2000). The standard NIST SRM 987 gave a mean value for $^{87}Sr/^{86}Sr$ of 0.710248 ± 10 (1 SE; quoted value 0.710245: Ravikant 2006) during the period of analysis. The reported values for both the ratios ($^{143}Nd/^{144}Nd$ and $^{87}Sr/^{86}Sr$) are the mean of about 350 ratios, and the errors are the mean of the standard error. The routine blank analysis during the period of this study was same as that of Alam *et al.* (2017), which was <1 ng for Sm and Nd, and <8 ng for Sr. Values of $^{146}Nd/^{144}Nd = 0.7219$ and $^{88}Sr/^{86}Sr = 0.1194$ were used for the mass fractionation correction for Nd and Sr isotopic analyses, respectively.

Results

Element distribution

The silica content of the GMI rocks ranged from 47.68 to 51.27 wt%, the Al_2O_3 content from 12.45 to 17.45 wt% and the MgO content from 3.79 to 9.09 wt%. TiO_2, which is one of the least mobile minor oxides during post-crystallization events, had a concentration range from 1.05 to 2.95 wt%.

The proportion of Fe_2O_3 as the total iron oxide concentration ranged from 9.1 to 17.15 wt%, while P_2O_5 ranged from 0.11 to 0.22 wt% and CaO from 7.35 to 10.37 wt%. The Mg number [Mg# = 100Mg/(Mg + Fe^{2+})]$_{atomic}$ (Gill 2010) ranged from 36.31 to 59.86, with an average value of 47.71. These rocks are quartz-normative and show a restricted range in their normative composition, indicating their tholeiitic nature. Among the trace elements, the abundances of the compatible elements Cr and Ni have average values of approximately 90 (28–187 ppm) and 69 (26–138 ppm), respectively. Among the high field strength elements (HFSEs), Zr ranges from 78 to 200 ppm with an average value of 127 ppm, Y ranges from 16 to 37 ppm with an average of 25 ppm and Nb ranges from 2 to 25 ppm. The range for strontium is 179–327 ppm and for barium is 128–578 ppm.

Variation diagrams using major and trace elements show differentiation trends and the nature of the mafic magmatic rocks (Figs 4 & 5). Major oxides, such as SiO_2 and P_2O_5, display magmatic trends, increasing with decreasing MgO. However, the concentration of Al_2O_3 first increases with the decrease in MgO up to about 5 wt% MgO, and then there is a sharp drop in the Al_2O_3 content with further decreases in MgO. On the other hand, a sharp positive trend of Al_2O_3 with CaO is observed. Furthermore, the abundance of CaO decreases with increases in TiO_2 and Fe_2O_3 (Fig. 4). Compatible elements, such as Ni, display a positive correlation with MgO, whereas Zr increases with a decrease in MgO (Fig. 5). Other incompatible elements, such as TiO_2, Nb, Th and Y, display a positive correlation with increasing Zr (Fig. 5) abundances, thus exhibiting an igneous trend. Thus, we can say that the characteristics these rocks possess much of the primary igneous chemistry.

Alteration and element mobility

The GMI rocks of the Gwalior Basin have not undergone any significant deformation and metamorphism, even the associated sedimentary beds do not show any significant effect of metamorphism; therefore, major compositional changes are unlikely. Nevertheless, several factors controlled the elemental mobility and it is probable that there has been some element mobility; one factor could be their age. Various alteration processes (such as albitization, sericitization, epidotization, silicification and fluid interaction) may cause chemical changes. Major elements (Al, Fe, Ti and P), HFSEs (Y, Th, Nb, Ta, Zr and Hf), transition metals (Cr, Ni, Co, Sc and V) and the REEs (except Ce and Eu) are relatively immobile, and remain unaffected during various alteration processes and metamorphism (Sun & Nesbitt 1978; Jochum *et al.* 1991; Polat *et al.* 2002; Manikyamba *et al.* 2008; Said & Kerrich 2009); they also provide important clues to infer the source characteristics, petrogenesis and tectonic implication of these rocks (Arndt *et al.* 1989; Song *et al.* 2006). The major and trace elements when plotted against the relatively immobile element Zr display either positive or negative trends (Fig. 5) due to compositional variations related to fractionation or variable degrees of partial melting. The distinct positive trend of REE v. Zr (not shown here) and the observed smooth, slightly inclined and parallel REE pattern indicates their near-pristine characteristics and that the REE were not perturbed during post-crystallization alteration. Likewise, the geochemical characteristics of some major and trace element following igneous trends (Figs 4 & 5) also indicate that they preserved the primary igneous chemistry and have not suffered much by variable alteration processes.

Other chemical criteria are also tested to determining the possible element mobility of these suites of rocks. According to criteria suggested by Hashiguchi *et al.* (1983), the alteration index (AI = $MgO + K_2O/MgO + K_2O + Na_2O + CaO \times 100$) of the studied rocks are within the range of unaltered rocks (30.04–46.68), with only one sample (J-10) having a higher value of 51.06. High AI values (AI > 50) are produced by alteration processes such as chrolitization and sericitization, whilst albitization processes lower the AI values (AI < 30). Polat *et al.* (2002) estimated the extent of alteration on some volcanic rocks based on loss on ignition (LOI) and suggested that rocks whose LOI is >6 wt% have been subjected to significant alteration. The LOI of the GMI rocks of the Gwalior Basin ranges between 0.42 and 3.85 (Table 2). Thus, we may summarize that there was no serious impact of secondary alteration on this suite of rocks, except for some known mobile elements. Therefore, in this study, it is more appropriate to use the least mobile element, such as the HFSE and REE, to evaluate magma evolution and source characteristics.

Classification

On the total alkali ($Na_2O + K_2O$) v. silica (SiO_2) TAS (total alkali silica) classification diagram (Cox *et al.* 1979), the gabbros of the Gwalior Basin fall within the sub-alkaline basalt field with tholeiitic affinity (Fig. 6a). A similar picture is observed on the Zr + Y–Ti–Cr ternary diagram (Fig. 6b) (Davies *et al.* 1979), where they also exhibit a tholeiitic trend. Classification for GMI samples is again tested using ratios of incompatible elements (Zr, Ti, Nb and Y) (Winchester & Floyd 1977), and it is observed that the GMI samples fall in the field of sub-alkaline basalt to andesitic basalt composition (Fig. 6c). On Jensen's cationic ternary diagram (Fe + Ti–Al–Mg), we observed a full spectrum of rocks ranging

Table 2. *Major (wt%) and trace (ppm) element concentrations of GMI rocks of the Gwalior Basin*

Major (wt%)	JG-1	JG-2	JG-3	JG-4	JG-5a	JG 6	JG-7	JG 8	JG 9	JG 10	JG 11	JG-12	JG 13	JG 14	JG 15	JG 16
SiO_2	50.4	50.3	50	50.5	50.7	50.6	48.8	50.7	50.1	50.1	50.8	49.5	49.3	51.3	50.9	51
TiO_2	1.32	1.32	1.3	1.45	2.1	2.13	1.75	2.95	2.05	2.08	1.21	1.09	1.13	1.3	1.22	1.15
Al_2O_3	15.5	15.3	15.2	14.9	13.1	15	14.8	13.3	15.1	14.2	16.5	16.7	16.3	16.4	17.5	13.6
FeO_3	10.98	11.05	11.16	12.54	14.68	14.48	14.12	17.15	14.22	14.04	10.77	9.1	9.57	11.24	10.4	12.26
MnO	0.15	0.15	0.16	0.17	0.18	0.18	0.17	0.19	0.18	0.18	0.16	0.13	0.14	0.16	0.15	0.19
MgO	5.2	5.3	5.1	5.3	4.5	5.1	6.1	3.8	4.8	4.3	5.8	4.7	5	5.8	5.2	8.3
CaO	9.59	9.3	9.48	9.24	7.81	9.16	8.18	7.35	9.11	8.39	9.7	10.07	9.44	9.76	9.41	10.28
Na_2O	2.71	2.68	2.79	2.84	2.85	2.82	2.54	2.86	2.77	2.66	2.72	2.67	3.41	2.77	2.86	2.25
K_2O	0.72	0.86	0.77	0.82	1.01	0.72	0.91	1.08	0.7	1.04	0.84	0.82	1.16	0.81	1.03	0.54
P_2O_5	0.14	0.14	0.13	0.13	0.23	0.16	0.16	0.23	0.12	0.2	0.12	0.12	0.12	0.13	0.12	0.11
LOI	1.42	1.2	0.87	0.97	0.99	0.87	1.03	0.42	0.42	1.12	0.87	1.12	1.98	0.81	1.53	0.56
Sum	98.13	97.6	96.96	98.8	98.14	101.24	98.53	100.06	99.51	98.35	99.48	96.03	97.56	100.47	100.24	100.18
Trace (ppm)																
Cr	54	76	80	88	106	103	68	91	145	147	131	146	52	153	166	187
Ni	51	68	66	64	44	67	71	26	64	41	72	67	42	71	61	107
Co	39	43	47	49	52	64	47	57	64	50	48	44	33	48	46	59
Sc	26	28	31	32	33	38	28	32	39	34	33	30	23	33	32	44
Ga	19	21	21	18	23	24	19	25	22	23	19	25	18	19	23	16
Rb	19	26	21	21	33	23	30	34	20	31	24	30	38	22	32	15
Ba	179	214	196	183	304	213	215	322	193	281	185	207	179	181	249	128
Sr	229	256	249	212	225	247	237	217	247	236	241	327	245	231	284	179
Cu	30	37	47	40	42	69	33	43	63	33	24	29	23	28	28	24
Zn	110	92	120	94	131	129	88	155	121	125	86	88	69	91	97	90
Zr	88	105	99	95	179	127	104	200	106	162	93	111	80	94	107	78
Y	19	20	20	19	35	25	21	37	22	31	19	21	16	19	21	18
Pb	9	5	10	5	8	7	4	9	6	8	5	5	5	5	6	4
Th	3	3	3	3	6	4	3	7	3	5	3	4	2	3	3	2
Nb	12	12	11	11	10	15	5	18	9	12	10	11	9	6	11	8
U	1	1	1	1	1	1	0	1	1	1	1	1	0	1	1	0
V	224	237	268	290	375	554	274	476	534	338	249	253	198	263	261	280
REE (ppm)																
La	11.48	13.15	12.32	11.57	23.65	15.77	12.6	26.03	12.62	19.96	11.31	14.28	10.58	11.78	13.73	10
Ce	25.24	29.1	27	25.29	51.84	34.66	27.68	56.45	28.03	44.22	24.96	31.05	23.17	26.02	29.92	22.33
Pr	3.31	3.67	3.44	3.24	6.53	4.37	3.6	7.16	3.57	5.55	3.16	3.9	2.93	3.25	3.69	2.87
Nd	13.83	15.57	14.67	13.84	27.89	18.47	15.68	30.34	15.35	23.53	13.35	16.76	12.25	14.03	16.03	12.31
Sm	3.34	3.88	3.66	3.29	6.44	4.34	3.76	7.12	3.75	5.57	3.23	3.81	2.93	3.35	3.81	3.07
Eu	1.06	1.19	1.12	1.09	1.74	1.36	1.2	1.96	1.27	1.57	0.99	1.17	0.9	1.03	1.2	0.91
Gd	3.39	3.98	3.66	3.4	6.6	4.57	3.87	7.2	3.9	5.64	3.33	3.88	2.96	3.43	3.73	3.09
Tb	0.57	0.62	0.59	0.56	1.04	0.74	0.63	1.16	0.62	0.9	0.53	0.6	0.46	0.57	0.6	0.52
Dy	3.25	3.59	3.4	3.43	6.16	4.37	3.76	6.56	3.91	5.44	3.2	3.71	2.77	3.36	3.64	3.15
Ho	0.67	0.73	0.72	0.72	1.3	0.91	0.79	1.35	0.82	1.12	0.68	0.75	0.57	0.69	0.74	0.67
Er	1.81	2.02	1.9	1.88	3.42	2.38	2.02	3.61	2.17	2.9	1.76	2.08	1.5	1.87	1.98	1.79
Tm	0.26	0.29	0.28	0.29	0.49	0.36	0.31	0.54	0.32	0.45	0.26	0.3	0.22	0.28	0.3	0.26
Yb	1.71	1.85	1.83	1.76	3.09	2.22	1.85	3.42	1.96	2.63	1.6	1.83	1.38	1.63	1.78	1.61
Lu	0.24	0.26	0.26	0.24	0.45	0.32	0.26	0.47	0.28	0.38	0.22	0.27	0.19	0.24	0.26	0.23

from basaltic komatiite to high Mg-tholeiite to high Fe-tholeiite (Fig. 6d). Based on the criteria given by Erlank *et al.* (1988), the GMI rocks of the Gwalior Basin are again classified as low-Ti tholeiites (average value of $Ti/Y < 410$ and $Zr/Y < 6$). Therefore, on the basis of these classification schemes, it may be concluded that the GMI rocks are sub-alkaline tholeiitic gabbros.

Geochemical characterization

A range of incompatible trace elements arranged in order of incompatibility were normalized to the chondrite and primitive mantle values to know the relative abundance of the rocks relative to Primordial Earth (Thompson *et al.* 1984). The immobile nature of the trace elements, including the REEs, helps to understand their geochemical characteristics and to identify the original petrogenetic processes of these rocks (Taylor 1965; Winchester & Floyd 1977; Hanson 1980). In chondrite normalized (Sun & McDonough 1989) REE plots (Fig. 7a), the GMI samples are moderately fractionated [$(La/Yb)_N = 4.43$–5.60, with an average value *c.* 5.045] with small Eu anomalies ranging from 0.81 to 1.01 (average $Eu/Eu^* = 0.916$). The average normalized LREE [$(La/Sm)_N = 2.21(1.96$–$2.42)$] and HREE [$(Gd/Yb)_N = 1.74(1.59$–$1.82)$] indicates an enriched LREE pattern, which is more fractionated than HREE. The slight depletion of HREE [$(La/Yb)_N = 4.43$–5.60] probably indicates involvement of garnet in the source. The smooth, parallel and enriched REE pattern indicates the derivation of these rocks from similar enriched mantle sources. Varying degrees of partial melting cause large variations in incompatible trace elements, whereas variable degrees of fractionation lead to variations in the compatible elements (Ahmad & Tarney 1991). The abundance of

JG-17	JG-18	JG-19	JG-20	JG-21	JG 22	JG 22A	JG 23	JG 24	JG-25	JG 27	JG-29	JG 30	JG-33	JG 35	JG 36
50.2	50.1	50.3	51	50.9	49.8	48.2	49.7	49.3	49.9	50.3	49.1	50.4	49.7	50.2	50.2
1.09	1.05	1.11	1.49	1.75	1.64	1.46	1.49	1.41	1.68	1.7	1.62	1.61	1.76	1.73	1.64
14.7	15.9	16.7	15.8	15.8	15.5	15.1	15.7	15.6	15.3	15.6	14.7	15.9	14.7	15.7	15.8
10.82	10.08	9.71	12.18	13.48	13.07	13.96	12.66	12.2	13.39	13.33	12.89	13.01	13.42	13.25	12.99
0.17	0.15	0.14	0.17	0.18	0.17	0.18	0.16	0.16	0.17	0.17	0.17	0.17	0.17	0.18	0.17
7.1	6.1	5.1	5.3	4.6	5.1	6.3	5.2	5.2	4.9	4.7	4.5	5.1	4.9	4.8	4.7
9.79	10.37	9.84	9.18	8.36	8.76	7.65	8.6	8.78	8.54	8.52	8.98	8.92	8.53	8.68	9.03
2.41	2.56	2.7	2.85	2.85	2.7	2.86	2.74	2.82	2.77	2.74	2.56	2.93	2.51	2.54	2.64
0.85	0.72	1.01	1.02	1.45	1.19	1.15	1.28	1.21	1.25	1.33	1.1	1.48	1.37	1.23	1.12
0.11	0.11	0.12	0.16	0.18	0.16	0.12	0.17	0.15	0.17	0.18	0.17	0.15	0.15	0.14	0.17
0.75	0.98	1.25	1.44	1.74	1.86	2.75	2.45	1.99	1.93	1.43	1	1.82	1.26	1.46	1.32
98.01	98.19	98.02	100.63	101.26	99.93	99.63	100.18	98.72	99.95	100.02	96.85	101.52	98.51	99.91	99.77
119	134	135	183	105	169	74	148	138	128	110	131	157	114	87	163
85	82	49	72	64	73	77	87	91	62	68	80	67	89	79	75
51	50	36	55	51	49	49	52	52	46	49	52	45	59	51	50
37	36	25	36	34	30	31	32	32	30	31	34	28	37	33	32
18	21	19	23	24	20	20	21	20	20	21	23	19	24	23	21
27	23	30	31	61	52	60	53	49	49	51	37	47	66	54	38
197	177	183	239	389	171	256	182	177	192	236	249	171	264	251	193
225	273	254	267	282	240	262	267	267	236	248	252	245	281	227	247
23	28	22	32	42	35	39	40	40	34	36	43	33	40	44	35
85	217	80	109	129	107	229	94	145	94	109	205	93	122	117	108
81	87	83	129	153	117	102	121	114	117	130	144	118	151	140	130
17	19	17	26	29	23	21	24	23	24	25	27	23	30	27	25
4	8	4	6	8	8	7	5	11	5	5	10	5	12	9	6
2	3	3	4	4	3	3	4	3	3	3	4	3	4	4	4
7	10	9	14	11	4	14	13	7	8	8	11	12	18	18	9
0	1	0	1	1	1	1	1	1	1	1	1	1	1	1	1
261	258	204	320	326	283	282	278	281	287	299	329	268	346	312	298
10.13	11.14	10.7	16.78	18.54	13.88	12.34	14.64	13.62	14.09	15.42	16.7	13.66	17.57	16.51	15.41
22.21	24.55	23.44	36.87	40.44	30.92	27.24	32.31	30.53	31.4	34.09	37.27	30.82	39.74	36.75	34.11
2.76	3.07	2.92	4.65	5.16	3.93	3.5	4.14	3.87	4.03	4.34	4.7	3.86	5.07	4.7	4.34
11.92	13.36	12.44	19.47	21.71	17.12	14.91	17.78	16.68	17.34	18.84	20.48	16.71	21.85	20.22	18.47
2.93	3.21	3.05	4.6	5.31	4	3.85	4.17	4.05	4.12	4.53	4.9	3.98	5.24	5.04	4.45
0.94	1	0.93	1.41	1.61	1.23	1.23	1.26	1.21	1.27	1.37	1.49	1.2	1.55	1.49	1.29
3.01	3.3	3.05	4.62	5.36	4.14	3.87	4.33	4.05	4.21	4.56	4.96	4.03	5.3	5.12	4.51
0.49	0.55	0.48	0.76	0.88	0.66	0.63	0.69	0.65	0.68	0.72	0.78	0.65	0.85	0.83	0.72
2.95	3.22	2.88	4.46	4.98	4.09	3.64	4.06	4	4.1	4.4	4.75	3.95	5.07	4.65	4.37
0.61	0.66	0.6	0.92	1.05	0.82	0.75	0.84	0.79	0.84	0.9	0.98	0.82	1.04	0.99	0.92
1.63	1.77	1.59	2.44	2.86	2.16	2.01	2.24	2.15	2.2	2.41	2.55	2.14	2.79	2.57	2.41
0.24	0.26	0.24	0.35	0.41	0.32	0.29	0.34	0.32	0.32	0.36	0.38	0.32	0.41	0.38	0.36
1.49	1.61	1.44	2.15	2.6	2.02	1.87	1.97	1.98	1.99	2.2	2.35	1.93	2.48	2.48	2.16
0.21	0.23	0.2	0.31	0.37	0.29	0.27	0.29	0.28	0.29	0.31	0.35	0.28	0.36	0.35	0.31

LREE enrichment is approximately 40–110 times chondrite, indicating variable degrees of partial melting of the source. In primitive mantle (PM) normalized (Sun & McDonough 1989) multi-element plots, the marked enrichment of large-ion lithophile elements (LILEs) over HFSE and LREE is evident, with negative anomalies for Nb and P (Fig. 7b) suggesting an affinity with continental tholeiites (Hergt *et al.* 1991). The observed distinct positive Pb anomalies for all the samples probably indicate a crustal influence. Ba is slightly enriched in a few samples, although the majority show negative anomalies. Nb and P indicate the influence of subcontinental lithosphere on the mantle-derived magma (Tarney 1992). The small anomalies observed in Eu and Sr (Fig. 7a, b) may indicate the minor role of plagioclase fractionation.

The Nd and Sr isotopic compositions of some representative samples of the GMI rocks from the Gwalior Basin are given in Table 3. Pyroxene and plagioclase were separated and a mineral–whole-rock isochron age for sample JG-13 is presented in Figure 8a. The three-point thermal ionization mass spectrometry (TIMS) mineral–whole-rock Sm–Nd isochron has a good spread in $^{147}Sm/^{144}Nd$ ratios and corresponds to an age of 2104 ± 23 Ma (Fig. 8a), with an initial $^{143}Nd/^{144}Nd = 0.509938 \pm 0.000023$ and $\varepsilon Nd^{t} = -0.90$ ($t = 2000$ Ma). Although we have more Nd data (Table 3), only one sample (JG-13) gave an isochron age within the analytical error (MSWD < 2.5), and the other samples could not fit a straight line. These ages are not reliable, so we are considering 2.1 Ga as the crystallization age of these mafic intrusives. This is the first Sm–Nd mineral-rock isochron age for the GMI bodies. The only other isotopic age available for GMI is a Rb–Sr age of 1830 ± 200 Ma for the mafic sills (Crawford & Compston 1969). The initial

Table 2. *Major (wt%) and trace (ppm) element concentrations of GMI rocks of the Gwalior Basin*

Major (wt%)	J-1	J-2	J-3	J-4	J-5	J-6	J-7	J-8	J-9	J-10	J-11	J-12	J-13	J-14	J-15	J 16
SiO_2	48.2	47.8	47.8	47.9	48.2	47.8	47.8	48.6	49.9	47.9	50.1	50.2	50.3	50.6	47.7	50
TiO_2	1.59	1.46	1.34	1.72	1.67	1.49	1.61	1.39	1.81	1.58	1.81	1.77	1.7	1.76	1.67	1.71
Al_2O_3	13.8	14.6	14.3	13.6	13.8	13.9	13.9	12.9	13.6	12.5	13.9	14.4	14.2	14.4	13.1	14.1
FeO_3	13.96	12.45	13.43	14.21	14.04	14.01	13.98	14.43	13.36	15.31	13.29	12.91	12.79	12.94	14.27	12.8
MnO	0.18	0.18	0.17	0.18	0.17	0.18	0.2	0.2	0.17	0.18	0.17	0.17	0.17	0.17	0.18	0.17
MgO	6.9	7	7.2	6.1	6.6	6.9	6.7	8.2	4.9	9.1	4.9	4.6	4.5	4.3	7.5	4.8
CaO	7.98	7.71	8.01	7.78	7.6	7.74	8.17	8.19	8.56	7.67	8.53	8.62	8.46	8.55	8.25	8.73
Na_2O	2.31	2.76	2.58	2.31	2.5	2.74	2.33	2.51	2.53	1.74	2.59	2.81	2.78	2.85	2.17	2.64
K_2O	1.25	1.21	1.33	1.18	1.21	1.14	1.08	1.17	0.93	0.77	0.99	1.09	1.12	1.08	0.82	0.93
P_2O_5	0.13	0.14	0.11	0.14	0.14	0.13	0.13	0.13	0.17	0.19	0.17	0.17	0.18	0.19	0.14	0.17
LOI	2.78	3.71	3.1	3.41	3.85	2.85	2.7	2.16	2.71	1.99	1.14	2.11	2.92	2.58	3.05	3.16
Sum	99.1	99.06	99.29	98.5	99.77	98.88	98.59	99.91	98.61	98.83	97.54	98.85	99.13	99.4	98.87	99.19
Trace (ppm)																
Cr	50	33	32	57	51	28	49	28	65	38	66	65	90	62	56	72
Ni	75	90	90	94	89	80	86	106	47	138	51	50	50	46	104	48
Co	52	47	46	60	55	46	60	55	46	61	50	47	45	44	57	44
Sc	31	26	23	35	33	25	35	25	31	26	32	31	30	31	32	29
Ga	19	20	18	24	23	18	24	18	21	20	23	21	20	20	22	20
Rb	38	47	34	46	45	29	40	34	28	30	31	31	31	29	30	27
Ba	578	446	372	481	536	458	422	520	274	367	315	253	263	237	302	245
Sr	246	244	214	316	267	201	289	264	229	211	247	245	225	223	255	228
Cu	39	39	35	41	39	30	44	54	35	37	46	38	32	32	44	39
Zn	151	70	104	109	87	103	125	128	112	84	136	111	119	105	89	107
Zr	102	107	83	140	147	99	127	100	145	150	162	126	125	152	120	139
Y	20	21	17	27	25	19	25	21	27	26	29	26	25	25	24	25
Pb	8	5	6	7	5	9	8	11	7	5	7	7	7	6	4	6
Th	3	3	2	3	3	3	3	2	5	4	5	4	4	5	3	4
Nb	14	15	11	20	17	13	17	7	16	12	9	19	12	16	7	18
U	1	1	0	1	1	0	1	0	1	1	1	1	1	1	1	1
V	308	256	218	351	321	237	334	248	309	247	347	310	284	324	301	291
REE (ppm)																
La	12.39	12.52	9.91	15.91	14.15	10.96	14.17	12.38	17.95	16.09	19.5	16.85	16.53	16.76	13.4	16.46
Ce	26.75	27.63	21.92	35.48	31.84	24.62	30.97	27.49	39.21	35.47	42.13	36.05	35.94	36.25	29.98	35.5
Pr	3.44	3.6	2.84	4.59	4.18	3.22	4.06	3.6	4.99	4.6	5.31	4.56	4.55	4.53	3.91	4.52
Nd	15.01	15.7	12.29	19.73	18.33	13.78	17.53	15.49	21.15	19.78	22.19	19.36	19.41	19.45	16.97	19.06
Sm	3.75	3.92	3.08	4.97	4.52	3.52	4.51	3.77	5.1	4.85	5.4	4.67	4.78	4.54	4.23	4.58
Eu	1.27	1.25	1.01	1.6	1.5	1.12	1.44	1.25	1.43	1.49	1.56	1.4	1.38	1.36	1.34	1.35
Gd	3.99	3.91	3.1	5.15	4.7	3.57	4.65	3.96	5.16	4.91	5.59	4.86	4.77	4.73	4.34	4.76
Tb	0.64	0.65	0.52	0.83	0.75	0.57	0.75	0.64	0.84	0.8	0.87	0.76	0.77	0.75	0.72	0.76
Dy	3.69	3.7	3.03	4.93	4.44	3.36	4.4	3.69	4.78	4.45	5.03	4.48	4.39	4.31	4.06	4.28
Ho	0.77	0.77	0.61	0.99	0.92	0.67	0.9	0.75	0.99	0.93	1.03	0.91	0.9	0.88	0.85	0.9
Er	2.04	2.02	1.65	2.63	2.4	1.82	2.41	1.98	2.62	2.46	2.79	2.42	2.42	2.38	2.25	2.34
Tm	0.31	0.3	0.24	0.39	0.36	0.28	0.36	0.29	0.39	0.36	0.4	0.36	0.35	0.36	0.33	0.35
Yb	1.88	1.81	1.5	2.47	2.28	1.7	2.25	1.83	2.45	2.26	2.62	2.34	2.2	2.27	2.12	2.19
Lu	0.27	0.26	0.22	0.35	0.33	0.25	0.31	0.26	0.35	0.33	0.37	0.32	0.32	0.32	0.3	0.32

$^{87}Sr/^{86}Sr$ (t = 2000 Ma) for the majority of samples ranges from 0.69276 to 0.70819, with one sample (JG-22A) having a lower value of 0.64248 and another (JG-22) having a higher value (0.7234). Average present-day measured values of $^{87}Sr/^{86}Sr$ were 0.7211 in the continental crust and 0.7045 in the mantle (Rollinson 1993). The average initial $^{87}Sr/^{86}Sr$ ratio of the GMI samples excluding the outliers is approximately 0.7033. This indicates that at 2 Ga, the melts formed from the mantle. $^{147}Sm/^{144}Nd$ measured values range from 0.1042 to 0.1302. Variability in the values of isotopic ratios clearly indicates heterogeneity in the mantle sources.

Depleted Nd model ages (T_{DM}) (extraction ages of the protoliths from the mantle sources) for the GMI show a wide range from 2.6 to 1.7 Ga (Table 3) with an average of 2.0 Ga. This indicates that the Nd isotope composition of the GMI rocks experienced their extraction from the mantle sources between 2.6 and 1.7 Ga. The εNd^t (t = 2000 Ma) for the GMI rocks varies from −7.69 to +3.91 with an average value of −0.17, indicating near-chondritic values ranging from depleted to enriched source. This reflects that that the mantle sources were heterogeneous: that is, with variable enrichment and depletion histories. The growth of the $^{143}Nd/^{144}Nd$ ratios were calculated at any time t, and are presented along with the curve lines of depleted mantle (DM) and chondritic uniform reservoir (CHUR) (Fig. 8b).

Petrogenesis

Geochemical variability (Figs 4 & 5) suggests that the magma evolution was controlled to some extent by fractional crystallization. This suggestion is also supported by low to moderate Mg# (36.31–59.86).

J 17	J-18	J-19	J-20	J-21	J-22	J-23	J-24	J-25	J-26	J-27	J-28	J-29	J-30	J-31	J-32	J-33	J-34
50.2	49.8	50	50.2	50.4	50.4	48.8	50.1	48.6	50.6	47.9	50.5	49.2	49.7	50.7	50.3	48.5	48.9
1.77	1.52	1.66	1.62	2.64	2.13	1.76	1.63	1.7	1.77	1.65	1.73	1.63	1.81	1.81	1.85	1.78	1.81
14.4	14.1	14.4	14.1	13.2	13.5	14.4	14	14	14.2	13.8	14.9	14.7	14.6	14.1	14.1	14.3	14.2
13.01	12.92	12.95	12.79	15.88	14.19	14.01	12.73	13.8	13.02	13.53	12.83	13.37	13.99	13.3	13.26	13.94	14.02
0.17	0.18	0.17	0.17	0.18	0.17	0.17	0.17	0.17	0.17	0.17	0.17	0.17	0.17	0.17	0.17	0.17	0.17
5	5.3	5.3	5.4	4.1	4.2	5.4	4.9	6.6	4.7	7	4.1	5.4	5.4	4.8	4.4	5.2	5.3
8.72	8.82	9.02	9.19	7.95	8.15	8.47	8.54	8.26	8.56	8.19	8.64	8.42	8.73	8.61	8.31	8.63	8.62
2.83	2.6	2.63	2.69	2.77	2.68	2.44	2.69	2.37	2.72	2.37	3.09	2.62	2.71	2.72	2.76	2.46	2.62
0.93	0.71	0.88	0.81	1.01	1.07	1.16	1.03	1.05	1.08	0.97	1.12	1.12	0.81	1.1	1.14	0.93	0.86
0.16	0.13	0.16	0.15	0.21	0.22	0.16	0.17	0.15	0.18	0.15	0.18	0.15	0.16	0.18	0.19	0.16	0.16
2.95	0.89	2.13	1.13	0.85	0.83	2.68	0.87	2.54	0.89	2.74	1.02	2.39	1.41	2.16	1.14	1.9	1.95
100.09	97.01	99.32	98.29	99.25	97.59	99.46	96.84	99.23	97.88	98.49	98.29	99.14	99.43	99.57	97.62	97.94	98.63
91	54	65	74	68	69	58	70	61	67	43	95	77	71	56	70	58	72
68	60	68	75	39	40	69	55	87	48	80	56	86	77	55	48	68	84
57	55	56	60	62	47	48	48	58	47	48	49	50	60	50	50	51	64
37	36	37	39	37	30	31	33	33	31	27	32	32	39	34	32	32	41
26	24	22	25	27	22	21	22	23	23	20	22	22	26	23	24	23	28
32	24	27	26	35	31	43	30	43	31	33	34	41	34	34	36	31	41
265	233	232	228	341	277	321	262	288	275	241	251	306	233	299	288	266	312
284	270	241	265	247	212	230	239	253	239	262	249	253	272	252	266	239	300
48	50	51	57	57	41	37	36	44	37	37	36	38	44	40	37	39	47
120	119	110	124	145	125	113	111	120	114	102	127	134	114	124	123	109	151
161	139	116	140	189	159	130	136	130	151	136	132	122	159	145	163	127	155
29	24	24	28	36	30	25	26	26	27	23	27	25	31	29	29	26	33
7	6	5	6	10	6	6	7	6	7	5	8	8	5	7	7	5	9
5	5	4	4	6	5	3	4	3	5	3	4	3	4	5	5	3	4
7	19	7	11	25	11	18	15	18	15	16	16	13	17	20	22	19	20
1	1	1	1	1	1	1	1	1	1	1	1	1	1	1	1	1	1
380	314	365	379	595	352	305	301	317	317	256	310	307	364	334	348	314	392
18.78	16.07	14.76	18.24	24.69	20.37	14.13	17.29	15.19	18.53	13.6	17.45	14.71	17.73	19.66	19.97	14.67	18.23
41.05	34.04	32.24	39.62	53.76	44.49	31.61	37.38	33.46	40.64	30.48	38.45	33.04	39.98	43.43	44.04	33.28	40.97
5.24	4.33	4.17	5.07	6.79	5.58	4.15	4.76	4.45	5.14	3.98	4.89	4.3	5.18	5.44	5.58	4.37	5.34
21.9	18.07	17.49	21.43	28.54	23.37	17.99	20.05	19.23	21.64	16.94	20.8	18.45	22.63	22.88	23.04	18.85	23.34
5.5	4.41	4.28	5.23	6.92	5.68	4.46	4.85	4.88	5.08	4.17	4.96	4.58	5.73	5.52	5.49	4.84	5.97
1.61	1.39	1.29	1.49	1.94	1.63	1.36	1.44	1.51	1.51	1.41	1.5	1.48	1.68	1.57	1.61	1.47	1.81
5.55	4.44	4.44	5.42	6.86	5.73	4.58	5.02	4.89	5.28	4.39	5.12	4.62	5.87	5.57	5.61	4.90	6.06
0.87	0.71	0.71	0.87	1.1	0.91	0.74	0.79	0.79	0.84	0.7	0.82	0.75	0.95	0.87	0.90	0.80	0.97
5.07	4.22	4.12	4.96	6.27	5.21	4.34	4.55	4.73	4.78	4	4.58	4.37	5.41	5.19	5.06	4.57	5.74
1.07	0.86	0.87	1.03	1.31	1.06	0.88	0.92	0.97	0.97	0.82	0.96	0.91	1.11	1.07	1.05	0.94	1.16
2.86	2.29	2.33	2.72	3.49	2.9	2.37	2.46	2.58	2.62	2.2	2.51	2.42	2.93	2.83	2.81	2.54	3.11
0.42	0.35	0.34	0.4	0.51	0.41	0.35	0.36	0.36	0.39	0.32	0.38	0.35	0.44	0.43	0.41	0.38	0.45
2.6	2.15	2.16	2.54	3.26	2.63	2.16	2.34	2.33	2.48	2.02	2.39	2.25	2.75	2.57	2.64	2.35	2.95
0.38	0.32	0.3	0.36	0.46	0.39	0.32	0.33	0.34	0.35	0.28	0.34	0.32	0.41	0.38	0.38	0.34	0.41

The slightly linear trend between MgO and SiO_2 also indicates that fractionation has played an important role. The drop in MgO with an increase in SiO_2 for samples up to 50 wt% SiO_2 (Fig. 4) indicates olivine-dominated fractionation in the more mafic magmas. However, the overall trend (Fig. 4) indicates that olivine was not the only fractionating phase and other phases were involved during the evolution of the magmas. Fractionation of olivine increases the CaO composition, which then decreases with the fractionation of clinopyroxene. Plagioclase and pyroxene can control both CaO and Al_2O_3. Positive correlation between Al_2O_3 v. CaO and negative correlation of SiO_2 with MgO indicates fractionation of clinopyroxene, olivine and plagioclase (Fig. 4). Further, negative correlation between TiO_2 and Fe_2O_3 against CaO (Fig. 4) also supports the fractionation of phases, such as plagioclase and clinopyroxene. The fractionation phases involved are consistent with the fractionation of a gabbro assemblage (olivine–pyroxene–plagioclase). Major oxides plotted with some trace elements (Fig. 5) also support this interpretation. The positive correlation between Ni and MgO, and between TiO_2 and Zr, and the negative correlation between Zr and MgO (Fig. 5) also indicate the fractionation of olivine + clinopyroxene ± plagioclase. The fractionation of olivine, clinopyroxene and plagioclase is again supported in Figure 9a, b, where a gabbro fractionation trend is followed. However, the feeble Eu anomalies (0.81–1.01) in the REE patterns (Fig. 7a) and the high Sr content (179–327, average *c.* 249) negate any significant role for plagioclase fractionation. In Figure 4, the Al_2O_3 content increases with decreasing MgO to about 5 wt% MgO, and then there is a sharp drop in the Al_2O_3

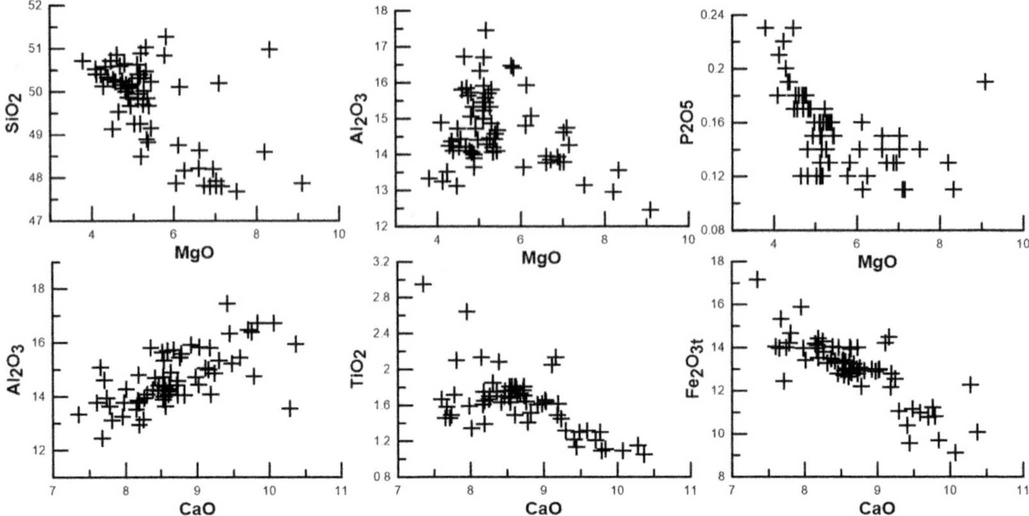

Fig. 4. Major element bivariate diagram where the oxides are plotted against MgO and CaO, showing a more or less linear trend.

content with further decreases in MgO. This is consistent with plagioclase accumulation peaking at about 5 wt% MgO. Likewise, the lack of negative correlation between TiO_2 against Zr (Fig. 5) and the negative correlation of P_2O_5 and MgO (Fig. 4) further suggest that the role of fractionation of accessory minerals, such as Fe–Ti oxides and apatite, was insignificant.

Incompatible trace elements ratios do not fractionate significantly during a moderate degree of crystal fractionation but are affected by variable degrees of partial melting of the mantle sources (Ahmad & Tarney 1991). The incompatible trace element ratio Zr/Y of the samples was plotted against Y (Fig. 9b), with the fractionation trends for amphibole, clinopyroxene and olivine + plagioclase also shown.

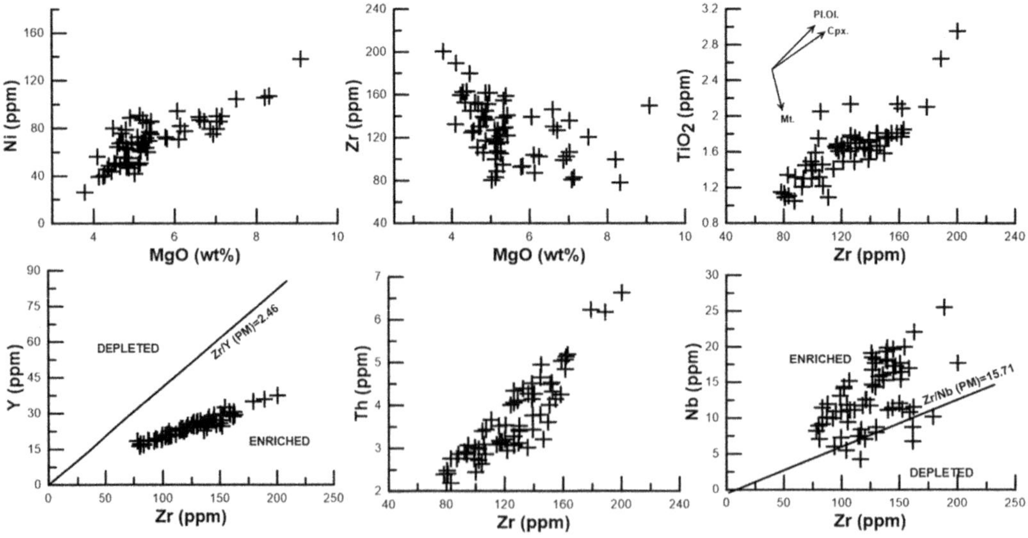

Fig. 5. Bivariate plots of some of the major oxides against trace, and trace v. trace of GMI samples showing the less mobile nature of these elements, indicating that these suites of rocks preserved much of the primary igneous chemistry. Primordial mantle (PM) ratio values shown in Y and Nb v. Zr are after Sun & McDonough (1989).

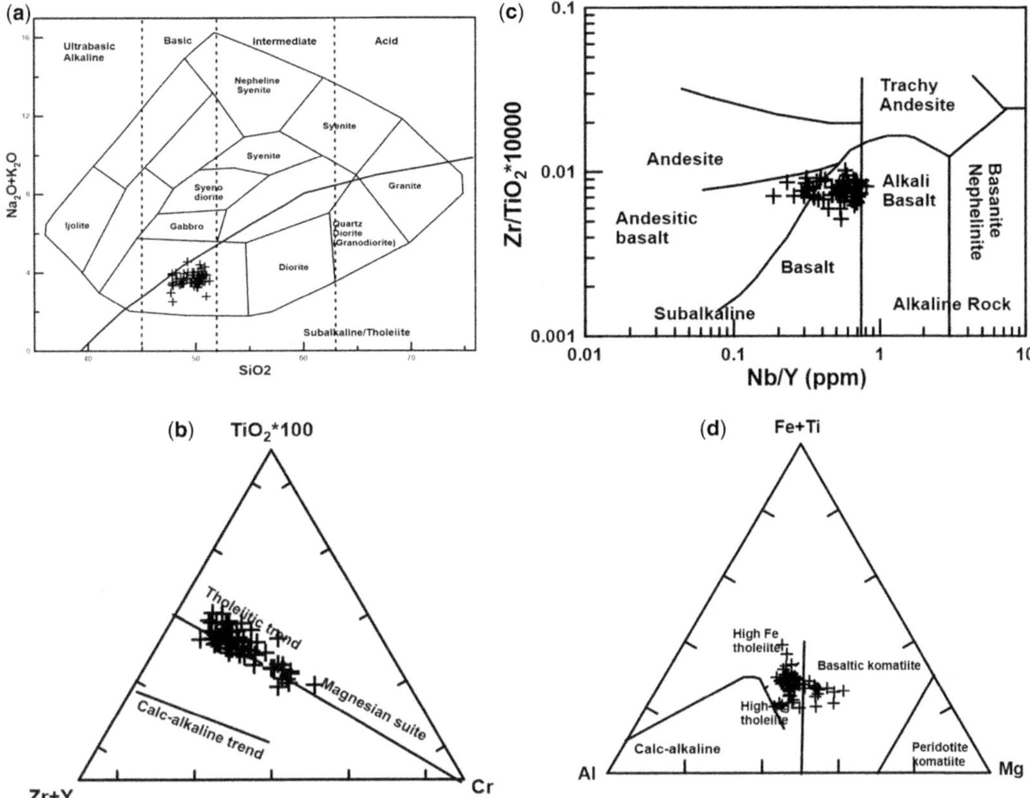

Fig. 6. Classification diagrams. (**a**) Total alkali v. silica (TAS) diagram (after Cox *et al.* 1979). (**b**) Zr + Y–TiO_2*100–Cr ternary diagram (Davies *et al.* 1979) showing the tholeiitic trend for GMI samples. (**c**) HFSEs Zr/Ti v. Nb/Y ratio plots (Winchester & Floyd 1977). (**d**) Cationic ternary diagram of Fe + Ti–Al–Mg (after Jensen 1976) for the mafic magmatic rocks of Gwalior Basin.

The fractional crystallization vectors trends are after Floyd (1993). In Figure 9b, we observed nearly flat to a slight increase in Zr/Y ratios with increasing Zr content, indicating more or less clinopyroxene and olivine + plagioclase fractionation. The slight increase in Zr/Y ratios with variable Zr in the GMI samples is consistent with evolution of the magmas by fractional crystallization of parental melts derived by varying degrees of partial melting of the same or similar mantle source(s). Again in this figure, two mantle sources represented by melting curves I and II are shown (Fig. 9b: after Drury 1983). Curve I represents the melting curve with proportions of 60% olivine + 20% orthopyroxene + 10% clinopyroxene + 10%plagioclase, and curve II represents 60% olivine + 20% orthopyroxene + 10% clinopyroxene + 10% garnet (Sun & Nesbitt 1977). The GMI samples fall in-between the two curves (I and II), but are closer to curve II (Fig. 9b) allowing for a clinopyroxene and olivine + plagioclase fractionation trend; these observed variations may reflect variable degrees and varying depth (with garnet) of partial melting of a common or a similar source, followed by fractionation of the gabbro assemblages.

The degree of mantle partial melting was examined based on the Ni v. Zr petrogenetic model diagram (Fig. 10: after Rajamani *et al.* 1985; Condie *et al.* 1987). The mantle source in the figure is a lherzolite (11 ppm Zr and 2000 ppm Ni) at a depth of 80–100 km (Condie *et al.* 1987); curve I (batch melting curve) and curves II, III and IV (olivine fractionation curves) are after Rajamani *et al.* (1985). The GMI samples fall around olivine fractionation curve III, but between curves I and IV, suggesting their derivation from a melt generated by about 7–19% (<20%) melting of a mantle source followed by fractional crystallization of olivine up to 35–50% (Fig. 10). In the Ce v. Nd plot (Fig. 11a), the wide variation observed in both incompatible elements again indicates that the magmas were generated by variable degrees of partial melting, in the range from approximately 6 to 18%. In Figure 11a, the chondrite ratio

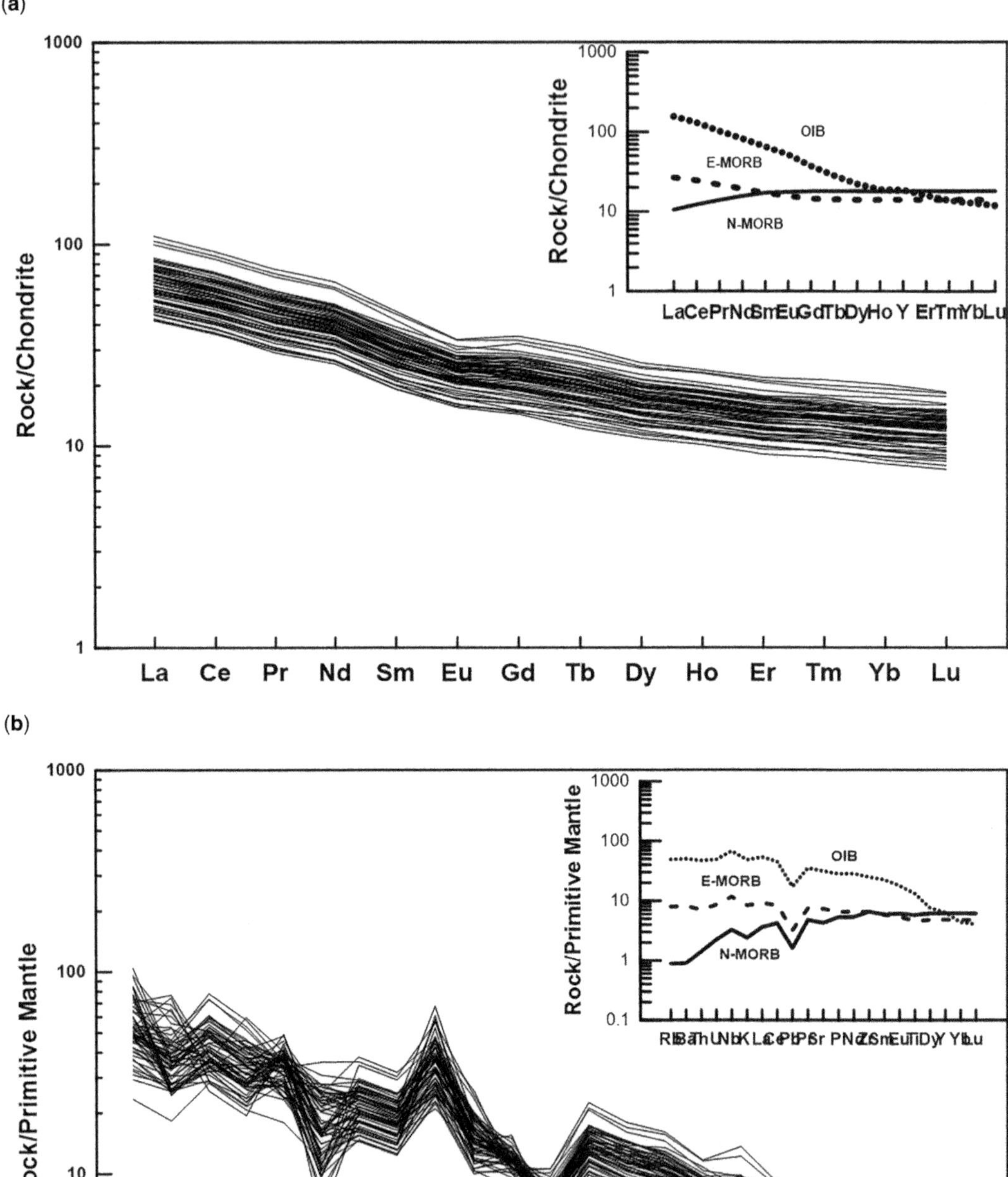

Fig. 7. (**a**) Chondrite-normalized REE diagram for GMI samples. (**b**) Primitive-mantle-normalized Spidergram diagram. Inset shows the REE and multi-element diagram of ocean island basalt (OIB), enriched mid-ocean ridge basalt (E-MORB) and normal mid-ocean ridge basalt (N-MORB: normalizing value after Sun & McDonough 1989).

Table 3. *Sm Nd and Sr isotopic data for whole-rock and minerals (plagioclase and pyroxene) of some representative samples for the GMI rocks of the Gwalior Basin*

Sample ID	Nd (ppm)	Sm (ppm)	$^{147}Sm/^{144}Nd$	$^{143}Nd/^{144}Nd$	Error	ε_{Nd} ($t=0$)	T_{DM} (Ga)	ε_{Nd} ($t = 2$ Ga)	f (Sm/Nd)	$^{87}Sr/^{86}Sr$*	Error
JG-1WR	6.23	1.07	0.1	0.51	7	−30.56	2.160	−2.14	−0.47	0.710551	8
JG-2WR	5.85	1.17	0.12	0.51	9	−31.82	2.648	−7.70	−0.39	0.712152	11
JG-4WR	6.19	1.25	0.12	0.51	4	−27.30	2.297	−3.41	−0.38	0.711761	5
JG-6WR	6	1.12	0.11	0.51	3	−23.01	1.770	3.15	−0.42	0.711439	12
JG-10WR	7.63	1.5	0.12	0.51	6	−24.69	2.018	−0.09	−0.39	0.714245	6
JG-11WR	6.04	1.14	0.11	0.51	3	−22.78	1.776	3.02	−0.42	0.711809	6
JG-13WR	5.67	1.14	0.12	0.51	8	−30.28	2.548	−6.41	−0.38	0.716437	7
JG-15WR	5.47	1.06	0.12	0.51	4	−23.29	1.862	1.87	−0.40	0.713206	8
JG-22WR	6.86	1.35	0.12	0.51	4	−23.50	1.917	1.16	−0.40	0.729187	52
JG-22AWR	5.89	1.27	0.13	0.51	3	−25.86	2.393	−4.13	−0.34	0.725101	13
JG-33WR	7.89	1.57	0.12	0.51	4	−26.72	2.219	−2.51	−0.39	0.721965	6
JG-35WR	6.82	1.3	0.12	0.51	6	−22.86	1.798	2.71	−0.41	0.719499	4
J-3WR	5.23	1.05	0.12	0.51	4	−24.01	2.020	−0.10	−0.38	0.716439	8
J-11WR	9.23	1.92	0.13	0.51	2	−24.34	2.140	−1.46	−0.36	0.714021	9
J-12WR	6.19	1.18	0.12	0.51	3	−23.58	1.853	2.01	−0.41	0.714723	7
J-16WR	7	1.26	0.11	0.51	7	−23.25	1.721	3.92	−0.44	0.713160	9
J-23WR	7.04	1.36	0.12	0.51	4	−22.61	1.797	2.71	−0.41	0.715340	11
J-32WR	6.9	1.28	0.11	0.51	3	−23.87	1.816	2.57	−0.43	0.715262	7
J-33WR	7.23	1.52	0.13	0.51	2	−20.99	1.881	1.47	−0.35	0.712653	9
JG-1Plag	1.66	0.35	0.13	0.51	13	−20.9	2.75			0.71	
JG-2Plag	2.55	0.54	0.13	0.51	11	−18.4	2.56			0.71	
JG-4Plag	2.38	0.49	0.12	0.51	14	−19.5	2.56			0.71	
JG-13Plag	1.04	0.21	0.12	0.51	7	−20	2.49			0.72	
JG-1Pyx	15.41	4.32	0.17	0.51	7	−7.5	3.02			0.71	
JG-2Pyx	14.94	4.17	0.17	0.51	5	−6.3	2.76			bdl	
JG-4Pyx	15.95	4.4	0.17	0.51	6	−7.2	2.8			0.71	
JG-13Pyx	16.02	4.4	0.17	0.51	4	−7.8	2.85			0.72	

*$^{87}Sr/^{86}Sr$ are measured values and $^{88}Sr/^{86}Sr = 0.1194$ is used for correction. bdl-below detection limits.

line and different extent of melting are after Hanson (1980), and the abundances of REE are from Sun & McDonough (1989). Thus, from various bivariate plots, it is evident that the magmas for the GMI gabbros were generated by variable degrees of partial melting of garnet-bearing mantle followed by fractionation of a gabbroic mineral assemblage.

Source and tectonic setting

The enriched nature of the GMI rocks is evident from the plot of Zr–Y (Fig. 5), Nb–Zr (Fig. 5) and Ce–Nd (Fig. 11a). The enrichment of the LILE and LREE relative to primitive mantle (PM) and mid-ocean ridge basalt (MORB) in spider diagrams (Fig. 7b) is a characteristic feature of continental basalts. Such enrichment has been ascribed to either crustal contamination (Dupuy & Dostal 1984; Arndt & Jenner 1986) or derivation from an enriched source (Weaver & Tarney 1983). Even a very low extent of partial melting of the primitive or depleted mantle source can generate enriched melt (Ahmad & Tarney 1993). So, in order to test the role of the crustal contamination or mantle metasomatism of the source of the GMI rocks, the plot of Y/Nb v. Zr/Y (Fig. 11b; Ahmad & Tarney 1993) was plotted. Since Nb and Zr are more incompatible than Y, mantle metasomatism processes will increase the Zr/Y ratio and decrease the Y/Nb ratio, and mantle-derived samples that were subsequently contaminated by crust will fall close to the crust–mantle mixing line (Ahmad & Tarney 1993). The majority of the GMI samples plot close to the enrichment trend expected for mantle sources (Fig. 11b); thereby, indicating that the GMI samples were probably derived from the partial melting of mantle sources that were enriched relative to PM, although a few samples with a low Nb content plot close to the mixing line. This interpretation against significant crustal contamination is again confirmed by the Ce–Nd plot (Fig. 11a). In Figure 11a, GMI samples follow a common trend, intersecting the origin but above the chondrite ratio line. If the line or the trend of the samples intersects the Nd axis, it indicates that the melt suffered crustal contamination, whilst the uncontaminated or insignificantly crustally contaminated rocks give a trend that intersects the origin (Ahmad & Tarney 1991). Another probable

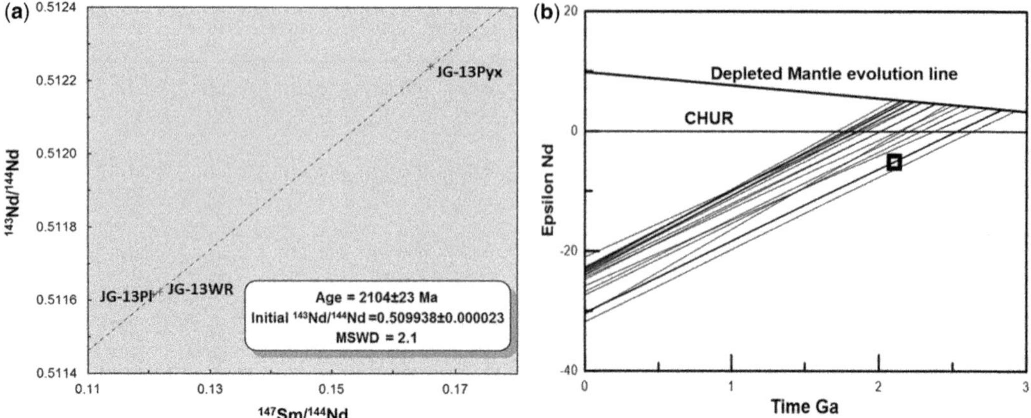

Fig. 8. (**a**) Sm–Nd mineral and whole-rock isochron diagram for the gabbroic sills (JG-13) of the Gwalior Basin. (**b**) ε_{Nd} v. time (Ga) and Nd evolution (ε_{Nd}) curves for samples for GMI samples; evolution curves for depleted mantle and CHUR are also shown. $\varepsilon_{Nd}(t)$ calculated at 2.1 Ga for the sample(JG-13) shows a negative value.

explanation for the enrichment of LREEs relative to chondrite is that the magmas were contaminated before or during fractional crystallization (Ahmad & Tarney 1991). To further envisage the composition of the mantle source(s) and the degree of melting of the mantle source(s), the Sm/Yb ratios v. Sm plot (Fig. 12) is used (Zhao & Zhou 2007). The GMI samples have higher Sm/Yb ratios than the spinel lherzolite melting curve, but fall on the spinel + garnet lherzolite melting curve. This indicates that the GMI were derived from the melt produced by approximately 7–20% (<30%) partial melting of a spinel + garnet lherzolite mantle source (Fig. 12).

The Archaean Bundelkhand Craton witnessed development/formation of Palaeoproterozoic sedimentary basins along its NW and SW fringes. The Gwalior and Bijawar basins were developed along the rifted and fracture zones due to lithospheric stretching along its NW and SW periphery (Ramakrishnan & Vaidyanadhan 2008). This was followed

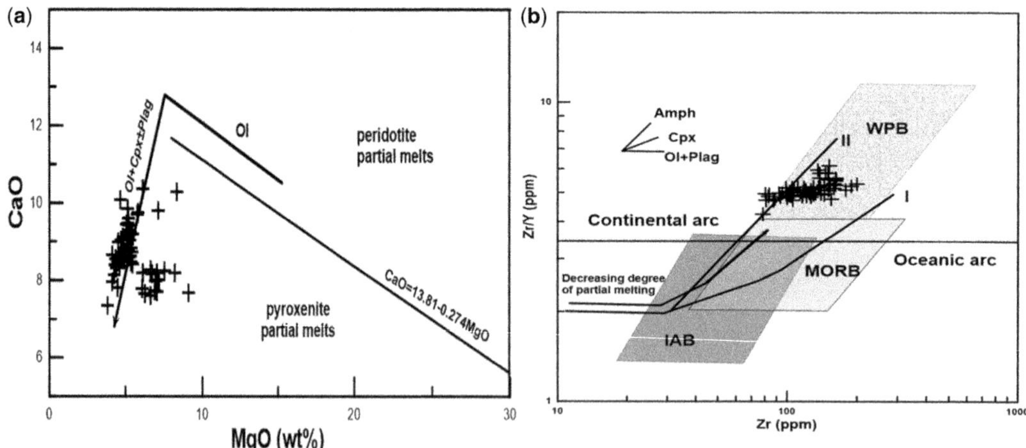

Fig. 9. Variation diagram. (**a**) CaO v. MgO plot for GMI rocks where they follow the fractionation trend of gabbro (Ol + Cpx ± Plag) assemblages. The line separates the field of magmas produced by partial melting of the pyroxenite and peridotite source; the liquid line of descent for magmas (primary) that crystallize gabbro is also shown (after Herzberg & Asimow 2008). (**b**) Binary plot of Zr v. Zr/Y for GMI rocks. In the figure, melting curves I and II are after Drury (1983), and fractional crystallization vectors trends are after Floyd (1993). The tectonic setting field boundary of island arc basalt (IAB), mid-oceanic ridge basalt (MORB) and within-plate basalt (WPB) (Pearce & Norry 1979) are also given where GMI samples fall in the field of WPB.

by emplacement of mafic magmatic rocks in both the basins. The geological and geochemical signatures of the mafic magmatic activity in both the basins depict similar tectonic environments in a rift setting (Pandey *et al.* 2012). The presence of rift-related Palaeoproterozoic–Mesoproterozoic intracratonic basins (Gwalior and Bijawar) in the Archaean Bundelkhand Craton are supportive evidence for the evolution of these basins in an extensional tectonic setting. Lithospheric stretching is evident by the presence of regional faults and the extension of the basin. This is also supported by the absence of compressional features, such as folds and thrusts. Other supportive evidence for the evolution of the basins in an extensional tectonic setting are the presence of authigenic minerals, such as feldspar in shales (Absar *et al.* 2009), no deformation observed along the basin margins and insignificant traces of metamorphism in the sediments (Siever 1983; Robinson 1987). The GMI rocks were emplaced following the bedding planes and fabric of the Gwalior sediments, and the terrain was uplifted due to magma emplacement. It is widely accepted that the Bundelkhand Craton stabilized at 2.5 Ga (Stein *et al.* 2004). A major thermal event affecting the Earth's mantle during early to mid-Proterozoic times, leading to extension in the northern Indian Shield has also been reported (Mallikharjuna Rao 2004). Model ages (2.6–1.7 Ga) indicate that their mantle sources had a protracted evolution and melting, which may have been triggered by a mantle plume in an extensional intra-continental rift tectonic setting. The presence of the associated continental-derived sediments (shales and sandstones) and geochemical evidence do support the emplacement of these mafic intrusives into a rift setting. In terms of a plate tectonic setting, using the discrimination diagram of Pearce & Norry (1979), most of the GMI samples plot exclusively in the 'within-plate' field (Fig. 9b).

Discussion

From the above petrogenetic interpretation, an enriched source for the GMI rocks most probably due to mantle metasomatism is evident, with crustal contamination being insignificant. Isotopically, they show near-chondritic sources, ranging from a depleted to an enriched source based on εNd^t (t = 2000 Ma) values; however, geochemical characteristics suggest an enriched character. Ahmad *et al.* (2008) pointed out the possibilities of an enriched mantle reservoir during early–middle Proterozoic times in the northern Indian Shield. They also suggested the presence of subcontinental lithospheric mantle (SCLM) in the northern Indian lithosphere, which comprised a relatively fertile mantle during

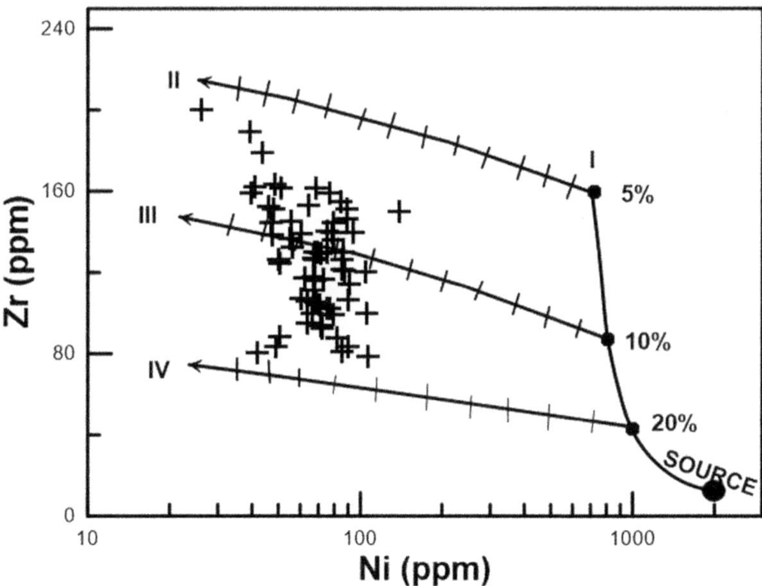

Fig. 10. Trace element modelling diagram of Ni v. Zr (Condie *et al.* 1987). Batch melting curve I, at 1500°C (1 atm equivalent), with degrees of melting (in percentage) and the melting relationship of the mantle lherzolite source is assumed to be 11 ppm Zr and 2000 ppm Ni (Rajamani *et al.* 1985), and curves II, III and IV are olivine fractionation curves (in percentage) with olivine removal noted in 5% increments.

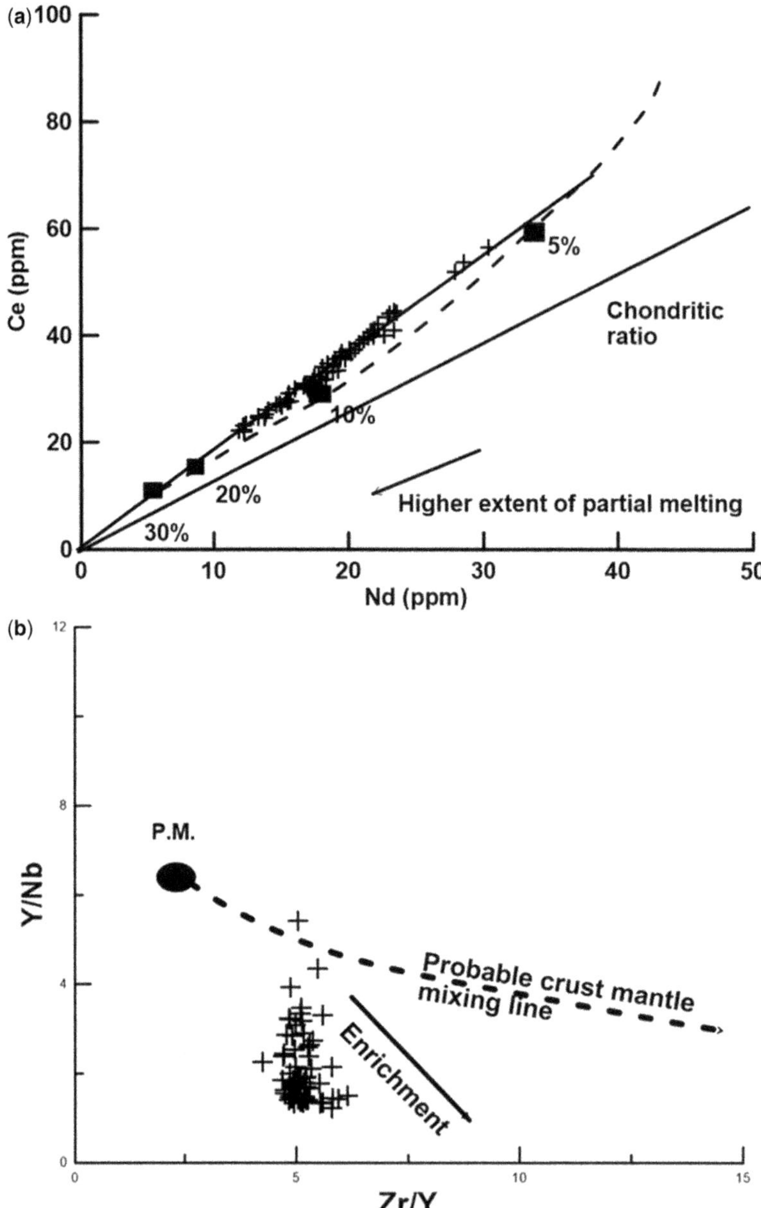

Fig. 11. (**a**) Partial melting diagram of Nd v. Ce (after Ahmad & Tarney 1991). The chondrite ratio line and calculated different extents of melting for the garnet lherzolite source (after Hanson 1980) are shown in this diagram; abundances of REE are from Sun & McDonough (1989). (**b**) Bivariate ratio plots of Zr/Y v. Y/Nb of GMI rocks falling near the enrichment trend, indicating their derivation from enriched sources rather than contamination of crustal components of their parental melt.

the Palaeoproterozoic–Mesoproterozoic (Ahmad & Tarney 1991; Ahmad *et al.* 2008). SCLM is thought to be variably enriched in incompatible elements, possibly due to the addition of melt or fluids or both (Weaver & Tarney 1983; Horan *et al.* 1987). The addition of fluids enriched the source in incompatible elements, whereas melt addition caused enrichment in both the major and incompatible trace elements (Ahmad & Rajamani 1991). It is possible that the source of the GMI samples is similar to

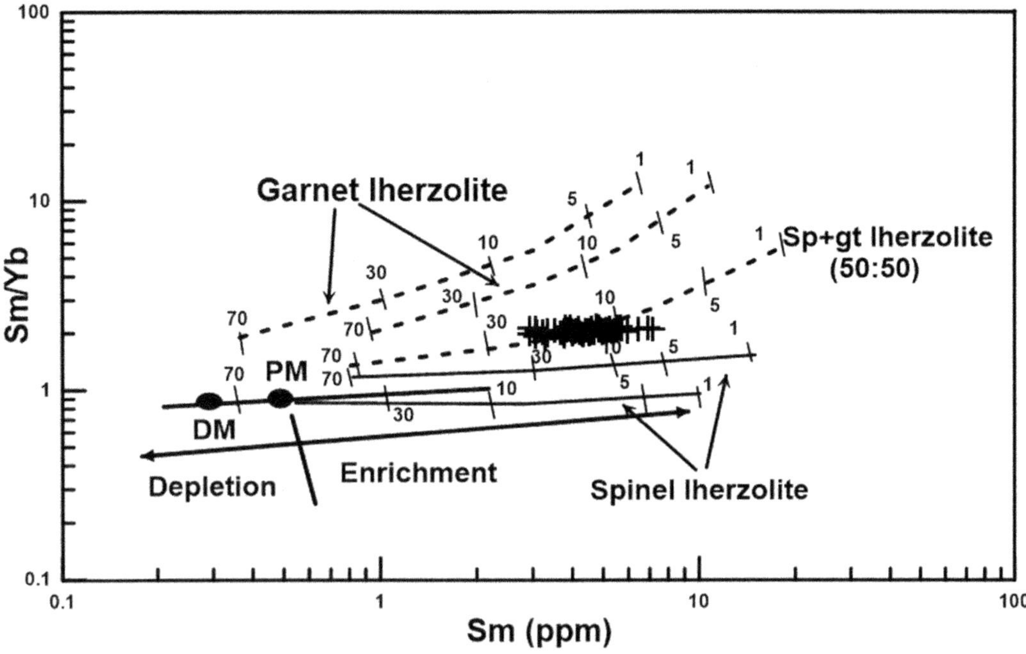

Fig. 12. Plots of Sm/Yb v. Sm (Zhao & Zhou 2007) for the GMI samples. Mantle curves for spinel lherzolite, Sp + gt lherzolite and garnet lhrzolite with both PM (primitive mantle: Sun & McDonough 1989) and DM (depleted mantle: after McKenzie & O'Nions 1991) compositions are after Aldammaz *et al.* (2000).

the SCLM source present in the northern Indian lithosphere during that time.

Figure 13 compares samples thought to be derived from the SCLM source in the northern Indian lithosphere (Ahmad & Tarney 1991) with the average value of the GMI samples. The unusually striking similarities between them, following all the crests and troughs in the pattern (Fig. 13a, b), indicates a similar SCLM source in north India as a source for the GMI during the Palaeoproterozoic. Mondal *et al.* (2008) suggested the lithospheric extension as the common driving force for such similar mafic magmatic rocks in the northern part of India. Lithospheric extension and rifting in response to plume activities and broadly similar widespread mafic magmatism during the Palaeoproterozoic–Mesoproterozoic is reported in many parts of northern India, such as from Aravalli, Garhwal, Mandi-Darla and the Delhi Supergroup in Ranakpur-Desuri (Ahmad & Tarney 1994; Ahmad *et al.* 1999, 2008; Mondal & Ahmad 2001 and references therein). Basal Aravalli rocks located on the western side of the study area also indicate active volcanism within the period from *c.* 2.3 to 1.8 Ga (Ahmad *et al.* 2008). Similar to this, the Lesser Himalayan region also experienced extensive magmatism during 2.0–1.8 Ga (Miller *et al.* 2000; Ahmad *et al.* 2008). The plausible aspect of the large-scale generation of melt (Large Igneous Province) and emplacement covering tens of thousands of square kilometres in the northern part of India in the Bundelkhand province is discussed (Sensarma *et al.* 2010). Iron enrichment (FeO: 9–16 wt %), tholeiites and higher abundances of incompatible trace elements are some of the common geochemical signatures observed. It has been common to argue that a major thermal or energy source is required to provide the energy to melt the mantle on a sufficently large scale to generate such widespread mafic magmatism (dyke swarms and LIPs). Mantle plumes are thought to be the energy source required for this melting of the mantle, thinning of the lithosphere and, eventually, rifting, thus producing magmas (White & Mckenzie 1989). Therefore, all such mantle-plume-related activities, such as the formation of basins, magmatism and rifting, are common geological processes that have occurred since at least Early Proterozoic times (Ahmad & Rajamani 1991). From the above discussion, the possibilities of long-lived enriched mantle sources, associated with the plume activity, and the existence of widespread mafic magmatism in north India during the Proterozoic are evident. It is also possible that all this magmatism may have been generated at about the same time, triggering the crustal extension by the upwelling of a mantle plume which caused large-scale extension and possibly giving rise to the LIPs in this part of the northern Indian Shield.

Such evidence of mafic magmatic activity and LIPs (*c.* 1.9 Ga), which are thought to have been associated with crustal extension due to plume activity, is quite conspicuous throughout the central and southern part of Peninsular India (French *et al.* 2008; Meert *et al.* 2011; Mishra 2011; Srivastava & Gautam 2012, 2015; Khanna *et al.* 2013). Some of the already reported LIPs from the southern part of Indian Shield are the 2370 Ma Bangalore LIP (Dharwar Craton), the Mahbubnagar LIP (Dharwar Craton), where gabbro and dolerite dyke are dated at 2173 ± 64 Ma (Sm–Nd), the Shimoga and NE Dharwar dolerite dykes (2180 ± 100 Ma, U–Pb baddeleyite and zircon), and the the southern Bastar–Cuddapah LIP, such as the Pulivendla sill (1885.4 ± 3.1 Ma, U–Pb baddeleyite: Dharwar Craton) from the Cuddapah Basin and the BD-2 swarms (1891–1883 Ma, U–Pb baddeleyite) from the Bastar Craton (Pandey *et al.* 1997; Halls *et al.* 2007; Ernst & Srivastava 2008; French *et al.* 2008). Sensarma (2007) reported a bimodal LIP of 2500 Ma from the central Indian Shield and this is represented by the Dongargarh Group of rocks. The widespread presence of synchronous intrusives in the Indian Peninsula confirms the existence of a plume or superplume at that time (French *et al.* 2008). The contiguous nature of the Archaean cratons of central (Bastar), northern (Bundelkhand) and southern (Dharwar) India are also reported, based on the reconstructed apparent polar paths of the widespread mafic magmatic rocks in these cratons (Meert 2002, 2003; Pradhan *et al.* 2008, 2010, 2012; Radhakrishna *et al.* 2013*a*, *b*). Such records of magmatic pulses (LIPs and dyke swarms) represent important tools for assembling the pieces of supercratons and eventually leading to the identification and reconstruction of supercontinents (Bleeker 2003, 2004; Ernst *et al.* 2005).

Geodynamic implications and conclusions

In several Precambrain terranes worldwide, there is an increasing indication that the time period between 2.1 and 1.8 Ga is dominated by many major events, such as global accretion, collisional orogeny and global rifting (Zhao *et al.* 2002, 2004). The time period 2.2–2.0 Ga is marked as a global rifting period and 2.0–1.8 Ga is marked as a global crustal amalgamation period (Zhao *et al.* 2002). The record of occurrence of such igneous events across the globe is an important parameter to mark the assembly and break-up of the various supercontinents (Bleeker 2003; Ernst *et al.* 2013; Nance *et al.* 2014 and references therein). Some of the known supercontinents that existed before Pangea (*c.* 260 Ma) are Gondwana (0.6 Ga), Rodinia (*c.* 1.0–0.7 Ga), Columbia (*c.* 1.8–1.3 Ga), Kenorland (Late Archaean to *c.* 2.0 Ga) and probably the earliest supercontinent Ur (*c.* 3 Ga) (Rogers 1996; Rogers & Santosh 2002, 2003, 2009; Ernst *et al.* 2013; Meert 2014; Nance *et al.* 2014; Srivastava & Gautam 2015). The rhythm of intraplate mantle melting through time recorded from the magmatic pulses of Earth's history helps in identifying LIPs (Ernst & Buchan 1997; Bleeker 2004). The record of intracratonic basin formation due to lithospheric stretching, emplacement of gabbro sills and dykes, plume-rift-induced magmatism, global magmatism and rifting events between 2.5 and 2.1 Ga associated with the break-up of the Kenorland supercontinent is well documented (Lauri *et al.* 2012; Strand & Köykkä 2012). There is also a report of similar LIPs preserved in other Archaean cratons of the world, like the Kalahari (Kaapvaal/Zimbabwe) Craton in southern Africa, and the Superior and Slave cratons in North America (French *et al.* 2008). Meert *et al.* (2011) also suggested the contemporaneous nature of mafic magmatic activity in Indian cratons with several worldwide cratons (North China, Baltica, Laurentia, Australia, Siberia, and the Kaapvaal and Zimbabwe, southern Africa) during that time (*c.* 1.9 Ga). Considering that these widespread mafic activities and LIPs are of broadly similar age, preserved in Indian cratons and several worldwide cratons, the possibilities of the contemporaneous nature of these rocks and the Precambrian Kenorland supercontinent need to be understood. The initial break-up stages of the Kenorland supercontinent took place at *c.* 2.5–2.1 Ga (Lauri *et al.* 2012; Strand & Köykkä 2012). Based on the crystallization age (2.1 Ga), it is possible that the mafic magmatism in the Gwalior Basin was formed in a rift environment that was contemporary with the break-up of the Kenorland supercontinent, as the GMI rocks are synchronous with this supercontinent. The observed Nd model or extraction ages (2.6–1.7 Ga) and the crystallization age of 2.1 Ga of the GMI samples also correspond with the global distribution of the 2.1–1.8 Ga orogenic activity. Thus, on the basis of the available geological, geochemical and geochronological data, it is possible that the mafic magmatism in this part of central Indian Shield (Gwalior Basin) may be related to the Kenorland supercontinent or contemporaneous with the global 2.1–1.8 Ga major orogeny activity (Zhao *et al.* 2004). However, more precise age data are required for the studied intrusive (GMI) rocks to throw more light on the construction of supercontinents.

The Gwalior mafic intrusive rocks of the Proterozoic sedimentary Gwalior Basin show geochemical features of Proterozoic continental tholeiites. Geochemically, the GMI samples are sub-alkaline tholeiitic in nature and gabbroic in composition. The protolith of the GMI samples was derived from

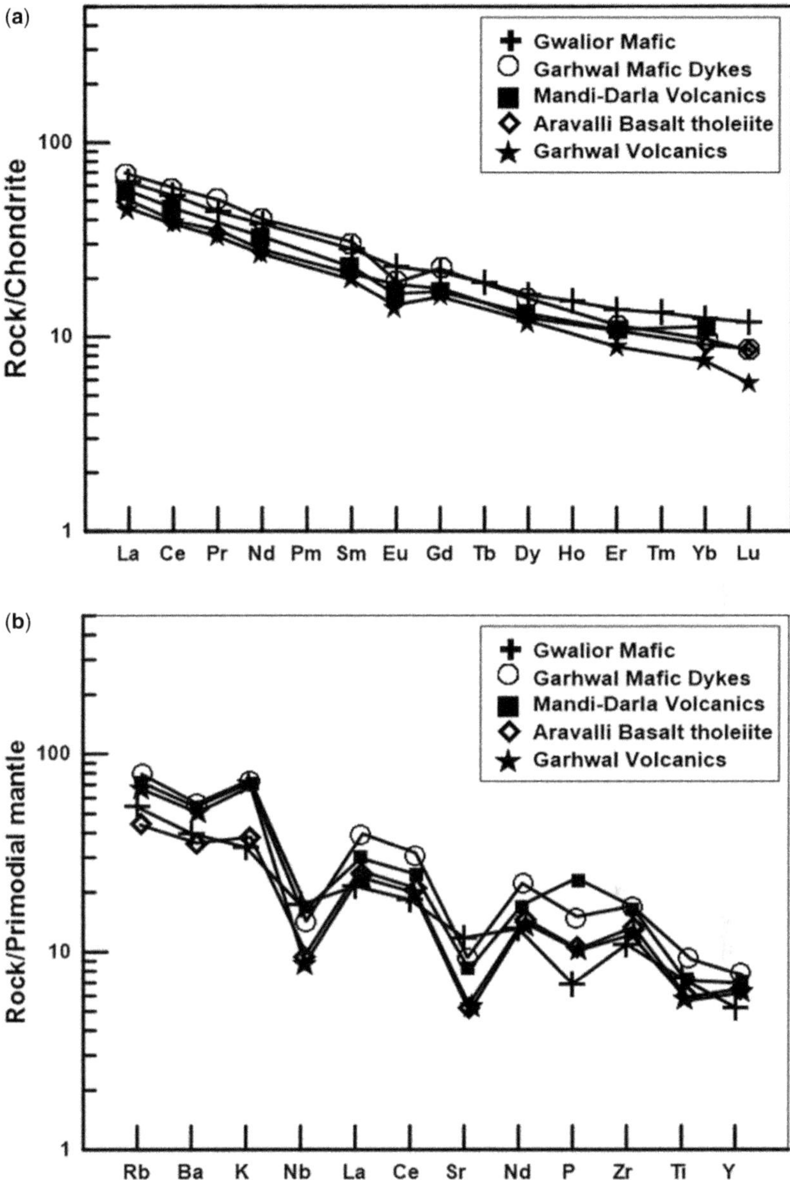

Fig. 13. Comparative diagrams. (**a**) Chondrite-normalized REE pattern and (**b**) primitive-mantle-normalized multi-element pattern of average GMI rocks with other mafics of broadly similar ages from the northern Indian Craton (Garhwal mafic dykes, Mandi-Darla volcanics, Aravalli basalt tholeiite, Garhwal volcanics: data from Ahmad & Tarney 1991 and references therein). Normalizing values are after Sun & McDonough (1989).

variably enriched mantle sources with no significant crustal contamination. The lithospheric mantle source was probably enriched by metasomatism. The GMI samples were formed by variable degrees of partial melting (5–18%; <20%) of mantle sources, followed by fractionation of clinopyroxene–olivine–plagioclase (gabbroic assemblages). Geological and geochemical evidence are consistent with the continental rift setting environment for these suites of rocks. The widespread mafic activity of broadly similar ages associated with large-scale crustal extension indicates the presence of an upwelling plume beneath the northern Indian Plate at this time. The age of the GMI (2.1 Ga) corresponds to global magmatic activity (2.1–1.8 Ga), mantle plume activity related to crustal extension and widespread magmatism associated with the break-up of the Kenorland supercontinent.

We are particularly grateful to staff at IIT Roorkee (especially Ms Poornima Saini) for assistance with the isotopic analysis. Dr Satyanarayan and Dr Subramanian, scientists in NGRI, are especially thanked for helping with the ICP-MS analysis and giving valuable suggestions. We would also like to thank the Head of the Department of Geology for providing all the facilities. The financial support provided by CAS (RFSMS), UGC is also greatly acknowledged. We also gratefully acknowledge Professor Sarajit Sensarma for inviting us to submit this paper for this special publication, and anonymous reviewers for their constructive reviews, comments and helpful suggestions, which have helped considerably in improving the manuscript.

References

ABSAR, N., RAZA, M., ROY, M., NAQVI, S.M. & ROY, A.K. 2009. Composition and weathering conditions of Palaeoproterozoic upper crust of Bundelkhand Craton, Central India: records from geochemistry of clastic sediments of 1.9 Ga Gwalior Group. *Precambrian Research*, **168**, 313–329.

ACHARYYA, S.K. 2003. The nature of Mesoproterozoic Central Indian tectonic zone with exhumed and reworked older granulites. *Gondwana Research*, **6**, 197–214.

AHMAD, T. & RAJAMANI, V. 1991. Geochemistry and petrogenesis of the basal Aravalli volcanics near Nathdwara, Rajasthan, India. *Precambrian Research*, **49**, 185–204.

AHMAD, T. & TARNEY, J. 1991. Geochemistry and petrogenesis of Garhwal Volcanics: implications for evolution of the N. Indian lithosphere. *Precambrian Research*, **50**, 69–89.

AHMAD, T. & TARNEY, J. 1993. North Indian Proterozoic volcanics, products of lithosphere extension: geochemical studies bearing on the lithospheric derivation rather than crustal contamination. *In*: CASSHYAP, S.M. (ed.) *Rifted Basins and Aulacogens, Geological and Geophysical Approach*. Gyanodaya Prakashan, Nainital, India, 130–147.

AHMAD, T. & TARNEY, J. 1994. Geochemistry and petrogenesis of late Archaean Aravalli volcanics, basement enclaves and granitoids, Rajasthan. *Precambrian Research*, **65**, 1–23.

AHMAD, T., MUKHERJEE, P.K. & TRIVEDI, J.R. 1999. Geochemistry of Precambrian mafic magmatic rocks of the western Himalaya, India: petrogenetic and tectonic implications. *Chemical Geology*, **160**, 103–119.

AHMAD, T., DRAGUSANU, C. & TANAKA, T. 2008. Provenance of Proterozoic Basal Aravalli mafic volcanic rocks from Rajasthan, Northwestern India: Nd isotopes evidence for enriched mantle reservoirs. *Precambrian Research*, **162**, 150–159.

ALAM, M.A., CHOUDHARY, A.K., MOURI, H. & AHMAD, T. 2017. Geochemical characterization and petrogenesis of mafic granulites from the Central Indian Tectonic Zone (CITZ). *In*: HALLA, J., WHITEHOUSE, M.J., AHMAD, T. & BAGAI, Z. (eds) *Crust–Mantle Interactions and Granitoid Diversification: Insights from Archaean Cratons*. Geological Society, London, Special Publications, **449**, 207–229, https://doi.org/10.1144/SP449.1

ALDAMMAZ, E., PEARCE, J.A., THIRWALL, M.F. & MITCHELL, J.G. 2000. Petrogenic evolution of late Cenezoic, post coolisional volcanism in western Antolia, Turkey. *Journal of Volcanology and Geothermal Research*, **102**, 67–95.

ARNDT, N.T. & JENNER, G.R. 1986. Crustally contaminated komatiites and basalts from Kambalda. Western Australia. *Chemical Geology*, **56**, 229–255.

ARNDT, N.T., TEIXEIRA, N.A. & WHITE, W.M. 1989. Bizarre geochemistry of komatiites from the Crixas greenstone belt, Brazil. *Contributions to Mineralogy and Petrology*, **101**, 187–197.

BAJPAI, M.P. 1935. The Gwalior traps from Gwalior, India. *Journal of Geology*, **43**, 61–75.

BALARAM, V. & GNANESHWAR RAO, T. 2003. Rapid determination of REEs and other trace elements in geological samples by microwave acid digestion and ICP-MS. *Atomic Spectroscopy*, **24**, 206–212.

BALARAM, V., MATHUR, R., BANAKAR, V.K., HEIN, J.R., RAO, C.R.M., GNANESWARA RAO, T. & DASARAM, B. 2006. Determination of the platinum-group elements and gold in manganese nodule reference samples by nickel sulphide fire-assay and Te-coprecipitation with ICP-MS. *Indian Journal of Marine Science*, **35**, 7–16.

BICKLE, M.J. 1986. Implications of melting for stabilisation of the lithosphere and heat loss in the Archaean. *Earth and Planetary Science Letters*, **80**, 314–324.

BLEEKER, W. 2003. The late Archean record: a puzzle in ca. 35 pieces. *Lithos*, **71**, 99–134.

BLEEKER, W. 2004. Taking the pulse of planet Earth: a proposal for a new multidisciplinary flagship project in Canadian solid Earth sciences. *Geoscience Canada*, **31**, 179–190.

BUCHAN, K.L., MERTANEN, S., PARK, R.G., PESONEN, L.J., ELMING, S.A., ABRAHAMSEN, N. & BYLUND, G. 2000. Comparing the drift of Laurentia and Baltica in the Proterozoic: the importance of key palaeomagnetic poles. *Tectonophysics*, **319**, 167–198.

CHAKRABORTY, P.P. 2006. Outcrop signatures of relative sea level fall on a siliciclastic shelf: examples from the Rewa Group of Proterozoic Vindhyan basin. *Journal of Earth System Sciences*, **115**, 23–36.

CHAKRABORTY, P.P. & PAUL, P. 2014. Depositional character of a dry-climate alluvial fan system from Palaeoproterozoic rift setting using facies architecture and palaeo-hydraulics: example from the Par Formation, Gwalior Group, central India. *Journal of Asian Earth Sciences*, **91**, 298–315.

CHAKRABORTY, P.P., DEY, S. & MOHANTY, S.P. 2010. Proterozoic platform sequences of Peninsula India: implication towards basin evolution and supercontinent assembly. *Journal of Asian Earth Sciences*, **39**, 589–607.

CONDIE, K.C., BOBROW, D.J. & CARD, K.D. 1987. Geochemistry of Precambrian mafic dykes from the Southern Superior Province. *In*: HALLS, H.C. & FAHRIG, W.F. (eds) *Mafic Dyke Swarms*. Geological Association of Canada, Special Papers, **34**, 95–108.

COX, K.G., BELL, J.D. & PANKHURST, R.J. 1979. *The Interpretation of Igneous Rocks*. Allen and Unwin, London.

Crawford, A.R. & Compston, W. 1969. The age of the Vindhyan System of Peninsular India. *Quarterly Journal of the Geological Society, London*, **125**, 351–371, https://doi.org/10.1144/gsjgs.125.1.0351

Davies, J., Grant, R.W. & Whitehead, R.E.S. 1979. Immobile trace elements and Archean volcanic stratigraphy in the Timmins mining area, Ontario. *Canadian Journal of Earth Sciences*, **16**, 305–331.

Drury, S.A. 1983. The Petrogenesis and tectonic setting of Archaean metavolcanics from Karnataka State, South India. *Geochimica et Cosmochimica Acta*, **47**, 317–329.

Drury, S.A. 1984. A Proterozoic intracratonic basin, dyke swarms and thermal evolution in South India. *Journal of the Geological Society of India*, **25**, 437–444.

Dupuy, C. & Dostal, J. 1984. Trace element geochemistry of some continental tholeiites. *Earth and Planetary Science Letters*, **67**, 61–69.

Erlank, A.J., Duncan, A.R. et al. 1988. A laterally extensive geochemical discontinuity in subcontinental Gondwana lithosphere. *In*: Gomes, C.B. et al. (eds) *Proceedings of the V Conference on Geochemical Evolution of the Continental Crust, Poços de Caldas, Brazil*, International Association of Geochemistry and Cosmochemistry. Paleopublications, Eagle I.A., U.S.A., 1–10.

Ernst, R.E. & Buchan, K.L. 1997. Giant radiating dyke swarms: their use in identifying pre-Mesozoic large igneous provinces and mantle plumes. *In*: Mahoney, J. & Coffin, M. (eds) *Large Igneous Provinces: Continental, Oceanic, and Planetary Volcanism*. American Geophysical Union, Geophysical Monograph Series, **100**, 297–333.

Ernst, R.E. & Buchan, K.L. 2001. The use of mafic dike swarms in identifying and locating mantle plumes. *In*: Ernst, R.E. & Buchan, K.L. (eds) *Mantle Plumes: Their Identification through Time*. Geological Society of America, Special Papers, **352**, 247–265.

Ernst, R.E. & Srivastava, R.K. 2008. *Speculations on the Proterozoic Large Igneous Province (LIP) Record of India*. Large Igneous Provinces Commission, http://www.largeigneousprovinces.org/08jun

Ernst, R.E., Buchan, K.L. & Campbell, I.H. 2005. Frontiers in large igneous province research. *Lithos*, **79**, 271–297.

Ernst, R.E., Bleeker, W., Soderlund, U. & Kerr, A.C. 2013. Large igneous provinces and supercontinents: toward completing the plate tectonic revolution. *Lithos*, **174**, 1–14.

Floyd, P.A. 1993. Geochemical discrimination and petrogenesis of alkalic basalt sequences in part of Ankara melange, central Turkey. *Journal of the Geological Society, London*, **150**, 541–550, https://doi.org/10.1144/gsjgs.150.3.0541

French, J.E., Heaman, L.M., Chacko, T. & Srivastava, R.K. 2008. 1891–1883 Ma southern Bastar–Cuddapah mafic igneous events, India: a newly recognized large igneous province. *Precambrian Research*, **160**, 308–322.

Gill, R. 2010. *Igneous Rocks and Processes: a Practical Guide*. John Wiley & Sons, Chichester, UK.

Halls, H.C., Kumar, A., Srinivasan, R. & Hamilton, M.A. 2007. Paleomagnetism and U–Pb geochronology of eastern trending dykes in the Dharwar craton, India: feldspar clouding, radiating dyke swarms and the position of India at 2.37 Ga. *Precambrian Research*, **155**, 47–68.

Hanson, G.N. 1980. Rare earth elements in petrogenetic studies of igneous systems. *Annual Review of Earth and Planetary Sciences*, **8**, 372–406.

Hashiguchi, H., Yamada, R. & Inoue, T. 1983. Practical application of low Na2O anomalies in footwall acid lava for delimiting promising areas around the Kosaka and Fukazawa Kuroko deposits, Akita Prefectural, Japan. *In*: Ohmoto, H. & Skinner, B.J. (eds) *The Kuroko and Related Volcanogenic Massive Sulfide Deposits*. Economic Geology Monographs, **5**, 387–394.

Hawkesworth, C.J., Dhuime, D., Pietranik, A.B., Cawood, P.A., Kemp, A.I.S. & Storey, C.D. 2010. The generation and evolution of the continental crust. *Journal of the Geological Society, London*, **167**, 229–248, https://doi.org/10.1144/0016-76492009-072

Hergt, J.M., Peate, D.W. & Hawkesworth, C.J. 1991. The petrogenesis of Mesozoic Gondwana low-Ti flood basalts. *Earth and Planetary Science Letters*, **105**, 134–148.

Herzberg, C. & Asimow, P.D. 2008. Petrology of some oceanic island basalts: PRIMELT2.XLS software for primary magma calculation. *Geochemistry, Geophysics, Geosystem*, **9**, Q09001, https://doi.org/10.1029/2008GC002057

Horan, M.F., Hanson, G.N. & Spencer, K.J. 1987. Pb and Nd isotope and trace elements constraints on the origin of basic rocks in an Early Proterozoic Igneous Complex, Minnesota. *Precambrian Research*, **37**, 323–342.

Jensen, L.S. 1976. *A New Cation Plot for Classifying Subalkaline Volcanic Rocks*. Ontario Division of Mines, Miscellaneous Papers, **66**.

Jochum, K.P., Arndt, N.T. & Hoffman, A.W. 1991. Nb–Th–La in komatiites and basalts: constraints on komatiites petrogenesis and mantle evolution. *Earth and Planetary Science Letters*, **107**, 272–289.

Joshi, K.B., Bhattacharjee, J. et al. 2017. The diversity of granitoids and plate tectonic implications at the Archaean–Proterozoic boundary in Bundelkhand Craton, Central India. *In*: Halla, J., Whitehouse, M.J., Ahmad, T. & Bagai, Z. (eds) *Crust–Mantle Interactions and Granitoid Diversification: Insights from Archaean Cratons*. Geological Society, London, Special Publications, **449**, 123–157, https://doi.org/10.1144/SP449.8

Kale, V.S. & Phansalkar, V.G. 1991. Purana basins of Peninsular India. A review. *Basin Research*, **3**, 1–36.

Khanna, T.C., Sai, V.V.S., Zhao, G.C., Rao, D.V.S., Krishna, A.K., Sawant, S.S. & Charan, S.N. 2013. Petrogenesis of mafic alkaline dikes from the *c.* 2.18 Ga Mahbubnagar Large Igneous Province, Eastern Dharwar Craton, India: geochemical evidence for uncontaminated intracontinental mantle derived magmatism. *Lithos*, **179**, 84–98.

Lauri, L.S., Mikkola, P. & Karinen, T. 2012. Early Paleoproterozoic felsic and mafic magmatism in the Karelian province of the Fennoscandian shield. *Lithos*, **151**, 74–82.

Longjam, K.C. & Ahmad, T. 2012. Geochemical characterization and Petrogenesis of Proterozoic Khairagarh Volcanics: implication for Precambrian crustal evolution. *Geological Journal*, **47**, 130–143.

McKenzie, D. & O'Nions, R.K. 1991. Partial melting distributions from inversion of rare earth element concentrations. *Journal of Petrology*, **32**, 1021–1091.

Mallikharjuna Rao, J. 2004. The wide spread 2 Ga dyke activity in the Indian shield – evidences from Bundelkhand Mafic Dyke Swarm, Central India and their tectonic implications. *Gondwana Research (Gondwana Newsletter Section)*, **7**, 1219–1228.

Malone, S.J., Meert, J.G. *et al.* 2008. Paleomagnatism and detrital zircon geochronology of the upper Vindhyan sequence, Son Valley and Rajasthan, India: a ca.1000 Ma closure age for the Purana basins? *Precambrian Research*, **164**, 137–159.

Manikyamba, C., Kerrich, R., Khanna, T.C., Krishna, A. K. & Satyanarayanan, M. 2008. Geochemical systematics of komatiite–tholeiite and adakite–arc basalt associations: the role of a mantle plume and convergent margin in formation of the Sandur Superterrane, Dharwar Craton. *Lithos*, **106**, 155–172.

Meert, J.G. 2002. Paleomagnetic evidence for a Paleo-Mesoproterozoic supercontinent Columbia. *Gondwana Research*, **5**, 207–216.

Meert, J.G. 2003. A synopsis of events related to the assembly of eastern Gondwana. *Tectonophysics*, **362**, 1–40.

Meert, J.G. 2014. Strange attractors, spiritual interlopers and lonely wanderers: the search for Pre-Pangaean supercontinents. *Geoscience Frontiers*, **5**, 155–166.

Meert, J.G., Pandit, M.K. & Pradhan, V.R. 2011. Preliminary report on the paleomagnetismof 1.88 Ga dykes from the Bastar and Dharwar cratons, Peninsular India. *Gondwana Research*, **20**, 335–343.

Miller, C., Klotzli, U., Fran, W., Thoni, M. & Grasemann, B. 2000. Proterozoic crustal evolution in the NW Himalaya (India) as recorded by circa 1.80 Ga mafic and 1.84 Ga granitic magmatism. *Precambrian Research*, **103**, 191–206.

Mishra, D.C. 2011. Long hiatus in Proterozoic sedimentation in India: Vindhyan, Cuddapah and Pakhal basins – a plate tectonics model. *Journal of the Geological Society of India*, **77**, 17–25.

Mondal, M.E.A. & Ahmad, T. 2001. Bundelkhand mafic dykes, central Indian shield: implications for the role of sediment subduction in Proterozoic crustal evolution. *The Island Arc*, **10**, 51–67.

Mondal, M.E.A., Goswamy, J.N., Deomurari, M.P. & Sharma, K.K. 2002. Ion microprobe $^{207}Pb/^{206}Pb$ ages of zircons from the Bundelkhand massif, northern India: implications for crustal evolution of the Bundelkhand–Aravalli proto-continent. *Precambrian Research*, **117**, 413–419.

Mondal, M.E.A., Chandra, R. & Ahmad, T. 2008. Precambrian mafic magmatism in Bundelkhand Craton. *Journal of the Geological Society of India*, **72**, 113–122.

Murthy, N.G.K. 1987. Mafic dyke swarms of the Indian shield. *In*: Halls, H.C. & Fahrig, W.F. (eds) *Mafic Dyke Swarms*. Geological Association of Canada, Special Papers, **34**, 393–400.

Murthy, N.G.K. 1995. Proterozoic mafic dykes in southern peninsular India: a review. *In*: Devaraju, T.C. (ed.) *Mafic Dyke Swarms of Peninsular India*. Geological Society of India, Memoirs, **33**, 81–98.

Nance, R.D., Murphy, J.B. & Santosh, M. 2014. The supercontinent cycle: a retrospective essay. *Gondwana Research*, **25**, 4–29.

Naqvi, S.M. & Rogers, J.J.W. 1987. *Precambrian Geology of India*. Oxford University Press, New York.

Nisbet, E.G., Cheadle, M.J., Arndt, N.T. & Bickle, M.J. 1993. Constraining the potential temperature of the Archaean mantle: a review of the evidence from komatiites. *Lithos*, **30**, 291–307.

Pandey, B.K., Gupta, J.N., Sarma, K.J. & Sastry, C.A. 1997. Sm–Nd, Pb–Pb and Rb–Sr geochronology and petrogenesis of the mafic dyke swarm of Mahbubnagar, South India: implications for Paleoproterozoic crustal evolution of Eastern Dharwar Craton. *Precambrian Research*, **84**, 181–196.

Pandey, U.K., Sastry, D.V.L.N., Pandey, B.K., Madhuparna, R., Rawat, T.P.S., Rajeeva, R. & Shrivastava, V.K. 2012. Geochronological (Rb–Sr and Sm–Nd) studies on intrusive gabbros and dolerite dykes from parts of northern and central Indian cratons: implications for the age of onset of sedimentation in Bijawar and Chattisgarh basins and uranium mineralisation. *Journal of the Geological Society of India*, **79**, 30–40.

Patranabis-Deb, S., Bickford, M.E., Hill, B., Chaudhari, A.K. & Basu, A. 2008. SHRIMP ages of zircon in the uppermost tuff in Chhattisgarh Basin in central India require up to 500 Ma adjustments in Indian Proterozoic stratigraphy. *Journal of Geology*, **115**, 407–416.

Pearce, J.A. & Norry, M. 1979. Petrogenetic implications of Ti, Zr, Y and Nb variations in volcanic rocks. *Contributions to Mineralogy and Petrology*, **69**, 33–47.

Pharaoh, T.C., Beckinsale, R.D. & Richard, D. (eds). 1987. *Geochemistry and Mineralization of Proterozoic Volcanic Suites*. Geological Society, London, Special Publications, **33**, http://sp.lyellcollection.org/content/33/1

Polat, A., Hofmann, A.W. & Rosing, M.T. 2002. Boninite-like volcanic rocks in the 3.7–3.8 Ga Isua greenstone belt, West Greenland: geochemical evidence for intra-oceanic subduction zone processes in the early Earth. *Chemical Geology*, **184**, 231–254.

Polat, A., Appel, P.W.U. & Fryer, B.J. 2011. An overview of the geochemistry of Eoarchean to Mesoarchean ultramafic to mafic volcanic rocks, SW Greenland: implications for mantle depletion and petrogenetic processes at subduction zones in the early Earth. *Gondwana Research*, **20**, 255–283.

Pradhan, V.R., Pandit, M.K. & Meert, J.G. 2008. A cautionary note on the age of the Paleomagnetic pole obtained from the Harohalli Dyke swarms, Dharwar Craton, Southern India. *In*: Srivastava, R.K., Shivaji, Ch. & Chalapathi Rao, V. (eds) *Indian Dykes: Geochemistry, Geophysics and Geochronology*. Narosa Publishing House, New Delhi, India, 339–351.

Pradhan, V.R., Meert, J.G., Pandit, M.K., Kamenov, G., Gregory, L.C. & Malone, S.J. 2010. India's changing place in global Proterozoic reconstructions: a review of geochronologic constraints on key paleomagnetic poles from the Dharwar, Bundelkhand and Marwar Cratons. *Journal of Geodynamics*, **50**, 224–242.

Pradhan, V.R., Meert, J.G., Pandit, M.K., Kamenov, G. & Mondal, E.A. 2012. Tectonic evolution of the Precambrian Bundelkhand craton, central India: insights from paleomagnetic and geochronologic studies on the mafic dyke swarms. *Precambrian Research*, **198–199**, 51–76.

RADHAKRISHNA, B.P. 1987. Introduction. *In*: RADHAKRISHNA, B.P. (ed.) *Purana Basins of Peninsular India*. Geological Society of India, Memoirs, **6**, 1–15.

RADHAKRISHNA, B.P. 1989. Suspect Tectono-stratigraphic Terrain elements in the Indian Subcontinent. *Journal Geological Society of India*, **34**, 1–24.

RADHAKRISHNA, T., CHANDRA, R., SRIVASTAVA, A.K. & BALASUBRAMONIAN, G. 2013*a*. Central/Eastern Indian Bundelkhand and Bastar cratons in the Palaeoproterozoic supercontinental reconstructions: a palaeomagnetic perspective. *Precambrian Research*, **226**, 91–104.

RADHAKRISHNA, T., KRISHNENDU, N. & BALASUBRAMONIAN, G. 2013*b*. Palaeoproterozoic Indian shield in the global continental assembly: evidence from the palaeomagnetism of mafic dyke swarms. *Earth-Science Reviews*, **126**, 370–389.

RAJAMANI, V., SHIVKUMAR, K., HANSON, G.N. & SHIREY, S.B. 1985. Geochemistry and petrogenesis of amphibolites from the Kolar Schist Belt, South India: evidence for ultramafic magma generation by low percent melting. *Journal of Petrology*, **26**, 92–123.

RAMAKRISHNAN, M. & VAIDYANADHAN, R. (eds). 2008. *Geology of India, Volume 1*. Geological Society of India, Bangalore, India.

RAMCHANDRA, H.M., MISHRA, V.P. & DESHMUKH, S.S. 1995. Mafic dykes in the Bastar Precambrian: study of the Bhanupratappur–Keskal mafic dyke swarm. *In*: DEVARAJU, T.C. (ed.) *Mafic Dyke Swarms of Peninsular India*. Geological Society of India, Memoirs, **33**, 183–207.

RAVIKANT, V. 2006. Utility of Rb–Sr geochronology in constraining Miocene and Cretaceous events in the eastern Karakoram, Ladakh, India. *Journal of Asian Earth Sciences*, **27**, 534–543.

ROBINSON, D. 1987. Transition from diagenesis to metamorphism in extensional and collision settings. *Geology*, **15**, 866–869.

ROGERS, J.J.W. 1996. A history of continents in the past three billion years. *Journal of Geology*, **104**, 91–107.

ROGERS, J.J.W. & SANTOSH, M. 2002. Configuration of Columbia, A Mesoproterozoic supercontinent. *Gondwana Research*, **5**, 5–22.

ROGERS, J.J.W. & SANTOSH, M. 2003. Supercontinents in Earth history. *Gondwana Research*, **6**, 357–368.

ROGERS, J.J.W. & SANTOSH, M. 2009. Tectonics and surface effects of the supercontinent Columbia. *Gondwana Research*, **15**, 373–380.

ROLLINSON, H. 1993. *Using Geochemical Data: Evaluation, Presentation, Interpretation*. Longman, London.

SAID, N. & KERRICH, R. 2009. Geochemistry of coexisting depleted and enriched Paringa Basalts, in the 2.7 Ga Kalgoorlie Terrane, Yilgarn Craton, Western Australia: evidence for a heterogeneous mantle plume event." *Precambrian Research*, **174**, 287–309.

SENSARMA, S. 2007. A bimodal large igneous province and the plume debate. *In*: FOULAGAR, G.R. & JURDY, D.M. (eds) *Plates, Plumes, and Planetary Processes*. Geological Society of America, Special Papers, **430**, 831–840.

SENSARMA, S., MARTIN, A., SINGH, V.K. & BANERJEE, S. 2010. The Bundelkhand granite massif: a Paleoproterozoic plutonic large igneous province. *In*: BHATT, S.C. & SINGH, S.P. (eds) *Proceedings of the National Symposium on Geology and Mineral Resources of Bundelkhand Craton (GMRB)*. Bundelkhand University, Jhansi, 27.

SIEVER, R. 1983. Burial history and diagenetic reaction kinetics. *American Association of Petroleum Geologist Bulletin*, **67**, 684–691.

SONG, X.Y., ZHOU, M.F., KEAYS, R.R., CAO, Z.M., SUN, M. & QI, L. 2006. Geochemistry of the Emeishan flood basalts at Yangliuping, Sichuan, SW China: implications for sulphide segregation. *Contributions to Mineralogy and Petrology*, **152**, 53–74.

SRIVASTAVA, R.K. & GAUTAM, G.C. 2012. Early Precambrian mafic dyke swarms from the Central Archaean Bastar Craton: geochemistry, petrogenesis and tectonic implications. *Geological Journal*, **47**, 144–160.

SRIVASTAVA, R.K. & GAUTAM, G.C. 2015. Geochemistry and petrogenesis of Paleo-Mesoproterozoic mafic dyke swarms from northern Bastar craton, central India: geodynamic implications in reference to Columbia supercontinent. *Gondwana Reaearch*, **28**, 1061–1078.

STEIN, H.J., HANNAH, J.L., ZIMMERMAN, A., MARKEY, R.J., SARKAR, S.C. & PAL, A.B. 2004. A 2.5 Ga porphyry Cu–Mo–Au deposit at Malanjkhand, central India: implicationsfor Late Archean continental assembly. *Precambrian Research*, **134**, 189–226.

STRAND, K. & KÖYKKÄ, J. 2012. Early Paleoproterozoic rift volcanism in the eastern Fennoscandian Shield related to the breakup of the Kenorland supercontinent. *Precambrian Research*, **214–215**, 95–105.

SUN, S.S. & MCDONOUGH, W.F. 1989. Chemical and isotopic systematics of oceanic basalts: implications for mantle composition and processes, *In*: SAUNDERS, A. D. & NORRY, M.J. (eds) *Magmatism in the Ocean Basins*. Geological Society, London, Special Publications, **42**, 313–345, https://doi.org/10.1144/GSL.SP.1989.042.01.19

SUN, S.S. & NESBITT, R.W. 1977. Chemical heterogeneity of the Archaean mantle composition of the bulk earth and mantle evolution. *Earth and Planetary Science Letters*, **35**, 429–448.

SUN, S.S. & NESBITT, R.W. 1978. Geochemical regularities and genetic significance of ophiolitic basalts. *Geology*, **6**, 689–693.

TANAKA, T., TOGASHI, S. ET AL. 2000. JNdi-1: a neodymium isotopic reference in consistency with La Jolla neodymium. *Chemical Geology*, **168**, 279–281.

TARNEY, J. 1992. Chapter 4. Geochemistry and significance of mafic dyke swarms in the Proterozoic. *In*: CONDIE, K. C. (ed.) *Proterozoic Crustal Evolution*. Elsevier, Amsterdam, 151–179.

TAYLOR, S.R. 1965. The application of trace element data to problem in petrology. *Physics and Chemistry of the Earth*, **6**, 133–213.

THOMPSON, R.N., MORRISON, M.A., HENDRY, G.L. & PARRY, S.J. 1984. An assessment of relative roles of crust and mantle genesis: an elemental approach. *Philosophical Transactions of the Royal Society of London, A*, **310**, 549–590.

WEAVER, B.L. & TARNEY, J. 1983. Chemistry of the sub continental mantle inferences from Archean and Proterozoic dykes and continental flood basalts. *In*: HAWKERWARTH, C.J. & NORRY, M.J. (eds) *Continental Basalt and Mental Xenoliths*. Shiva Publications, Nantwich, UK, 209–229.

WHITE, R. & MCKENZIE, D. 1989. Volcanism at rifts. *Scientific American*, **261**, 62–71.

WINCHESTER, J.A. & FLOYD, P.A. 1977. Geochemical discrimination of different magma series and their differentiation products, using immobile elements. *Chemical Geology*, **20**, 325–344.

ZHAO, G.C., CAWOOD, P.A., WILDE, S.A. & SUN, M. 2002. Review of global 2.1–1.8 Ga orogens: implications for a pre-Rodinia supercontinent. *Earth-Science Reviews*, **59**, 125–162.

ZHAO, G.C., SUN, M., WILDE, S.A. & LI, S.Z. 2004. A Paleo–Mesoproterozoic supercontinent: assembly, growth and breakup. *Earth-Science Reviews*, **67**, 91–123.

ZHAO, J.H. & ZHOU, M.F. 2007. Geochemistry of Neoproterozoic mafic intrusions in the Panzhihua district (Sichuan Province, SW China): implications for subduction related metasomatism in the upper mantle. *Precambrian Research*, **152**, 27–47.

Index

Page numbers in *italics* denote Figures. Page numbers in **bold** denote Tables.